ACS SYMPOSIUM SERIES 476

Emerging Technologies for Materials and Chemicals from Biomass

Roger M. Rowell, EDITOR
U.S. Department of Agriculture

Tor P. Schultz, EDITOR
Mississippi State University

Ramani Narayan, EDITOR
*Michigan Biotechnology Institute
and
Michigan State University*

Developed from a symposium sponsored
by the Cellulose, Paper, and Textile Division
at the 200th National Meeting
of the American Chemical Society,
Washington, D.C.,
August 26–31, 1990

American Chemical Society, Washington, DC 1992

Library of Congress Cataloging-in-Publication Data

Emerging technologies for materials and chemicals from biomass /
  Roger M. Rowell, editor, Tor P. Schultz, editor, Ramani Narayan, editor.

  p.   cm.—(ACS Symposium Series, 0097–6156; 476)

  "Developed from a symposium sponsored by the Cellulose, Paper, and Textile Division at the 200th National Meeting of the American Chemical Society, Washington, D.C., August 26–31, 1990."

  Includes bibliographical references and indexes.

  ISBN 0–8412–2171–5

  1. Biomass chemicals—Congresses. 2. Biomass fuel—Congresses.

  I. Rowell, Roger M. II. Schultz, Tor P., 1953–   . III. Narayan, Ramani, 1949–   . IV. American Chemical Society. Cellulose, Paper, and Textile Division. V. American Chemical Society. Meeting (200th: 1990: Washington, D.C.) VI. Series.

TP248.B55E44 1991
661'.8—dc20   91–36048
                                       CIP

The paper used in this publication meets the minimum requirements of American National Standard for Information Sciences—Permanence of Paper for Printed Library Materials, ANSI Z39.48–1984. ∞

Copyright © 1992

American Chemical Society

All Rights Reserved. The appearance of the code at the bottom of the first page of each chapter in this volume indicates the copyright owner's consent that reprographic copies of the chapter may be made for personal or internal use or for the personal or internal use of specific clients. This consent is given on the condition, however, that the copier pay the stated per-copy fee through the Copyright Clearance Center, Inc., 27 Congress Street, Salem, MA 01970, for copying beyond that permitted by Sections 107 or 108 of the U.S. Copyright Law. This consent does not extend to copying or transmission by any means—graphic or electronic—for any other purpose, such as for general distribution, for advertising or promotional purposes, for creating a new collective work, for resale, or for information storage and retrieval systems. The copying fee for each chapter is indicated in the code at the bottom of the first page of the chapter.

The citation of trade names and/or names of manufacturers in this publication is not to be construed as an endorsement or as approval by ACS of the commercial products or services referenced herein; nor should the mere reference herein to any drawing, specification, chemical process, or other data be regarded as a license or as a conveyance of any right or permission to the holder, reader, or any other person or corporation, to manufacture, reproduce, use, or sell any patented invention or copyrighted work that may in any way be related thereto. Registered names, trademarks, etc., used in this publication, even without specific indication thereof, are not to be considered unprotected by law.

PRINTED IN THE UNITED STATES OF AMERICA

ACS Symposium Series

M. Joan Comstock, *Series Editor*

1992 ACS Books Advisory Board

V. Dean Adams
Tennessee Technological
   University

Alexis T. Bell
University of California—Berkeley

Dennis W. Hess
Lehigh University

Mary A. Kaiser
E. I. du Pont de Nemours and
   Company

Gretchen S. Kohl
Dow-Corning Corporation

Bonnie Lawlor
Institute for Scientific Information

John L. Massingill
Dow Chemical Company

Robert McGorrin
Kraft General Foods

Julius J. Menn
Plant Sciences Institute,
   U.S. Department of Agriculture

Marshall Phillips
Office of Agricultural Biotechnology,
   U.S. Department of Agriculture

A. Truman Schwartz
Macalaster College

Stephen A. Szabo
Conoco Inc.

Robert A. Weiss
University of Connecticut

Foreword

THE ACS SYMPOSIUM SERIES was founded in 1974 to provide a medium for publishing symposia quickly in book form. The format of the Series parallels that of the continuing ADVANCES IN CHEMISTRY SERIES except that, in order to save time, the papers are not typeset, but are reproduced as they are submitted by the authors in camera-ready form. Papers are reviewed under the supervision of the editors with the assistance of the Advisory Board and are selected to maintain the integrity of the symposia. Both reviews and reports of research are acceptable, because symposia may embrace both types of presentation. However, verbatim reproductions of previously published papers are not accepted.

Contents

Preface ix

1. **Biomass (Renewable) Resources for Production of Materials, Chemicals, and Fuels: A Paradigm Shift** 1
   Ramani Narayan

   LIGNOCELLULOSIC MATERIALS AND COMPOSITES

2. **Opportunities for Lignocellulosic Materials and Composites** 12
   Roger M. Rowell

3. **Opportunities for the Cost-Effective Production of Biobased Materials** 28
   Helena L. Chum and Arthur J. Power

4. **Lignocellulosic–Plastic Composites from Recycled Materials** 42
   John Youngquist, George E. Myers, and Teresa M. Harten

5. **Compatibilization of Lignocellulosics with Plastics** 57
   Ramani Narayan

6. **Mechanical Properties of Surface-Modified Cellulose Fiber–Thermoplastic Composites** 76
   R. G. Raj and B. V. Kokta

7. **Dimensionally and Ultraviolet-Stable Composites from Lignocellulosics** 88
   David V. Plackett

8. **Thermal Plasticization of Lignocellulosics for Composites** 98
   Hideaki Matsuda

9. **Activation and Characterization of Fiber Surfaces for Composites** 115
   Raymond A. Young

v

10. Liquefaction of Lignocellulosics in Organic Solvents and Its Application 136
    Nobuo Shiraishi

    BIOPOLYMERS: ALLOYS, DERIVATIVES, AND BLENDS

11. Low-Cost Uses of Lignin 146
    Robert A. Northey

12. New Developments in Cellulosic Derivatives and Copolymers 176
    David N.-S. Hon

13. Emerging Polymeric Materials Based on Starch 197
    William M. Doane, Charles L. Swanson, and George F. Fanta

14. Monomers and Polymers Based on Mono-, Di-, and Oligosaccharides 231
    Stoil K. Dirlikov

15. Cellulose Ethers: Self-Cross-linking Mixed Ether Silyl Derivatives 265
    Arjun C. Sau and Thomas G. Majewicz

16. Polymeric Materials from Agricultural Commodities 273
    S. F. Thames and P. W. Poole

17. Emerging Polymeric Materials Based on Soy Protein 299
    Thomas L. Krinski

18. Enzymatic Treatments of Pulps 313
    Thomas W. Jeffries

    CHEMICALS AND FUELS FROM BIOMASS AND WASTES

19. Chemicals and Fuels from Biomass: Review and Preview 332
    Irving S. Goldstein

20. The Promise and Pitfalls of Biomass and Waste Conversion 339
    Helena L. Chum and Arthur J. Power

21. Potential for Fuels from Biomass and Wastes 354
    K. Grohmann, C. E. Wyman, and M. E. Himmel

22. **Chemicals from Pulping: Product Generation from Pulping Residuals**.. 393
    David E. Knox and Philip L. Robinson

23. **Feedstock Availability of Biomass and Wastes**............................... 410
    J. W. Barrier and M. M. Bulls

24. **Pyrolysis of Agricultural and Forest Wastes**................................. 422
    D. S. Scott, J. Piskorz, and D. St. A. G. Radlein

25. **Plants as Sources of Drugs and Agrochemicals**............................ 437
    James D. McChesney and Alice M. Clark

Author Index 452

Affiliation Index 452

Subject Index 452

Preface

TO DESCRIBE BIOMASS as an emerging material and source of chemicals may seem redundant because wood has been used as a material from early times and as a source of chemicals for several hundred years. However, the definition of "emerge" is "to rise, to become apparent, and to evolve." In this sense, biomass as a source of material and chemicals is still emerging in the 1990s and will probably continue to do so well into the future.

The term "biomass" is actually a little misleading. We should really use the terms "phytomass", "photomass", "solarmass", or "photosynthetic mass" because this book does not include biomass such as bone, protein, lipids, and other biological components. This book covers plant-based resources including wood, agricultural crops and residues, grasses, and components from these sources.

Until about the 1920s, we relied almost completely on biomass and coal for our materials and chemicals. In the early 1900s, a new material emerged called petroleum, which has since become our major source of chemicals and plastic materials. At the same time, modern metal, glass, and chemical technologies were emerging to give us materials such as high carbon and stainless steel, structural aluminum alloys, organometallics, ceramics, and various plastics and adhesives. At a time when plastics, metals, and glass were emerging, the market share of biomass-derived materials decreased.

Because of the oil embargo of the 1970s, the Iraq War of 1990–1991, increased demand from third world countries for modern materials, changes in the economics of competing materials, a re-awareness of our environment, and global interest in recycling, there is a renewed interest in biomass utilization. After thousands of years of use, biomass is once again emerging as a source of materials and chemicals.

Acknowledgments

For their financial support of the symposium upon which this book is based, we thank the following organizations: the American Chemical Society Petroleum Research Fund, the Georgia-Pacific Corporation, the

Texaco-Exploration and Production Division, the CSI/Laport-Timber Division, and the American Chemical Society Cellulose, Paper, and Textile Division.

We also thank these organizations for release time and clerical assistance: the U.S. Department of Agriculture's Forest Products Laboratory, the Michigan Biotechnology Institute, and the Mississippi Forest Products Laboratory.

The organization of a symposium and the publication of a book require the efforts of many people. We thank them all, especially the ACS Cellulose, Paper, and Textile Division; the authors; and the patient and professional staff of the ACS Books Department.

ROGER M. ROWELL
U.S. Department of Agriculture Forest Service
Forest Products Laboratory and Forestry Department
University of Wisconsin
Madison, WI 53705

TOR P. SCHULTZ
Mississippi Forest Products Laboratory
Mississippi State University
Mississippi State, MS 39762

RAMANI NARAYAN
Michigan Biotechnology Institute and
Michigan State University,
Lansing, MI 48910

May 14, 1991

Chapter 1

Biomass (Renewable) Resources for Production of Materials, Chemicals, and Fuels
A Paradigm Shift

Ramani Narayan

Michigan Biotechnology Institute and Michigan State University, Lansing, MI 48909

> There is a resurgence of interest in abundant, renewable, biomass resources as precursors for the production of polymeric materials, organic chemicals, and fuel. This heightened awareness and sense of urgency toward biomass utilization is driven by 1) environmental concerns, and the need for environmentally compatible polymers and chemicals. 2) the need to be weaned away from our dependance on non renewable imported petroleum feedstocks, and 3) utilization of the nation's abundant, renewable agricultural feedstocks in new non-food, non-feed uses.

Industry has always been driven by three paramount factors in the development of products and processes. They are:
- Productivity
- Efficiency and Functionality
- Economics

To meet these criteria for production of polymeric materials, organic chemicals and fuel we have come to rely heavily on a single non-renewable resource, namely oil, to the total exclusion of other resources. Other factors, namely:
- sustainability of oil resources and dependance on foreign imports,
- the impact on the environment, both before and after use (waste management),
- the folly of depending on only one resource for all our needs,

have not been taken in to consideration during the development of a product or process. The need and importance of taking these additional factors in to consideration has been the subject of frequent debates as during the energy crisis of the 70's and more recently during the Persian Gulf crisis. However, as normalcy returns, the questions are swept aside and we go back to business as usual. This has resulted in an increasing dependance on foreign oil and, therefore, more vulnerable to supply disruptions and price manipulation. Our oil imports have risen sharply since the 80's, and U.S. dependence on foreign oil is close to the 50% mark - half from a politically unstable region (Figure 1).

As Daniel Yergin says in his recent book about oil "The Price" (Simon and Schuster, 1991), "Petroleum remains the motive force of the industrial society and the life- blood of civilization that it helped create. It also remains an essential element in national power, a major factor in world economics, a critical focus for war and conflict, and a decisive force in international affairs." Today, with a victorious Persian Gulf war behind us and oil being pumped out of the Persian Gulf as never before - 2.2 million barrels a day (1), the continued use of oil as the feedstock for polymeric materials, organic chemicals and fuel seems even more enshrined.

The Paradigm Shift

In spite of the above seeming complacency about oil as the primary feedstock, there is a climate of change in industry. The driver for this change is the growing concerns about the environment. Therefore, industry is being asked to reevaluate its priority relative to productivity, efficiency and economics and add three more factors to the equation. These are:
- Resource conservation - the utilization of renewable resources as opposed to non-renewable oil resources; and more importantly,
- The effect of processes and products on the global environment -- compatibility of products and processes with the environment;
- Waste management - disposal of waste in an environmental and ecologically sound manner - issues of recyclability and biodegradability.

The key word is "Environmentalism," and environmentalism will be the biggest business and industry issue of the 90's and beyond. This "greening" of the industry resulted in key international companies and industrial organizations meeting in Rotterdam to endorse a set of principles and a charter that will commit them to environmental protection into the 21st century (2). One hundred and fifty companies including some major U.S. chemical concerns and more than thirty-five organizations adopted the business charter for sustainable development. In support of this document 16 principles developed by the Paris-based International Chamber of Commerce (ICC) were adopted. According to ICC the principles are designed to place environmental management high on corporate agendas and to encourage policies and practices for carrying out operations in environmentally sound ways. The U.S. affiliate of ICC, the U.S. Council for International Business, sees the principles "as the culmination of building American corporate awareness of the importance of sustainable development." Some of the key principles of the charter are:
- Develop and operate facilities and undertake activities with energy efficiency, **sustainable use of renewable resources** and waste generation in mind
- Conduct or support research on the impact and ways to minimize the impacts of raw materials, products or processes, emissions and wastes

- Modify manufacturing, marketing, or use of products and services to prevent serious or irreversible environmental damage; develop and provide products and services that do not harm the environment
- Contribute to the transfer of environmentally sound technology and management methods.

Biomass derived renewable polymer resources can play a major role under this new environmental climate. Clearly, the processes, products and technologies adopted and developed using renewable resources must be compatible with the environment. Furthermore, the wastes generated should be recycled or transformed into environmentally benign products. Youngquist and co-workers, in Section 2 of the book, describe work on lignocellulosic plastic composites made from recycled plastics. The idea ties in biomass renewable resource utilization with recycling of the post- consumer plastic waste - a major waste management issue.

Thus, the twin issues of environmental responsibility and resource conservation (moving away from non-renewable oil feedstocks) has led to a resurgence of interest in renewable biomass resources as the precursor for polymeric materials, organic chemicals and fuels. The chapters in the book present an overview of the emerging technologies, products and processes that are capitalizing on this renewed interest and enthusiasm for biomass derived polymeric materials and chemicals (Figure 2).

One of the major questions raised about biomass utilization is its effect on global warming. The consumption of biomass resources would mean that there will be less biomass available to fix CO_2 emmissions. There is also the question of CO_2 released in the atmosphere during combustion or in other waste disposal schemes such as composting. These are important points, since environmental considerations are the driver for the present change. However, by producing biomass at a sustainable rate (continuous replenishment of biomass utilized with new growth), the CO_2 consumed during photosynthesis should balance the amount of CO_2 released in processing biomass. Thus, biomass utilization would make no net contribution to the CO_2 in the atmosphere, and so it would have negligible impact on global warming. Figure 3 illustrate these concepts and shows some of the energy and profits from the biomass utilization cycle being ploughed back into replantation of biomass. Thus, biomass resources, if properly managed, can contribute to a sustainable resource and environment base. However, one must, also, carefully address other environmental consequences of biomass utilization including air pollution, residues, ash, and depletion of cell nutrients to make biomass a acceptable feedstock. These must, therefore, be studied carefully and appropriate standards and regulations developed. Life cycle or Cradle to Grave analysis must be performed on the emerging biomass technologies. This analysis is a holistic environmental and energy audit (accounting procedure) that focuses on the entire life cycle of a product, from raw material production to final product disposition, rather than a single manufacturing step or environmental emission.

Figure 1. US dependance on foreign oil.

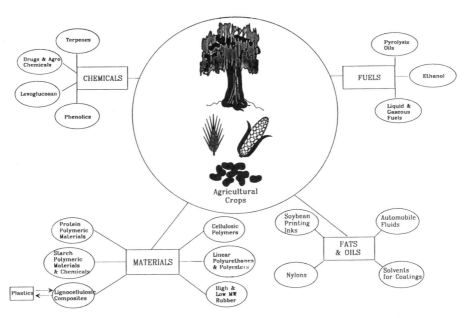

Figure 2. Emerging materials, chemicals, and fuels from biomass.

Chum and Power's chapter on "Opportunities for the Cost Effective Production Biobased Materials" and "The Potential and Pitfalls of Biomass & Wastes"; Grohaman and coworkers chapter on "Potential for fuels from Biomass and Wastes"; Barrier and Bull's chapter on "Feedstock Availability of Biomass & Wastes, and Goldsteins chapter provides details on these issues. The avaialability of biomass feedstocks, and the potential economic impact of biomass utilization is discussed later in this chapter.

U.S. national policy seems moving in the direction of producing industrial materials from biomass resources. The provisions of the 1990 Farm Bill (Food Agricultural, Conservation and Trade Act of 1990, XVI Subtitle G) is to expedite the development and market penetration of industrial products that use agricultural materials. The act also provides for R&D and commercialization assistance. The U.S. Department of Agriculture (USDA) through the Agriculture Research Service (ARS) has R&D programs to produce new industrial products from agricultural raw materials. Doane and coworkers review some of this work in their chapter. The USDA Forest Products Research Service is looking at utilizing lignocellulosic biomass resources to produce new composite materials and other industrial products (see Section 2). The Department of Energy through their Advanced Industrial Materials Program in the Office of Conservation and Renewable Energies have a small R&D program on new industrial products from renewable resources (see Chum's chapters in this book).

Biomass Derived Materials

Biomass derived materials are being produced at substantial levels. For example, paper and paperboard production is around 139 billion lbs, and biomass derived textiles production around 1.2 million tons (see Chum's chapter). However, biomass use in production of plastics, coatings, resins and composites is negligible. These areas are dominated by synthetics derived from oil and represent the industrial materials of today.

Lignocellulosics-Composites. Wood is the oldest known composite material -- flexible cellulosic fibers assembled in an amorphous matrix of lignin and hemicellulosic polymer. It continues to find extensive use in the construction industry. A lignocellulosic composite such as particleboard and fiberboard are lignocellulosic fibers embedded in a thermoplastic or thermoset matrix. Lignocellulosic materials such as wood flour are used as inexpensive filler in phenolic resins and thermoplastics. However, the cost, availability, renewability, and recyclability of lignocellulosics offer the potential to expand their market use into the large volume, low-cost, high performance structural composites -- the automobile and durable goods market (Figure 4) (3, 4). Section 2 of the book is devoted to this area, and presents emerging technologies that can provide biomass derived low-cost high performance materials that competes with the totally synthetic materials. Rowell's chapter showcases in detail the opportunities and requirements for lignocellulosic materials and composites. Youngquist deals with the effective utilization of recycled plastics by combination with lignocellulosics. The

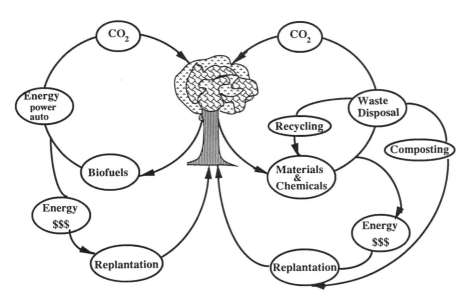

Figure 3. Impact of biomass utilization on global warming.

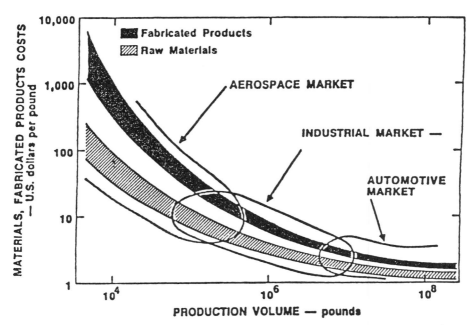

Figure 4. Major markets for structural composites. (Reproduced from ref. 4. Copyright 1988 American Chemical Society.)

compatibilization of lignocellulosics with synthetic polymers to provide new thermoplastic alloys with improved functionality and properties is discussed by Narayan. The properties of such new composite materials is discussed in chapters by Raj & Kokta, and Plackett. The ability to turn lignocellulosics into thermoplastics with the desired properties and ease of processing is described by Matsuda. The new world of lignocellulosic liquefaction is introduced by Shiriashi.

Biopolymers and Derivatives. The production of only five of the leading thermoplastics (low and high density polyethylene, polystyrene, polypropylene, and polyvinyl chloride) totalled 48 billion pounds in 1990. The replacement of these synthetic polymers in part or full would create new large volume markets for biomass polymers. Cellulosic derivatives (Chapters by Hon, and Majewics & Sau), starch and modified starches (Doanne and co-workers), soy protein polymers (Krinski), monomers and polymers based on oligosaccharides (Dirlikov), lignin polymers (Northey) are potential contenders for the markets dominated by the synthetic polymers.

Thermoplastics made from synthetic oil are today formulated to be strong, light weight, durable and bioresistant. They are resistant to biological degradation in the environment because the natural microbial population does not have polymer specific enzymes capable of degrading and using most man-made synthetic polymers. In addition, the hydrophobic character of most synthetic plastics inhibits natural enzymatic activities and the low surface area of the plastic with the inherent high molecular rates compounds the problem further. It is this durability, light weight and indestructibility that makes these plastics materials of choice for many packaging and consumer goods applications but also creates problems when they enter the waste streams. Plastic litter and errant medical waste scar landscape, foul our beaches, and pose a serious hazard to marine life. Nationwide, between 40-60% of beach debris is plastic. An additional 10-20% is expanded polystyrene foam (5). It has been estimated that 50-80% of materials washing ashore will remain undegraded in the environment, i.e., they are persistent and recalcitrant, and not readily broken down by the elements to become a part of the natural carbon cycle of the ecosystem (6, 7). As a result there are mounting concerns over the disposal of persistent disposable and non-degradable plastics that are often and perhaps not always fairly singled out as the major culprit (8).

This leads us to the concept of designing and engineering new biodegradable polymeric materials, materials that are plastics, i.e. strong, light weight, easily processable, energy efficient, excellent barrier properties, disposable (mainly for reasons of hygiene and public health) yet break down under appropriate environmental conditions just like its organic (lignocellulose) counterpart. It also includes developing new concepts and technologies for handling our waste, and also creating waste disposal infrastructures in tune with the natural carbon cycle. The rationale for biodegradable polymers and its role in waste management has been discussed in detail by Narayan (4,5).

Thus, new market opportunities for biopolymers and derivatives, such as described in section 3 of the book, presents itself in areas such as single use, disposal, short life packaging materials, service ware items and disposal non-wovens. An estimated 30% of synthetic polymers from non-renewable oil feedstocks totalling 16.5 billion pounds annually are used in these applications. Marine plastics is another category that lends itself to utilization of biodegradable material concepts. These include fishing gear such as drift net traps and packaging materials such as plastic sheets, strapping, shrinkwrap, polystyrene foam products and domestic trash such as plastic bags, bottles and beverage ring containers. Indeed, major corporations such as Warner-Lambert (Novon product line), National Starch and Chemical, Archer Daniels & Midland (ADM), and more recently the Ecological Chemical Products Company (ECOCHEM -- a joint venture of DuPont and ConAgra) have announced production of polymeric materials derived from renewable biomass resources solely with such an environmental theme.

Biofuels and Chemicals

As the only renewable technology to produce liquid transportation fuels, biofuels have the potential to displace the crude oil being imported today. In doing so, it could reduce the environmental problems because they consume as much carbon dioxide (a greenhouse gas) in their growth and production as they release in their use. They also produce few, if any, sulfur compounds. This is an important factor in a world that is learning about the greenhouse effect and becoming concerned about the diminishing quality of the environment. Because biofuel conversion processes produce little ash and few if any sulfur compounds, biofuels contribute to an improved environment. In using municipal waste to produce energy, biofuels technology is helping to solve a growing environmental problem. it also reduces the need for landfills by using industrial and municipal wastes. The U.S. Department of Energy (DOE) Biofuels Municipal Waste Technology (BMWT) program focuses on five pathways for biofuel production. Their combined energy potential is estimated at approximately 17 quads or 20% of current U.S. energy consumption. Approximately 3 quads (3-5%) are currently generated from wood and waste alone representing roughly the same contribution made by nuclear energy (4-5%) or hydropower (4%). More than eighty facilities nationwide burn municipal solid wastes (MSW) and refuse derived fuel (RDF) to provide heat for industrial processes and for electricity generation. Meanwhile, starch and sugar crops are being converted to 850 MM gallons of fuel ethanol each year for use in gasoline blends where it raises the octane rating and reduces carbon monoxide emissions. K. Grohmann and co-workers and Barrier and Bulls review this very important biofuels area in Section 4 of this book.

While the primary thrust of this book is on production of polymeric materials, composites and chemicals from lignocellulosics and its polymer constituents primarily cellulose and lignin, other biomass resources such as starches, oilseeds, industrial crops such as guayule, soybean and many more can also serve as feedstocks. Some of these have been covered in the book -- polymeric materials from starches, rubber and other materials from guayule, materials based on soy protein, plants as sources for drugs and chemicals. Some have not been covered, again testifying to the versatility and flexibility of biomass resources. Of these mention must be made of the preparation of nylons -- nylon 9, and nylon 13, 13 from the erucic acid of crambe or industrial rapeseed oil (9), the use of soybean oil for petrochemical resins in printing inks, and vernonia oil, a natural epoxidized vegetable oil, as a replacement for conventional solvents in alkyd and epoxy coatings (10). Vegetable oils are used in lubricants, and industrial rapeseed oil is used as a supplement in automatic tramsmission fluid.

Biomass Feedstocks

The abundance of renewable biomass resources can be illustrated from the fact that the primary production of biomass estimated in energy equivalents is 6.9×10^{17} kcal/year(11). Mankind utilizes only 7% of this amount, i.e 4.7×10^{16} kcal/year, testifying to the abundance of biomass resources. In terms of mass units the net photosynthetic productivity of the biosphere is estimated to be 155 billion tons/year (12) or over 30 tons per capita and this is the case under the current conditions of non-intensive cultivation of biomass. Forests and crop lands contribute 42 and 6%, respectively, of that 155 billion tons/year. The world's plant biomass is about 2×10^{12} tons and the renewable resources amount to about 10^{11} tons/year of carbon of which starch provided by grains exceeds 109 tons (half which comes from wheat and rice) and sucrose accounts for about 108 tons. Another estimate of the net productivity of the dry biomass gives 172 billion tons/year of which 117.5 and 55 billion tons/year are obtained from terrestrial and aquatic sources, respectively (13).

It is estimated that U.S. agriculture accounts directly and indirectly for about 20% of theGNP by contributing $ 750 billion to the economy through the production of foods and fiber, the manufacture of farm equipment, the transportation of agricultural products, etc. It is also interesting that while agricultural products contribute to our economy with $ 40 billion of exports, and each billion of export dollars creates 31, 600 jobs (12) (1982 figures), foreign oil imports drains our economy and makes up 23% of the U.S. trade deficit (U.S. Department of Commerce 1987 estimate)

Forests cover one third of the land in the 48 contiguous states (759 MM acres) and commercial forests make up about 500 MM acres. Fortunately, we are growing trees faster than they are being consumed, although sometimes the quality of the harvested trees is superior to those being planted. Agriculture uses about 360 MM acres of the 48 contiguous states, and this acreage does not include idle crop lands and pastures. These figures clearly illustrate the potential for biomass utilization in the U.S.

Conclusions

The conversion of biomass resources to polymeric materials, chemicals and fuels can:

- Spur new sustainable industrial development with energy efficiencies and economies
- Mitigate environmental concerns and provide for a sustainable environment
- Reduce our dependance on foreign oil imports and the drain on our economy
- Contribute to the growth and development of our economy

The absurd situation of the U.S. government artificially limiting the agricultural production capabilities of this country will and can be eliminated, if utilization of renewable biomass resources toward addressing the problems of stable resource base and ecologically sound waste management practices be adopted. Rather than subsidies and PIK (pay in kind) program that compensates farmers for a reduction in cultivated land (by giving them credit with crops held in government storage) we should be using our biomass potential as a feedstock for materials, chemicals and fuel besides food.

The time is right for a greater national commitment toward biobased materials, and chemicals. New emerging technologies that are showcased in this book have created products and processes that are competitive with oil based synthetics, and contribute to a sustainable environment. A recent Chemical & Engineering News article (14) chronicles very nicely this push for new materials and chemicals from biomass.

Literature Cited

1. News Item *C& E News*, July 17, **1991**, pp. 20.
2. News Item, *C & E News*, April 8, **1991**, pp 4.
3. Chum, H. L.,In *Assessment of Biobased Materials*, Solar Energy Research Institute (SERI), Report No. SERI/TR-234-3610, 1989.
4. Fishman, N., "Abstracts of Papers", 196th National Meeting of the American Chemical Society, Los Angeles, Calif., Sept. 1988, CME 1.
5. Alaska Sea Grant Rep[ort No. 88-7 on "Workshop on Fisheries, Generated Marine Debris and Derelict Fishing Gear" February 9-11, 1988.
6. Narayan, R., *Kunstoffe*, **1989**, 79 (10), 1022.
7. Narayan, R., *INDA Nonwoven Res.*, **1991**, 3, 1.
8. News Item, *Modern Plastics,* Waste Solutions, April, 1990.
9. Van Dyne, D. L., and Blase, G. M., *Biotechnol Prog.*, **1990**, 6, 273.
10. Dirlikov, S., *ACS Symp Ser.*, **1990**, 433, 176.
11. *Primary Productivity of the Biosphere*, Lieth, H., and Whittaker, H. R., Eds., Springer Verlag, 1975.
12. Institute of Gas Technology, Symposium on "Clean Fuels from Biomass and Wastes", Orlando, Florida, 1977.
13. Szmant, H. H., Industrial Utilization of Renewable Resources, Technomic Publishing Co., Lancaster, Basel, 1986.
14. Borman, S., *C & E News*, September 10, **1990**, pp 19.

RECEIVED July 29, 1991

LIGNOCELLULOSIC MATERIALS AND COMPOSITES

Chapter 2

Opportunities for Lignocellulosic Materials and Composites

Roger M. Rowell

Forest Products Laboratory and Forestry Department, U.S. Department of Agriculture Forest Service, University of Wisconsin, Madison, WI 53705

High performance lignocellulosic composites with uniform densities, durability in adverse environments, and high strength can be produced by using fiber technology, bonding agents, and fiber modification to overcome dimensional instability, biodegradability, flammability, and degradation caused by ultraviolet light, acids, and bases. Products with complex shapes can also be produced using flexible fiber mats, which can be made by nonwoven needling or thermoplastic fiber melt matrix technologies. The wide distribution, renewability, and recyclability of lignocellulosics can expand the market for low-cost materials and high performance composites. Research is being done to combine lignocellulosics with materials such as glass, metals, plastics, inorganics, and synthetic fibers to produce new materials that are tailored for end-use requirements.

Lignocellulosic substances contain both cellulose and lignin. Sources of lignocellulosics include wood, agricultural residues, water plants, grasses, and other plant substances. In general, lignocellulosics have been included in the term biomass, but this term has broader implications than that denoted by lignocellulosics. Biomass also includes living substances such as animal tissue and bones. Lignocellulosic materials have also been called photomass because they are a result of photosynthesis.

A material is usually defined as a substance with consistent, uniform, continuous, predictable, and reproducible properties. An engineering material is defined simply as any material used in construction. Wood and other lignocellolosics have been used as "engineering materials" because they are economical, low in processing energy, renewable, and strong. In some schools of thought, however, lignocellulosics are not considered materials because they do not have consistent, predictable, reproducible, continuous, and uniform properties. This is true for solid lignocellulosics such as wood, but it is not necessarily true for composites made from lignocellulosics. A lignocellulosic composite is a reconstituted product made from a combination of one or more substances using some kind of a bonding agent to hold the components together. The best known lignocellulosic composites are plywood, particleboard, fiberboard, and laminated lumber.

For the most part, materials and composites can be grouped into two basic types: price-driven, for which costs dictate the markets, and performance-driven, for which properties dictate the markets. A few materials and composites might be of both types, but usually these two types are very distinct. In general, the wood industry has produced the price-driven type. Even though the wood industry is growing, it has lost some market share to competing materials such as steel, aluminum, plastic, and glass. Unless something is done to improve the competitive performance of lignocellulosics, future opportunities for growth will be limited.

Performance-driven materials and composites in the wood industry range from high performance–high cost, to low performance–low cost. Figure 1 shows three examples of industrial uses for materials and composites. The aerospace industry is very concerned about high performance in their materials and composites, and this results in high cost and low production. The military is also concerned about performance, but their requirements are not as strict as those of the aerospace industry. Thus, the rate of production of military-type materials and composites is higher than that of aerospace materials and the cost is lower. In contrast, the wood industry depends on a high production rate and low cost. As a result, wood materials and composites have low performance.

Lignocellulosic materials can provide the wood industry with the opportunity to develop high performance materials and composites. The chemistry of lignocellulosic components can be modified to produce a material that has consistent, predictable, and uniform properties. As will be discussed later, properties such as dimensional instability, biodegradability, flammability, and degradation caused by ultraviolet light, acids, and bases can be altered to produce value-added, property-enhanced composites.

Even though wood is one of the most common "materials" used for construction, information on wood and other lignocellulosics is not included in many university courses in materials science. Wood and other lignocellulosics are perceived as not being consistent, uniform, and predictable in comparison to other materials. Most materials science classes deal mainly with metals and, to a lesser extent, with ceramics, glass, and plastics, depending on the background of the professor. The general public is also reluctant to accept lignocellulosic composites. In the early days of wood composites, products such as particleboard developed a bad reputation, mainly because of failures in fasteners and swelling under moist conditions. This led to public skepticism about the use of wood composites. The markets that were developed for wood composites were all price-driven, with few claims for high performance. This is still true, for

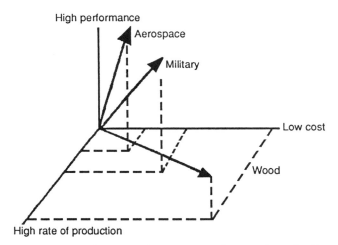

Figure 1. Comparison of cost, performance, and rate of production for various types of composites.

the most part, although such wood composite products do have high performance. For example, the high performance of expensive furniture is a result of composite cores.

The perception that wood is not a material as defined earlier is easy to understand. Wood is anisotropic (different properties in all three growing directions of a tree); it may contain sapwood, heartwood, latewood, earlywood, juvenile wood, reaction wood, knots, cracks, splits, and checks; and it may be bent, twisted, or bowed. These defects occur in solid wood but they need not exist in wood composites. The smaller the size of the components in the composite, the more uniform the properties of the composite. For example, chipboard is less uniform than flakeboard, which is less uniform than particleboard, which is less uniform than fiberboard. However, fiberboard made from lignocellulosic fiber can be very uniform, reproducible, and consistent, and it is very close to being a true "material".

The perception that lignocellulosics are the only substance with property shortcomings is not entirely accurate. Table I shows the properties of several commonly used engineering materials. Wood and other lignocellulosics swell as a result of moisture, but metals, plastics, and glass also swell, as a result of increases in temperature. Lignocellulosics are not the only substances that decay. Metals oxidize and concrete deteriorates as a result of moisture, pH changes, and microbial action. Lignocellulosics and plastic burn, but metal and glass melt and flow at high temperatures. Lignocellulosics are excellent insulating substances; the insulating capacity of other materials ranges from poor to good. Furthermore, the strength-to-weight ratio is very high for lignocellulosic fibers when compared to that for almost every other fiber. Given these properties, lignocellulosics compare favorably to other products.

Composition of Lignocellulosics

To understand how wood and other lignocellulosics can become materials in the materials science sense, it is important to understand the properties of the components of the cell wall and their contributions to the composite.

Wood and other lignocellulosics are three-dimensional, naturally occurring, polymeric composites primarily made up of cellulose, hemicelluloses, lignin, and small amounts of extractives and ash (Figure 2). Cellulose is a linear polymer of D-glucopyranose sugar units (actually the dimer, cellobiose) linked in a beta configuration. The average cellulose chain has a degree of polymerization of about 9,000 to 10,000 units. There are several types of cellulose in the cell wall. Approximately 65 percent of the cellulose is highly oriented, crystalline, and not accessible to water or other solvents (1). The remaining cellulose, composed of less oriented chains, is only partially accessible to water and other solvents as a result of its association with hemicellulose and lignin. None of the cellulose is in direct contact with the lignin in the cell wall.

The hemicelluloses are a group of polysaccharide polymers containing the pentose sugars D-xylose and L-arabinose and the hexose sugars D-glucose, D-galactose, D-mannose, and 4-0-methylglucuronic acid. The hemicelluloses are

A

B

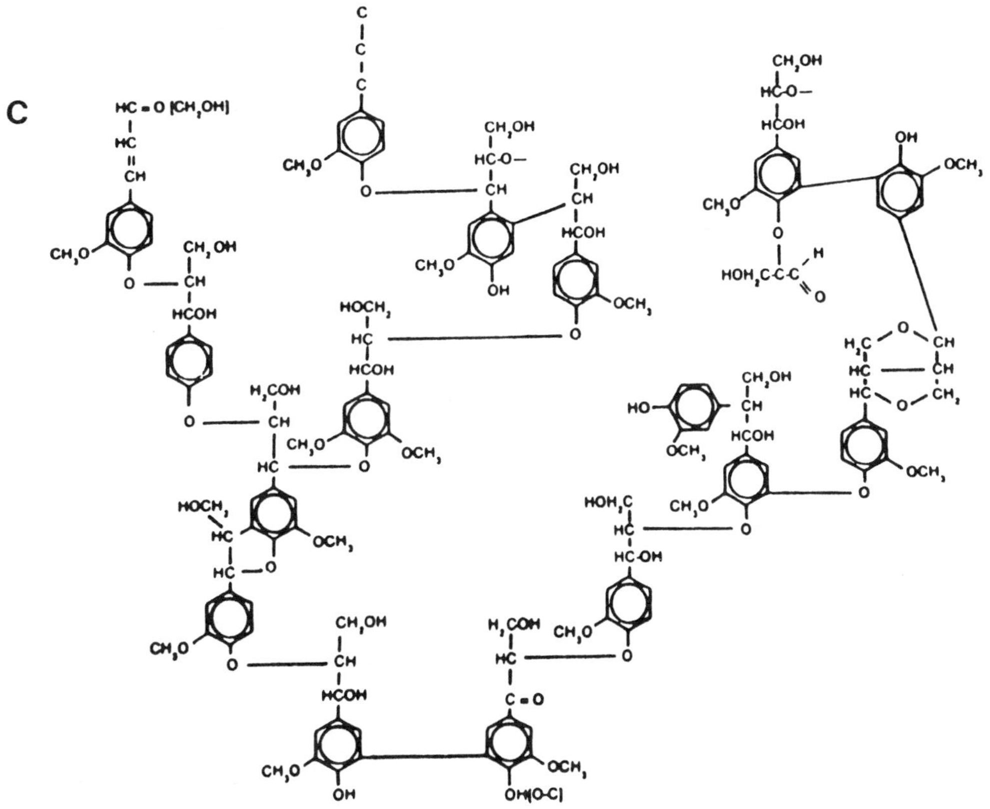

Figure 2. Major polymer components of wood: A, partial structure of cellulose; B, partial structure of a hemicellulose; C, partial structure of a softwood lignin.

Table I. Properties of Engineering Materials

Material	Insulation	Strength to Weight Ratio	Degradation			Swelling	Determination
			Thermal	UV Light	Acid		
Lignocellulosics	Good	High	Yes (fire)	Yes[a]	Yes[b]	Yes (moisture)	Yes[c]
Metals	Poor	Low	Yes (melt)	No	Yes	Yes (temp)	Yes[b]
Plastics	Poor to good	Fair	Yes (fire)	Yes/No	Yes/No	Yes (temp)	No
Glass	Poor	Low	Yes (melt)	No	No	Yes (temp)	No
Concrete	Poor	Low	No	No	Yes	No	Yes[d]

[a] Limited to surface.
[b] Oxidation.
[c] Caused by organisms.
[d] Caused by moisture.

not crystalline, but are highly branched with a much lower degree of polymerization than cellulose. Hemicelluloses vary in structure and sugar composition depending on the source.

Lignins are composed of nine carbon units derived from substituted cinnamyl alcohol; that is, coumaryl, coniferyl, and syringyl alcohols. Lignins are highly branched, not crystalline, and their structure and chemical composition is a function of their source. Lignins are associated with the hemicelluloses and plays a role in the natural decay resistance of the lignocellulosic substance.

The remaining mass of lignocellulosics consists of water, organic soluble extractives, and inorganic materials. The extractive materials are primarily composed of cyclic hydrocarbons. These vary in structure depending on the source and play a major role in natural decay and insect resistance and combustion properties of lignocellulosics. The inorganic portion (ash content) of lignocellulosics can vary from a few percent to over 15 percent, depending on the source.

The strongest component of the lignocellulosic resource is the cellulose polymer. It is an order of magnitude stronger than the lignocellulosic fiber, which, in turn, is almost another order of magnitude stronger than the intact plant wall. The fiber is thus the smallest unit that can be used to produce high-yield lignocellulosic composites. Using isolated cellulose requires a loss of over 50 percent of the cell wall substance.

Most lignocellulosics have similar properties even though they may differ in chemical composition and matrix morphology. Table II shows the chemical composition and Table III the dimensions of some common lignocellulosic fibers.

Table II. Chemical Composition of Some Common Fibers

Type of Fiber	Chemical Component (percent)				
	Cellulose	Lignin	Pentosan	Ash	Silica
Stalk fiber					
Straw					
Rice	28–36	12–16	23–28	15–20	9–14
Wheat	29–35	16–21	26–32	4.5–9	3–7
Barley	31–34	14–15	24–29	5–7	3–6
Oat	31–37	16–19	27-38	6–8	4–6.5
Rye	33–35	16–19	27–30	2–5	0.5–4
Cane					
Sugar	32–44	19–24	27–32	1.5–5	0.7–3.5
Bamboo	26–43	21–31	15–26	1.7–5	0.7
Grass					
Esparto	33–38	17–19	27–32	6–8	—
Sabai	—	22.0	23.9	6.0	—
Reed					
Phragmites communis	44.75	22.8	20.0	2.9	2.0
Bast fiber					
Seed flax	47	23	25	5	—
Kenaf	31–39	15–19	22–23	2–5	—
Jute	45–53	21–26	18–21	0.5–2	—
Leaf fiber					
Abaca (Manila)	60.8	8.8	17.3	1.1	—
Sisal (agave)	43–56	7–9	21–24	0.6–1.1	—
Seed hull fiber					
Cotton linter	80–85	—	—	0.8–1.8	—
Wood					
Coniferous	40–45	26–34	7–14	<1	—
Deciduous	38–49	23–30	19–26	<1	—

The cell wall polymers and their matrix make up the cell wall and in general are responsible for the physical and chemical properties of lignocellulosics.

Properties of Lignocellulosic Composites

The properties of lignocellulosic composites can be modified to be consistent, predictable, reproducible, continuous, and uniform. Before we discuss how lignocellulosic properties can be modified, it is important to describe how the components of lignocellulosic composites interact.

Table III. Dimensions of Some Common Lignocellulosic Fibers

Type of Fiber	Fiber Dimension (mm)		
	Length	Average Length	Width
Cotton	10–60	18	0.02
Flax	5–60	25–30	0.012–0.027
Hemp	5–55	20	0.025–0.050
Manila hemp	2.5–12	6	0.025–0.040
Bamboo	1.5–4	2.5	0.025–0.040
Esparto	0.5–2	1.5	0.013
Cereal straw	1–3.4	1.5	0.023
Jute	1.5–5	2	0.02
Deciduous wood	1–1.8	—	0.03
Coniferous wood	3.5–5	—	0.025

Interaction of Chemical Components. Figure 3 shows how the components of lignocellulosics interact in various processes. Lignocellulosics change dimensions with changing moisture content because the cell wall polymers contain hydroxyl and other oxygen-containing groups that attract moisture through hydrogen bonding (1,2). The hemicelluloses are mainly responsible for moisture sorption, but the accessible cellulose, noncrystalline cellulose, lignin, and surface of crystalline cellulose also play major roles. Moisture swells the cell wall, and the lignocellulosic expands until the cell wall is saturated with water. Beyond this saturation point, moisture exists as free water in the void structure and does not contribute to further expansion. This process is reversible, and the lignocellulosic shrinks as it loses moisture.

Lignocellulosics are degraded biologically because organisms recognize the carbohydrate polymers (mainly the hemicelluloses) in the cell wall and have very specific enzyme systems capable of hydrolyzing these polymers into digestible units. Biodegradation of the high molecular weight cellulose weakens the lignocellulosic cell wall because crystalline cellulose is primarily responsible for the strength of the lignocellulosic (3). Strength is lost as the cellulose polymer undergoes degradation through oxidation, hydrolysis, and dehydration reactions. The same types of reactions take place in the presence of acids and bases.

Lignocellulosics exposed outdoors undergo photochemical degradation caused by ultraviolet light. This degradation takes place primarily in the lignin component, which is responsible for the characteristic color changes (4). The lignin acts as an adhesive in lignocellulosic cell walls, holding the cellulose fibers together. The surface becomes richer in cellulose content as the lignin degrades. In comparison to lignin, cellulose is much less susceptible to ultraviolet light degradation. After the lignin has been degraded, the poorly bonded carbohydrate-rich fibers erode easily from the surface, which exposes new lignin to further degradative reactions. In time, this "weathering" process causes the wood surface to become rough and can account for a significant loss in surface fibers.

Lignocellulosics burn because the cell wall polymers undergo pyrolysis reactions with increasing temperature to give off volatile, flammable gases. The hemicellulose and cellulose polymers are degraded by heat much before the lignin (4). The lignin component contributes to "char" formation, and the charred layer helps insulate the lignocellulosic from further thermal degradation.

Modification of Properties. Although solid wood may never be accepted by the academic community as a true engineering material because of its inconsistent properties, lignocellulosic composites conceivably can be made to conform to the definition of a material. The process of taking wood apart and converting it into fiber removes knots, splits, checks, cracks, twists, and bends from the wood. The mixture of sapwood, heartwood, springwood, and summerwood results in a consistent composite.

If the fiber is reconstituted using a high performance adhesive, very uniform lignocellulosic composites can be produced. However, properties such as dimensional instability, flammability, biodegradability, and degradation caused by acids, bases, and ultraviolet radiation are still present in the composite, which will restrict its use for many applications. If the properties of the composite do not meet the end-use performance requirements, then the question is what must be changed, improved, and redesigned so that the composite does meet end-use expectations? Although such changes are regularly made in the textile, plastic, glass, and metal industries, they are rarely made in the lignocellulosic industry.

Because the properties of lignocellulosics result from the chemistry of the cell wall components, the basic properties of a lignocellulosic can be changed by modifying the basic chemistry of the cell wall polymers. Chemical modification technology can greatly improve the properties of lignocellulosics (5–8). Dimensional stability can be greatly improved by "bulking" the cell wall either with simple bonded chemicals or by impregnation with water-soluble polymers (6). For example, acetylation of the cell wall polymers using acetic anhydride produces a lignocellulosic composite with a dimensional stability of 90 to 95 percent as compared to that of a control composite (7,8). Many different types of lignocellulosic fibers have been acetylated. When acetyl content resulting from the acetylation of many different types of fibers is plotted as a function of reaction time, all data points fit a common curve (Figure 4) (9). A maximum weight percent gain of about 20 percent was reached in 2 h of reaction time, and an additional 2 h increased the weight gain only by about 2 to 3 percent. Without a strong catalyst, acetylation using acetic anhydride alone levels off at approximately 20 weight percent gain for the softwoods, hardwoods, grasses, and water plants. This means that this simple procedure can be used to impart dimensional stability to many different types of mixed lignocellulosic fibers without the need for separation before reaction because all the fibers react at the same rate.

Sorption of moisture is mainly due to hydrogen bonding of water molecules to the hydroxyl groups in the cell wall polymers. Replacing some

Biological Degradation
  Hemicelluloses > > > Accessible Cellulose > Non Crystalline Cellulose > > > > Crystalline Cellulose > > > > > Lignin

Moisture Sorption
  Hemicelluloses > > Accessible Cellulose > > > Non Crystalline Cellulose > Lignin > > > Crystalline Cellulose

Ultraviolet Degradation
  Lignin > > > > > Hemicelluloses > Accessible Cellulose > Non Crystalline Cellulose > > > Crystalline Cellulose

Thermal Degradation
  Hemicelluloses > Cellulose > > > > > Lignin

Strength
  Crystalline Cellulose > > Non Crystalline Cellulose + Hemicelluloses + Lignin > Lignin

Figure 3. Cell wall polymers responsible for lignocellulosic properties.

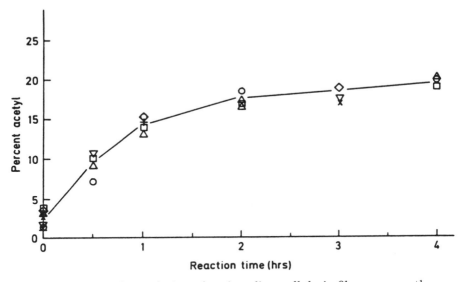

Figure 4. Rate of acetylation of various lignocellulosic fibers. o, southern yellow pine; □, aspen; △, bamboo; ◊, bagasse; ×, jute; +, pennywort; ▽, water hyacinth.

hydroxyl groups on the cell wall polymers with acetyl groups reduces the hygroscopicity of the lignocellulosic material.

The reduction in equilibrium moisture content (EMC) at 65 percent relative humidity of acetylated fiber compared to unacetylated fiber, as a function of the bonded acetyl content, yields a linear plot (Figure 5) (*9*). Even though the points shown in Figure 5 came from many different lignocellulosic fibers, they all fit a common line. A maximum reduction in EMC was achieved at about 20 percent bonded acetyl. Extrapolation of the plot to 100 percent reduction in EMC would occur at about 30 percent bonded acetyl. Because the acetate group is larger than the water molecule, not all hygroscopic hydrogen-bonding sites are covered.

The fact that EMC reduction as a function of acetyl content is the same for many different lignocellulosic fibers indicates that this technology can be used to improve dimensional stability of a diverse mixture of lignocellulosic fibers without separation.

Biological resistance can be improved by several methods. Bonding chemicals to the cell wall polymers increases resistance as a result of lowering the EMC point below that needed for microorganism attack and changing the conformation and configuration requirements of the enzyme–substrate reactions (*3*). Acetylation has also been shown to be an effective way of slowing or stopping the biological degradation of lignocellulosic composites. Toxic chemicals can also be added to the lignocellulosics to stop biological attack. This is the focus of the wood preservation industry (*4*).

Resistance to ultraviolet light radiation can also be improved by bonding chemicals to the cell wall polymers, which reduces lignin degradation, or by adding polymers to the cell matrix, which helps hold the degraded fiber structure together so water leaching of the undegraded carbohydrate polymers cannot occur (*4*).

The fire performance of lignocellulosic composites can be greatly improved by bonding fire retardants to the lignocellulosic cell wall (*4*). Soluble inorganic salts or polymers containing nitrogen and phosphorus can also be used. These chemicals are the basis of the fire retardant wood treating industry.

The strength properties of a lignocellulosic composite can be greatly improved in several ways (*10*). Lignocellulosic composites can be impregnated with a monomer and polymerized *in situ* or impregnated with a preformed polymer. In most cases, the polymer does not enter the cell wall and is located in the cell lumen. Mechanical properties can be greatly enhanced using this technology (*4*). For example, composites impregnated with methyl methacrylate and polymerized to weight gain levels of 60 to 100 percent show increases in density of 60 to 150 percent, compression strength of 60 to 250 percent, and tangential hardness of 120 to 400 percent compared to untreated controls. Static bending tests show increases in modulus of elasticity of 25 percent, modulus of rupture of 80 percent, fiber stress at proportional limit of 80 percent, work to proportional limit of 150 percent, and work to maximum load of 80 percent, and, at the same time, a decrease in permeability of 200 to 1,200 percent.

Future of Lignocellulosic Composites

Lignocellulosics will be used in the future to produce a wide spectrum of products ranging from very inexpensive, low performance composites, to expensive, high performance composites. The wide distribution, renewability, and recyclability of lignocellulosics can expand the market for low-cost composites. Fiber technology, high performance adhesives, and fiber modification can be used to manufacture lignocellulosic composites with uniform densities, durability in adverse environments, and high strength.

Products having complex shapes can also be produced using flexible fiber mats, which can be made by textile nonwoven needling or thermoplastic fiber melt matrix technologies. The mats can be pressed into any desired shape and size. Figure 6 shows a lignocellulosic fiber combined with a binder fiber to produce a flexible mat. The binder fiber can be a synthetic or natural fiber. A high performance adhesive can be sprayed on the fiber before mat formation, added as a powder during mat formation, or be included in the binder fiber system. Figure 7 shows a complex shape that was produced in a press using fiber mat technology. Within certain limits, any size, shape, thickness, and density is possible. With fiber mat technology, a complex part can be made directly from a lignocellulosic fiber blend; the present technology requires the formation of flat sheets prior to the shaping of complex parts.

Many different types of lignocellulosic fibers can be used to make composites. Wood fiber is currently used as the major source of fiber, but other sources can work just as well. Problems such as availability, collection, handling, storage, separation, and uniformity will need to be considered and solved. Some of these are technical problems whereas others are a matter of economics. Currently available sources of fiber can be blended with wood fiber or used separately. For example, large quantities of sugar cane bagasse fiber are available from sugar refineries, rice hulls from rice processing plants, and sunflower seed hulls from oil extraction plants.

All of this technology can be applied to recycled lignocellulosic fiber as well as virgin fiber, which can be derived from many sources. Agricultural residues, all types of paper, yard waste, industrial fiber residues, residential fiber waste, and many other forms of waste lignocellulosic fiber base can be used to make composites. Recycled plastics can also be used as binders in both cost-effective and value-added composites. Thermoplastics can be used to make composites for nonstructural parts, and thermosetting resins can be used for structural materials.

More research will be done to combine lignocellulosics with other materials such as glass, metal, inorganics, plastic, and synthetic fibers to produce new materials to meet end-use requirements (*10*). The objective will be to combine two or more materials in such a way that a synergism between the materials results in a composite that is much better than the individual components.

Wood–glass composites can be made using the glass as a surface material or combined as a fiber with lignocellulosic fiber. Composites of this type can have a very high stiffness to weight ratio.

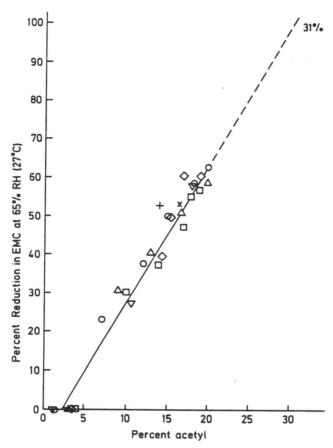

Figure 5. Reduction in equilibrium moisture content as a function of bonded acetyl content for various acetylated lignocellulosic fibers. o, southern yellow pine; □, aspen; △, bamboo; ◊, bagasse; ×, jute; +, pennywort; ▽, water hyacinth.

Figure 6. Lignocellulosic fiber mixed with a synthetic fiber to form a non-woven fiber mat.

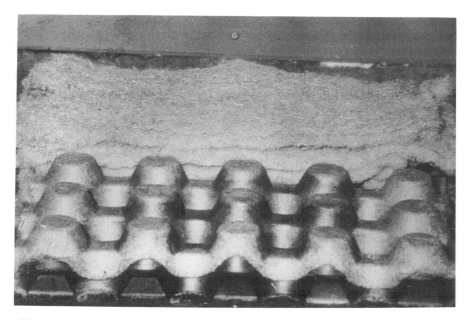

Figure 7. Fiber mats can be molded into any desired thickness, density, size, and shape.

Metal films can be overlayed onto lignocellulosic composites to produce a durable surface coating, or metal fibers can be combined with fiber in a matrix configuration in the same way metal fibers are added to rubber to produce wear-resistant aircraft tires. A metal matrix offers excellent temperature resistance and improved strength properties, and the ductility of the metal lends toughness to the resulting composite. Another application for metal matrix composites is in the cooler parts of the skin of ultra-high-speed aircraft. Technology also exists for making molded products using perforated metal plates embedded in a phenolic-coated wood fiber mat, which is then pressed into various shapes.

Lignocellulosic fibers can also be combined in an inorganic matrix. Such composites are dimensionally and thermally stable, and they can be used as substitutes for asbestos composites.

Lignocellulosics can also be combined with plastic in several ways. In one case, thermoplastics are mixed with lignocellulosics and the plastic material melts, but each component remains as a distinct separate phase. One example of this technology is reinforced thermoplastic composites, which are lighter in weight, have improved acoustical, impact, and heat reformability properties, and cost less than comparable products made from plastic alone. These advantages make possible the exploration of new processing techniques, new applications, and new markets in such areas as packaging, furniture, housing, and automobiles.

Another combination of lignocellulosics and plastic is lignocellulosic–plastic alloys. At present, the plastic industry commonly alloys one polymer to another. In this process, two or more polymers interface with each other, resulting in a homogeneous material. Lignocellulosic–plastic alloys are possible through grafting and compatibilization research.

Combining lignocellulosics with other materials provides a strategy for producing advanced composites that take advantage of the enhanced properties of all types of materials. It allows the scientist to design materials based on end-use requirements within the framework of cost, availability, renewability, recyclability, energy use, and environmental considerations.

Literature Cited

1. Stamm, A.J. *Wood and Cellulose Science*; The Ronald Press Co.: New York, 1964.
2. Rowell, R.M.; Banks, W.B. *Gen. Tech. Rep. FPL–GTR–50*; USDA Forest Service, Forest Products Laboratory, Madison, WI, 1985. 24 pp.
3. Rowell, R.M.; Esenther, G.R.; Youngquist, J.A.; Nicholas, D.D.; Nilsson, T.; Imamura, Y.; Kerner–Gang, W.; Trong, L.; Deon, G. In *Proc. IUFRO wood protection subject group*, Honey Harbor, Ontario, Canada; Canadian Forestry Service, 1988, pp. 238–266.
4. Rowell, R.M., Ed. *The Chemistry of Solid Wood*, Advances in Chemistry Series No. 207; American Chemical Society, Washington, DC, 1985.
5. Rowell, R.M.; Konkol, P. *Gen. Tech. Rep. FPL–GTR–55*; Forest Products Laboratory, Madison, WI, 1987, 12 pp.
6. Rowell, R.M.; Youngs, R.L. *Res. Note FPL–0243*; Forest Products Laboratory, Madison, WI, 1981, 8 pp.
7. Rowell, R.M. Commonwealth Forestry Bureau, Oxford, England, *6 (12)*, pp. 363–382, 1983.
8. Rowell, R.M. In *Composites: Chemical and Physicochemical Aspects*, Vigo, T.L.; Kinzig, B.J., Eds., Advances in Chemistry Series, American Chemical Society: Washington, DC, in press.
9. Rowell, R.M.; Rowell, J.S. In *Cellulose and Wood-Chemistry and Technology*, Schuerch, C., Ed. *Proc. 10th Cellulose Conference*, Syracuse, NY, May 1988. John Wiley and Sons, Inc.: New York, pp. 343–353, 1989.
10. Youngquist, J.A.; Rowell, R.M. In *Proc. 23rd International Particleboard/Composite Materials Symposium*, Maloney, T.M., Ed., Pullman, WA, April 1989.

RECEIVED March 4, 1991

Chapter 3

Opportunities for the Cost-Effective Production of Biobased Materials

Helena L. Chum[1] and Arthur J. Power[2]

[1]Chemical Conversion Research Branch, Solar Energy Research Institute, 1617 Cole Boulevard, Golden, CO 80401
[2]Arthur J. Power and Associates, Inc., 2360 Kalmia Avenue Boulder, CO 80304

> There are a number of evolving technologies that convert wood, wood waste, agricultural residues, and recycled fibers into new biobased materials such as composites that incorporate additional plastics and other materials. The use of non-woven technologies to produce mats that can be molded into shaped parts is very common in the composites industry. The direct use of the thermally treated fibers is possible; however, chemically modified fibers improve the chemical, physical, and biological properties significantly. To illustrate the potential of biobased fibers in composites, a technoeconomic analysis of wood acetylation is presented. Worldwide trends in the use of chemically modified wood products are described in relation to their possible market penetration relative to the current products.

This contribution is based on a few recent reviews of biobased materials (*1-3*), literature and patent searches, analysis of market trends of the North American forest products industry (e.g., *4*) compared to those of a few other countries (e.g., *1,5*), and a detailed technoeconomic assessment of one option of chemical modification of fibers, the acetylation of wood. Several chapters in this book are also excellent sources of current literature and trends, principally for the use of recycled fibers. This chapter discusses current utilization trends for value-added forest products, such as those produced from chemical modifications. Some of these processes are becoming technically feasible and are beginning to penetrate the marketplace, primarily in Japan.

New products need to be cost-effective and produced in an environmentally benign way. The product properties must also be reproducible, regardless of possible variations in the quality of the incoming feedstocks, a characteristic that impacts wood natural polymer feedstocks more than petroleum-derived plastics. A deep understanding of the materials industry (both petroleum-derived and forest-based) in the particular country or region is essential, since the overall assessment of the effects of the new process (or product) needs to be carried out. To displace a high-sales, low-cost material produced in a process that operates profitably in an integrated industry is difficult. In most cases, the capital has been depreciated already and the plants can continue to operate profitably for a reasonable time. One needs to search for opportunities in the industry, such that the materials that we want to produce from biomass or wastes can either be substantially less expensive (both in capital and operating costs) or provide some special market advantages to the investing company. The risk of the new technology needs to be largely reduced, for instance, through joint funding between government and industry, to a point where industry can invest in a new venture and operate the plant profitably.

Cradle-to-Grave Materials Cycle Analysis

The materials cycle includes drilling/mining/harvesting raw materials such as oil, rock, coal, sand, or renewable resources. Through extraction, refining, or processing, these raw materials are converted into various bulk materials such as chemicals, metals, cement, fibers, and paper. These in turn can be made into engineering materials such as plastics, alloys, ceramics, crystals, and textiles through appropriate processing. The engineering materials are then fabricated into products, devices, structures, and machines, through design, manufacture, and assembly. The materials have to be disposed after their useful service life. At each step of the materials cycle residues are generated, which need to be minimized.

The materials cycle thus includes all associated emissions to the biosphere from each individual step on the way to the consumer products as well as post-consumer use. In the disposal area, many alternatives exist, such as landfill, reuse, recycle, and recovery of materials and energy. Many of these strategies are still under development.

Analyses of the complete life cycle of the product are becoming more common, albeit yet imprecise, since there are uncertainties in many of the assumptions used (e.g., references 6,7). Uncertainties exist in data gathering on the processes themselves (substantial proprietary data). In addition, the existing processes are not static, and the improvements have to be taken into account in a timely manner. Discrepancies between results of current comparative analyses by different industrial groups can be explained by using technologies at different developmental stages. The product recycle rate that can be achieved is another variable, as is the degree of landfill degradation that will result if that were the chosen option for disposal of that product.

A total life cycle analysis that includes the raw materials, process, and all associated environmental costs is desirable before reaching the decision on what is the best material for a given application. Society may move toward such a decision-making process, different from the current low (or acceptable) cost to the producer or consumer. Any analysis based on our current knowledge will only provide a partial answer, which needs to be updated periodically as new knowledge is developed, since these issues affect major industrial sectors, municipalities, and government in the United States.

Better-defined calculation methodology will continue to emerge. The size of the envelope of the cradle-to-grave analysis is important, and it should be recognized that we know progressively less about the second, third, and higher order impacts, which need to be considered as the envelope gets bigger. Thus, it is more difficult to quantify effects such as greenhouse and ozone layer damage. Nevertheless, all of these factors influence market decisions on what products to make, how to dispose of them, or better yet, how to design a product life cycle in such a way that it moves from one useful application into another, until its final environmentally acceptable disposal. Thus, reutilization of waste fibers through other applications that maintain the material in useful life for another 5-10 years, and that are followed by an environmentally acceptable disposal (or yet another use) are important. It is, however, abundantly clear that our society needs and will continue to need inputs of virgin renewable materials, synthetic plastics, and their combinations.

Use of renewable resources for materials will certainly have very little impact on the carbon cycle provided that 1) the energy use in the production process is small and not fossil fuel derived, and 2) the environmental impact of the process is small and minimizes energy/materials input. All materials have finite useful life cycles, but the released carbon from renewable resources can be fixed again in a sustainable way by the planted forests, in a short time. Therefore, the use of renewables coupled with a sustainable environmentally sound forest production and management program can have a significant positive effect on the carbon cycle, reducing the net increase in atmospheric carbon dioxide.

Plastics Industry

The plastics industry is complex. It is intimately related to the chemicals/petroleum refining industries, which responded to the globalization of world markets by becoming international in scope. The chemicals industry is operating profitably in the United States where many of the chemicals operations have consolidated over the past five years (*8*), with a concomitant substantial increase of foreign investments in this sector of the economy and an associated trade surplus. The plastics industry is moving today towards value-added operations, which could displace, in the materials area, a number of the conventional wood-based materials in the buildings and construction area, and in the commodities area, low-value plastics production.

The total resins/plastics volume (includes miscellaneous plastic products) was 58 billion lb in 1988 in the United States. In the past 50 years of the domestic plastics industry, a growth of more than 50 billion lb has been achieved, primarily because of the production of better quality and more durable plastics (*9*). The value of these shipments was $87 billion in 1988 (*10*). Key applications include packaging (31%), construction and buildings (22%), and consumer and institutional products (10%). The U. S. organic chemicals industry produces about 320 billion lb yearly with a value of about $110 billion (*11*).

Energy profiles for plastics (*12*), which include raw materials extraction, processing into refined materials, and fabrication into finished products, range from 40,000 Btu/lb for high density polyethylene (HDPE), polypropylene (PP), polyvinyl chloride (PVC), urea formaldehyde (UF), and resins to 90,000 Btu/lb for nylon-6,6. Plastics are lightweight and finished products consume a relatively low amount of energy per unit.

Forest-products Based Industries

The paper and paperboard industry in the United States produced 130 billion lb in 1986, with a total sales value of $75 billion (*13*). In 1988, the apparent per capita consumption of these products was 700 lb in the United States versus an average world figure of 100 lb. An additional penetration of 9 billion lb was achieved by 1988, thus bringing the U.S. production to 139 billion pounds. This constitutes a significant fraction of the worldwide production of 453 billion lb (*14*).

The production of lumber, plywood and veneer, and panels in similar units was 131 billion lb (*15,16*), with a value of shipments of $56 billion (1986). Thus, the forest-products industry expressed as the sum of these two industry segments shipped well over 260 billion lb in products (much more than the equivalent raw materials to manufacture them) and had a combined value of more than $130 billion.

Energy profiles for wood products (*12*), which include raw materials extraction, processing into refined materials, and fabrication into finished products, are 4,000- 10,000 Btu/lb for pulpwood, veneers, recycled boxboard, and furniture, in order of increasing energy of manufacture; 19,000 Btu/lb for unbleached kraft paper; and 25,000 Btu/lb for bleached kraft paper. Wood is a lightweight material and its density depends on the species; wood products can be very lightweight, compared to metals and ceramics.

Synthetics and Renewables

On a per pound basis, simple plastics costs range from $0.30-$0.80; wood costs range from $0.005-$0.02 for wood waste, to $0.05-$0.07 for some fibers, to $0.18- $0.30 for pulp (February 1991 - pulp and paper company data and Chemical Marketing Reporter for a variety of individual plastics) It is not surprising, based on the low cost, that wood flour is a major filler for materials such as phenolic resins and many other plastic products (*15*). These filler applications do not take advantage of the wood-polymer composite fiber properties.

The specific product densities need to be taken into account when comparing energy consumption for their manufacture. Also, in comparing energy consumption and environmental impact, the recycle rate of both products has to be taken into account, as well as the amount of energy recovery and/or emissions production in the disposal processes. The U.S. market is driven mostly by the price/cost of the materials. The Japanese market, for instance, has some properties of the wood such as natural warmth and aesthetic value as market drivers for high-performance products based on improved wood materials.

Opportunities for market penetration by additional biobased materials exist, but good targets are more difficult to identify, and certainly involve producing value-added, performance-enhanced materials obtained with as little additional cost as possible. In the future, countries with less stringent environmental requirements for permitting and operating chemical/wood-based processes may become more attractive to some of these industries. However, at present, both the chemical and the broad forest products industries are indeed operating profitably in the United States (8,9). Overall, industry globalization considerations and waste minimization strategies need to be taken into account at the start of a new business venture in a materials area.

From the figures above, both the chemicals/plastics and the forest/pulp/paper industries have made significant progress and have reached the marketplace with a variety of products the consumers wanted (purchased). Competition will continue among these traditional industries. The plastics industries will continue to try to penetrate into traditional wood-products construction markets based on increased durability and special properties of products (10).

Composites

In many cases, the products that will reach the consumer are not simple plastics, but **composite materials**, which combine lightness and strength. Composites consist of high-strength fibers (e.g., glass, ceramic, or synthetic fibers) held together by a continuous phase or matrix (rubber, metal or ceramic, or plastic materials), which gives it definite shape and durable surfaces. In these materials the mechanical loads to which the structure is subjected during its use are supported by the fiber reinforcements. The matrix adds strength by transferring the load to the fibers, while protecting them from fracture. This is an area of growth for both synthetic and renewable products industries (2).

A recently disclosed example of such a matrix is the Louisiana Pacific proposal to substitute an adaptation of particle-board for whole lumber. The market force is the growing shortage of valuable timber and the eventual need to use wood chips from short-rotation forestry as the primary source of wood fiber. Such a development would expand markets for wood-gluing resins and related chemicals. Another example is Parallam, a product from MacMillan Bloedel Ltd. Strands of veneer (including imperfect materials) are mixed with adhesive, aligned, and then pressed together to give large beam structures of any desired length. This process edits out the weaknesses in the wood and averages different species properties that could be employed in the product manufacture (1).

At one extreme of the composites market are the very high-value, low-volume materials that meet stringent specifications of aerospace and defense industries. Composites for construction and automotive industries have major potential markets with large volumes, but, contrary to aerospace or industrial markets, require low-cost products (17,18).

Automobile weight reduction alone can lead to an increase in fuel economy of about 0.7 mpg per 100 lb saved in the weight of the vehicle (19). In 1986, 169 million automobiles, buses, and trucks consumed roughly 124 billion gallons of motor fuels (20). Thus the impact of weight reduction on overall fuel consumption can be very high. Between 1978 and 1984, approximately 16% of the total 36% increase in fleet fuel economy could be attributed to automobile weight reductions (19). As oil prices decrease, automobile downsizing strategies fail, so that materials substitution with high-performance, light-weight, and low-cost materials becomes one of the best strategies for lowering oil consumption in that sector through weight reduction.

Plastics composites increasingly penetrate the automotive industry because they provide manufacturing strategies for parts consolidation and increase design flexibility. Both features lead to increased production flexibility in an extremely competitive market which is undergoing globalization. In the past ten years, not only plastics composites using fiberglass as the prime reinforcing material but also those that employ wood and other renewable fibers as reinforcements have been used (2,21). A comparison of fiber properties of synthetic and wood materials is shown in Figure 1, which highlights wood fiber's lightweight and good compromise of tensile properties compared to E-type glass fibers principally when weight and costs are included: wood fibers at $0.15-$0.25/lb and fiberglass at $0.8-$1.0/lb (2).

A few strategies exploring wood fiber reinforcements are being used in commercial products or are in the developmental stage. Examples include the use of web technologies for mat formation (2,21), reviewed by Brooks in reference 22:
1) Ligna-Tock mats of wood fiber, textiles, and phenolic resin can be used for interior door panel substrates; these mats are pressed at about 450°F for 2.5 min, and yield shaped forms, with some restrictions as to the formed angles (e.g., hanging brackets cannot be formed). These materials are in the process of being used in Chrysler's automobile parts, and will be provided by Ligna Corp., Nashville, Illinois.
2) Fibrit Mat, which uses similar starting materials as described before, for use in a variety of parts, including hanging brackets, door panel substrates, and interior van molded parts (5 ft by 10 ft). These mats are more flexible compared to those above, and require less time at temperature for parts formation.
3) ASA Mats have been developed at Cadillac ASA, Troy, Michigan, but the development through that company ceased in 1989. These are air-lay non-woven mats, which employed fibers such as kenaf or heat-treated steam-exploded wood fibers (23) alone or in combination with other fibers. A wide variety of parts have been produced and tested, such as headliners, interior door panels, hood cores with

resin-impregnated glass faces, and interior door panels. These developments continue today at other companies (22,24).

Combinations of wood fibers and thermoplastics such as PP or PE are being used. D. A. I. Goring provided many insights to improve the utilization of wood fibers in these types of composite structures (25). They are:

o   Chemical and mechanical pulps are proving to be excellent fiber components of composites. For materials of equivalent stiffness, pulp/HDPE composites are proving to be superior to glass-filled HDPE on a cost-weight basis. However, low impact strength and compatibility have to be overcome.

o   Research in this area has been done mostly with pulps produced for papermaking, where much effort is put into making fibers flexible with hydrophilic surfaces. In the case of mechanical pulps, a large expenditure of energy is required. It is possible mechanically to produce stiff fibers coated with lignins at much lower energy consumption. Such pulps would be useless for papermaking but may prove to be the ideal fiber for a composite.

These technologies will provide an opportunity to use waste fibers and recycled fibers, as well as to develop additional market applications (see other chapters in this book and reference 26). The lifetime of the interior panels produced with some of the commercial technologies described above can be substantially improved if the fibers are chemically modified. Whether the cost of the chemical modification will drive these products outside of the price range of competitive automotive composites is still an open question. For that reason, a technoeconomic assessment of one such concept was developed. The preliminary results are reported here.

The Case Of Acetylated Wood

Chemical and biological resistance can be imparted to wood fibers through chemical modification. Products made thus far from acetylation fiber technology show 1) high dimensional stability both in the thickness and lineal directions; 2) a high level of rot resistance; 3) a low degree of thermal expansion; 4) smooth surfaces that do not require further sanding; 5) a uniform density throughout the product wall; 6) no increase in toxicity of the wood; 7) high strength--both wet and dry; 8) a high degree of UV radiation stability; and 9) no change in flammability. Acetylated veneers can be pressed along with the acetylated fiber mat to yield veneer-faced fiber-backed products (27).

A technoeconomic assessment of the acetylation of wood process (28) was made, based on the process scheme shown in Figure 2. The financial results of the study are summarized in Tables I and II, which present the estimated capital and processing costs for each of the two general classes of feedstocks. The process based on particle feedstocks has the lowest fixed investment, about 50% of the level

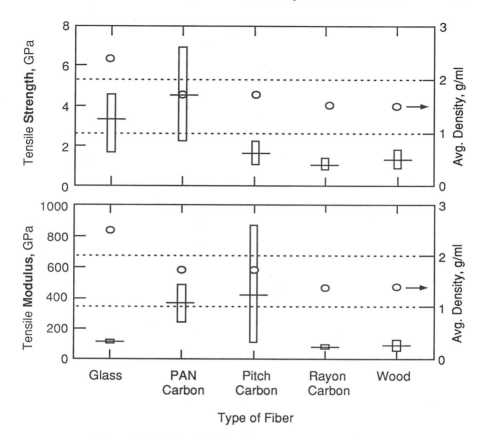

Figure 1. Reinforcing Fiber Properties Comparison
Tensile <u>Strength</u> or <u>Modulus</u> and Density

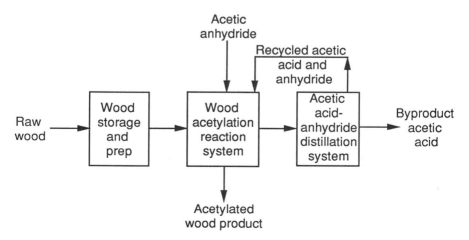

Figure 2. Process Flow Diagram Acetylated Wood

TABLE I
CAPITAL INVESTMENT SUMMARY
For the production of 500,000,000 pounds per year
of acetylated wood
Figures given in millions of dollars

	Particles	Fibers
Major Equipment Cost	3.78	7.65
Installation Costs	4.53	9.18
Engineering/Contingencies	4.15	8.41
Offsites	3.46	7.02
Total Fixed Investment	**15.92**	**32.26**
Working Capital	14.13	14.13
Total Utilized Investment	**30.05**	**46.39**

TABLE II
ESTIMATED PROCESSING COSTS

In the production of 500,000,000 pounds per year of acetylated wood. Wood feedstock assumed at zero cost. Add 1.7 cents/lb for a feed cost of $40/dry ton to the processing costs to account for that feed cost. Capital recovery factor is 15% of the total fixed investment and the plant life is assumed to be 20 years.

Figures given in cents per pound of product

	Particles	Fibers
Raw materials	14.4	14.4
Utilities	-	0.9
Operating Costs/Expenses	0.4	0.8
Byproduct credits (5.5)	(5.5)	
Processing Cost	**9.3**	**10.6**
Taxes/depreciation	0.4	0.5
Manufacturing cost	**9.7**	**11.1**

Conversion Factors:	To Convert	To	Multiply By
	¢/lb	¢/kilogram	2.205
	$/ton	$/tonne	1.12
	$/lb	$/kilogram	2.205

required for processing fibers. Particles processing is simpler in the solids-handling operations and requires much less specialized and expensive related equipment.

Processing costs are dominated by the net cost of the acetic anhydride raw material, which corresponds to the total acetic anhydride cost minus the acetic acid by-product cost (14.4 - 5.5 = 8.9 cents per pound of the product), which is assumed to be the same for these feedstocks. This element of cost is 96% in the case of fibers. The latter case is higher-cost principally because it requires more operators in the more complex solids-handling systems, with their expensive equipment. In both cases, acetic anhydride was assumed to be purchased at $0.22/lb and acetic acid was credited at $0.10/lb.

Processing costs rise approximately linearly with increasing feed moisture content and wood cost as illustrated in Table III. Note that processing costs are much more sensitive to wood cost than to feed moisture content. Additional moisture leads to excess consumption of acetic anhydride, which is converted into low-value acetic acid.

TABLE III
PROCESSING COST VERSUS WOOD COST AND MOISTURE CONTENT

In the production of 500,000,000 pounds per year of acetylated wood

Processing Cost, ¢/lb	Influence of Feed Moisture Weight Percent		Influence of Wood Cost $/Dry Ton	
	Particles	Fibers	Particles	Fibers
8	3.0	-	-	-
10	7.5	3.0	5	-
12	9.5	7.7	55	15
14	12.5	10.0	105	65
16	13.5		120	

Acetylated Wood Commercialization Trends

The costs of wood acetylation appear to be acceptable for their use in composite products, principally if the starting fibers have low cost, as is the case with wood waste, recycled fibers, and agricultural residues. There remains to be demonstrated in longer term testing that the compatibility of the acetylated wood fibers with the other components of the composites is acceptable and reproducible.

While there has been an interest in the United States both at government and industrial sectors to continue the development of technologies for cost-effective wood acetylation (29), most of the interest in this area has been in other countries, such as Sweden and Japan, with the latter leading the trend towards commercialization of products (5). For instance, Daiken Trade and Industry Co., Ltd. is beginning to commercialize acetylated wood under the name of α-wood for a number of higher value applications. Examples include: flat plate for large speaker systems, exploring the good dimensional stability afforded by the chemical modification; bath tubs and bathroom doors, which exploit the considerably slower biodegradation of acetylated wood compared to the unmodified wood; kitchen applications; and wood plastic composites for heavy traffic walkways (30). At least nine companies are currently exploring resin reinforced wood in Japan from patent searches (5). Applications considered are indeed those afforded by the increased performance of the material which are of higher value. On a comparative basis, there seems to be less industrial interest by U.S. companies in these technologies. However, the wood acetylation appears to be technically feasible, and reasonably cost effective for selected applications. Whether additional substantial niche markets can be found remains to be seen; this market-driven research is largely in the hands of the private sector.

Conclusions

There are many future opportunities in the area of biobased materials and principally composites (1-3,25,27) which are worthwhile to pursue. Examples are summarized as follows:

o   Explore non-pulp/paper fiber technologies for the production of renewable fiber reinforcements. Include wood, new fiber feedstocks (e.g., kenaf), and agricultural residues, as well as recycled fibers. Processes designed for multiple feedstocks may have greater flexibility and could achieve lower costs.

o   Add value through chemical modification of less cost than the increased value. Therefore, it is important to carry out technoeconomic assessment of the chemical modification at an early stage.

o   Understand the chemistry of the chemical modification and its relationship to property buildup. Avoid repetition of empirical developments of the past, which did not advance the field because they did not establish the property relationship.

o   Use evolving powerful analytical techniques that allow better understanding of the materials interfaces and properties.

o   Develop individual fibers/crystallites, ideal for composite reinforcement, through innovative concepts for fiber producing domains that have not been thoroughly investigated. Bacterial celluloses may provide at least insight in the research directions necessary to achieve appropriate crystallite domains from wood. Bacterial celluloses are also proving to be very interesting materials themselves.

o   Establish multidisciplinary teams with polymer/wood/ agriculture/chemical industry components to define target materials. The markets are worldwide; develop appropriate regional technologies.
o   Establish properties/parts (products) philosophy for renewable fibers in composites for selected markets.
o   Design integrated product life cycle -- understand chemical/mechanical/ physical properties that affect aging, recovery, and reuse of the product. A cradle to grave analysis of the product is necessary.

Acknowledgements

This work describes part of the strategy and rationale of the U.S. Department of Energy's Advanced Industrial Materials (AIM) Program, in the element Lightweight and Biobased Material now called Organic Polymeric Materials - Synthesis and Disposal by Design. Over its past four years, the program has had the support of many outstanding DOE Program Managers: Drs. James Eberhardt, Stanley Wolf, and Marvin Gunn. At the start, it was part of the Energy Conversion and Utilization Technologies Division of DOE. This program element is managed for DOE through the Solar Energy Research Institute, with the help of several people, including the ORNL researchers involved in the program, and in particular, Dr. P. Angelini and J. Carpenter. The program participants need to be acknowledged for their contribution to the development of the rationale and numerous discussions on how to achieve our program goals: R. Evans, W. G. Glasser, K. Grohmann, M. Himmel, D. K. Johnson, L. Mathias, L. Moens, R. Narayan, C. Rivard, R. Rowell, G. Tesoro, and R. Young. In addition, the participation of many individuals in the community is gratefully acknowledged. They continue to help us direct our efforts to the most important areas: R. M. Brown, D. Braunstein, Jr., S. H. Hunter Brooks, H. M. Chang, W. Daly, N. Fishman, D. A. I. Goring, B. Gunnesin, M. Hearon, H. Hergert, J. Hyatt, S. Loud, N. Lewis, E. Malcolm, G. Maffia, B. Rasmussen, M. Rutenberg, K. V. Sarkanen, W. Surber, and V. Stannett. The suggestions and recommendations from the AIM Guidance and Evaluation Board are gratefully acknowledged, in particular, those of Dr. R. Isaacson. N. Greer from the SERI Technical Library performed much of the research to uncover the statistical data used in this paper. This information is quite difficult to obtain since the industry is reported in a fragmentary way. Her special efforts are gratefully acknowledged.

Literature Cited

1. *Assessment of Biobased Materials*; Chum, H. L., Ed.; Solar Energy Research Institute: Golden, CO, 1989; SERI/TR-234-3610. Note: The *Assessment of Biobased Materials* is being reprinted by Noyes Data Corporation, 1991.
2. Chum, H. L. "Structural Materials for the Automobile of the Future -- Composite Materials," In *Assessment of Biobased Materials;* Chum, H. L.,

Ed.; Solar Energy Research Institute: Golden, CO, 1989; SERI/TR-234-3610, and references therein.
3. Zadorecki, P.; Michell, A. J. *Polym. Compos.* **1989**, *10*(2), 69, and references therein.
4. *Markets 90-94, The Outlook for North American Forest Products*; Ransom, K. L. Ed.; Widman Management Ltd./Miller Freeman Publications, Inc.: Vancouver, B.C./San Francisco, CA, 1990.
5. *New Development of Building Materials, Examples of New Techniques and New Products Developed in Japan*; Hongo, M., Ed.; Toray Research Center, Inc.: Tokyo, Japan, 1990.
6. Hocking, M. B. *Science* **1991**, *251*, 504.
7. Franklin Associates, Ltd. *Resource and Environmental Profile Analysis of Foam Polystyrene and Bleached Paperboard Containers*. Report to the Council for Solid Waste Solutions; Franklin Associates, Ltd.: Prairie Village, KS, 1990.
8. Dosher, J. R. "Outlook for Foreign Investments in United States Petrochemicals," In *CMRA Washington and the Chemical Industry - Partners in World Trade*, Washington, D.C., Chemical Marketing Research Association: New York, December 1989.
9. Alper, J.; Nelson, G. L. *Polymeric Materials*; American Chemical Society: Washington DC, 1989.
10. Society for the Plastics Industry, Inc. *Facts and Figures of the U.S. Plastics Industry*, SPI: Washington, DC, 1989.
11. Busche, R. M. *Appl. Biochem. Biotechnol.* **1989**, *20/21*, 655.
12. Bider, W. L.; Seitter, L. E.; Hunt, R. G. *Total Energy Impacts of the Use of Plastics Products in the United States*; Franklin Associates, Ltd.: Prairie Village, KS, 1985.
13. Cavaney, R. *Pulp Pap Int.* **1989**, *31*, July, 37.
14. Cavaney, R. *Pulp Pap Int.*, **1987**, *29*, July, 89.
15. Ulrich, A. H. "U. S. Timber Production, Trade, Consumption, and Price Statistics 1950-86." U.S. Department of Agriculture, Forest Service, Miscellaneous Publication No. 1460, June 1988.
16. *Millwork, Plywood, and Structural Wood, Not Elsewhere Classified*: 1990; U.S. Department of Commerce, Bureau of Census, 1987 Census of Manufactures MC87-I-24B, Industry Series. (Industries 2431, 2434, 2435, 2436, and 2439), issued April 1990.
17. Manson, J. A.; Sperling, L. H. *Polymer Blends and Composites*, Plenum: New York, 1977.
18. Fishman, N. Presented at the 196th Meeting of the American Chemical Society, Los Angeles, CA, September, 1988; Abstract CME 1.
19. Kulkarni, S. V. *State-of-the-Art Reviews in Selected Areas of Materials for Energy Conservation*, Carpenter, Jr., J. A., Ed.; ORNL/CF-83/291, Oak Ridge National Laboratories: Oak Ridge, Tennessee, 1984, pp 185-219.
20. U.S. Statistics; Federal Highway Administration: National Transportation and Safety Board, 1986.

21. *Web Processing and Converting Technology and Equipment*; Satas, D., Ed.; Van Nostrand: New York, New York, 1984.
22. Brooks, S. H. W. In *Proceedings of the 1990 Nonwovens Conference*, TAPPI: Atlanta, GA, p 87.
23. Chornet, E.; Overend, R. P. *Fractionation of Lignocellulosics, Centre Quebecois de Valorisation de la Biomasse*; Sainte-Foy: Quebec, Canada, 1986; Chapter 3.
24. Brooks, S. H. W. "Web Products." In *Proceedings of the Forest Products Research Conference on Opportunities for Combining Wood with Nonwood Materials*; (Madison, Wisconsin), October 4-6, 1988; Brooks, S. H. W. private communications to H. L. Chum, 1988-1989; Brooks, S. H. W. "Self-supporting Moldable Fiber Mat," PCT International Patent WO82/01507, May 13, 1982, and references therein.
25. Goring, D. A. I. In *Assessment of Biobased Materials*; Chum, H. L., Ed.; SERI/TR-234-3610, Solar Energy Research Institute: Golden, CO, 1989, Chapter 5 and references therein.
26. Brooks, S. H. W. In *Proceedings of the Disposing of Disposables Conference*, Washington DC, Association of the Non-Woven Fibers Industry: Cary, NC, 1989.
27. Rowell, R. M.; Young, R. A. In *Assessment of Biobased Materials*; Chum, H. L., Ed.; SERI/TR-234-3610, Solar Energy Research Institute: Golden, CO, 1989, Chapter 2 and references therein.
28. Rowell, R. M.; Simonson, R.; Tillman, A.-M. European Patent Application 85850268.5, 1985; U. S. Patent 4,804,384 1989.
29. USDA Forest Products Laboratory Programs, U.S. DOE Advanced Industrial Materials Program, and DARPA, are a few examples of sponsored government programs. House, C. B.; Leichti, R. J. U. S. Patent 4,388,378, 1983.
30. Examples of patents assigned to Daiken Kogyo KK on wood chemical modification, especially acetylation: JP Patent 63056403, 1988; JP Patent 62236702, 1987; JP Patent 62225553, 1987; U.S. Patent 4,592,962, 1986; JP Patent 58007310, 1983; JP Patent 58007308, 1983; JP Patent 63199604, 1988; JP Patent, 63056402, 1988; JP Patent 61040104, 1986; JP Patent, 61035208, 1986.

RECEIVED March 28, 1991

Chapter 4

Lignocellulosic—Plastic Composites from Recycled Materials

John Youngquist[1], George E. Myers[1], and Teresa M. Harten[2]

[1]Forest Products Laboratory, U.S. Department of Agriculture Forest Service, One Gifford Pinchot Drive, Madison, WI 53705–2398
[2]U.S. Environmental Protection Agency, 26 West Martin Luther King Drive, Cincinnati, OH 45268

> Reductions are urgently needed in the quantities of municipal solid waste (MSW) materials that are currently being landfilled. Waste wood, waste paper, and waste plastics are major components of MSW and offer great opportunities as recycled ingredients in wood fiber-plastic composites. With cooperation from the U.S. Environmental Protection Agency, a research and development program has been initiated at the Forest Products Laboratory and University of Wisconsin–Madison to investigate the processing, properties, and commercial potential of composites containing these recycled ingredients. Two processing technologies are being employed—melt blending and nonwoven web. Some past research studies are briefly reviewed to illustrate the behavior of wood fiber-polyolefin composites.

Use of the word "waste" projects a vision of material with no value or useful purpose. However, technology is evolving that holds promise for using waste or recycled wood and plastics to make an array of high-performance products that are, in themselves, potentially recyclable. Preliminary research at the USDA Forest Service, Forest Products Laboratory (FPL), and elsewhere (1,2) indicates that recycled plastics such as polyethylene, polypropylene, or polyethylene terephthalate can be combined with wood fiber waste to make useful reinforced thermoplastic composites. Advantages associated with these composite products include lighter weight and improved acoustic, impact, and heat reformability properties—all at a cost less than that of comparable products made from plastics alone. In addition, previous research has shown that composite products can possibly be reclaimed and recycled for the production of second-generation composites (3).

This chapter not subject to U.S. copyright
Published 1992 American Chemical Society

Since the turn of this century, the United States has been dubbed "the throw-away society" (4), generating approximately 50 percent of the world's solid and industrial waste. If present trends continue, this nation's solid-waste stream will increase from 157.7×10^6 t in 1986 to 192.7×10^6 t by the year 2000 (5). It is increasingly difficult to dispose of the growing volume of municipal solid wastes (MSW) in landfills because most people will not tolerate MSW in their neighborhood. In the next 15 years, 75 percent of our landfills will be closed, and by the year 2000, this nation will be short 56×10^6 t per year of disposal capacity. As political and conservation pressures increase, it is anticipated that the recovery of recyclable materials from the solid-waste stream will increase from 16.9 to 23.9×10^6 t by the year 2000 (5).

In many uses, wood fiber-plastic composites can be opaque, colored, painted, or overlaid. Consequently, recovered fibers or resins used in these composites do not require the extreme cleaning and refinement needed when they are to be used as raw materials for printing paper or pure plastic resins. This fact greatly reduces the cost of wood-fiber plastic composites as raw materials and makes composite panels or molded products an unusually favorable option for the recycling of three of our most visible and troublesome classes of MSW—newspapers, waste wood, and plastic bottles.

Research at the FPL on wood fiber-plastic composites focuses on two very different manufacturing technologies—melt blending (for example, extrusion and injection molding) and air laid or nonwoven web. A host of new natural fiber-synthetic plastic fiber products can be made because of the increased processing flexibility inherent in both of these technologies. These products can be produced in various thicknesses, from a thin material of 3 mm to structural panels up to several centimeters thick. A large variety of applications are possible because of the many alternative configurations of the product. A list of potential products could include

1. storage bins for crops or other commodities,
2. furniture components, including both flat and curved surfaces,
3. automobile and truck components,
4. paneling for interior wall sections, partitions, and door systems,
5. floor, wall, and roof systems for light-frame construction, and
6. packaging applications, including containers, cartons, and pallets.

Taken together, these two processing techniques provide options for balancing performance properties and costs, depending upon the product application under consideration.

The overall goal of this paper is to illustrate the potential that currently exists for manufacturing thermoformable composites from waste materials such as waste wood, paper, and plastics. We first discuss the availability of such waste materials from MSW streams and the desirability of developing the means to recycle them. We then describe how these composites are made and point out why materials from the MSW should be suitable ingredients. Next, we illustrate the properties of such composites by describing some recent research on the effects of composition and processing variables, using both virgin and recycled ingredients. We follow this with an outline of the research and development needed to convert wood fiber and plastics into durable products.

Municipal Solid Waste as a Source of Lignocellulosic Fiber and Plastics

A considerable amount of data are related to the inventory of the U.S. MSW stream (Table I). In 1986, paper and paperboard, wood, and plastics in the MSW stream accounted for approximately 65, 5.8, and 10.3×10^6 t, respectively. By the year 2000, these figures are expected to increase to 86.5, 6.1, and 15.7×10^6 t annually. In addition to the wood fiber in the MSW stream, vast quantities of low-grade wood, wood residues, and industry-generated wood waste in the form of sawdust, planer shavings, and chips are now being burned or otherwise disposed of.

Table I. Estimated Distribution of Materials in Municipal Solid Waste in 1986[a]

Source	Amount in municipal solid waste	
	(Percentage)	Weight ($\times 10^6$ t)
Paper and paperboard	35.6	56.1
Yard waste	20.1	31.7
Metals	8.9	14.0
Food waste	8.9	14.0
Glass	8.4	13.3
Plastics	7.3	11.5
Textiles	2.0	3.2
Wood	4.1	6.5
Rubber–leather	2.8	4.4
Miscellaneous inorganics	1.9	3.0
Total	100.0	157.7

[a]SOURCE: Adapted from ref. 6.

The data in Table I include all the residential waste products but not all the industrial waste materials. Data are available for the total volume or weight of certain wood-based products in the MSW stream, such as paper, packaging, and pallets, but only incomplete information is available for timber thinnings, leaves, industrial production wastes, bark, and sawdust. These latter categories of wood waste also represent potentially valuable sources of raw materials.

A number of problems are associated with the use of these waste materials, some of which include collection, analysis, separation, clean up, uniformity, form, and costs. Assuming that these problems can be overcome on a cost-effective basis, some of the resultant reclaimed materials should be useful ingredients for a range of valuable composites, from low-cost, high-volume materials to high-cost, low-volume materials for a wide range of end-use applications.

Source separation and recycling not only extend the life of landfills by removing materials from the MSW stream, but they also make available large volumes of valuable raw materials for use by industry in place of virgin resources. Industrial use of such materials reduces both costs for raw materials and the energy it takes to make a finished product (4). The main requirement is that the recycled ingredients meet the quality and quantity requirements of the consuming production operation.

Thermoformable Composites as Outlets for Waste Paper, Wood, and Plastics

Two general types of thermoformable composites exist, distinguished by their very different manufacturing processes. Both processes allow and require differences in composition and in the lignocellulosic component. The two processes used to produce thermoformable composites are melt blending and air laying or nonwoven mat formation.

A typical composition for a melt-blended composite is 40 to 60 weight percent wood flour or cellulose pulp fiber with a powdered or pelletized thermoplastic such as polypropylene or polyethylene. In the melt-blending process, the wood-based fiber or flour is blended with the melted thermoplastic matrix by shearing or kneading. Currently, the primary commercial process employs twin screw extruders for the melting and mixing; the mixture is extruded as sheets that are subsequently shaped by thermoforming into the final product. Limits on the melt viscosity of the mixture restrict the amount of fiber or flour to about 50 weight percent as well as the length of the fibers that can be used. In any event, fiber length is limited by fiber breakage as a result of the high shear forces during melt mixing.

In contrast, the nonwoven mat technology involves a room temperature air mixing of lignocellulosic fibers (or even fiber bundles) with fibers of the thermoplastic. The resultant mixture passes through a needling step that produces a low-density mat in which the fibers are mechanically entangled. That mat must then be shaped and densified by a thermoforming step. With this technology, the amount of lignocellulosic fiber can be greater than 90 weight percent. In addition, the lignocellulosic fiber can be precoated with a thermosetting resin; for example, phenol-formaldehyde. After thermoforming, the product possesses good temperature resistance. Because longer fibers are required, this product can achieve better mechanical properties than that obtained with the melt-blending process. In contrast, high wood fiber contents lead to increased moisture sensitivity.

It is virtually certain that virgin ingredients can be replaced by some recycled ingredients in melt blending and nonwoven mat formations for many applications. For example, the thermoplastic polymer might be totally or partially replaced by high-density polyethylene (HDPE) from milk bottles, polyethyleneterephthalate (PET) from beverage bottles, or even nonsegregated plastic mixtures from MSW. Large quantities of a variety of industrial waste plastics are also available and should be considered. The virgin lignocellulosic component might be replaced by fibers from waste paper or wood. These substitutions offer potential benefits in reducing both the MSW problem and the cost of the composite processes. In some cases, we can also reasonably expect the properties of the composite to be improved; for example, by substituting waste paper fibers for wood flour in the melt-blending process.

Currently, the primary application of the thermoformed composites, both melt blended and air laid, is for interior door panels and trunk liners in automobiles. As noted, additional large-volume, low-to-moderate cost applications are expected in areas such as packaging (trays, cartons), interior building panels, and door skins.

Recent Research on Wood Fiber-Thermoplastic Composites

The following is not intended to be a comprehensive review of recent research on wood fiber-thermoplastic composites. Instead, we simply illustrate the effects of some important composition and processing variables in the composite processes, including preliminary indications of the effects of recycled ingredients.

Composites Made by Melt Blending. The 1980s brought a resurgence of research into various aspects of melt-blended composites made from wood-based flour or fiber in virgin thermoplastic matrices. For example, Kokta and his colleagues have published numerous papers in this area, emphasizing improvements in the filler-matrix bond through coupling agents and grafting of polymers on cellulosic fiber surfaces (7–9). Klason and colleagues carried out extensive investigations on the effects of several polymer and fiber types and the influence of a variety of processing aids and coupling agents (10,11). Woodhams and others examined several types of pulp fiber in polypropylene (PP) and HDPE (12,13). Shiraishi and colleagues showed improvements in mechanical properties as a consequence of using high-molecular-weight maleated PP instead of normal PP (14,15). Finally, Maiti and Hassan measured the effects of wood flour on the melt rheology of PP (16).

At the FPL and the University of Wisconsin (UW), we have investigated in some detail the influence of a low-molecular-weight maleated PP (Eastman's Epolene E–43) (The use of trade or firm names in this publication is for reader information and does not imply endorsement by the U.S. Department of Agriculture of any product or service.) on the mechanical and physical properties of wood flour and PP–extruded composites. For example, Figure 1 shows that small amounts of E–43 produce small but statistically significant improvements in maximum flexural strength and that those effects are greater when smaller wood flour particles are present (17). Moreover, Figure 2 indicates that the major gains in strength are produced at less than 2 percent E–43 and that the optimum extrusion temperature for this property is about 210°C (18). In contrast, Figure 3 shows that notched impact energy is significantly reduced by both the E–43 and the high extrusion temperature (18). Other experiments indicated that the E–43 probably is not acting as a true coupling agent but instead has some effectiveness as a dispersing agent (19).

Publications are beginning to appear on the effect of recycled ingredients on the behavior of melt-blended lignocellulosic-polyolefin composites. Selke and colleagues showed that composites from aspen fiber and once-recycled blow-molding HDPE from milk bottles possessed essentially equivalent strength and

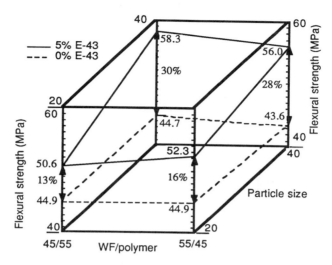

Figure 1. Effect of wood flour particle size, wood flour to polymer ratio, and Epolene E–43 on flexural strength of melt-blended composites. (WF is wood flour.) SOURCE: Adapted from ref. 17.

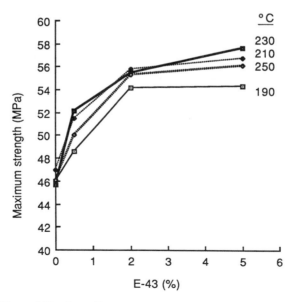

Figure 2. Effect of Epolene E–43 concentration and extrusion temperature on flexural strength of melt-blended wood flour-polypropylene composites. SOURCE: Adapted from ref. 18.

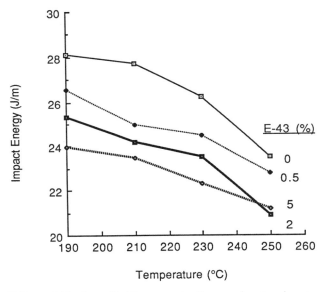

Figure 3. Effect of Epolene E-43 concentration and extrusion temperature on notched impact energy of melt-blended wood flour-polypropylene composites. SOURCE: Adapted from ref. 18.

modulus properties as those of composites made with virgin HDPE; however, impact energy was reduced (*1,2*). Woodhams and others found that composites made from PP and pulp fibers or fiberized newspaper (ONPF) possessed strength and impact properties that were very similar and apparently much superior to those for wood flour (WF)–PP systems (*13*). In preliminary work at the FPL, we compared the properties of WF–PP and WF–HDPE systems with an ONPF–HDPE composite (Figures 4 and 5). The differences between WF–PP and WF–HDPE are qualitatively consistent with expectations based on the lower strength and greater flexibility of HDPE relative to PP. Also, strength has been improved by substituting ONPF for WF.

Composites Made by Nonwoven Mat Technology. Numerous articles and technical papers have been written and several patents have been issued on both the manufacture and use of nonwoven fiber webs containing combinations of textile and cellulosic fibers. This technology is particularly well-known in the consumer products industry. For example, Sciaraffa and others (*20*) have been issued a patent for producing a nonwoven web that has both fused spot bonds and patterned embossments for use as a liner material for disposable diapers. Bither (*21*) has found that polyolefin pulps can serve as effective binders in nonwoven products. Many additional references could be cited in this area.

S. Hunter Brooks (*22*) has published a review of the history of technology development for the production and use of moldable wood products and air-laid, nonwoven, moldable mat processes and products. The first moldable wood product using the wet slurry process was developed by Deutche Fibrit during 1945 to 1946 in Krefeld, West Germany (*22*). A moldable cellulose composition containing pine wood resin was patented by Roberts (*23*) in 1955, and a process for producing molded products from this composition was patented by Roberts (*24*) in 1956. This composition consisted of a mixture of comminuted cellulose material and at least 10 percent of a thermoplastic pine wood resin derived from the solvent refining of crude rosin. Both of Roberts' patents were assigned to the Weyerhaeuser Company.

From 1966 to 1968, a series of patents (*25–29*) was issued to Caron and others and assigned to the Weyerhaeuser Company. These patents cover the use of a wood fiber-thermoplastic resin system in conjunction with a thermosetting resin system.

In the early 1970s, Brooks developed a process that produced a very flexible mat using a thermoplastic Vinyon fiber in combination with a thermosetting resin system (*22*). The mat was fed through an oven to melt and set the Vinyon fiber without affecting the setting of the thermosetting resin component. This process was patented in 1984 by Doerer and Karpik and was assigned to the Van Dresser Corporation (*30*).

Brooks also developed an interesting method of recycling waste cellulosic materials for the production of medium-density fiberboard and paper (*31*). After being shredded, sorted from other waste materials like plastic and metal, and steamed, the cellulosic fibers and fiber bundles are abraided under heat and pressure to break down any hydrogen bonds and to soften any lignin and other resins. The resultant cellulose fibers are then mixed with resin, formed

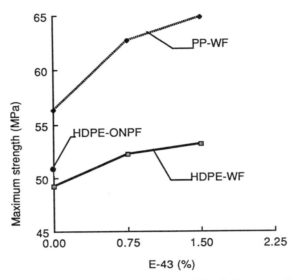

Figure 4. Flexural strength of several melt-blended composites at 50/50 weight ratio filler to polymer. (WF is wood flour, ONPF is newspaper fiber, PP is polypropylene, HDPE is high-density polyethylene.)

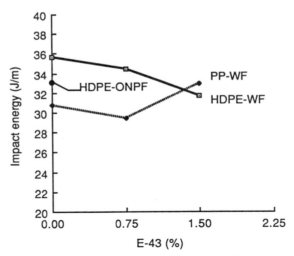

Figure 5. Notched impact energy of several melt-blended composites at 50/50 weight ratio filler to polymer. (WF is wood flour, ONPF is newspaper fiber, PP is polypropylene, HDPE is high-density polyethylene.)

into a mat, and consolidated under pressure to form flat fiberboard and paper products.

A general review of the opportunities for combining wood with nonwood materials was written by Youngquist and Rowell (*32*). This review included a discussion of the materials and the properties of composites consisting of wood-biomass, wood-metals, wood-plastics, wood-glass, and wood-synthetic fibers.

In a recently published paper, Youngquist and others (*33*) reported on the mechanical and physical properties of wood-plastic fiber composites made with an air-formed dry-process technology. This paper reported the effect of species, the ratios of WF to PP, and the type of plastic fiber or plastic fiber-thermosetting resin blends on the mechanical and dimensional stability properties of pressed panels having a density of 1 g/m^3.

Krzysik and Youngquist (*34*) reported on the bonding of air-formed wood-polypropylene fiber composites using maleated polypropylene as a coupling agent between the hydrophilic wood and the hydrophobic polyolefin materials.

Tables II and III present mechanical and physical property data obtained at FPL on nonwoven webs made into composites having densities of 1.0 and 1.2 g/m^3, respectively.

In many cases, the mechanical properties for each of the three formulations within a density grouping (e.g., 1.0 g/m^3) were statistically different from one another. As the panel densities increased, the level of magnitude of the mechanical properties generally increased correspondingly. Also note that the dimensional stability property levels for each of the formulations at the two density levels did not exhibit large variations when comparing one density level to another. In most cases, as each formulation changed, all properties, with a few exceptions, increased in that order. The use of a thermoplastic polyolefin-like polypropylene greatly improved the dimensional stability properties of the composite, compared to the polyester copolymer-containing composite. These results can probably be explained by the fact that the polypropylene melts, and to some extent, partially encapusulates the wood fibers. In all cases, the ability to absorb impact energy of the polyester copolymer-containing composite was superior to that of the polypropylene-containing composite. This can be attributed to the fact that the polyester maintains a fibrous matrix, whereas the polypropylene fibers melt and flow under heat and pressure. Phenolic resin, in combination with wood and polyester homopolymer fibers, forms a composite that has greatly improved mechanical properties compared to that of the other two composite formulations tested.

A number of preliminary trials have been conducted at the FPL, using recycled office wastepaper, shredded old newspapers, dry fiberized old newspapers, and fiberized demolition waste wood. The raw paper materials, which did not have the ink removed, were reduced to a suitable form using a number of reduction methods, as noted previously. The demolition waste was first sorted mechanically, and then manually, to remove nonwood materials, washed, and then fiberized using a pressurized refiner. These recycled wood-based fibers were then air mixed with virgin polyester or polypropylene, transferred by an air stream to a moving support bed, needled, and subsequently formed into a continuous, low-density mat of intertwined fibers. When polyester fibers were

Table II. Properties of Nonwoven Web Composite Panels[a]
(Density 1.0 g/m^3)

Property measured[c]	Formulation[b]		
	90H/10PE	90H/10PP	80H/10PE/10PR
Static bending strength (MPa)	23.3	25.2	49.3
Static bending modulus (GPa)	2.82	2.99	3.57
Tensile strength (MPa)	13.5	12.5	27.7
Tensile modulus of elasticity (GPa)	3.87	4.20	5.07
Internal bond (MPa)	0.14	0.28	0.81
Impact energy (J)	26.7	21.5	34.3
24-h water-soak thickness swell (percent)	60.8	40.3	21.8
24-h water-soak water absorbency (percent)	85.0	54.7	45.1
2-h water-boil thickness swell (percent)	260.1	77.5	28.2
2-h water-boil water absorbency (percent)	301.6	99.5	55.7
Linear expansion at 65 percent relative humidity (percent)[d]	0.38	0.25	0.76

[a]Averages connected by bars located below the numbers are not statistically different; averages not connected by bars are statistically different at 0.05 significance level.
[b]90H/10PE = 90 percent hemlock and 10 percent polyester copolymer, 90H/10PP = 90 percent hemlock and 10 percent polypropylene, and 80H/10PE/10PR = 80 percent hemlock, 10 percent polyester homopolymer, and 10 percent phenolic resin.
[c]Specimens were conditioned at 65 percent relative humidity and 20°C.
[d]Linear expansion values are based on 0 percent moisture content.

Table III. Properties of Nonwoven Web Composite Panels[a]
(Density 1.2 g/m³)

Property measured[c]	Formulation[b]		
	90H/10PE	90H/10PP	80H/10PE/10PR
Static bending strength (MPa)	36.5	36.8	76.4
Static bending modulus[d] (GPa)	4.59	3.81	5.42
Tensile strength (MPa)	20.2	17.8	42.2
Tensile modulus of elasticity (GPa)	6.56	6.36	7.80
Internal bond (MPa)	0.28	0.52	1.44
Impact energy[d] (J)	31.1	27.4	36.6
24-h water-soak thickness swell (percent)	57.1	42.7	23.6
24-h water-soak water absorbency (percent)	57.5	37.5	28.4
2-h water-boil thickness swell	245.0	88.3	32.4
2-h water-boil water absorbency	227.3	79.6	38.6
Linear expansion at 65 percent relative humidity (percent)[e]	0.33	0.20	0.86

[a] Values connected by bars located below the numbers are not statistically different; values not connected by bars are statistically different at 0.05 significance level.
[b] 90H/10PE = 90 percent hemlock and 10 percent polyester copolymer, 90H/10PP = 90 percent hemlock and 10 percent polypropylene, and 80H/10PE/10PR = 80 percent hemlock, 10 percent polyester homopolymer, and 10 percent phenolic resin.
[c] Specimens were conditioned at 65 percent relative humidity and 20°C.
[d] The only statistically different values were between the 90H/10PP and 80H/10PE/10PR formulations.
[e] Linear expansion values are based on 0 percent moisture content.

used, the wood fibers were first sprayed with a liquid phenolic resin, prior to web formation.

Panel-sized mats were cut from each roll. Six to eight mats were selected according to their weight and stacked. Demonstration prototype panels and molded shapes were then produced using a steam-heated press. These prototype samples appeared to be reasonably successful; therefore, a series of experiments will be conducted so that physical and mechanical properties can be determined on a number of different composites made from a variety of recycled wood raw materials.

Research and Development Needs

At the FPL and the UW, we are conducting a program aimed at developing technology to convert recycled wood fiber and plastics into durable products that are themselves recyclable, are environmentally friendly, and will remove the raw materials from the waste stream. This program is being conducted in cooperation with the U.S. Environmental Protection Agency. In support of this goal, we have defined a number of research and development needs. Some of these are as follows:

1. Processing methods—Processing methods must be improved or developed, first, for converting waste wood and waste plastics into forms suitable for subsequent melt-blending and nonwoven web processing and, second, for carrying out the actual processing into composites. In the melt-blending technology, for example, the necessary short fibers from waste paper (ONPF) possess very low-bulk density, and this creates difficulties in handling and in feeding to an extruder. Therefore, preblending or pelletizing steps may be desirable. In the nonwoven web process, for example, both the wood and plastic must be converted to fibers or at least to a long, slender form (paper strips or shreds). For either technology, the processing conditions required for good composite properties must be established for different ingredients or even for different forms of the same ingredients.
2. Database—A database is needed that systematically describes the effects of ingredient, formulation, and processing variables on the physical and mechanical properties of the composites. For example, variables must include, plastic and wood component specifications, ratio of plastic to wood components, and presence of additives such as coupling agents or dispersing aids. The properties of interest include those of the melt, such as melt viscosity, melt strength, thermoformability, and those of the final composite, such as strength, modulus, impact resistance, and moisture sensitivity.
3. Recyclability—It is necessary to establish the extent to which the composites formed with recycled ingredients can be recycled into similar products without undue loss in composite properties.
4. Industrial implementation—An essential aspect of the program is the identification of potential processing limitations and the practical utility of the products. This will require cooperation with industry processors and producers and extension of the database to confirm that composite systems are available to meet processing and product specifications.

Concluding Remarks

In any approach to recycling, the government and the private sector must be full cooperative partners. Government cannot logically mandate the increased use of recyclable materials without industry involvement because only the industrial sector has the technical knowledge and equipment to separate and process solid waste and to make useful, economically viable products from these materials. Industry is the market for recycled resources, and it must be a full partner in all aspects of the process.

As a society, we must take a broad look at our opportunities and the responsibilities that go with them. We must be concerned with the reliable performance of products and also the health and safety of those making and using the products. We must be concerned with the prudent use of renewable resources and reduce our use of products that will deplete our nation's or the world's resources. These limitations can be considered as constraints or as opportunities.

We believe that using recycled raw materials for wood-based composites presents tremendous opportunities for growth, for progress, and for further industry competitiveness in a world that is rapidly consuming many of our nonrenewable resources at an alarming rate.

Literature Cited

1. Selke, S.E.; Yam, K.L.; Gogoi, B.; Lai, C.C. *Abst. Cellulose, Paper, and Textile Div., ACS Meeting.* **1988**, June.
2. Yam, K.; Kalyankai, S.; Selke, S.; Lai, C.C. *ANTEC 1988.* **1988**, pp. 1809–1811.
3. Maldas, D.; Kokta, B.V. *Polym. Composites.* **1990**, *11(2)*, pp. 77–83.
4. New York Legislation Commission on Solid Waste Management. *New York legislative communication.* "The Economics of Recycling Municipal Waste." 1986.
5. Environmental Protection Agency. *Environmental Protection Agency Report.* "Characterization of Municipal Solid Waste in the United Stated 1960 to 2000." 1988.
6. Franklin Associates, Ltd. (Prairie Village, KS). *Environmental Protection Agency Report*, NTIS No. PB87-178323. July 25, 1988. "Characterization of municipal solid waste in the United States, 1960–2000."
7. Kokta, B.V.; Maldas, D.; Daneault, C.; Beland, P. *Polym. Composites*, **1990**. *11(2)*, pp. 84–89.
8. Maldas, d.; Kokta, B.V.; "Daneault, C. *J. Appl. Polym. Sci.* **1989**, *37*, pp. 751–775.
9. Maldas, d.; Kokta, B.V.; Raj, R.G.; Daneault, C. *Polymer.* **1988**, *29*, pp. 1255–1265.
10. Dalvag, H.; Klason, C.; Stromvall, H.-E. *Intern. J. Polym. Mater.* **1985**, *11*, pp. 9–38.
11. Klason, C.; Kubat, J.; Stromvall, H.-E. *Intern. J. Polym. Mater.* **1984**, *10*, pp. 159–187.
12. Woodhams, R.T.; Thomas, G.; Rodgers, D.K. *Polym. Eng. Sci.* **1984**, *24(15)*, pp. 1166–1171.

13. Woodhams, R.T.; Law, S.; Balatinecz, J.J. *Proc. Symp. on Wood Adhesives.* **1990**, Madison, WI, May 16–18.
14. Kishi, H.; Yoshioka, M.; Yamanoi, A.; Shiraishi, N. *Mokuzai Gakkaishi.* **1988**, *34(2)*, pp. 133–139.
15. Takase, S.; Shiraishi, N. *J. Appl. Polym. Sci.* **1989**, *37*, pp. 645–659.
16. Maiti, S.N.; Hassan, M.R. *J. Appl. Polym. Sci.* **1989**, *37*, pp. 2019–2032.
17. Myers, G.E.; Chahyadi, I.S.; Coberly, C.A.; Ermer, D.S. *Intern. J. Polym. Mater.*, in press.
18. Myers, G.E.; Chahyadi, I.S.; Gonzalez, C.; Coberly, C.A.; Ermer, D.S. *Intern. J. Polym. Mater.*, in preparation.
19. Kolosick, P.; Myers, G.E.; Koutsky, J.A. Unpublished data, 1990.
20. Sciaraffa, M.A.; Dhome, D.G.; Bogt, C.M. *U.S. Patent 4,333,979.* 1982.
21. Bither, P. *Proc. Air-Laid and Advanced Forming Conference.* Hilton Head Island, SC., Nov. 16–18, 1980.
22. Brooks, S. H. *Proceedings 1990 Tappi Nonwovens Conference.* 1990, pp. 87–108.
23. Roberts, J. R. *U.S. Patent 2,714,072.* 1955.
24. Roberts, J. R. *U.S. Patent 2,759,837.* 1956.
25. Caron, P.E.; Grove, G.A. *U.S. Patent 3,230,287.* 1966.
26. Caron, P.E.; Grove, G.A. *U.S. Patent 3,261,898.* 1966.
27. Grove, G.A.; Caron, P.E. *U.S. Patent 3,279,048.* 1966.
28. Caron, P.E.; Allen, G.D. *U.S. Patent 3,367,820.* 1968.
29. Caron, P.E.; Grove, G.A. *U.S. Patent 3,265,791.* 1966.
30. Doerer, R.P.; Karpik, J.T. *U.S. Patent 4,474,846.* 1958.
31. Brooks, S. H. *U.S. Patent 3,741,863.* 1973.
32. Youngquist, J.A.; Rowell, R.M. *Proc. 23rd.Washington State University Intern.Particleboard/Composite Mater.Symp.* Washington State University, Pullman, WA. 1989, pp. 141–157.
33. Youngquist, J.A.; Muehl, J.; Krzysik, A.; Tu Xin. In Wang, S. Y.: Tang, R. E., eds. *Proc. Joint Intern. Conf. on Processing and Utilization of Low-Grade Hardwoods and Intern. Trade of Forest-Related Products.* National Taiwan Univ. Wang, S.Y.; Tang, R.E., eds. 1990, pp. 159-162.
34. Krzysik, A.M.; Youngquist, J.A. *Intern. J. Adhesion and Adhesives*, in press.

RECEIVED July 22, 1991

Chapter 5

Compatibilization of Lignocellulosics with Plastics

Ramani Narayan

Michigan Biotechnology Institute and Michigan State University, Lansing, MI 48909

> The blending of lignocellulosic polymers with synthetic polymers leads to immiscible blends whose properties tend to be poor and undesirable. Tailor-made cellulose-polystyrene graft copolymers have been used as compatibilizers/interfacial agents to prepare cellulosic-polystyrene alloys and wood-plastic alloys. The graft copolymers function as emulsifying agents and provide for a stabilized, fine dispersion of the polystyrene phase in the continuous phase of the cellulosic matrix. Transmission electron microscopy and thermal analyses was used as evidence for formation of these compatibilized cellulosic blends (alloys).

An important aspect of today's polymeric materials industry involves the blending of commercially available materials. The objective is to prepare new materials by blending two or more unique polymers to obtain desirable combinations of characteristics imparted by its components, while maintaining an optimized relationship of cost to performance. The blending/alloying of two polymers can produce a homogeneous one-phase material, or result in a two-phase morphology. The presence of a stabilized two-phase morphology is often desirable because it can be organized into a variety of structures. Variations in morphology, imparted by varying domain structures, generally lead to significant changes in physical properties and subsequent end-uses.

The driving force for blends and alloys comes from:
1. Poor economics of new polymer and polymer production, and
2. The need for new materials whose performance-cost ratios can be closely matched to specific applications.

It has been predicted that the worldwide market for polymer blends and alloys will reach a level of 2.4 billion lbs by 1994, at an annual growth rate of 8.3% from 1.6 billion lb in 1989 (1). From being relatively unknown on a broad commercial basis in 1980, it is reported that for every five pounds of resin sold about one pound is a blend or alloy (2). As many as 1000 technical papers are published and another

1000 patents awarded on the subject each year. Several excellent books and reviews have been written (3 - 5). It is, therefore, rather surprising that lignocellulosic and other biomass materials have not seen much use as one of the components of the blend/alloy systems. This paper reviews the use of lignocellulosic biopolymers in blends and alloys with synthetic polymers, with emphasis on our work in compatibilization of lignocellulosics with plastics.

Blends and Alloys

The terms blends and alloys are often used interchangeably possibly because of the convenience of semantics equating the two concepts. While the term "blend" is a general term for the mixture of two or more polymers, the term "alloy" is generally used to describe a specific type of blend, namely a "compatibilized blend" that offers a unique combination or enhancement in properties.

Miscible Blends. Completely miscible blends are relatively uncommon. They exhibit single phase behavior, are thermodynamically compatible, form a molecular solution, give rise to a single glass transition, and are generally optically transparent. There is a smooth variation of properties with composition. An example of a miscible polymer blend system includes the well-known poly(phenylene oxide) (PPO) - polystyrene (PS) blends commercialized by General Electric (GE) under the trade name NORYL. In these blends PS imparts processability to a relatively difficult-to-process PPO. Other examples include poly(vinyl chloride) (PVC) - poly(methylmethacrylate) (PMMA) blends, and poly(styrene-co-acrylonitrile) (SAN) - PVC blends. Miscible polymer blends are useful to overcome specific problems such as processability, as in the case of the PPO-PS blend. Heat distortion, hardness, tensile, creep are some of the additional properties that can be added by blending in another polymer. The goal is to fine tune the properties of a particular polymer to meet specific application needs.

Miscible Cellulosic Blends. In the lignocellulosic area, cellulose acetate is reported (6) to yield miscible blends with polystyrene phophonate esters, and poly(vinyl pyridine) (PVP). A wholly amorphous blend with a single T_g is obtained from CA-polystyrene phosphonate ester systems when annealed above the glass transition temperature of CA (7). Another cellulose derivative, namely nitrocellulose, is miscible with poly(caprolactone) over the full composition range and clearly shows a single T_g at any composition (8). It is more difficult to prepare cellulose-polymer blends than polymer-polymer blends due to the small number of common solvents available (blends cannot be prepared from the melt because cellulose degrades before softening). Total miscibility has never been observed, but miscibility at certain compositions has been shown to occur in many instances(911).

Immiscible Blends. Most polymer blends fall in to this category. The phases undergo gross segregation with minimal interfacial contact between the two phases resulting in poor mechanical properties. Strength and toughness values are minimal and are lower for the blend than for any of the pure components (12-14). This is a direct consequence of their incompatibility arising from negligible entropy of mixing and typical positive heat of mixing. Such blends generally have little value for end-use applications. Examples of such blends are PS/PE, PPO/PE, Nylon/PPO, and Cellulosics/Synthetic polymers.

Immiscible but Compatibilized Blends - Alloys. Immiscible blends which have been "compatibilized" generally exhibit a two-phase stabilized morphology, in which one of the phases is either finely dispersed or co-interpenetrating on a microscopic scale. These alloys show two glass transition temperatures and are opaque if the phases are large enough. The improved compatibility can lead to a dramatic improvement in properties usually because of control and stabilization of the morphology. Ultimately the nature of the domain structures dictates the physical properties and performance of these systems. The extent of interfacial adhesion between the dispersed and continuous phases also contributes to the properties. In the field of polymer blends, such compatibilized mixtures of polymers are referred to as polymer alloys. Concerning the original meaning of the term "alloy" it must be noted for the puritans amongst us that in the field of metallurgy, from which it was borrowed, this usually refers to miscible blends showing a smooth variation of properties with composition. The key concept in preparing polymer alloys with improved properties is the use of compatibilizers or interfacial agents. Block and graft copolymers of the form A-B have been used by the polymer industry as compatibilizers or interfacial agents to improve interfacial adhesion, reduce interfacial tension, and provide for a stabilized and ordered morphology. Understanding the role compatibilization plays is key to conceptualizing not only the performance of the materials, but also how they differ from blends.

Examples of commercial polymer alloys include (15):
- ABS/nylon containing alloys which are about 10 times stronger than the simple blend
- PC/ABS containing grades of alloys for exterior auto parts
- PC/PBT containing alloys for engine-rack cradles
- Poly(phenylene ether)/polyamide containing alloys which combine the chemical resistance of nylon and the creep resistance and toughness of PPE, with an especially high heat deflection temperature.
- ABS/PVC containing alloys for applications ranging from seat components for mass transit systems to providing materials with improved melt flow and thermal stability.

Block and Graft Copolymers

Pure block and graft copolymers by themselves are also two-phase systems in which gross segregation of the two phases is prevented because the component polymers are chemically linked. This results in microphase separation in which the microphases exist in a unique and ordered domain morphology, conferring unique and altered physical properties to the product. In particular, studies of extensive morphology and physical properties have been reported for block copolymer systems (16). Considerable amount of work, pioneered by Stannett (17) and Arthur (18) on the synthesis & characterization of cellulosic graft copolymers, has been done. However large scale industrial applications have eluded these polymers, and the problems preventing large scale use have been discussed by Stannett (12,19). Cellulosic block copolymers have been prepared by Gilbert's group (20,21). D. N.-S. Hon discusses in detail new developments in cellulosic derivatives and copolymers in another chapter in this book. In this paper the discussion will be restricted to cellulosic blends and alloys as defined earlier and illustrated below.

Figure 1 illustrates the concept of blends and alloys. Thus, when A and B polymers are mixed in unmodified form, the resultant product does not maintain the property characteristics of either A or B homopolymers and there is a loss in property as shown by the curve 1 in Figure 1 (Immiscible blend). If the A and B polymers give rise to a miscible polymer blend, the properties of the blend will be composition dependant and follow the line 2. Curve 3 represents the synergism of a compatibilized blend or A/B-alloy, in which enhancement and a unique balance of properties is achieved. As discussed earlier the synergism observed by using the compatibilization approach is due to improved interfacial adhesion and reduced interfacial tension between the phases, as well as the formation of a stabilized and finely dispersed phase morphology.

Very little work has focussed on alloys in which one of the blend components is a lignocellulosic polymer. Glasser and coworkers (22) have investigated polymer blends of hydroxypropyl lignin (HPL) with polyethylene(PE), ethylene vinyl acetate copolymer (EVA), poly(methyl methacrylate) (PMMA), and with poly(vinyl alcohol) (PVA). The blends produced a two-phase morphology, with the PE-HPL system behaving as a immiscible blend system. However, incorporation of vinyl acetate groups (ethylene-vinyl acetate copolymer) resulted in a compatibilized blend (alloy) with improved tensile strength. Stannett and coworkers pioneered some work on cellulosic block copolymers and its use in cellulosic alloys (23), and recently, we have been working on cellulosic alloys (24, 25) which is discussed in detail in this paper.

Cellulosic Alloys

The blending of lignocellulosic polymers with synthetic polymers generally leads to immiscible blends. As discussed earlier, the properties of such blend tend to be poor and undesirable. We have prepared cellulosic-PS alloys using tailor-made cellulose acetate (CA) - PS graft copolymers as compatibilizers. The graft copolymers function as emulsifying agents and provide for a stabilized, fine dispersion of the PS phase in the continuous phase of the CA matrix.

Synthesis of Tailor-Made Graft Copolymers. We have reported the synthesis of cellulose-PS graft copolymers with precise control over the molecular weight of the PS graft, the degree of substitution, and the backbone-graft linkage (26 - 28). The key step in the synthesis is the preparation of the carboxylate polymer anion, and its nucleophilic displacement reaction with mesylated cellulose acetate. Figure 2 outlines the synthetic scheme which was adopted. Recently, we have simplified the synthesis as outlined in Figure 3 by preparing anhydride-terminated PS instead of carboxylate-terminated PS which reacts directly with the -OH groups in CA, thereby

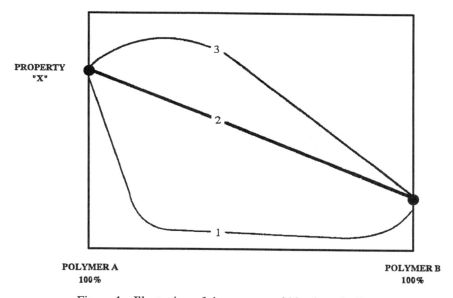

Figure 1. Illustration of the concept of blends and alloys.

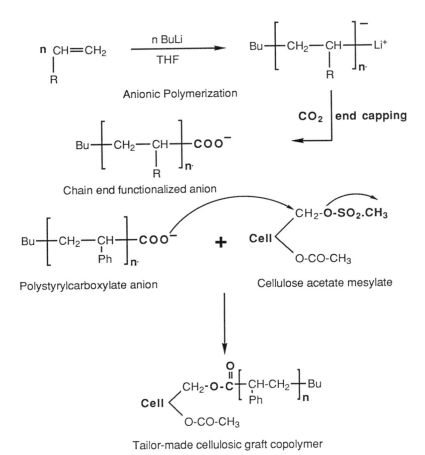

Figure 2. Preparation of tailor-made cellulose-polystyrene graft copolymers.

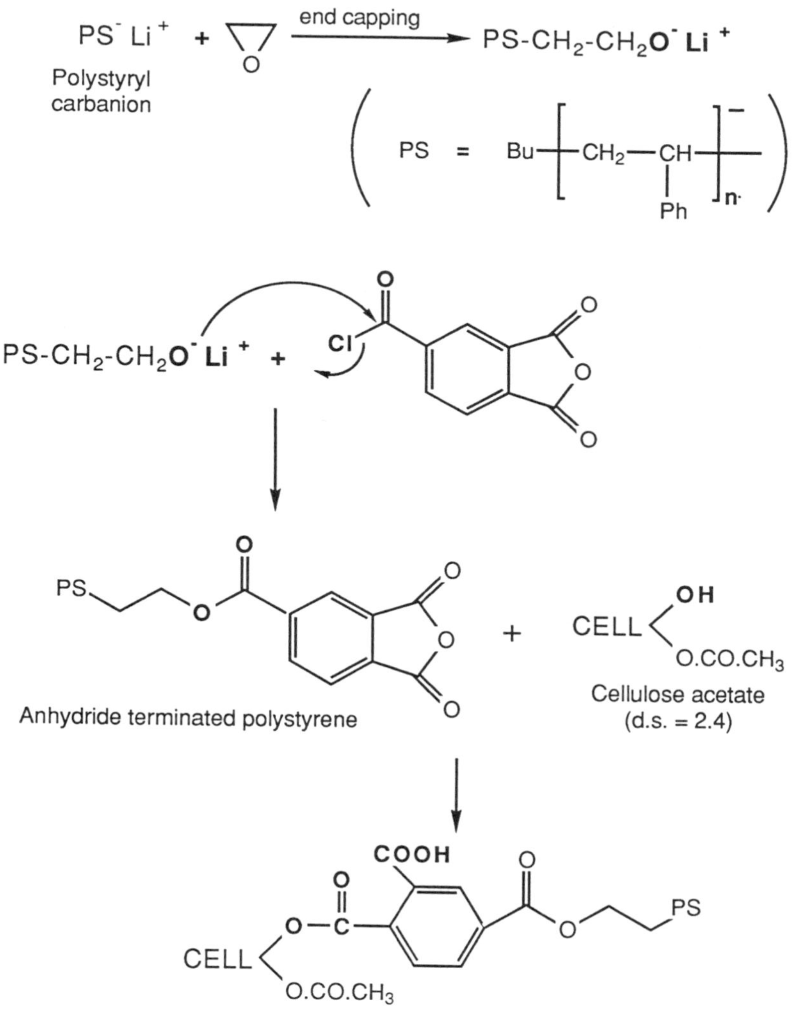

Figure 3. Reaction of anhydride terminated polystyrene with cellulose acetate.

eliminating an additional CA mesylation step. The anhydride terminated PS is prepared in excellent yields by reaction of the ethylene oxide end capped polystryl carbanion with trimellitic anhydride acidchloride (Narayan, R, and Chen, Z., X., unpublished work). Differential scanning calorimetry (DSC) of the graft copolymers has revealed the existence of two glass transitions (Figure 4). The lower T_g at 102 °C corresponds to the PS phase and the higher T_g at 180 °C corresponds to the CA phase. The presence of a two-phase structure that would localize at the interphase, without dissolving in either of the two phases is an essential criteria for the graft copolymer to function as a compatibilizer. As will be shown later, these graft copolymers function effectively as compatibilizers in the preparation of CA-PS alloys with finely dispersed and stabilized PS domains.

Using well-characterized low molecular weight lignins of narrow polydispersity, tailor-made lignin-PS graft copolymers have also been synthesized (18) following the same synthetic methodology as used in the preparation of CA-g-PS outlined in Figure 2. The lignin graft copolymers can function as compatibilizers in preparing blends of kraft lignins ($ 0.10-0.30/lb) with PS ($ 0.55-.74/lb) leading to new lignin-PS alloys (29).

Cellulose Acetate - Polystyrene Alloys. DSC scans of CA-g-PS - PS alloys (MW of PS = 20,000) of different compositions are given in Figure 5. These alloys show a single broad T_g around 92 °C corresponding to the PS phase. No apparent composition-dependant T_g was observed for the blends located between the T_g of PS in the graft copolymer and the T_g of PS homopolymer. This suggests that complete miscibility of the PS chains in the homopolymer and the graft copolymer is not occurring. However, because the T_g of PS in the graft copolymer cannot be distinguished from the homopolymer the observed Tg indicates good interfacial mixing is taking place.

Transmission Electron Microscopy Studies (TEM). The ability of the CA-g-PS to function as a compatibilizer to improve adhesion between the PS and CA phases and provide for stabilized, finely dispersed PS domains in the CA matrix is illustrated by TEM studies on the following samples:

- RN-1: 80 CA/20 PS blend, no graft copolymer added
- RN-2: 80 CA/20 PS blend, 10% CA-g-PS added
- RN-3: 75 CA/25 PS, graft copolymer

Film samples were sectioned at room temperature using a DuPont MT5000 microtome and diamond knives. Sections of 70-90 nm nominal thickness were used. Some were treated with RuO_4 vapor for 3 hr. Comparison of stained and unstained images showed that while RuO_4 slightly enhanced the image contrast, it was essential for evaluation of these samples. TEM images were obtained using a JEOL 2000FX microscope operated at 120 or 200 KV accelerating voltage and recorded on sheet

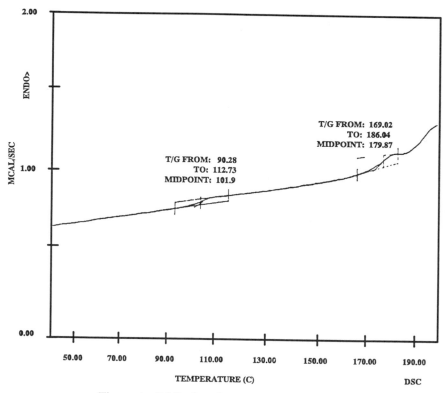

Figure 4. DSC of graft copolymer (CA-g-PS).

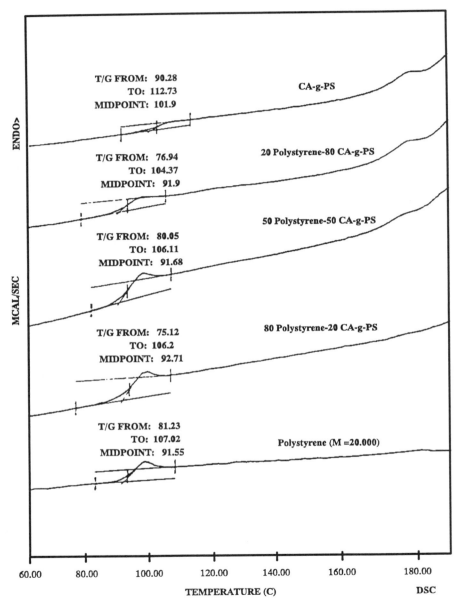

Figure 5. DSC of graft copolymer (CA-g-PS) and polystyrene homopolymer blends.

film. In sample RN-1 microdebonding of the PS phase from the CA matrix was caused by the microtoming process, indicating poor interfacial adhesion between the phases. Figure 6 shows the TEM image of RN-1 where the large domain sizes of PS is evident as dark spheres. A debonded PS phase is also readily seen. Figure 7 shows the TEM image of sample RN-2 which contains 10% CA-PS graft copolymer, but has the same overall composition as sample RN-1, i.e. 80 CA/20 PS. The interfacial adhesion appears to be much improved, and there is a decrease in the upper limit of the PS domain size from 15 microns in sample RN-1 to 5 microns in sample RN-2. This demonstrates that addition of a tailor-made graft copolymer having the requisite two-phase morphology functions effectively as a compatibilizer to improve adhesion between the CA and PS phases and provides for a stabilized and finely dispersed PS domain in the CA matrix. The TEM image of the CA-g-PS graft copolymer sample (RN-3) having a 75 CA/25 PS composition, is shown in Figure 8. The incompatible segments of the graft copolymer undergo microphase separation that is restricted to molecular dimensions with the microspheres existing in an ordered domain morphology. Some degree of membrane character is visible although more detailed studies are needed to draw further conclusions about the morphology.

Cellulose Acetate-Polystyrene-maleic anhydride (SMA) Alloys. Earlier, we had shown the preparation of a cellulose-polystyrene graft copolymer by reacting a anhydride terminated PS directly with cellulose acetate (Figure 3). The chemistry involved the reaction of the free hydroxyl groups on CA with the anhydride group on the PS resulting in the formation of a half ester. Using similar chemistry, we have prepared CA-SMA alloys by the direct reactive extrusion of cellulose acetate with a commercial random copolymer of polystyrene and maleic anhydride (supplied by ARCO Chemical Company, maleic anhydride content of 7% and a weight average molecular weight of 100,000) (Figure 9) (30). The advantages associated with reactive polymer extrusion are manyfold (31 - 33), the chief among them being the elimination of the use of solvents, lower equipment and energy costs, improved process and product control. CA and SMA were extruded through a Killion single screw extruder (length/diameter ratio of 24:1). Composition and screw speed were varied to optimize the reaction conditions. Extrudate samples were extracted with toluene in a soxhlet apparatus for 48 hr to remove unreacted SMA. DSC scan (Figure 10A) showed the presence of two glass transitions. The first T_g was at 115 ºC, which corresponds to the T_g of SMA, and the second one is at 185 ºC. The CA melting endotherm at 217 ºC was also observed. Figure 10B shows the DSC scan for a binary blend of CA homopolymer (75%) and SMA homopolymer (25%) with no graft copolymer. The ratio of CA to SMA is the same as in the alloy CA+CA-g-SMA. The most obvious difference between the two scans is the depression of the CA melting point peak by 14 ºC from 232 ºC in pure CA (Figure 10A) as opposed to only a 5 ºC depression in the blend with no graft copolymer (Figure 10B). A depressed melting point is a result of imperfections in the crystalline structure. It is logical to assume that the presence of the graft copolymer is interfering with the CA crystalline structure. In addition to a depressed CA melting point, the CA alloy has a

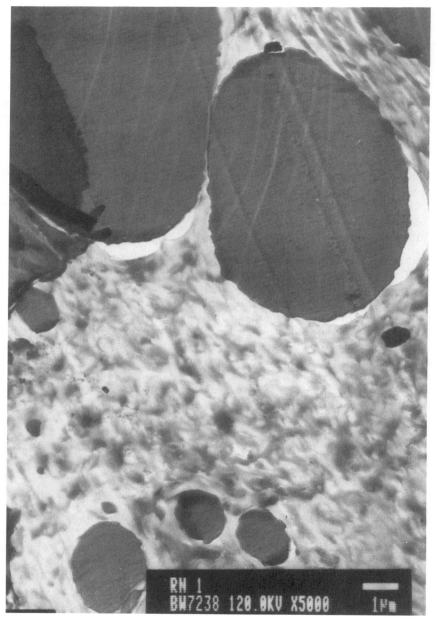

Figure 6. TEM of CA(80%)-PS(20%) blend with no addition of graft copolymer.

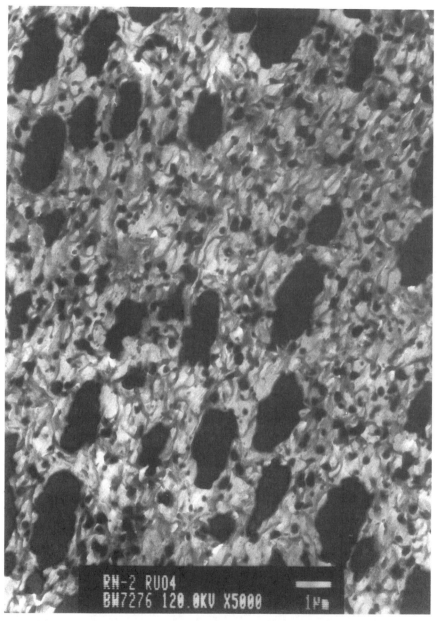

Figure 7. TEM of CA-PS alloy with 10% CA-g-PS. (Overall composition CA=80% and PS=20%).

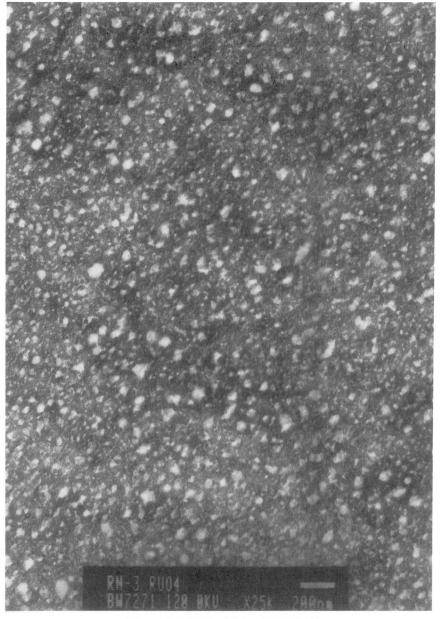

Figure 8. TEM of CA-g-PS (75:25).

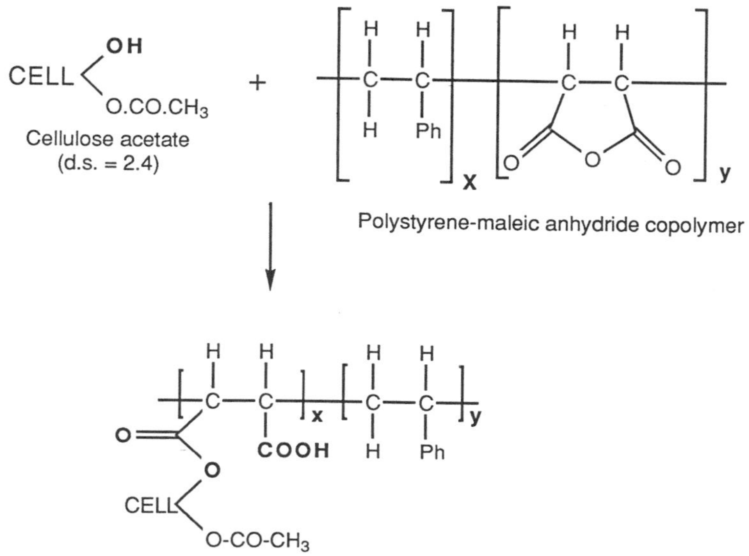

Figure 9. Grafting Reaction of cellulose acetate (CA) and polystyrene-maleic anydride (SMA) copolymer.

slightly lower T_g (T_g = 185 °C) than the straight blend (T_g = 190 °C.). TGA and elemental analysis of the toluene insoluble extrudate showed that the weight percent of SMA present was 16%. The TEM of the extrudate is shown in Figure 11. This compatibilized blend (alloy) of CA + CA-g-SMA gave the finest domain texture as compared to the CA-PS system with submicron domains of the SMA copolymer finely dispersed in the continuous CA matrix.

Wood-Plastic Alloys (Composites)

We were interested in the design and development of wood-plastic alloys, specifically wood-PS alloys (composites). Again, the major problem in developing this new material system is the inherent incompatibility of the two components the hydrophobic polystyrene and the polar wood-adhesive matrix. Two different cellulose-polystyrene graft copolymers prepared as described in Figure 2 (see Fig. 12) as compatibilizers to improve the bonding between the polystyrene and the wood-adhesive matrix. Cell-g-PS with a PS content of 58% and a molecular weight of 6250, and a cellulose-PS crosslinked graft copolymer with a PS content of 64%, and a molecular weight of 10,900 were used in the study (34). Two-ply veneer shear specimens were used to study how the graft copolymers, polystyrene, and a commercial phenolic resin interact when combined with wood. Analyses (according to ASTM D2339-82) showed that the graft copolymer had a favorable influence on the bonding of the polystyrene to the wood. The average bond strength was only 334 [Standard Deviation (SD) = 181] psi when only PS and the wood-resin matrix was present. By using the graft copolymers the bond strengths increased to 658 (SD=228) and 819 (SD=185) psi respectively. Forty-five test specimens were used for each bond strength determination.

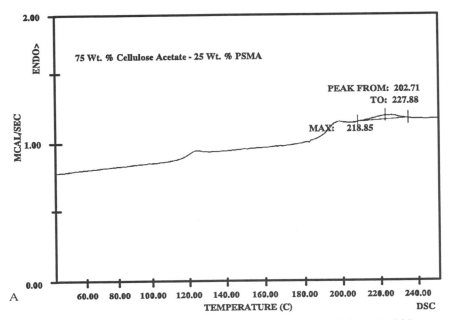

Figure 10A. DSC of cellulose acetate-styrene maleic anhydride copolymer graft copolymer (CA-g-SMA; 75 wt. % CA - 25 wt % SMA).

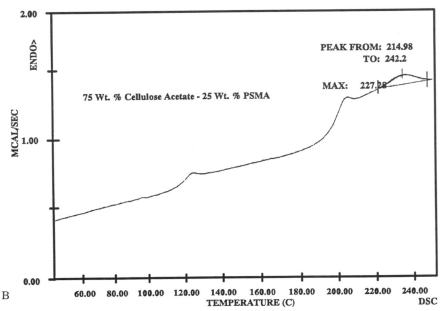

Figure 10B. DSC of cellulose acetate - styrene maleic anhydride blend (75 wt. % CA - 25 wt % SMA).

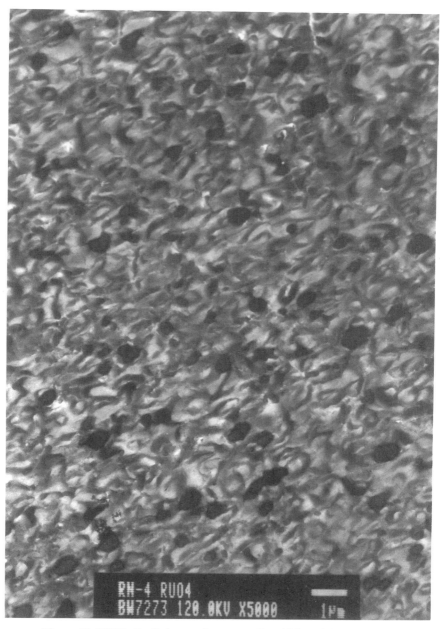

Figure 11. TEM of cellulose-polystyrene alloy (CA + CA-g-SMA).

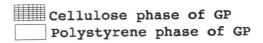
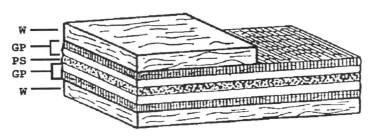

Figure 12. Exploded view of the use of CA-g-PS to compatibilize bonding of PS to wood -- wood plastic alloys.

Conclusions.

Alloying of lignocellulosics with synthetic polymers (thermoplastics) offers an excellent opportunity to tap in to the growing, and relatively new, polymer blends and alloys markets . Design and engineering of tailor-made and well defined compatibilizers is the key to synthesizing these new lignocellulosic-synthetic polymer alloy materials. The compatibilizer provides for the creation of a stabilized, finely dispersed two-phase morphological system with synergistic improvement in properties. This report summarizes the results of using graft copolymers as compatibilizers to prepare new cellulosic alloys, and is, hopefully, the beginning of many more compatibilized lignocellulosic-synthetic polymer alloy systems with commercial potential.

Literature Cited

1. News Item, *Elastomerics*, **August 1990**, 26(8), 14
2. News Item, *Plastics Engineering*, **November 1988**, pp 25
3. Paul, D. R. , and Newman S., eds., *Polymer Blends, Vol. 1&2*, Academic Press, New York, 1978
4. Utracki, L. A., Guest Ed., *Polym Eng. Sci.*, vol 30(17), 1990
5. Mason J. A., and Sperling L. H., Polymer Blends and Composites, Plenum Publishing Corp., New York, 1976
6. Cabasso, I., *in Cellulose and Wood - Chemistry and Technology*, Editor Schuerch, C., Wiley Interscience, 1989, pp 1361
7. Gardiner, E., and Cabasso, I., *Polymer*, **1987**, 28, 2052
8. Jutier, J. J., Lemieux, E., and Prud'homme, R. E., *J. Polym. Sci., Polym. Phys.*, **1988**, 26, 1313

9. Nishio, Y., and St-John Manley, R., Macromolecules, 1988, 21, 1270.
10. Nishio, Y., Roy, S. K., and St-John Manley, R., Polymer, 1987, 28, 1385
11. Prud'homme, R. E., in Wood Processing and Utilization, Eds., J. F. Kennedy, G. O. Phillips, P. A. Williams, Ellis Horwood Limited, Chichester, U.K., 1989, pp 221
12. Milkovich, R. M., Chiang, M. T.; U.S. Patent 3, 786, 116, 1974
13. Perry, E.; *J. Appl. Polym. Sci.*, **1964**, 8, 2605
14. Feldman, D., Rusu, M.; *Eur. Polym. J.*, **1974**, 10, 41
15. News Item, *Modern Plastics*, **October 1988**, pg 62
16. Meier D. J., in *Polymer Blends and Mixtures*; Editors, Walsh, D. J. , Higgins, J. S., Maconnachie, A., Matinus Nijhoff Publishers, Boston, 1985, pp 173
17. Stannett, V.,*ACS Symp. Ser.*, **1982**, 187, 1, and references cited therin
18. Arthur, J. C. Jr., *ACS Symp. Ser.*, **1982**, 187, 21, and references cited therin
19. Stannett, V., in *Cellulose - Structural and Functional Aspects*, Editors, Kennedy, J. F., Phillips, G. O., and Williams, P. A., Ellis Horwood Limited, Chichester, U.K., 1989, pp 19
20. Kim, S. L., Stannett V. T. and Gilbert, R. D. *J. Macromol. Sci., - Chem.*, **1976**, A10, 1967.
21. Amick, R., Gilbert, R. D., and Stannett, V. T., *Polymer*, **1980**, 21, 648.
22. Ciemniecki S. L., and Glasser, W. G., *ACS Symp Ser.*, **1989**, 397, 452
23. Penn, Gilbert, and Stannett, V *J. Polymer Sci.*, **1967**, A1, 5, 1341
24. Narayan, R.; Biermann, C.J.; Hunt, M. O.; Horn, D. P.; *ACS Symp. Ser.*, **1989**, 385, 337
25. Narayan, R.; Neu R. P.; *Materials Research Soc., Symp. Proc.*, **1990**, 197, 55
26. Biermann, C. J., Chung J. B., and Narayan, R.; *Macromolecules*, **1987**, 20(5), 954
27. Biermann, C. J., and Narayan, R.; *Polymer*, **1987**, 28, 2176
28. Narayan, R.; *in Cellulose and Wood - Chemistry and Technology*, Editor Schuerch, C., Wiley Interscience, 1989, pp 945
29. Narayan, R., Stacy, N., Ratcliff, M, and Chum, H.; *ACS Symp Ser.*, **1989**, 397, 476
30. Neu R. P. M.S Thesis, Purdue University, October 1989.
31. Tucker , C. S., and Nichols, R. J., *Plast. Eng.*, May 1987, pp 27-30
32. Frund Z. N. Jr., *Plast. Comp.*, **1986**, 9(15), 24.
33. Tzoganakis, C., *Advan. Polym. Tech.*, **1989**, 9(4), 321
34. Narayan, R., Biermann, C. J., Hunt, M. O., and Horn, D. P., *ACS Symp. Ser.*, **1989**, 385, 337

RECEIVED July 29, 1991

Chapter 6

Mechanical Properties of Surface-Modified Cellulose Fiber–Thermoplastic Composites

R. G. Raj and B. V. Kokta

Centre de Recherche en Pâtes et Papiers, Université du Quebec à Trois-Rivières, Quebec G9A 5H7, Canada

> Recent attention has examined the use of natural organic fillers, such as cellulose fibers, as fillers/reinforcing agents in elastomers and thermoplastic polymers. These fibers, when used in a polymer matrix, cannot function as an effective reinforcement system due to poor adhesion at the fiber-matrix interface. Cellulose fibers also tend to aggregate and thus the fibers do not disperse well in a hydrophobic polymer matrix. The objective of this study was to improve the suitability of cellulose fibers in a hydrophobic polymer matrix (polyethylene) through the use of various processing aids/coupling agent. Stearic acid and mineral oil were used as additives and maleated ethylene as a coupling agent. Tensile strength and modulus of the composites increased with the fiber concentration, largely because of improved fiber dispersion (with stearic acid) at higher filler concentrations. Increased mechanical properties of the composites due to improved compatibility between the fiber and matrix was also achieved. Maleated ethylene improved the adhesion between the fiber and polymer matrix. The rule of mixture equation was used to calculate tensile modulus of the composites and these values compared with the experimental results. Factors affecting the modulus of the composites are discussed.

Many studies have described the use of cellulose or wood fibers as a filler/reinforcing agents in thermoplastic polymer matrices (1-6). These fibers are relatively cheap and lightweight (lower density) compared to inorganic fillers. In addition, the biodegradable nature of cellulosic fillers offers a potential solution to the growing waste disposable problem. In spite of these advantages, wood or cellulose fillers have not found much use in thermoplastics. It has been argued that certain drawbacks such as incompatibility to the hydrophobic polymer matrix, tendency to form aggregates, poor resistance to moisture, and the limitation of processing temperature greatly reduce the potential of cellulose and/or wood as a filler/reinforcement. Many approaches have been described in the literature to improve adhesion at the polymer-matrix interface. One of the methods used to improve compatibility between the filler and matrix is chemical grafting. By attaching a suitable polymer segment to the surface of a filler with a similar solubility parameter as the polymer matrix, a significant improvement in bonding between the fiber and matrix can be achieved. For example, the polymerization of methyl methacrylate on sawdust improved the physical and mechanical properties of the composite (7). Aspen and birch pulps grafted with polystyrene and incorporated into the polystyrene matrix gave a 40% increase in mechanical properties as compared to unfilled polystyrene (8).

Modification of the filler-matrix interface has also been attempted by the addition of various additives or coupling agents during processing. Dalvag et al. (9) used maleic anhydride modified propylene to improve strength and ductility of polypropylene (PP)-wood and cellulose flour composites. Cellulose fibers treated with vinyl chloride, a plasticizer and an isocyanate produced better adhesion with polyvinyl chloride (10). High density polyethylene (HDPE) filled with silane A-174 treated chemithermomechanical pulp (CTMP) of aspen fibers produced higher tensile strength and modulus (11). An increase in tensile and impact strength was reported when rosin was used in PP-wood flour composites (12).

Present Technology

Cellulose fibers in the form of paper have been used with interleaving thermoplastic polymer films, after hot pressing, to obtain laminates (13). McKenzie and Yuritta (14) reported that pretreating wood fibers with urea formaldehyde, followed by sheet formation and hot-press lamination with polyethylene, gave a product with very good retention of wet tensile strength. However, the above fabrication system is not practical for the manufacture of thick products. Better bonding between wood fiber and PP matrix was achieved by gamma irradiation of wood fiber (15). Improvements in the mechanical strength of wood fiber-filled thermoplastic composites can also be achieved with the use of coupling agents (4,11). The polymer is chemically bonded to the filler particle by a coupling agent, which improves interface adhesion. The degree of adhesion depends on the type of polymer, coupling agent, and filler combination. Formation of a strong adhesive bond between the polymer matrix and cellulose fiber gives increased strength and stiffness to the composite material. The factors which can influence the mechanical properties of short fiber-filled thermoplastics are summarized in Table I.

Table I. Factors Affecting Mechanical Properties of Short Fiber Composites

Tensile strength:

- strong interface
- low stress concentrations
- fiber orientation

Tensile modulus:

- high fiber aspect ratio
- fiber wetting
- fiber concentration

Impact strength:

- ductile matrix
- energy absorption

Both the matrix and fiber properties are important to improve mechanical properties of the composite. Tensile strength is more sensitive to matrix properties, while fiber properties are more important for modulus. However, a balance between the fiber and matrix properties is required to achieve good impact strength. When short discontinuous cellulose fibers are used in thermoplastics, it is very important to reduce the fiber-to-fiber interaction to achieve a uniform distribution of fiber in the matrix. The surface characteristics of the reinforcing fiber are important in the transfer of stress from the matrix to the fiber. The pretreatment of cellulose fibers with a suitable additive prior to incorporating the fibers with the polymer matrix aids dispersion and significantly improves the mechanical properties of the composite (16,17). Many types of wood pulps are available (thermomechanical, chemithermo-mechanical and chemical); the physical and mechanical properties of these fibers vary widely depending upon the pulping process. Thus, modification of the fiber surface needs to be specific for a particular polymer matrix. In an earlier work, Quick (18) reported that softwood kraft pulp fiber treated with ethylene-acrylic acid co-polymer had good reinforcing effect on low density polyethylene, but the same fiber did not perform well with PP and polystyrene (PS). Hydrolytic pretreatment of cellulose fibers with oxalic acid improved the homogeneity and mechanical properties of PP, HDPE and PS which contained various amounts of bleached pulps (19).

Current work

Processing aids/coupling agents. The performance of fiber-filled plastic is greatly influenced by the degree of fiber

dispersion within the polymer matrix. Aspen fiber with an average aspect ratio (L/D) of 12 was used as a reinforcing fiber. Maleic anhydride modified HDPE (DuPont Canada; melt index 13.5 dg/min.) was used as a matrix. Two different processing aids and a coupling agent were used to improve the dispersion of fiber as well as the compatibility of the fiber with the polymer matrix;
a) stearic acid (Aldrich)
b) mineral oil (Sigma)
c) maleated ethylene (Epolene C-18, Eastman Kodak)
The compounding of predried polymer and fiber (0-40% by weight) was done at 170°C in a two-roll mill. In a typical mixing process, about half of the polymer-fiber mixture was added to the heated co-rotating rolls. After mixing for 3 minutes, the processing aids (0-2% by weight of fiber) and the remaining polymer-fiber mixture was slowly added and mixed throughly to obtain the wetting of fiber by the polymer. The above mixture was compression molded in a Carver laboratory press to obtain tensile specimens (ASTM D-638, Type V). After heating of the mold for 15 min. at 160°C (pressure 3.2 MPa), the samples were slowly cooled to room temperature with the pressure maintained during the process. Tensile properties of the composites were measured with an Instron model 4201. The full-scale load was 500 N and the cross-head speed was 10 mm/min. The test results were automatically calculated by a HP86B computing system using the Instron 2412005 General Tensile Test Program. Six specimens were tested in each case and the average was reported. The coefficients of variation of the reported properties were: Tensile strength, 1.2-6.9%; Elongation, 3.4-7.6%; Tensile modulus, 2.5-6.7%; and Izod-Impact strength, 4.9-8.6%.

The additive effect on tensile strength of HDPE-cellulose fiber composites (at 10 and 30% fiber concentrations) is presented in Table II. The concentration of additive was 1% by weight of fiber. The results show an increase in tensile strength and modulus with the use of stearic acid, mineral oil, and maleated ethylene. Tensile strength of the composite (at 30% fiber content) with stearic acid increased to 35.2 MPa as compared to 29.8 MPa for the control. The results show that stearic acid is highly effective as a dispersant; i.e., reducing fiber-to-fiber interaction. Normally, improved fiber dispersion results in higher strength values (9). An increase in tensile strength and modulus was also observed with mineral oil. When compared to control, tensile strength increased by 13% at 30% fiber concentration. It was observed that mineral oil functions as a lubricant which is adsorbed by the fiber, and this facilitates the disentanglement of individual fibers (17). However, the best improvement in tensile strength and modulus was achieved with maleated ethylene; i.e., tensile strength increased from 29.8 MPa (control) to 36.1 MPa at 30% fiber concentration. This may be attributed to improved bonding between the fiber and matrix. The coupling reaction (ester linkage) between the maleated ethylene and the hydroxyl groups of cellulose thus provides a means to improve the bonding between fiber and matrix (20).

Table II. Tensile properties of HDPE-cellulose fiber composites

Processing aids	Fiber (% wt.)	Tensile strength (MPa)	Elongation (%)	Tensile modulus (GPa)
control	10	23.7 (1.2)	8.6 (4.5)	1.16 (3.6)
	30	29.8 (2.3)	6.0 (3.6)	1.60 (4.9)
stearic acid	10	26.6 (1.4)	8.0 (4.9)	1.28 (4.7)
	30	35.2 (6.2)	6.0 (3.8)	1.86 (4.5)
mineral oil	10	26.5 (3.5)	8.6 (5.0)	1.26 (3.9)
	30	33.8 (4.7)	6.1 (2.8)	1.84 (6.5)
maleated ethylene	10	25.9 (2.7)	8.2 (2.1)	1.29 (3.0)
	30	36.1 (3.7)	6.0 (5.3)	1.94 (5.8)

( ) values represent coefficients of variation.

Tensile modulus of the composites which contained additives improved as compared to the control (Table II). With maleated ethylene, the modulus increased from 1.60 GPa to 1.94 GPa at 30% fiber concentration. The data show that stearic acid and mineral oil were also effective in improving the modulus of the composites. To realize the maximum stiffness potential of the fiber, a good contact is essential between the matrix and fiber phases so that the load can be transferred effectively from the matrix to the fiber. Tensile modulus is also affected by the concentration of fiber in the composite. This is evident from the data, where the modulus increased steadily with a rise in fiber concentration; i.e., 1.29 GPa at 10% fiber concentration to 1.94 GPa at 30% fiber concentration. While the addition of the fiber increased the stiffness of the composite, it had a negative effect on elongation. The results show a decrease in elongation with the increase in fiber concentration. Unlike tensile strength and modulus, additives had no influence on elongation. Earlier studies have shown a similar reduction in elongation with the increase in fiber concentration (11).

Figures 1 and 2 show the effect of additive concentra-

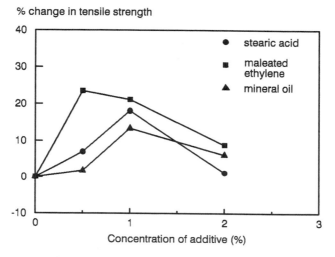

Figure 1. Effect of additive concentration on tensile strength of HDPE-cellulose fiber composites.

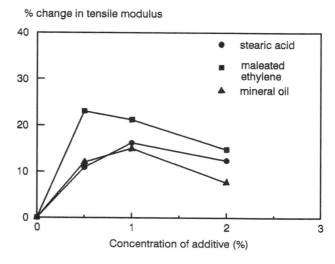

Figure 2. Effect of additive concentration on tensile modulus of HDPE-cellulose fiber composites.

tion on tensile strength and modulus of composites with 30% fiber concentration. The concentration of stearic acid, mineral oil, and maleated ethylene varied from 0 to 2% (based on fiber weight). The results show that the % change in tensile strength (compared to control) is significant at low additive concentration (Figure 1). Also, the increase in tensile strength is affected by the type of additive used. Maleated ethylene seems to be most effective at improving tensile strength, with a 25% increase in strength observed at 0.5% concentration. But in the case of stearic acid and mineral oil, maximum increase in strength was observed at 1% concentration. A further increase in additive concentration resulted in a strength decrease. A similar trend was observed in the case of tensile modulus (Figure 2). The loss in strength and modulus at the higher concentration of additives may be due to the plasticizing effect on the polymer matrix.

Figure 3 shows the Izod-impact strength (un-notched) as a function of fiber concentration of HDPE-cellulose fiber composites containing untreated, stearic acid-treated, and maleated ethylene-treated fibers. Impact strength steadily decreased with an increase in fiber concentration. Compared to untreated fiber composites, maleated ethylene-treated fibers produced slightly higher impact strength. For good impact strength, an optimum bonding level is necessary. Good bonding may produce poor impact strength because a crack can propagate rapidly from the matrix through a fiber and into the matrix again if the interface between fiber and matrix resists separation. On the other hand, if the fibers are not bonded strongly with the matrix, they may separate easily from the matrix and can divert the crack by absorbing its energy. The degree of adhesion, fiber pull-out, and a mechanism to absorb energy are some of the parameters which can influence the impact strength of short fiber-filled composites (4).

Fiber dispersion. Surface modification of cellulose fiber is important to promote high fiber loading in the polymer matrix and to improve polymer compatibility and property enhancement. Cellulosic fibers tend to form aggregates in a polymer matrix due to their hydrophilic nature. The degree of fiber dispersion depends on the interfacial characteristics of the polymer and fiber. The ability to break down fiber aggregates is sharply reduced by poor wetting of the fiber surface by the polymer. The results show that the addition of 1.0% stearic acid during the compounding of discontinuous cellulose fibers greatly improves the fiber dispersion in the polymer matrix. When compared to untreated fibers (no additive) as shown in Figure 4, the number of aggregates is greatly reduced in stearic acid-treated fiber composites (Figure 5). The stearic acid significantly reduces fiber-to-fiber interaction, and as a result, the fibers are better dispersed in the polymer matrix. This is demonstrated by the increase in

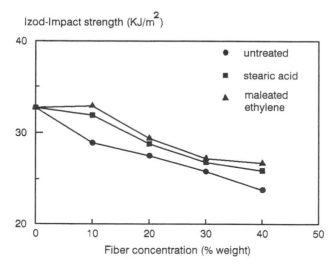

Figure 3. Effect of fiber concentration on Izod-impact strength (un-notched) of HDPE-cellulose fiber composites.

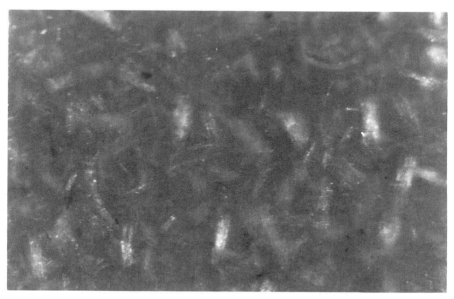

Figure 4. Fracture surface of HDPE-untreated cellulose fiber Composite (30% fiber weight). magnification 190 X

tensile strength and modulus of the composite. The mixing time is often increased to achieve good dispersion. However, repeated mixing tends to break the fiber due to high shear forces. Delvag, et al. *(9)* found that the degree of fiber length reduction during compounding increased proportionally with the mixing time. Additives help to reduce the mixing time to minimize the fiber breakage during the compounding process.

Young's modulus of the short fiber-filled composites may vary depending on the orientation of the fibers, the angle between the fiber axis, and the test direction. It has been suggested that a high degree of orientaion can be achieved by milling or extrusion *(16)*. Several factors can influence the degree of orientation: fiber aspect ratio, viscosity and flow gradient of the polymer matrix, and the fiber concentration. Figure 6 shows some of the possible types of fiber orientations in short discontinuous fiber composites. The first case (a) represents fibers which are randomly oriented in a plane; the orientation factor k is ≤ 0.33. If the fibers are crossed at 90° and tested in either of the two fiber directions as in case (b), the orientation factor k is ≥ 0.5. However, the ideal case is (c) where the fibers are fully aligned in the direction of test (k is 1.0). In such a case as (c), maximum strength and modulus of the composite would be realized. In practice, however, case (b) is most likely.

The modulus of the composite can be calculated from the rule of mixtures equation:

$$E_c = k V_f E_f + E_m (1 - V_f) \tag{1}$$

where E_c is modulus of the composite, E_f and E_m are moduli of the fiber and matrix respectively, V_f is the volume fraction of fiber, and k is the orientation factor. Figure 7 shows the relationship between the orientaion factor, volume fraction of the fiber, and tensile modulus of the composite. From Eq. 1, the k values were calculated for the fibers treated with stearic acid. The data show that as k varies from 0.27 ($V_f = 0.13$) to 0.47 ($V_f = 0.26$), the modulus value calculated using Eq. 1 agrees well with the experimental value at lower concentration of fiber. However, at higher fiber concentrations, the experimental values were lower than predicted perhaps because of poor fiber orientation.

Future needs

Additional research is needed before cellulose fiber composites can compete with high performance composites such as glass, graphite, and kevlar fiber composites. New technologies need to be developed for the use of cellulose-based materials as a source of reinforcement. Microfibrillated cellulose may provide a stiffer reinforcing fiber for the composites *(21)*. The reaction Injection

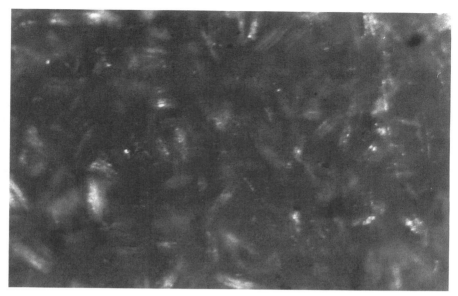

Figure 5. Fracture surface of HDPE-stearic acid treated cellulose fiber composite (30% fiber weight). magnification 190 X

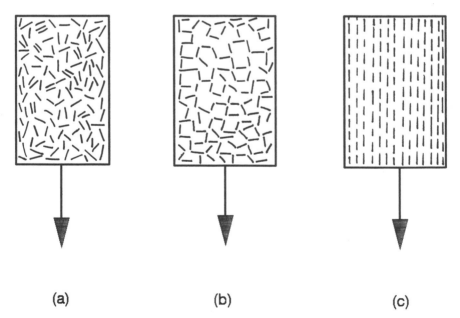

(a) (b) (c)

Figure 6. Possible types of fiber orientations in short discontinuous fiber composites.

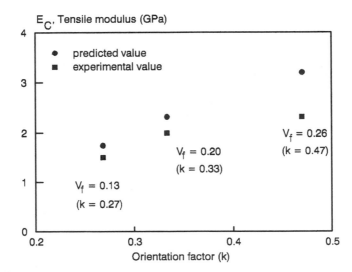

Figure 7. Relation between the orientation factor, fiber concentration and tensile modulus of the composite.

Molding (RIM) process can be used to improve the fiber wetting. In this process, two liquid components-one consisting of fibers suspended in monomer and the other consisting of the catalyst and further monomer-are injected into a mold at a suitable temperature to obtain the final product (22). Different reinforcing elements such as single pulp fiber, microfibrils, and crystallites with varying elastic modulus (10-250 GPa) can be derived from wood (23). Unfortunately, no technology is currently avilable to obtain pure crystallites of cellulose. Zadorecki and Michell have discussed some possibile processes for the fabrication of much stiffer celluose fiber composites using non-conventional technologies (24).

Literature Cited

1. Seymour, R.B. Popular Plastics **1978**, 23, 27-31.
2. Nakajima, Y. Jap. Pat. Kokai 127 632, **1981**.
3. Coran, A.Y.; Patel, R. U.S. Patent 4,323,625, **1982**.
4. Woodhams, R.T.; Thomas, G.; Rodgers, D.K. Polym. Eng. and Sci. **1984**, 24, 15, 1166-1171.
5. Coutts, R.S.P.; Campbell, M.D. Composite, **1979**, 10, 228.
6. Goettler, L.A. Polym. Compos. **1983**, 4, 249-260.
7. Moustafa, A.B.; Abd El-hady; Ghanem, N.A. Die Angewwandte Makromolekulare Chimie **1980**, 85, 91-105.

8. Kokta, B.V.; Chen, R.; Daneault, C.; Valade, J.L. Polym. Compos. **1983**, 4, 229-232.
9. Dalvag, H.; Klason, C.; Strömvall, H.-E. Intern. J. Polymeric Mater. **1985**, 11, 9-38.
10. Goettler, L.A. U.S. Patent 4,376,144, **1983**.
11. Raj, R.G.; Kokta, B.V.; Daneault, C. J. of Appl. Polym. Sci. **1989**, 37, 1089-1103.
12. Lightsey, G.; Short, P.H.; Kalasinsley, K.S.; Mann, L. J. of Mississipi Acad. Sci. **1979**, 24, 76-83.
13. Michell, A.J.; Vaughan, J.E.; Willis, D. J.Appl. Polym. Sci., **1978**, 22, 2047-2058.
14. McKenzie, A.W.; Yuritta, J.P. Appita **1979**, 32, 460-465.
15. Czvikovszky, T.; Tapolcai, I. Proc. of 5th symposium on Radiation chemistry, Budapest, **1982**, 785-792.
16. Boustany, K; Coran, A.Y. U.S. Patent 3,697,364, **1972**.
17. Hamed, P. U.S. Patent 3,943,079, **1976**.
18. Quick, J.R. Paper presented at 169 th ACS National Meeting, Philidelphia, **1975**.
19. Boldizer, A.; Klason, C.; Kubat, J. Intern. J. Polymeric Mater. **1987**, 11, 229-262.
20. Gaylord, N.G., U.S. Patent 3,645,939, **1972**.
21. Herrick, F.W.; Casebier, R.L.; Hamilton, J.K.; Sandberg, K.R. J. Appl. Polym. Sci: App. Polym. Symp. **1983**, 37, 797-805.
22. Zadorecki, P; Abbas, K.B. Polym. Compos. **1985**, 6, 162-171.
23. Jeronimidis, G. "Wood, One of Nature's Challenging Composites," in The mechanical Properties of Biological Materials, Eds. Vincent, J.F.V.; Currey, J.D., Cambridge University Press, Cambridge, **1980**.
24. Zadorecki, P; Michell, A. Polym. Compo. **1989**, 10, 2, 69-77.

RECEIVED June 4, 1991

Chapter 7

Dimensionally and Ultraviolet-Stable Composites from Lignocellulosics

David V. Plackett

Forest Research Institute, Private Bag 3020, Rotorua, New Zealand

> Composite products in the form of panels, molded articles, and structural beams can potentially be made from a wide variety of lignocellulosic raw materials. Adding value through the introduction of enhanced dimensional and ultraviolet (UV) light stability can open up new uses and new markets for such products. Dimensional stabilisation is possible through a variety of methods including heat treatment or chemical modification of raw materials or final products. The chemical modification route also offers improved resistance to the effects of UV light and water and this has been demonstrated in yellowing trials and outdoor weathering experiments with acetylated lignocellulosic composites.

Lignocellulosics such as wood, wheat straw, bagasse, and flax can be used, either alone or in combination with adhesives, as raw materials for a wide range of composite products. Of these materials, wood is the most commonly used at present with world production recorded at about 109 million m^3 of plywood, particleboard, and fiberboard in 1985 (*1*). Annual production of these wood products has increased at an average rate of 0.9% per annum since 1975. Medium-density fiberboard (MDF) production capacity in particular has been increasing rapidly in recent years.

Composites from lignocellulosic materials such as wood are now well accepted as panel materials offering predictable, reliable properties and ease of installation during construction. Wood composite beams made from veneer strands (Parallam) (*2*) or crushed juvenile wood (Scrimber) (*3*) are also now entering the marketplace to compete with laminated veneer lumber (LVL) for high-value structural uses. Like solid wood, composite materials have certain physical characteristics that require improvement if new generation materials with enhanced properties are to be developed. New dimensionally stable composites, whether from wood or other lignocellulosics, could find application in high humidity indoor or exposed outdoor situations where the use of existing composites is limited because water absorption can lead to dimensional changes and decay. Dimensional instability in wood-based composites, as an example, results both from the hygroscopic nature of the wood components (reversible swelling) and from the release of stresses induced by the pressing process (irreversible swelling). Possible methods for dimensionally stabilising wood products have been listed (*4, 5*) and include:
  1. Application of surface coatings.

2. Reduction of hygroscopicity by chemical reaction.
3. Cross-linking the substrate with reactive chemicals.
4. Bulking the material with impregnated chemicals.
5. Mechanical stabilisation by lamination.

The last of these choices is practiced with plywood where alternate veneers are laid at 90° to each other and good dimensional stability is obtained.

Methods for dimensionally stabilising composites by chemical modification of the lignocellulosic raw material offer opportunities to introduce improvement in other properties (6) and chemical modification procedures may be particularly suited to composite material production.

Using wood as the primary example, this chapter reviews methods for dimensionally stabilising composites based on heat treatment and contrasts this approach with chemical modification. UV stability of the product can also be gained through chemical modification procedures and this may have particular significance for veneer-based products in decorative uses or in applications exposed to the weather. The needs for future research in dimensional and UV stabilisation of lignocellulosic composites are summarised.

Heat Treatment as a Route to Dimensional Stabilisation of Composites

Wood-based composites are subjected to varying temperatures at different stages of processing from the raw material. For example, in production of MDF, the fiber receives low-pressure steam treatment at 100°C before proceeding to the digester at 170-180°C and then on to the refining process. The refined fiber is mixed with an adhesive and dried, typically at 160°C, before mat formation and processing. In particleboard, the raw wood particles are not exposed to temperatures above ambient before pressing. In both MDF and particleboard production, pressing occurs between heated platens at temperatures of 160-200°C for a time dependent on adhesive type and desired product thickness. Products have been examined to assess whether general differences in dimensional stability exist between particleboard and MDF. Suchsland et al. (7) showed that MDF appeared more resistant to surface deterioration caused by water soaking and redrying. In contrast, Watkinson and Van Gosliga (8) found that the two product types responded similarly in terms of the effect of moisture content on thickness swelling and linear expansion. Clearly, it is difficult to separate the influence of fiber or particle temperatures during processing from the impact that differences in exposed surface fiber or particle area would have on properties such as moisture sorption and dimensional stability.

A number of researchers have investigated the possible enhancement of wood dimensional stability through heat treatment. In the United States, a product known as Staybwood was developed during the 1950s and involved heating wood to 150-300°C in the absence of air. Staybwood showed a 40% improvement in dimensional stability but had reduced strength and abrasive resistance. It was a natural extension of this concept to assess whether dimensional stabilisation through heat treatment could be applied to composite wood products. Suchsland et al. (9) produced dry-formed binderless hardboard from refined Masonite pulp that had been prepared using hardwood chips heated to 192-240°C for two to four minutes. The purpose of their research was to determine what effect increasing steam pressure, and hence temperature, would have on mechanical and physical properties of hardboard. The results indicated that most board properties were enhanced when steam pressures up to 500 psi were used, but that properties such as modulus of elasticity (MOE), modulus of rupture (MOR), and linear expansion were less satisfactory at pressures higher than 500 psi. The fines content of the pulp was thought to be an important variable in bond and bending strength development.

The possibility of producing dimensionally stabilised composites by steam treatment of particles or fibers has recently been discussed by Hsu et al. (10). They considered how the forces responsible for disintegration of wood composites on exposure to high humidity could be minimised. The concept of thermochemical treatment of wood raw material with steam was developed on the basis that some chemical breakdown of the hemicellulose components could increase the compressibility of the wood material and might also soften the wood cell wall leading to greater fiber flexibility. Hsu et al. (10) manufactured samples of aspen waferboard and mixed pine/spruce particleboard both by conventional industrial practice and also by exposing each raw material to saturated steam at 1.55 MPa for various preselected times before drying. Although this sort of treatment resulted in some darkening of wood wafers or particles, the thickness swelling of phenol-formaldehyde (PF) resin-bonded waferboard on water soaking was dramatically reduced when compared with control material. Steam treatments lasting three minutes were sufficient to produce this improvement and longer steam exposure times gave only a marginal further increase in stability. The MOE and MOR of waferboard after an accelerated aging procedure were improved by steam pretreatment of the wafers. The thickness swelling and linear expansion of urea-formaldehyde (UF) resin-bonded particleboard were also significantly reduced by the steam pretreatment of particles even at very short steam treatment times (i.e., one to two minutes). Pressing times were reduced for both product types, presumably because of enhanced wafer or particle compressibility. Chemical analyses of steam-treated wood chips showed a time-dependent degradation of carbohydrate polymers occurring during steaming. A reduction in the molecular weight of lignin can occur in steaming (11) and it is possible that a reduction in the softening point of lignin, as well as the partial hydrolysis of hemicelluloses, could contribute to improved composite dimensional stability. The mild steam treatments proposed by Hsu et al. (10) were not thought to change total lignin or cellulose contents and were not found to impair the bending strength of composite wood panels.

Heat post-treatment has been shown to be effective as a means of dimensional stabilisation of phenolic-bonded particleboard or hardboard (12, 13). However, heating times have been generally considered too long for commercial practice. Hsu et al. (14) have developed a faster process dependent on heating products to temperatures higher than the glass transition temperatures of carbohydrates and lignin (e.g., 230-250°C) (15). Research has shown that board heating can be an effective way of dimensionally stabilising wood composites and, since analyses indicated little effect on chemical composition of boards, stabilisation is thought to arise from physical rather than chemical changes in wood components at temperatures above the softening points of carbohydrate components and lignin (16). A reduction in the accessibility and availability of hydroxyl groups can also be induced by heat treatment (17).

Processes involving steam injection into mats of wood fibers or particles during pressing have now been developed for commercial production of thick wood panel products. In New Zealand, a radiata pine product consisting of surface layers of MDF with a strandboard core is being produced in thicknesses up to 100 mm by a steam-injection process (18). In an early report on steam-pressing and its application to phenolic-bonded particleboard, Shen (19) demonstrated that 25-mm-thick maple particleboard could be produced in one minute at a steam pressure of 300 psi contrasting with a minimum pressing time of 15 minutes using conventional heated platens at the same temperatures. It was notable that steam-pressed boards were superior to conventionally pressed boards in regard to dimensional stability. For example, thickness expansion and water absorption in steam-pressed boards after 24-hour sample soaking were respectively about one-third and one-half of the values for conventionally pressed samples. Springback (permanent swelling after samples had been boiled in water for two hours and then reconditioned to equilibrium at 21°C and 65% relative humidity) for steam-pressed boards was 2-5% while for conventional boards it was 21-33%. More recently, Price and Geimer (20) prepared isocyanate-bonded oak waferboard at a thickness of 25 mm and noticed better dimensional stability than with conventionally pressed panels. Enhancement of dimensional stability was also found with phenolic-bonded panels.

Shen (21) described a procedure in which lignocellulosics may be bonded without using adhesives. In his process, lignocellulosic material is exposed to steam in the temperature range 120-280°C and the resulting fiber material can be hot-pressed without adhesive to form a panel product. Shen suggested that process conditions led to the conversion of hemicelluloses to low-molecular weight carbohydrates which then polymerise to form wood bonding agents *in situ*. It has also been suggested that hemicellulose degradation products may act in a bulking manner to enhance stability. By this method it is claimed that wood and agricultural residues such as cereal straw, bagasse, cotton stalks, and rice husks can potentially be converted to exterior grade panels. The process is claimed to be suitable for a variety of products including S2S hardboard, MDF, and waferboard. Claimed process advantages are the need for less energy in refining, the production of easily-machinable materials with hard and smooth surfaces, and superior product dimensional stability. Disadvantages of the process are the dark color of the final product, a loss of 5-10% of the material mass as volatiles and, perhaps most importantly for practical purposes, the need for higher steam pressures and press temperatures than those commonly available in panel pressing operations. Also, unlike synthetic resin-bonded materials in which adhesive content can be easily manipulated to give satisfactory properties, the Shen binderless process depends very much on process optimisation to obtain adequate bond strength. There has been some interest in the process in the Peoples' Republic of China where it has been seen as a possible outlet for a wide range of agricultural residues.

An alternative commercial process (22) involves treatment of wood furnish with steam at 300-450 psi and 210-238°C for one to four minutes in a continuous process. Results of dimensional stability tests on UF-bonded particleboard produced from raw material treated by this process indicated a dimensional stability enhancement by a factor of three over conventionally produced particleboard.

In summary, the literature indicates that improvements in dimensional stability of composite lignocellulosic materials are achievable through heat treatment, preferably under wet conditions at temperatures in the 180-240°C range. Post treatment of the finished product is also possible but is limited by potential degradation of UF resins. Evidence suggests steam-injection pressing can also be used as a method of controlling dimensional stability. Much scope remains for examining improvements in the proposed heat treatment techniques to produce dimensionally stable composites under conditions that will be technically and economically acceptable to industry. At the same time, a greater understanding of the impact of thermochemical treatments on raw materials with or without adhesive addition could allow further potential developments to be defined. In the literature on enhancing composite dimensional stability through heat treatments, frequent reference is made to the darkening of the wood material and there is no suggestion that color stabilisation has been achieved. However, uniform color is an important attribute in certain markets and this aspect may also need addressing in future research on heat treatment of fiber.

Enhancement of Lignocellulosic Composites through Chemical Modification

Dimensional stabilisation of composite wood products through chemical modification of the raw material has been widely reported with particular emphasis on opportunities for acetylation processes. Acetylation involves the replacement of accessible hydroxyl groups in the lignocellulosic structure with acetate groups by reaction of the substrate with agents such as acetic anhydride in liquid or vapor form or ketene gas. In theory, weight gains up to 30% through acetylation should be possible; however, in practice, weight gains of 20-22% are more common. The difference is thought to arise because not all hydroxyl sites on lignin, cellulose, or hemicelluloses are accessible to the treatment chemical.

Rowell *et al.* (*23*) described a simple acetylation procedure in which acetic anhydride may be used without a co-solvent or catalyst. For example, batches of pine or birch particles were dipped in acetic anhydride for one minute at 20°C, drained of excess reagent, heated at 120°C for various periods, and then oven-dried for 12 hours. The dimensional stability of melamine-urea-formaldehyde (MUF) resin-bonded particleboard made from acetylated material was compared with that of boards incorporating unmodified particles and the same resin. Acetylation weight gains of 7-10% gave anti-shrink efficiencies (ASE) of about 40%, while acetylation weight gains of 18-20% gave ASE values of about 70%. Acetylation reduced both the rate and the final extent of board thickness swelling in water. Control particleboards made from unmodified particles swelled about 50% in five days, while board samples prepared from particles acetylated to 23% weight gain swelled by only about 5%. Similar results have been obtained with phenolic-bonded waferboard prepared from acetylated radiata pine (*24*).

Imamura *et al.* (*25*) prepared low-density particleboards from acetylated or unmodified particles of *Albizzia falcata*, a fast-growing tropical timber species, using PF or isocyanate adhesives. The target specific gravity was 0.5 in air-dry condition for both control and acetylated particleboards. Rate of thickness swelling was less for isocyanate-bonded boards manufactured from untreated control particles than for PF-bonded controls. Rate of thickness swelling in acetylated board samples was significantly less than in controls and was about the same with both adhesive types. Interestingly, MOR in bending was 20-25% less in boards produced from particles acetylated to 18% weight gain than the corresponding value for untreated control boards. However, under wet conditions, acetylated boards had higher MOR values than untreated control boards.

Rowell (*26*) has summarised the benefits to be gained through acetylation of a wide range of lignocellulosics. Materials that have been studied include softwoods, hardwoods, bamboo, bagasse, jute, pennywort, and water hyacinth. When the influence of acetylation on moisture sorption of materials is examined by plotting the percent reduction in equilibrium moisture content (EMC) at 65% relative humidity as a function of acetyl content for a range of lignocellulosics, a straight line plot is obtained. This plot (Figure 1) highlights the potential for dimensionally stabilising a wide variety of lignocellulosic materials through acetylation and indicates that there is a common factor controlling cell wall stability. It is known that isolated cellulose is not easily acetylated, acetate is found mostly in the lignin and the hemicelluloses, and the lignin and hemicellulose contents of the various materials differ quite markedly. It seems reasonable to assume therefore that acetylation is controlling the material moisture sensitivity due to lignin and hemicelluloses but may not be reducing the moisture sensitivity of the cellulose polymer.

Rowell and Rowell (*27*) acetylated spruce chips, measured the EMC of the chips at various relative humidities and then processed some of the chips in a laboratory refiner before re-measuring EMC values. The EMC of the acetylated chips was higher at all conditions of relative humidity than the EMC of fiber that had been acetylated after fiberising untreated chips. These results suggested that new moisture sorption sites had been created in the refining process. Direct acetylation of fiber, rather than acetylation of chips followed by refining, would therefore be the most efficient means of reducing the EMC of the raw material as a step towards producing a dimensionally stabilised wood composite.

Chemical modification processes other than acetylation have been reviewed (*28*) and in some cases represent alternative routes to dimensionally stabilised composite materials. Rowell (*29*) has reviewed methods of wood cell wall stabilisation, the extent of stabilisation and also how much stability is needed. In addition to acetylation, other single-site reactive chemicals that may be used to chemically modify lignocellulosics include other anhydrides, methylating agents, alkyl chlorides, and various aldehydes (*30*). Polymerisation reactions with epoxides (*31*) and isocyanates (*32*) have been studied. Cross-linking reactions with formaldehyde have also been proposed as routes to dimensional stabilisation of wood or paper (*33-36*). A formaldehyde treatment resulting in 7% weight gain in wood material was reported to give 90% stabilisation efficiency as measured by water soaking (*33*). Higher weight gains do not increase stabilisation efficiency and this suggests that the mechanism of dimensional stabilisation may be based on the accessibility of cell wall sites where cross-linking can occur. Formaldehyde cross-linking also occurs most readily in the presence of a Lewis acid catalyst (e.g., $ZnCl_2$) and this can lead to reductions in material strength and surface darkening. Formaldehyde cross-linking of wood without catalyst is possible but much lower weight gains are achieved.

Of the various approaches to chemical modification of lignocellulosic materials, acetylation has been most widely studied because the chemicals used (e.g., acetic anhydride) are readily available, relatively non-toxic, and relatively inexpensive. Disadvantages are the requirement for corrosion resistance in all plant equipment and the need to remove acetic acid by-product from treated material. It is the latter factor that makes acetylation as a process more suitable for composite products than for solid products. Treatment of the flakes or fibers required for composite production allows easy removal of acetic acid by-product. Acetylation of wood has been established as a commercial process by Daiken Kogyo Ltd of Okayama, Japan (Sadachi, M. personal communication, 1989) and an in-depth analysis of process economics for particles, fibers, or veneers is also now available (*Power, A.J. and Associates, Inc., unpublished report*).

Acetylation offers benefits in addition to dimensional stabilisation. In particular, the ultra-violet (UV) light stability of materials can be improved. For example, Feist *et al.* (*in press*) showed that the surface of acetylated aspen eroded about 50% slower than the surface of unmodified aspen in accelerated weathering experiments. In outdoor tests, acetylated aspen flakeboard remained light yellow in color whereas unmodified aspen flakeboard turned dark orange to light grey over the same period.

Weathering of wood is caused by the combined effects of light (both visible and UV), water, heat, and abrasion by dirt and other particles. In most climates, UV radiation and the stresses imposed by cycles of surface wetting and drying are likely to be the most important factors in wood weathering. In general, wood modification procedures have been designed to enhance dimensional stability and durability rather than UV stability and so it was of interest to observe how different chemically modified woods might respond to weather. Feist and Rowell (*37*) found that chemical modification of southern yellow pine with butylene oxide or butyl isocyanate was ineffective in providing enhanced weather resistance as measured by wood surface erosion or wood sample weight loss during accelerated weathering. However, lumen filling by treatment with methyl methacrylate

(MMA) followed by MMA curing *in situ* within the wood cells was effective in reducing surface erosion when compared with unmodified or chemically modified pine. It was suggested that the MMA polymer probably reduces water uptake by blocking routes for water entry and retards subsequent leaching of products from photodegradation of wood material at the surface. Feist and Rowell (*37*) also found that isocyanate or epoxide modification of the wood cell wall followed by lumen filling with MMA improved both dimensional stability and weather resistance. Weight losses as a result of 800 hours' accelerated weathering of pine that had received such two-step treatments were at least 50% lower than those for material that had received only chemical modification treatment with isocyanate or epoxide. As expected, UV light without water exposure produced very low weight losses for all samples and, where surface erosion did occur in weathering experiments, earlywood eroded much more rapidly than the denser latewood bands.

Feist *et al.* (*in press*) found that acetylation of aspen followed by MMA treatment was more effective than either treatment alone in reducing the rate and extent of sample swelling in water and in reducing the rate of surface erosion during weathering. Aspen samples that had been treated with MMA alone had swelled to the same extent as untreated controls after one hour suggesting that the weather resistance introduced by MMA treatment may result from polymerised MMA acting as an agent to bind degraded fibers to the surface rather than impeding water entry. It has also been suggested that the MMA polymer could reduce photochemical degradation of wood by offering an alternative pathway for energy flow when wood is exposed to light. Chemical analyses of acetylated aspen wood surfaces after accelerated weathering showed that acetyl content had been reduced (e.g., 17.5% acetyl content before weathering compared with 12.8% acetyl content in the outer 0.5 mm of the wood after 700 hours' accelerated weathering). Acetylated aspen showed chemical changes in lignin and cellulose similar to those occurring in untreated wood but there were differences between acetylated and untreated samples in terms of hemicellulose content. It was suggested that the reduction in weathering of acetylated wood may be a result of acetylation of both lignin and hemicelluloses.

Feist *et al.* (*in press*) acetylated aspen fiber to 15% weight gain and prepared PF-bonded aspen fiberboard from the material. Acetylated fiberboard showed less thickness swelling and smoother board surfaces after weathering than boards prepared from unmodified fibers.

In New Zealand, research at the Forest Research Institute (FRI) (*unpublished data*) has demonstrated the long-term color stability (Figure 2) of acetylated radiata pine veneer and the improved performance of exterior paints, stains, or clear finishes when applied over acetylated veneer. Surfaces of acetylated radiata pine veneer showed signs of mildew growth within three months of initial exposure at 45°, north-facing in New Zealand, indicating that some surface protection might be necessary in practice. The nature of mildew growth on acetylated wood surfaces and the long-term performance of exterior finishes on acetylated wood are presently being investigated in detail at FRI.

Summary, and Future Research Needs

Lignocellulosics such as wood, bagasse, wheat straw, flax, and a wide range of other materials can potentially be reduced to particle, wafer, or fiber form and then hot-pressed to create composite building materials. Particleboard, waferboard, and MDF are already well established wood products. Production of MDF, in particular, is growing rapidly and this is likely to continue. Use of other lignocellulosics also has the potential to consume materials that currently might be regarded as agricultural residues.

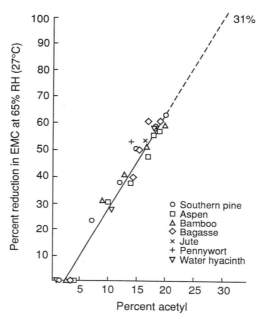

Figure 1. Reduction in equilibrium moisture content (EMC) as a function of bonded acetyl content for various acetylated lignocellulosic materials. (Reproduced from ref. 26.)

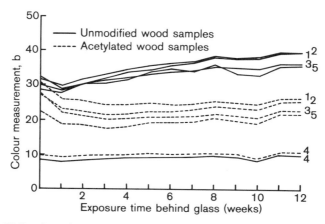

Figure 2. Yellowing of acetylated (21-22% weight gain) and unmodified radiata pine veneer with various surface finishes as a function of exposure time behind glass.
(1 = alkyd-urethane varnish, 2 = moisture-curing urethane varnish, 3 = acrylic varnish, 4 = bleaching stain with alkyd-urethane varnish topcoat, 5 = unfinished).

Increased use of composite products might be possible if dimensional and UV light stability could be introduced into lignocellulosic raw materials. For material in particulate or fiber form, two potential ways of introducing dimensional stability are either by heat treatment (of the raw material or the final product) or by chemically modifying the lignocellulosic. There is considerable evidence that heat treatment is effective and commercial processes have been established (e.g., 22). However, two factors still limiting the development of heat stabilisation procedures are the darkening of the material or product surfaces and the need to develop process conditions that are more compatible with current industry practices.

Chemical modification, particularly acetylation, has the advantage that lignocellulosics can be modified to enhance not only dimensional stability but also durability (i.e., decay and insect resistance) and UV light stability. Obtaining improvement in a wide range of material properties with a one-shot treatment has great appeal but the process costs for acetylation of fiber, for example, will probably require targeting of high-value, low-volume end uses in the first instance. In Japan, a company has been established to produce acetylated wood veneer-based products such as outdoor sign boards and components for Japanese wooden bathtubs and this is the first large application of chemical modification technology applied to wood. Veneer-based products such as plywood have built-in dimensional stability because of the alternating directions in which adjacent veneers are laid; however, improved surface properties and finishing characteristics can be introduced through the use of acetylated veneers. Research suggests that the decorative value of veneer can be enhanced through acetylation and that long-term color stability of veneer for indoor uses can be achieved.

Future research needs for chemical modification as a route to dimensionally and UV light-stable composites include the continued evaluation of new chemical modifiers and an improved understanding of the mechanism by which acetylation can introduce these properties to lignocellulosic material. The latter area, and particularly the inter-relationship between dimensional and UV stabilisation in a chemical sense, is the subject of joint research between the US Forest Products Laboratory at Madison, Wisconsin, and FRI in New Zealand.

Literature Cited

1. Food and Agricultural Organisation of the United Nations Production Yearbook. 1985, Volume 38, FAO Statistics Series No. 61.
2. Barnes, D. Proceedings of the 1987 Marcus Wallenberg Foundation Symposium. 1988, Söderhamn, Sweden.
3. Wilson, G. New Zealand Forest Industries. 1989, 40.
4. Bains, B.S. Report No. 12 for Alberta Forest Resource Development Agreement. 1986.
5. Bariska, M. Paper presented at the Three Nations Conference on Forest Products. 1978, Klagenfurt, Austria.
6. Rowell, R.M.; Youngquist, J.A.; Montrey, H.M.. Forest Products Journal. 1988, 38 (7/8), 67.
7. Suchsland, O.; Lyon, D.E.; Short, P.E. Forest Products Journal. 1978, 28(9), 45.
8. Watkinson, P.J.; Van Gosliga, N.L. Proceedings of the second Pacific Timber Engineering Conference. Auckland, New Zealand, 1989; Volume 1, 171.
9. Suchsland, O.; Woodson, G.E.; McMillin, C.W. Forest Products Journal. 1987, 37 (11/12), 65.
10. Hsu, W.E.; Schwald, W.; Schwald, J.; Shields, J.A. Wood Science and Technology. 1988, 22, 281.

11. Chua, M.G.S.; Wayman, M. Canadian Journal of Chemistry. 1979, 57, 1141.
12. Suchsland, O.; Enlow, R.C. Forest Products Journal. 1968, 18(8), 24.
13. Klinga, L.O.; Back, E. L. Forest Products Journal. 1964, 14, 425.
14. Hsu, W.E.; Schwald, W.; Shields, J.A. Wood Science and Technology. 1989, 23, 281.
15. Back, E.L. Holzforschung. 1987, 41, 247.
16. Spatt, H.A. ACS Symposium Series. 1977, 43, 193.
17. Hillis, W.E. Wood Science and Technology. 1984, 18, 281.
18. Lawler, C. Australian Forest Industries Journal. 1987, 12, 31.
19. Shen, K.C. Forest Products Journal. 1973, 23 (3), 21.
20. Price, E.W.; Geimer, R.L. In Composite board products for furniture and cabinets - innovations in manufacturing and utilisation; Hamel, M.P., Ed.; Greensboro, N.C., 1986, 65-71.
21. Shen, K.C. In Proceedings of the Composite Wood Products Symposium; Burton, R.J.; Tarlton, G.L.; Eds.; Rotorua, New Zealand, 1990, 105-107.
22. Taylor, J.D. In Proceedings of the 21st Washington State University Symposium; Pullman, Washington, 1987; 237-251.
23. Rowell, R.M.; Tillman, A.-M.; Simonson, R. Journal of Wood Chemistry and Technology. 1986, 6(3), 427.
24. Rowell, R.M.; Plackett, D.V. New Zealand Journal of Forestry Science. 1988, 18(1), 124.
25. Imamura, Y.; Subiyanto, B.; Rowell, R.M.; Nilsson, T. Wood Research. 1989, 76, 49.
26. Rowell, R.M. In Proceedings of the Composite Wood Products Symposium; Burton, R.J.; Tarlton, G.L., Eds.; Rotorua, New Zealand, 1990, 57-67.
27. Rowell, R.M.; Rowell, J.S. In Proceedings of the 10th cellulose conference, Schuerch, C., Ed.; Syracuse, N.Y., 1989, 343-355.
28. Rowell, R.M. Commonwealth Forestry Bureau Review, Oxford, England, 1983, 363-382.
29. Rowell, R.M. In Wood Science Seminar 1: Stabilisation of the wood cell wall; Suchsland, O., Ed.; East Lansing, MI, 1988, 53-63.
30. Rowell, R.M. ACS Symposium Series. 1984, 207, 175.
31. Rowell, R.M.; Ellis, W.D. USDA Forest Service Research Paper FPL 451, 1981.
32. Rowell, R.M.; Ellis, W.D. ACS Symposium Series. 1981, 172, 263.
33. Stamm, A.J. Tappi. 1959, 42(1), 39.
34. Stamm, A.J. Tappi. 1959, 42(1), 45.
35. Caulfield, D.F. In Wood Science Seminar 1: Stabilisation of the wood cell wall; Suchsland, O., Ed.; East Lansing, MI, 1988, 87-98.
36. Minato, K.; Norimoto, M. Mokuzai Gakkaishi. 1985, 31(3), 209.
37. Feist, W.C.; Rowell, R.M. ACS Symposium Series. 1982, 187, 349.

RECEIVED May 2, 1991

Chapter 8

Thermal Plasticization of Lignocellulosics for Composites

Hideaki Matsuda

Research Laboratory, Okura Industrial Company, Ltd., 1515 Nakatsu-cho, Marugame, Kagawa-ken 763, Japan

Chemical modifications of wood with dicarboxylic acid anhydrides and epoxides are useful methods for obtaining plasticized crosslinked woods. By addition reaction of the anhydrides with hydroxyl groups of wood, esterified woods bearing carboxyl groups are obtained. When the esterified woods are allowed to react with bisepoxide under hot-pressing, crosslinking and plasticization of wood components occur simultaneously, to give reddish brown, yellowish brown, or blackish brown plasticized crosslinked wood boards. Further, polymerizable oligoester chains can be introduced into wood by oligoesterification reactions of wood with the anhydrides and polymerizable monoepoxides. Products of this reaction consist of the oligoesterified woods bearing polymerizable oligoester chains and viscous liquids consisting mainly of polymerizable free oligoesters not linked with the wood matrix. The products, when subjected to hot-pressing, give plasticized crosslinked wood boards. In this case, the free oligoesters work as a plasticizer for the wood components and are combined, by the crosslinking, with the oligoesterified woods, resulting in the formation of the network structure. These boards exhibit outstanding properties depending on the anhydride and the epoxide.

Effective utilization of unused woods such as twiggy woods and periodically thinned woods, and of wood meal, chips, etc., by-produced as industrial wastes, has not yet been practiced well and development of more effective utilization processes is desired. Therefore, active investigations have been carried out for this purpose.

However, it was not attempted until recently that wood, as a whole, is chemically modified, with a view to providing properties not observed for the original wood (1). Shiraishi et al. (2-8) have recently reported that wood becomes thermally meltable by esterification such as lauroylation and stearoylation, and also by etherification such as benzylation. In this case, the thermal fluidity becomes higher as the carbon number of the acylation agent increases. Generally, lignin is belived to be a three dimensional, phenolic molecule of complex structure and ultrahigh molecular weight. However, from the fact that the chemically modified woods are thermally meltable, Shiraishi says that lignin might also be regarded as a highly branched, linear high polymer with large branches (7). Of other studies on chemical modification of wood by using its hydroxyl groups, those on the improvement of wood with isocyanates recently became actual (9,10).

However, there was little work on the method for efficiently introducing active functional groups into wood, and also on utilization of the obtained functional group-bearing woods.

Recently, reaction of commercially easily available dicarboxylic acid anhydrides with hydroxyl groups of wood was investigated. It was found that carboxyl group-bearing esterified woods could efficiently be obtained by the addition reaction (esterification) of the wood with the anhydrides, as shown by Scheme 1 (11,12). Further,

Wood-OH + R(CO)$_2$O ⟶ Wood-OOC-R-COOH   Scheme 1

properties of the esterified woods were investigated and preparation of novel wood-based polymers was attempted by the addition esterification using the introduced active carboxyl groups. In this series of studies, it was found that novel plasticized crosslinked woods having properties not observed for the original wood could be obtained. That is, novel advantageous methods for thermal plasticization of lignocellulosics for composites were found. This chapter reviews interesting results obtained in this series of studies.

Introduction of Carboxyl Groups into Wood

Esterification Reaction. The acylation of wood with higher aliphatic acid chlorides, from caproyl to stearoyl chloride, proceeds easily in a nonaqueous cellulose solvent (N_2O_4-N,N-dimethylformamide (DMF) solvent) (4,5). Further, the reaction can be conducted in a trifluoroacetic anhydride (TFAA) - higher aliphatic acid system at 30°C or 50°C (TFAA method) or in a higher aliphatic acid chloride - pyridine - DMF system at 100°C (chloride method) (13).

In DMF or dimethyl sulfoxide, which have a high swelling ability for wood, the esterification reaction of Scheme 1 proceeds at room temperature, and the anhydrides add to the wood by ring-opening of the anhydride group, giving esterified woods bearing pendant carboxyl groups (11). In this case, wood meal is used as the original wood sample. Maleic anhydride (MA), phthalic anhydride (PA), and succinic anhydride (SA) are the anhydrides used. After the reaction, the products are obtained by washing with acetone and water, and then subjecting to Soxhlet extraction with acetone. The esterified woods thus obtained are denoted by W·MA, W·PA, and W·SA, respectively, depending on the anhydride used. However, the acids such as maleic acid, phthalic acid, and succinic acid do not react with the wood. The degree of esterification can be evaluated by weight increase, acid value, and saponification value of product (11,12,14). When the carboxyl group of monoester derived from the anhydride further reacts with remaining hydroxyl group of wood, the monoester converts into diester. The content of the monoester is obtained from the acid value, and the content of the diester from the difference between the acid value and the saponification value. Throughout this chapter sum of the monoester content and the diester content will be referred to as ester content.

In esterified woods prepared in a solvent, the ester content agrees well with the monoester content, indicating that the anhydride has added to the wood in the form of monoester. In the infrared (IR) spectra of the esterified woods, they exhibit a sharp absorption band in the range of 1,720 - 1,750 cm^{-1}, attributable to -COOH and -COO-. In addition, W·MA shows a peak at 1,640 cm^{-1}, due to -CH=CH-.

It is noteworthy that, even in the absence of solvent, the esterification reaction proceeds easily at high temperatures (12). Figure 1 shows the effect of reaction temperature on weight increase of esterified wood meal for the reaction of wood meal with SA without solvent. The reaction was conducted by heating the mixture of 2 g of wood meal and 70 g of SA with stirring. The reaction time was 3 hr. The esterificatin reaction without catalyst begins to occur above ca. 60°C; above ca. 80°C, the progress of the reaction becomes remarkable. Interestingly, the reaction proceeds even below the melting point (120°C) of the SA, that is, in the solid state of SA. This is very advantageous from the industrial standpoint.

The presence of a catalyst such as Na_2CO_3 accelerates the esterification reaction further. For example, the esterified wood obtained at 160°C in the presence of Na_2CO_3 shows a very high weight increase of 118 %. This value corresponds to ca. 12 moles of SA/1000 g of wood.

The esterification reactions without solvent, as compared with those in the presence of a solvent, are industrially advantageous and give esterified woods with a wide range of monoester contents (12). Therefore, the esterified woods prepared without solvent are described in the following sections.

It has been found that an increase in the ester content in the esterified woods leads to decreases in hygroscopicity and in initial weight loss temperature (15).

<u>Thermal Plasticity of Esterified Woods.</u> The acylated woods prepared by the TFAA method and the chloride method are thermally meltable (4,5). For example, the lauroylated wood can be molded into transparent sheet by hot-pressing at 140°C under a pressure of ca. 15.0 MPa. Further, Morita and Sakata recently reported that a chemically modified wood by cyanoethylation exhibits thermal flow and that the thermal fluidity and the solubility in organic solvent of the cyanoethylated wood are considerably improved by chlorination (16-18).

The carboxyl group-bearing esterified wood meal can be molded into yellowish or reddish brown, plasticized sheets by hot-pressing at 160°C, 55.9 MPa for 10 min (19). At lower pressures and temperatures, the fluidity decreases. Thus, the plasticity is lower than that of the lauroylated wood. This is considered due to the introduced polar carboxyl groups. Table I shows the thermal plasticity of W·SA prepared at various reaction temperatures, in the presence of Na_2CO_3. When the reaction temperature is below ca. 100°C, the ester content agrees well with the monoester content, indicating

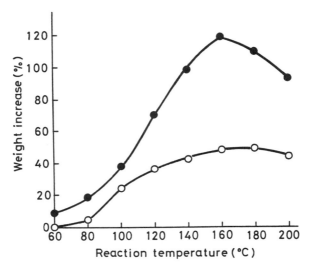

Fig. 1 Effect of reaction temperature on weight increase of esterified wood meal for reaction of wood meal (2 g) and SA (70 g) without a catalyst (○) or in the presence of a catalyst (Na_2CO_3: 0.2 g) (●).

that SA has added to the wood in the form of monoester, that is, carboxypropionyl group. However, above ca. 100°C not only the mono-

Table I. Thermal Plasticity of Esterified Woods Based on SA

Wood Meal Esterified[a]			
Reaction Temperature (°C)	Ester Content[b] (%)	Monoester Content[b] (%)	Thermal plasticity[c]
60	9.0	9.0	Poor
80	14.8	15.6	Intermediate
100	28.5	24.5	Good
120	60.8	51.0	Good
140	86.4	69.4	Good
160	105.2	84.1	Good
180	99.3	75.7	Poor
200	90.6	67.5	Poor

[a] Obtained by reaction of 2 g wood meal with 70 g SA in the presence of 0.2 g of Na_2CO_3. Reaction time = 3 hr.
[b] Values reported in ref. 19.
[c] Hot-press condition = 160°C, 55.9 MPa , 10 min. Sheet thickness = 0.5 mm.

ester content but also the diester content shows a tendency to increase with increase in reaction temperature.

An increase in the monoester content results in increased thermal plasticity. However, as the reaction temperature increases, the plasticity decreases at a given temperature, probably due to the increase in the diester content, leading to a decrease in the thermal fluidity by crosslinking.

Addition Reaction of Carboxyl Group-Bearing Esterified Woods

Addition Reaction with Epoxides. The carboxyl groups introduced into the wood are reactive with epoxide groups. When the esterified woods are subjected to the addition reaction with phenyl glycidyl ether (PGE), the epoxide group in PGE adds to the carboxyl group in the esterified wood to form an ester linkage, as shown by Scheme 2 (20). In this case, other epoxides such as allyl glycidyl ether

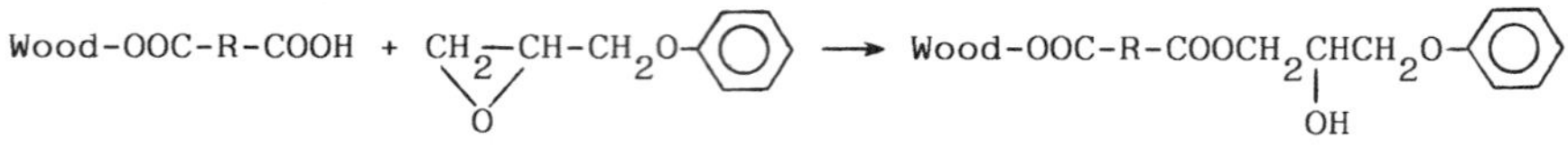

$$\text{Wood-OOC-R-COOH} + \underset{O}{CH_2-CH-CH_2O}-\langle O \rangle \longrightarrow \text{Wood-OOC-R-COOCH}_2\underset{OH}{CHCH_2}O-\langle O \rangle$$

Scheme 2

(AGE) and glycidyl methacrylate (GMA) also react with the esterified woods (21).

The epoxide-adducted esterified woods also can be molded by the hot-pressing into reddish brown, yellowish brown, or yellow plasticized sheets (22). Furthermore, the addition of the epoxide results in an improvement in the moisture resistance of the molded sheets.

Alternately Adding Esterification Reactions. When the epoxide-adducted esterified woods are further allowed to react with the anhydride and the epoxide at high temperatures, alternately adding esterification reactions occur, to produce oligoesterified woods, as shown by Scheme 3 (20).

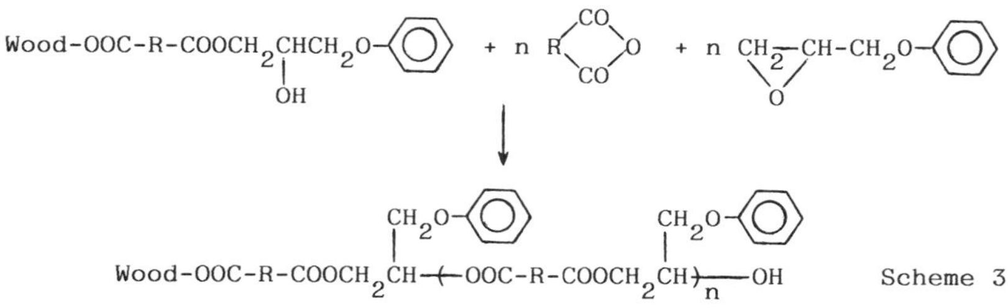

$$\text{Wood-OOC-R-COOCH}_2\underset{OH}{CHCH_2}O-\langle O \rangle + n\,R\underset{CO}{\overset{CO}{<}}O + n\,\underset{O}{CH_2-CH-CH_2}O-\langle O \rangle$$

$$\downarrow$$

$$\text{Wood-OOC-R-COOCH}_2\underset{|}{CH}\!\!\leftarrow\!\!\text{OOC-R-COOCH}_2\underset{|}{CH}\!\!\xrightarrow{}_{n}\!\!OH$$
$$\qquad\qquad\qquad CH_2O-\langle O \rangle \qquad CH_2O-\langle O \rangle$$

Scheme 3

Crosslinking Reaction of Esterified Woods with Bisepoxide

Crosslinking Reaction Accompanying Plasticization of Wood. Crosslinking reactions of the esterified woods with bisphenol A diglycidyl ether (BADG) having two epoxide groups proceed smoothly at high temperatures (23). In this case, the epoxide groups in BADG add to the carboxyl groups in the esterified woods to produce ester linkages, resulting in a formation of crosslinks between the esterified woods via BADG. Further, under higher pressures, the wood components plasticize to give reddish brown, yellowish brown, or blackish brown, crosslinked wood boards whose surfaces are smooth, glossy, and plastic-like. The following conditions have been found to be suitable for the preparation of plasticized crosslinked wood boards: first step, 150 - 170°C, 1.80 MPa, 10 - 40 min; second step, 180 - 190°C, 18.0 - 27.0 MPa, 20 - 60 min.

Properties of Plasticized Crosslinked Wood Boards. Plasticized crosslinked wood boards of various wood contents can be prepared as follows. First, esterified wood meal with a desired wood content and BADG are blended so that acid value and epoxide value of the system might become equal, and then the system is subjected to the hot-pressing. Wood contents (true wood contents) of 60 - 70 % are desirable to obtain plasticized crosslinked wood boads with high water resistance.

Table II shows the physical and other properties of the plasticized crosslinked wood boards based on various esterified woods. The

Table II. Physical and Other Properties of Plasticized Crosslinked Wood Boards

Physical and Other Properties[a]	W·(25.7)MA - BADG[b]	W·(25.1)SA - BADG[b]	W·(30.4)PA - BADG[b]	BADG -DSA[c]	BADG -PA[c]
WC (%)	59	62	60		
FS (MPa)	86.8	70.5	77.6	93.1	110.3
CS (MPa)	178.0	183.5	198.7	73.1	151.6
IS (J/m)	16	15	16		
RH (M scale)	93	95	116		
HDT (°C)	81	114	116	66-70	110-152
WA (%)	0.81	0.50	0.35		
THSW (%)	0.76	0.41	0.25		
LSW (%)	0.04	0.02	0.01		

[a] WC = wood content; FS = flexural strength; CS = compressive strength; IS = impact strength; RH = Rockwell hardness; HDT = heat distortion temperature; WA = water absorption; THSW = thickness swelling; LSW = linear swelling.
[b] Adapted from ref. 23.
[c] Adapted from ref. 24. DSA = Dodecylsuccinic anhydride.

wood contents are ca. 60 %. Generally, these boards exhibit properties which are much superior to those of usual woody boards such as fiber boards and particle boards. The board based on MA exhibits the highest flexural strength of 86.8 MPa. Meanwhile, compressive strength, hardness, and heat distortion temperature (HDT) are the highest in the board based on PA and are 198.7 MPa, 116, and 116°C, respectively. It is characteristic of these plasticized crosslinked wood boards that they show very high compressive strength. In addition, the PA-based wood board has higher water resistance than the MA and SA-based ones. This is considered to be attributable to the high hydrophobicity of the phenyl ring.

Furthermore, in Table II are also shown the properties of the representative cured resins of the usual BADG - anhydride systems (24). The above plasticized crosslinked wood boards are greatly superior to these cured resins in compressive strength, although the former are inferior to the latter in flexural strength.

Introduction of Polymerizable Oligoester Chains into Wood and Crosslinking

Oligoesterification Reaction. Polymerizable oligoester chains can be introduced into wood by the oligoesterification reaction of wood with the anhydrides and the epoxides such as AGE or GMA, as shown by Scheme 4 (25,26). This route is an extention of the reaction of

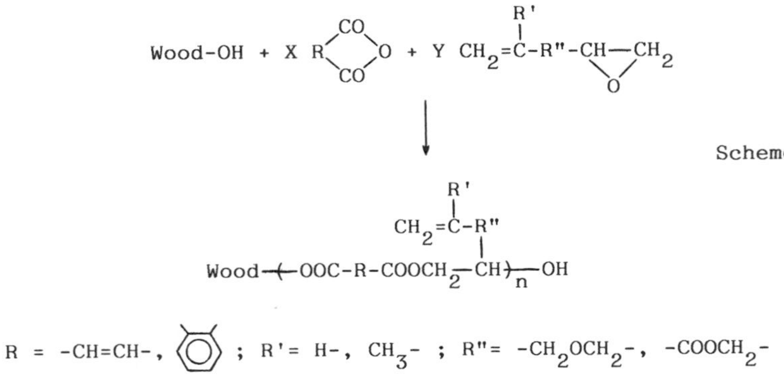

$$\text{Wood-OH} + X \underset{\text{CO}}{\overset{\text{CO}}{R}}\!\!\!\!\diagdown\!\!\text{O} + Y\ CH_2=\underset{\text{R'}}{\overset{|}{C}}-R''-CH\!\!-\!\!CH_2$$
$$\underset{O}{\diagdown\diagup}$$

$$\downarrow$$

Scheme 4

$$\text{Wood}\!\!-\!\!\!\left(\!OOC\text{-}R\text{-}COOCH_2\text{-}\underset{\underset{R''}{\overset{|}{\underset{|}{C}}}=CH_2}{\overset{R'}{\overset{|}{CH}}}\right)_{\!\!n}\!\!\!-OH$$

$$R = -CH=CH-,\ \bigcirc\!\!\!\!-\ ;\ R' = H-,\ CH_3-\ ;\ R'' = -CH_2OCH_2-,\ -COOCH_2-$$

Scheme 3. The reaction of Scheme 4 without the process of isolating the intermediate esterified wood is advantageous from the industrial standpoint. The oligoesterified woods thus obtained are of interest in that they will be crosslinked at high temperatures and under high pressures, accompanying plasticization of the wood components, to give plasticized crosslinked woods.

In this type of oligoesterification, the following main reactions of the functional groups are considered:

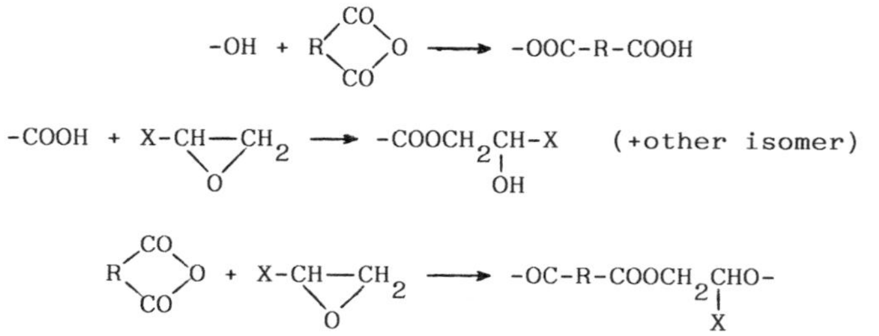

$$-OH + R\underset{CO}{\overset{CO}{<}}O \longrightarrow -OOC-R-COOH \qquad (1)$$

$$-COOH + X-CH\underset{O}{-}CH_2 \longrightarrow -COOCH_2\underset{OH}{CH}-X \quad (\text{+other isomer}) \qquad (2)$$

$$R\underset{CO}{\overset{CO}{<}}O + X-CH\underset{O}{-}CH_2 \longrightarrow -OC-R-COOCH_2\underset{X}{CHO}- \qquad (3)$$

First, the reaction of hydroxyl groups of wood with acid anhydride group, that is, reaction 1 should occur, to produce esterified wood bearing carboxyl groups. The carboxyl groups are starting points for chain extension. That is, to the carboxyl group thus produced, the epoxide group adds to form a new hydroxyl group, as shown by reaction 2. Next, the reaction of the hydroxyl group so produced with the acid anhydride group, that is, again reaction 1 occurs. Thus, these addition reactions are considered to repeat alternately, leading to the formation of oligoesterified woods, as shown by Scheme 4. Meanwhile, since the concentration of the acid anhydride groups is high in the initial stages of the reaction, the reaction of the acid anhydride group with the epoxide group (27), that is, reaction 3 is considered possible. However, this reaction results in structurally analogous free oligoester chains which are not linked with the wood matrix. In addition, active hydrogens contained in trace amounts of impurities existing in the reaction mixture would initiate the alternately adding esterification reactions. Also in this case, similar free oligoester chains are formed.

Figure 2 shows, as a typical example, the oligoesterification reaction of wood with PA and GMA. After 1 hr of the initial wood - PA reaction at 150 °C, GMA was gradually added over 15 min at 90 °C. Then, the mixture was stirred further at 90 °C. The reaction proceeds rapidly during the addition. The acid values are due to carboxyl groups and further to those produced by hydrolysis of anhydride groups in the determination. The NA acid values determined in nonaqueous medium are due to the carboxyl and the anhydride groups. The difference between the acid value and the NA acid value corresponds to concentrations of the anhydride groups. With increase in reaction time, acid value and epoxide value decrease and conversions reach nearly maximum conversions attainable after 7 - 8 hr. Anhydride groups are almost consumed after ca. 6 hr, indicatinng that the progress of the reaction in the latter stages is due mainly to reaction 2. The decrease of epoxide value is a little greater than that of acid value, even with a slight excess of GMA over PA in feed. This shows that etherification of epoxide occurred to a slight degree under the catalytic influence of the anhydride and carboxylic acid (28). PA is fairly reactive at 90°C, probably because GMA works as a solvent for PA in the system of the oligoesterification. For the oligoesterification reactions with MA, temperatures of 110 - 120°C have been found to give a convenient rate (25),(Matsuda, H., Okura Industrial Co., Ltd., Marugame, unpublished data).

By varying the feed weight ratio, reaction systems with various wood contents are obtained. Products of the reaction consist of acetone-insoluble and soluble parts (25,26). The insoluble parts are oligoesterified woods bearing polymerizable oligoester chains. The soluble parts which are viscous liquids consist mainly of polymerizable free oligoesters not linked with the wood matrix. The formation of the free oligoesters is considered due to the side reactions described above.

Effects of wood content in feed on acetone-soluble and insoluble parts, and weight increase of the insoluble part for the products were investigated for various systems (25,26),(Matsuda, H., Okura Industrial Co., Ltd., Marugame, unpublished data) and are summarized in Figure 3. The soluble part shows a tendency to decrease with increase in the wood content. However, a reverse trend is observed for the insoluble part. Meanwhile, the insoluble parts show weight increases (based on original wood) of ca. 5 - 65 %, which decrease with increase in the wood content. The weight increases are due to oligoester chains linked with the wood matrix. In addition, the insoluble and soluble parts exhibit slight residual acid values, indicating that they contain small amounts of residual terminal carboxyl groups. In the case of the wood - PA - GMA and wood - MA - GMA series, residual epoxide values of the soluble parts are not negligible (26),(Matsuda, H., Okura Industrial Co., Ltd., Marugame, unpublished data). The soluble parts are a mixture of free oligoesters (predominantly) and small amounts of unreacted GMA and dissolved

8. MATSUDA *Thermal Plasticization of Lignocellulosics for Composites* 109

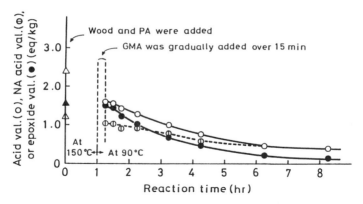

Fig. 2 Oligoesterification reaction of wood with PA and GMA. Feed weight ratio of wood : PA : GMA = 100 : 29.7 : 37.0. Mole ratio of PA : GMA = 1 : 1.3. (○), (◐), and (●) are initial acid value, NA acid value, and epoxide value, respectively, which were calculated on the assumption that wood, PA, and GMA were mixed at a time at the beginning of the reaction.

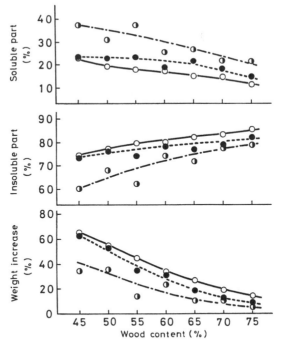

Fig. 3 Effect of wood content in feed on acetone-soluble part, insoluble part, and weight increase of insoluble part for products of oligoesterification reaction of wood with MA or PA and AGE or GMA. (○) wood–MA–AGE series; (◐) wood–PA–GMA series; and (●) wood–MA–GMA series.

oligoesterified wood components. GMA seems to have higher solvent power for wood components than AGE.

Crosslinking Accompanying Plasticization of Wood. The insoluble parts, that is, the oligoesterified woods do not show good thermoplastic properties. Similarly, in the case of the wood - MA - AGE series, when the products which have not been separated into the insoluble and the soluble parts, that is, the oligoesterified wood-containing mixtures are subjected to hot-pressing, a great part of the soluble part exudes from the system, and sufficient plasticization of the wood components is not observed. Meanwhile, in the case of the mixtures into which a catalytic amount of dicumyl peroxide (DCP) has been added, both at high temperatures and under high pressures, the wood components plasticize to give reddish or yellowish brown, crosslinked wood boards whose surfaces are smooth, glossy, and plastic-like (25). In this case, the exudation of the soluble parts is not observed, indicating that the free oligoester chains are combined, by the crosslinking, with the oligoesterified woods, resulting in the formation of the network structure. It is advantageous that the free oligoesters which are hardening work as a plasticizer for the wood components. The crosslinking is due largely to polymerization of the allylic double bonds; however, also, copolymerization of the allylic and the maleate double bonds would occur to some extent (27,29). The plasticization of the wood components is difficult for the products from wood - PA - AGE system (Matsuda, H., Okura Industrial Co., Ltd., Marugame, unpublished data). It is known that mere allylation or carboxymethylation does not render wood thermally meltable. However, by blending the allylated wood or carboxymethylated wood with appropriate synthetic polymers or low molecular weight plasticizers such as dimethyl phthalate or resorcinol, the wood components become thermally meltable (30).

On the other hand, in the case of the products of the wood - PA - GMA and wood - MA - GMA series, the wood components plasticize, under the hot-pressing, to give plasticized crosslinked wood boards even in the absence of radical initiator (26),(Matsuda, H., Okura Industrial Co., Ltd., Marugame, unpublished data). Methyl methacrylate is known to thermally polymerize even without radical initiator (31). Also, the methacrylate double bonds in the oligoester chains polymerize at high temperatures without radical initiator. This is industrially favorable. Meanwhile, in polymerization of the allyl group, the existence of radical initiator is absolutely necessary because of degradative chain transfer (32).

Properties of Plasticized Crosslinked Wood Boards. It is known that films obtained by the thermal plasticization of acetylated-butylated woods have tensile strengths of 41.0 MPa and benzylated wood films 27.7 - 40.6 MPa (33). Further, films from allylated wood - polyethylene and allylated wood - polypropylene (1 : 2) blends exhibit tensile strengths of 92.2 and 159.0 MPa, respectively and allylated wood - resorcinol formaldehyde film 87.3 MPa (30).

Table III shows the physical and other properties of the plasticized crosslinked wood boards obtained from the oligoesterified

Table III. Physical and Other Properties of Plasticized Crosslinked Wood Boards from Products of Wood - MA - AGE Series

Physical and Other Properties[a]	Wood content (%) of Plasticized Crosslinked Wood Board[b]						
	45	50	55	60	65	70	75
HDT (°C)	>220	>220	211	209	191	179	165
TS (MPa)	35.7	37.1	43.8	46.8	49.6	48.4	39.2
FS (MPa)	77.6	79.7	81.1	77.1	79.7	82.8	76.9
IS (J/m)	14	13	14	13	15	13	14
RH (M Scale)	112	112	112	109	109	106	104
CS (MPa)	227.9	220.1	216.6	215.5	192.5	180.3	158.3
WA (%)	1.32	1.45	1.59	1.98	2.67	2.65	3.37
THSW (%)	0.95	1.08	1.32	1.89	2.90	2.84	3.87
LSW (%)	0.24	0.23	0.22	0.23	0.21	0.20	0.23

[a] HDT, FS, IS, RH, CS, WA, THSW, and LSW are the same as in Table II. TS = tensile strength.
[b] Obtained by hot-pressing at 150°C, 27.0 MPa, 30 min.

wood-containing mixtures of the wood - MA - AGE series (25). They exhibit HDT above 165°C, which increase with decrease in the wood content. It should be noted that the HDT values are above 220°C at wood contents of 45 - 50 %. Tensile strength ranges from ca. 36.0 to ca. 50.0 MPa, showing a peak at a wood content of 65 %. Flexural and impact strengths are little influenced by the wood content, remaining around 80.0 MPa and around 14 J/m, respectively. Rockwell hardness shows values in the range of 104 - 112, and had a tendency to decrease with increase in the wood content. Compressive strength increases with decrease in the wood content and, at wood contents of 45 - 60 %, exhibits very high values of ca. 216 - 228 MPa.

As for water resistance, linear swelling is not affected by the wood content, showing almost constant values of ca. 0.2 %; however, water absorption and thickness swelling range from ca. 1.3 to ca. 3.4 %, and from ca. 0.9 to ca. 3.9 %, respectively, and increase with increase in the wood content.

On the other hand, the plasticized crosslinked wood boards of the wood - PA - GMA series exhibit outstanding properties in tensile strength (~ ca. 69 MPa), flexural strength (ca. 88 - ca. 100 MPa), and Rockwell hardness (ca. 120) (26). The boards of the wood - MA - GMA series show excellent properties in HDT, tensile and compressive strengths, which are superior to those of the other series (Matsuda, H., Okura Industrial Co., Ltd., Marugame, unpublished data).

The above crosslinkable mixtures are able to give, by compression molding or injection molding, various types of plasticized crosslinked wood samples as shown in Figure 4 (34).

Summary

As described above, the plasticized crosslinked woods having properties not observed for the original wood can be obtained from the carboxyl group-bearing esterified woods or the oligoesterified woods bearing the polymerizable double bonds in their oligoester chains by the crosslinking reactions accompanying the plasticization of the wood components. The chemical modifications of wood with the anhydrides and the epoxides are industrially advantageous, because the addition reactions, such as hydroxyl - anhydride reaction and carboxyl - epoxide reaction, are utilized, which produce no by-products to be removed from the system, and because no solvent is needed.

However, further research work is needed for commercializing the plasticized crosslinked woods. From the economical viewpoint, production by extrusion or injection molding would be more advantageous. In this case, further higher thermal fluidity of the mixture is required. For this purpose, pretreatment of wood by suitable solvents, steam explosion, etc. would be effective. Furthermore, investigations on resistances against weathering and biodeterioration are necessary. These are in progress.

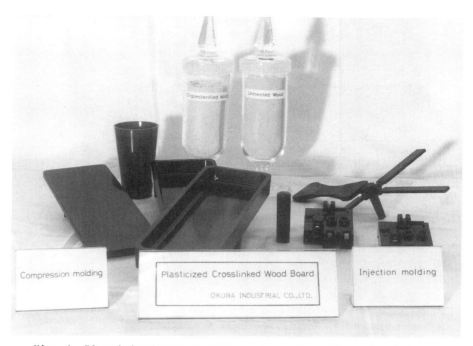

Fig. 4 Plasticized crosslinked wood samples. (Reproduced from ref. 34. Copyright 1990 American Chemical Society)

Literature Cited

1. Shiraishi, N. Nihon Setchaku Kyokaishi 1977, 13, 49.
2. Kawakami, H.; Shiraishi, N.; Yokota, T. Mokuzai Gakkaishi 1977, 23, 143.
3. Shiraishi, N.; Matsunaga, T.; Yokota, T. J. Appl. Polym. Sci. 1979, 24, 2347.
4. Shiraishi, N.; Matsunaga, T.; Yokota, T. J. Appl. Polym. Sci. 1979, 24, 2361.
5. Funakoshi, H.; Shiraishi, N.; Norimoto, M; Aoki, T.; Hayashi, S.; Yokota, T. Holzforschung 1979, 33, 159.
6. Shiraishi, N. Mokuzai Kogyo 1980, 35, 150.
7. Shiraishi, N. Mokuzai Kogyo 1980, 35, 200.
8. Shiraishi, N.; Aoki, T.; Norimoto, M.; Okumura, M. Chemtech 1983, 6, 366.
9. Steiner, P. R.; Chow, S.; Vagda, S. Forest Prod. J. 1980, 30, 21.

10. Rowell, R. M.; Ellis, W. D. in "Urethane Chemistry and Applications"; Edwards, Kenneth N., Ed.; ACS SYMPOSIUM SERIES No.172; ACS: Washington, D. C., 1981; 263-284.
11. Matsuda, H.; Ueda, M.; Hara, M. Mokuzai Gakkaishi 1984, 30, 735.
12. Matsuda, H.; Ueda, M.; Murakami, K. Mokuzai Gakkaishi 1984, 30, 1003.
13. Shiraishi, N.; Tsubouchi, K.; Matsunaga, T.; Yokota, T.; Aoki, T. Proc. - Annu. 30th Meet. Japan Wood Research Soc., 1980, 34.
14. Matsuda, H. Wood Sci. Technol. 1987, 21, 75
15. Matsuda, H.; Ueda, M. Mokuzai Gakkaishi 1985, 31, 103.
16. Morita, M.; Sakata, I. J. Appl. Polym. Sci. 1986, 31, 831.
17. Morita, M.; Shigematsu, M.; Sakata, I. Cellulose Chem. Technol., 1987, 21, 255.
18. Morita, M.; Sakata, I. Mokuzai Gakkaishi 1988, 34, 917.
19. Matsuda, H.; Ueda, M. Mokuzai Gakkaishi 1985, 31, 215.
20. Matsuda, H.; Ueda, M. Mokuzai Gakkaishi 1985, 31, 267.
21. Matsuda, H.; Ueda, M. Mokuzai Gakkaishi 1985, 31, 468.
22. Matsuda, H.; Ueda, M. Mokuzai Gakkaishi 1985, 31, 579.
23. Matsuda, H.; Ueda, M. Mokuzai Gakkaishi 1985, 31, 903.
24. "Kobunshi Kako Bessatsu 9: Epoxy Resins"; Saeki, K., Ed.; Kobunshi Kankokai: Kyoto, 1973; Vol. 22.
25. Matsuda, H.; Ueda, M.; Mori, H. Wood Sci. Technol. 1988, 22, 21.
26. Matsuda, H.; Ueda, M.; Mori, H. Wood Sci. Technol. 1988, 22, 335.
27. Fischer, R. F. J. Appl. Polym. Sci. 1963, 7, 1451.
28. Fisch, W.; Hofmann, W. J. Polym. Sci. 1954, 12, 497.
29. Urushido, K.; Matsumoto, A.; Oiwa, M. J. Polym. Sci. Polym. Chem. Ed. 1980, 16, 1081.
30. Shiraishi, N.; Goda, K. Mokuzai Kogyo 1984, 39, 329.
31. Tani, H. "Synthesized high molecular compounds"; Kotake, M., Ed.; Daiyukikagaku; Asakura Shoten: Tokyo, 1958; Vol. 22; 41.
32. Bamford, C. H.; Barb, W. G.; Jenkins, A. D.; Onyon, P. F. "The kinetics of vinyl polymerization by radical mechanisms"; Butterworths Scientific Publications: London, 1958; 46.
33. Shiraishi, N. "Advanced Technology and Perspective of Wood Chemicals"; R & D Report No.40; CMC: Tokyo, 1983; 271-285.
34. Worthy, W. C & EN, 1990, January 15, 19.

RECEIVED March 20, 1991

Chapter 9

Activation and Characterization of Fiber Surfaces for Composites

Raymond A. Young

Department of Forestry, University of Wisconsin, Madison, WI 53706

> Surface activation is a simple method for altering the surface properties of both natural and synthetic fibers. These treatments generally result in improved compatibility and provide superior bonding, adhesion and strength properties in composite structures. The activation methods described in this chapter are restricted to chemical and electric-discharge treatments of natural and synthetic fibers. To properly assess changes at the fiber surface due to surface activation it is necessary to employ the appropriate analytical methods for surface characterization. A very brief review of the methods available for surface characterization is given and three analytical methods applicable to fibers are described; namely, the Wilhelmy wetting method, ESCA and inverse phase chromatography.

The growth of composite materials has been more dramatic than that of even polymers and continued growth is expected in the future (1). This is due to the ability to "tailor" products to specific end-use applications by proper selection of the component materials. Fibers have been traditionally utilized in fiber-reinforced composites for many years. However, a new class of composites has been developed based almost entirely on fibrous materials as described in another chapter. Further development of these composites will require new methods for modification of the fiber surfaces to improve compatibility, bonding and strength properties. There are a variety of methods available for surface characterization of fibers as described below.

Surface Characterization of Fibers

Surface science has made great strides in recent years due to the development of powerful, sophisticated instrumentation (2). Many analytical tools are available for analysis of surfaces; however, it is important to consider the effective sampling depth of the analytical technique. Sampling depth of the measurement technique must be appropriate to the phenomenon under study. In Figure 1 it can be seen that bonding to surfaces and wettability involve only a few atomic layers, whereas, corrosion and

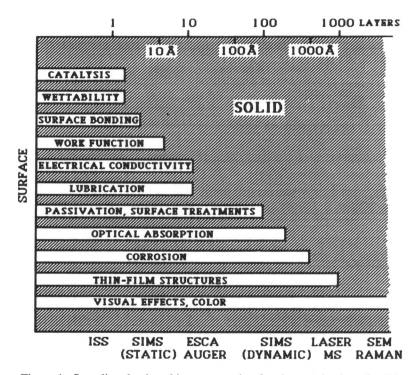

Figure 1. Sampling depth and instrumentation for characterization of solid surfaces (see Table I for description of acronyms). (Reproduced with permission of the National Academy of Science Press from ref. 3).

surface hardening treatments involve 10-1,000 atomic layers. Different instruments have different sampling depths; low energy ion scattering samples at 1-2 atomic layers, SIMS at 5 Å depth, Auger and ESCA at 20 Å, ion etching coupled with SIMS at 100 Å and laser mass spectrometry Raman microprobe and scanning electron microscopy (SEM) from 1,000 to 10,000 Å (1 μ). The acronyms and applications for the instrumental techniques are summarized in Table I *(2, 3)*. Most of the instrumental techniques give qualitative information on what is present at the surface, while, quantitative data is frequently desired. It is possible to obtain quantitative surface information with wettability and inverse gas chromatographic analyses. These methods provide excellent fiber surface information at a fraction of the cost of the more sophisticated instrumental methods. Wettability, inverse gas chromatography and ESCA analyses will be further described and their use for fiber surface characterization demonstrated.

Wetting. In the nineteenth century the Young-DuPré equation was developed to describe the interaction of a liquid with a solid surface as shown below.

$$\gamma_{LV} \cos \theta = \gamma_{SV} - \gamma_{SL} \tag{1}$$

The equation describes the conditions for wetting as a three phase equilibrium of solid (S), liquid (L) and vapor (V) in terms of surface free energies (γ). The angle (θ) is the contact angle and serves as a convenient means of visualizing the solid-liquid interaction. Generally a contact angle of zero is the condition for spreading while a contact angle other than 0 is a non-spreading condition. Whether or not a liquid spreads on a surface, there is always some wetting when a liquid comes in contact with a surface.

DuPré also developed the relationship for work of adhesion W_A;

$$W_A = \gamma_{LV} + \gamma_{SV} - \gamma_{SL} \tag{2}$$

Combining equations 1 and 2 gives

$$W_A = \gamma_{LV} + \gamma_{LV} \cos = \gamma_{LV}(1 + \cos \theta) \tag{3}$$

The measurement of the contact angle for flat surfaces is a relatively simple matter; however, for fine, anisotropic structures such as fibers this approach is quite problematical. Therefore an alternate method based on the Wilhelmy principle was developed. About forty years ago Collins *(5)* described the use of the Wilhelmy relationship to obtain fiber perimeters, using wetting liquids $(\theta = 0°)$ of known surface tensions. Knowing the liquid surface tension (γ_{LV}) and fiber perimeter (P), the force measurement provides the apparent contact angle according to the Wilhelmy equation given below.

$$F = P \gamma_{LV} \cos \theta \tag{4}$$

The work of adhesion is then

$$W_A = \frac{Fg}{P} + \gamma_{LV} \tag{5}$$

where g is the acceleration due to gravity.

The Wilhelmy technique was not widely utilized until sensitive microbalances were developed which allowed reproducible, quantitative force measurements on the fine natural and synthetic fibers *(5-8)*. The wetting force is obtained by measuring the apparent weight increase when the fiber contacts a liquid of known surface tension.

Table I. Surface Science Instrumentation

Method	Acronym	Source/Species	Application
Auger electron spectroscopy	AES	Medium-energy electrons/low-energy electrons	Elemental composition, some chemical state information
Diffuse reflectance IR Fourier transform (spectroscopy)	DRIFT	IR photons	Molecular chemical information
Electron energy loss spectroscopy	EELS	Electrons (1-10 eV)	Molecular identity, orientation and surface bonding of adsorbed molecules
Electron spectroscopy for chemical analysis, or X-ray photoelectron spectroscopy	ESCA (XPS)	X-rays/low-energy electrons	Elemental composition and chemical states Few atom layers sensitivity
Extended X-ray absorption fine structure spectroscopy	EXAFS	X-rays	Element-specific chemical states and structures, mm lateral resolution
Ion scattering spectroscopy	ISS	Low-energy ions	Elemental composition Outermost atom layer only; mm spatial resolution
Laser microprobe mass spectrometry	--	Visible, UV light	200-micron surface compositional mapping. Focused light-induced molecular desorption
Low-energy electron diffraction	LEED	Electrons (10-300 eV)	Atomic surface structure
Scanning electron microscopy	SEM	High-energy electrons	Surface morphology (5 nm image resolution)
Secondary ion mass spectroscopy	SIMS	Low-energy electrons	Elemental and molecular information. Outermost atomic layer only; 50 nm lateral resolution
Ultraviolet photoelectron spectroscopy	UPS	UV light/low-energy electrons	Elemental and molecular composition, chemical state details. A few atomic layers surface sensitivity; mm lateral resolution

SOURCE: Adapted from ref. 2 & 3.

The technique is now widely accepted as a very useful method for fiber surface characterization. Further details on this technique are given in the references (6-8).

A variety of secondary forces have been implicated in wetting and adhesion; these include dispersion (van der Waal's) forces, dipole-dipole interactions, hydrogen bonding, acid-base interactions, etc. Additional information relative to surface free energy can be obtained by evaluating the relative contributions of the secondary forces, namely, dispersive $\left(W_A^d\right)$, polar $\left(W_A^p\right)$, hydrogen bonding $\left(W_A^h\right)$ and acid-base $\left(W_A^{ab}\right)$ to the overall work of adhesion.

For interpretation, the work of adhesion has been simplified by considering only the two components, dispersion and polar forces,

$$W_A = W_A^d + W_A^p \tag{6}$$

However, Fowkes (9) has suggested that the polar forces are negligible and that hydrogen bonding is a form of acid-base interactions such that,

$$W_A = W_A^d + W_A^{ab} \tag{7}$$

If it is accepted that the dispersive components of the surface free energy interact according to a geometric mean square, then

$$W_A^d = 2\left(\gamma_S^d \, \gamma_{LV}^d\right)^{1/2} = \gamma_L(1 + \cos\theta) \tag{8}$$

Then, assuming the reduction of the surface energy of the solid surface due to the adsorption of the vapor of the probe liquid (π) is negligible, the dispersive component is obtained from the equation,

$$\gamma_S^d = \frac{\gamma_{LV}\,(1+\cos\theta)^2}{4} \tag{9}$$

Methylene iodide or tricresyl phosphate can be used as dispersive probe liquids to obtain γ_S^d. The dispersion force component of the advancing work of adhesion is determined from equation 9. The acid/base contributions to the work of adhesion can be determined using formamide as a Lewis base probe and from the equation 7 rewritten as equation 10 below (10),

$$W_A^{ab} = W_A - W_A^d \tag{10}$$

The respective polar and dispersive surface free energies can also be determined by measuring the fiber wetting in two probe liquids such as water and methylene iodide and solving simultaneous equations according to a reciprocal means approach proposed by Wu (11) below.

$$\gamma_{LV}\cos\theta = \gamma_{LV} + \frac{4\,\gamma_S^d\,\gamma_L^d}{\gamma_S^d + \gamma_L^d} + \frac{4\,\gamma_S^p\,\gamma_L^p}{\gamma_S^p + \gamma_L^p} \tag{11}$$

Inverse Gas Chromatography. In conventional gas chromatography the sorbents are modified to separate the components of liquid mixtures. With inverse gas chromatography (IGC) the known properties of fluids are utilized to characterize the surface properties of fibers.

A sample injected into a gas chromatography column which is not retained by the stationary phase will elute with a retention time t_m. A sample that interacts will have some retention time t_r, such that $t_r > t_m$. The retention volume of the sample corrected for the dead volume of the column is thus related to the volume flow rate, V, of the carrier gas by

$$V_R = (t_r - t_m)V \tag{12}$$

The viscosity of the gas passing between the column packing particles causes a finite pressure drop of the carrier gas across the column which necessitates a correction for gas compressibility. A derived correction factor j for the pressure drop is

$$j = \frac{3}{2} \frac{(P_i/P_0)^2 - 1}{(P_i/P_0)^3 - 1} \qquad (13)$$

where P_o and P_i are the outlet and inlet pressures, respectively.

The net retention volume V_N is then defined as,

$$V_N = j V_r = j(t_r - t_m) V \qquad (14)$$

and the specific retention volume is,

$$V_g = 273 V_N / WT \qquad (15)$$

where T is the column temperature and W the mass of the column loading (absorbent). Thus V_g is the elution volume of carrier gas corrected to 0°C, per gram of stationary phase in the column.

Since the retention volume is dependent on the probe concentration in the gas phase, sorption isotherms can be obtained according to,

$$V_N = WRT \left(\frac{\delta q}{\delta p} \right)_T \qquad (16)$$

where p is the partial pressure of the absorbate and q is the number of moles of absorbate per gram of absorbent. The amount of sorbed vapor is obtained by integration of equation 16,

$$q = \frac{1}{WRT} \int_0^p V_N \, d_p \qquad (17)$$

which when plotted against p for increasing values of V_N, gives the adsorption isotherm *(12-14)*. By assuming the peak height h is proportional to the partial pressure of the absorbate p, the latter can be calculated for a given detector response by a simple calibration procedure *(13)*.

Columns are prepared by pulling skeins of yarn into the stainless steel tubing. A variety of probes can be utilized to deduce the nature and extent of the solid/gas interactions. For example, for Lewis acids, t-butanol or chloroform; Lewis bases, t-butylamine or tetrahydrofuran; and amphoteric or neutral probes include ethyl acetate, pentane or octane *(14)*.

Electron Spectroscopy for Chemical Analysis (ESCA). Electron spectroscopy for chemical analysis (ESCA) is also referred to as x-ray photoelectron spectroscopy (XPS) by some investigators. Siegbahn and coworkers *(15)* developed the ESCA technique which provides a valuable method for the study of solid surfaces. The sample is irradiated with x-ray photons which collide with the inner shell electrons of atoms at the surface of the solid sample. The kinetic energy at the surface after the collision is equal to that of the photon minus the binding energy, E_b, of the surface atom electron. It is possible to determine the elements present at the surface of a solid because the binding energy is different for the electronic shells of every element. The energy emitted thus characterizes the elements present in the surface layer and the intensity of the signal gives the relative abundance. If the element under study is involved in a chemical bond, then the energy of the emitted electrons will be changed by a small but detectable amount. Thus, this chemical shift can indicate the types of chemical bonds in which the surface atoms are involved *(16)*.

The measured energy is only due to surface atoms because electrons emitted at depths greater than 1 to 5 nm lose their energy through collisions within the bulk of the solid. A number of investigators have measured and tabulated binding energies for carbon (1s) atoms in organic compounds. Dorris and Gray *(16)* have classified carbon atoms in lignocellulosic materials into four broad categories. In order of increasing chemical shifts, these are carbon atoms bonded:

- Class I.   only to carbon and/or hydrogen (C1),
- Class II. to a single oxygen, other than a carbonyl oxygen (C2),
- Class III. to two noncarbonyl oxygens, or to a single carbonyl oxygen (C3),
- Class IV. to a carbonyl and a noncarbonyl oxygen (C4).

Oxygen substitution on a neighboring carbon atom increases the binding energy; therefore, each of the four classes exhibits a number of different binding energies. Figure 2 shows the approximate range of binding energies for each group together with estimates of the range of peak widths at half height as determined by Dorris and Gray *(16)*. Considerable overlap of the ESCA peaks is expected.

Surface Modification of Fibers

There are many approaches possible for modification of fiber surfaces. It would be impossible to review all the possible modes of chemical reactions in this short chapter; therefore, this treatment will be restricted to those procedures generally referred to as "activation". The term activation implies increased reactivity and bonding of a surface but cannot be restricted to this constraint since in some cases the treatments may cause reduced reactivity. What is excluded from this chapter is direct reactions with reagents to form specific derivatives, such as esters, ethers, etc. A great deal has also been written about graft modification of polymers and fibers; therefore, graft modification will also not be covered. The activation approaches for discussion are through chemical treatments and application of electric discharges.

Chemical Activation. Most of the activation treatments applied to improve the reactivity and bonding of natural and synthetic polymers and fibers have been with oxidizing agents. Application to two types of substrates, polyolefins and lignocellulosics will be reviewed here.

Polyolefins. Brewis *(17)* evaluated surface activation of polyethylene with organic peroxides and chromic acid. The treatments with the organic peroxides involved immersion in a methylene chloride solution of the peroxide (5%) for 5 sec., removing the polyolefins and then heating the film in an oven. Although no changes were noted in the contact angles of the polyethylene after treatments with dicumyl or lauryl peroxides, the lap shear strength was doubled when the treated samples were heated at 120°C for 24 hours in a nitrogen gas. When the gas in the oven was air the strength was 3-5 times that of the control sample. The temperature of the treatment appears to be very important since no increase in bond strength was noted for the polyethylene when treated at 90°C in air for 16 hours, whereas the strength improved dramatically when heated at 120°C in air for 24 hours *(17)*.

Several investigators have noted increased bond strength of polyolefins after treatment with chromic acid *(17, 18)*. As shown in Table II, treatment with chromic acid results in an increase in polarity as measured by contact angle. There is also a considerable increase in the bond strength after the treatments. The surface activation appears to be relatively stable since there is no loss when a week interim is allowed before bonding of the oxidized samples. Based on the results in Table II it is probable that reactive functional groups such as carbonyl and carboxyl were produced at the polyolefin surface. Further surface analyses were not performed to verify these conclusions.

Table II. Effect of Chromic Acid on Wetting and Bonding of Polyethylene

Polymer[a]	Treatment	θ (degrees)	Lap Shear Strength (kg/cm^2)
LDPE	Control	99	11.0
	[b]Chromic acid (CA)	72	140.9
	[c]CA and bonded after 7 days	--	145.1
HDPE	Control	98	18.3
	Chromic acid	75	176.1
	CA and bonded after 7 days	--	167.7

[a] LDPE = low density polyethylene, HDPE = high density polyethylene
[b] Immersed in chromic acid for 1 hour, washed with distilled water, dried at 60°C for 1 hour under vacuum.
[c] As above but bonded after 7 days.

Chlorosulfonation has also been used for modification of polyethylene fibers to improve bonding with gypsum plaster (19). The fibers were immersed in a neat solution of chlorosulfonic acid, washed and dried. The treatment decreased the tensile strength but the Young's modulus was increased by more than fifty percent. The interfacial bond strength between the polyethylene and gypsum was improved over 4.8 times. The strength improvement was attributed to surface roughening of the polyethylene by the chlorosulfonic acid reaction. Previous investigators (20) noted fracture processes at surface irregularities such as kink and shear bands in the fiber after treatment with chlorosulfonic acid. This roughening effect may promote contact and bonding with gypsum plaster.

Lignocellulosic Materials. Surface activation of lignocellulosic materials with chemicals has been performed by a variety of investigators over the past 50 years. The various methods used for activation of lignocellulosic materials can be categorized as follows (21):
1. *Oxidation* - Nitric acid, periodate, peroxyacetic acid, hypochlorites, perchlorates, etc.
2. *Free Radical Generation Via Redox Reactions* - hydrogen peroxide and ferric ions.
3. *Acid Catalyzed Degradation and Condensation of Wood Polymers* - sulfuric acid.
4. *Base Treatment* - sodium hydroxide.

The distinction between oxidation, free radical generation, condensation and base effects cannot always be held rigidly since, for example, nitric acid acts as both an acid and oxidant, while hydrogen peroxide is both an oxidant and free radical generator. Zavarin (22) has reviewed the developments in wood surface activation. There has been renewed interest in activation techniques for molded fiber composite products.

Recent work by the author in collaboration with Professor John Philippou at the Aristotelian University in Greece (23) demonstrated that strong, water resistant fiber-based molded products could be produced by nitric acid activation of attrition milled whole-wood aspen fibers, with or without additives. Because of the greater surface area and compressibility of a fibrous dry-formed web, optimum results are produced for chemical activation.

Table III shows a comparison of aspen fiber board properties for a control prepared from untreated fibers, nitric acid activated fibers and phenol-formaldehyde bonded fibers. The boards from the untreated fibers show very poor strength and dimensional stability. However, nitric acid activation results in strong boards with dimensional stability much better than even the phenol-formaldehyde (PF) bonded boards. This difference between nitric acid activated and PF bonded boards is characteristic of the differences generally noted for the two bonding methods *(23)*.

Table III. Summary Comparison of Aspen Fiber Board Properties

Treatment	Density g/cc	Internal Bond kg/cm²	COLD WATER SOAK		2 HR Boil, % T.SW.
			2 HR, % T.SW.	24HR, % T.SW.	
Control	0.90	1.0	82.8	97.6	Dis.
Nitric Acid	1.00	5.4	15.8	28.4	46.0
Phenol-Formaldehyde	1.00	6.0	27.3	42.7	80.1

T.SW. = thickness swelling
Dis. = disintegrated

A variety of variables were evaluated for nitric acid bonding of aspen fiber boards. Although a longer reaction time for aspen fibers with nitric acid before bonding improved the internal bond strength of the fiber boards, no significant increase in the dimensional stability was noted beyond a two-hour nitric acid activation. The most dramatic improvements were noted for both strength (IB) and dimensional stability when the press temperature was raised (Figure 3). The higher temperatures most likely provide greater plasticization, surface contact and reaction rates for improved reactions between the fibers, resulting in better bonding and stronger boards. Significant improvements in dimensional stability as determined by thickness swell were noted up to 200°C *(23)*.

There are apparently two modes of dimensional stabilization. The first results from increased densification which gives improved bonding, holding the structure intact. The other mode of dimensional stabilization is due to modification and crosslinking of the cell wall of individual fibers such that the fibers do not expand in water and destroy the board structure. This auto-crosslinking apparently occurs with high temperature treatment of wood and fibers to impart dimensional stability by similar mechanisms *(24-26)*. The small effect of density on dimensional stability for nitric acid treated boards suggests the cell wall crosslinking mechanism is the most important for dimensional stability of these products.

When additives were incorporated with nitric acid treated aspen fibers, synergistic effects were noted. Both higher internal bond strengths and excellent dimensional stability were obtained for all boards activated with nitric acid and combined with furfuryl alcohol/maleic acid and/or lignin formulations. Ammonium lignosulfonate gave the best properties of the lignin formulations evaluated. The nitric acid treatment appears to provide compatibility with the lignin material. Thus, lignin preparations may be used as additives to a furfuryl alcohol/maleic acid system but only if preceded by nitric acid activation; otherwise, inferior boards are produced *(23)*.

A variety of techniques has been utilized to determine the mechanisms of nitric acid activation. Analyses were carried out on isolated wood polymers as well as on whole and ground wood samples. This work has already been reviewed and the reader is referred to previous publications for an in-depth discussion *(21, 22, 27-31)*.

To summarize the effects of nitric acid on wood, it appears that nitric acid activates wood by oxidation to primarily carboxyl groups, nitration of mainly aromatic lignin units, and hydrolysis of wood polymers to low molecular weight moieties even at room temperature. It is probable that the nitric acid also swells and plasticizes the

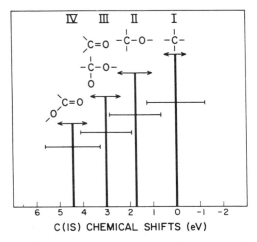

Figure 2. Range of binding energies for (1s) carbon-oxygen bonds. Upper and lower arrows indicate expected range of peak positions and peak widths at half height, respectively. (Adapted from ref. 16).

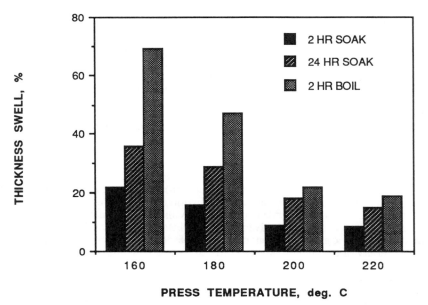

Figure 3. Percent thickness swell versus press temperature for boards produced from nitric acid activated aspen fibers.

wood surface to some extent. Extractives at the wood surface are likely solubilized and removed from the wood surface by this treatment. These effects are summarized in the schematic shown in Figure 4a&b *(31)*.

Treatment of wood surfaces with sodium hydroxide will also activate the wood (Base Activation) so that autohesive bonds can be formed by bonding under conditions similar to those used for nitric acid. Dry shear strengths as high as 2,900 psi were obtained by activation of wood with 3N sodium hydroxide; but only low wet strength was realized *(31)*.

The mechanisms for bonding with base are similar in many respects to those with nitric acid. The caustic swells and plasticizes the wood and probably hydrolyzes the wood polymers to some extent, although the degradation products would be distinctly different. The alkali also solubilizes and removes the extractives from the wood surface. In contrast to nitric acid treatment, however, a more reactive alkali cellulose (hemicellulose), and lignin would be formed which could enhance the reactivity of the wood surface. It was also speculated in a previous publication that sodium hydroxide treatment creates a much larger porous network at the wood surface that could greatly enhance interactions with applied adhesives. The combined effects of aqueous, sodium hydroxide on wood surfaces are shown schematically in Figure 4c *(31)*.

Activation by Electric Discharge. Activation of material surfaces can be accomplished in an ionized gas produced by gaseous electrical discharges. In a discharge, free electrons gain energy from an imposed electric field and lose this energy through collisions with neutral molecules. Particle bombardment energies of less than 15 ev are developed in a corona discharge at atmospheric pressure, but covalent bonds can be broken with much less energy. The transfer of energy to the molecules leads to the formation of new species including metastable species, atoms, free radicals and ions *(32-34)*.

An ionized gas is technically termed a plasma if the ionized gas has equal concentrations of positive and negative charge carriers. The properties of the plasma are dependent on the type of electric discharge; in plasma chemistry the glow discharges are of primary interest. The lack of equilibrium between the electron and gas temperature makes it possible to obtain a plasma in which the gas temperature is near ambient while the electrons have sufficient energy to rupture molecular bonds. Thus the glow discharges are well suited for initiating chemical reactions with thermally sensitive materials *(32-34)*.

Corona discharges are special cases of plasmas produced with a low frequency a.c. discharge. A corona is characterized by high electric field strength and high pressure (1 atm); a cool plasma of low ionic density. Low pressure plasmas, produced with high frequency discharges, provide an even cooler plasma environment than the low frequency excitation *(32-34)*.

Plasmas have been utilized for modification and grafting of polymer and fiber surfaces. Plasmas are ideally suited for surface modification since the effect is limited to the superficial surface layers of the substrate. Thus modified fibers generally retain their physical and mechanical properties with 99% of the fiber weight remaining unaffected by the treatment. However, prolonged plasma treatments (5-10 minutes) can have a detrimental effect on the mechanical properties of fibers *(35)*. Under prolonged exposure, the presence of oxygen in the polymer structure makes the polymer more susceptible to the plasma, while, the presence of nitrogen has the opposite effect *(32)*. However, most surface modification treatments do not require exposure times greater than one minute, therefore the fiber properties are not adversely affected by the treatment.

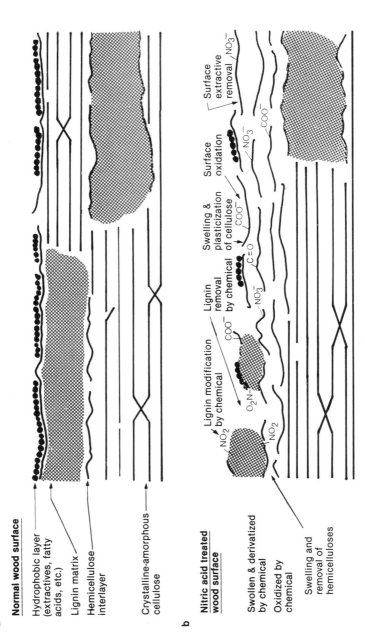

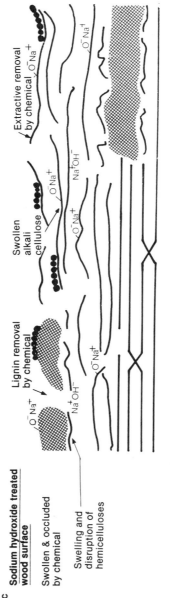

Figure 4. Schematic depiction of the surface of a) untreated wood; b) nitric acid activated wood; c) base activated wood.

Surface Modification by Plasma. The principal changes brought about by exposure of a fiber to a plasma are the chemical composition of the surface, wettability and/or the polymer molecular weight at the surface. A primary effect of plasma treatment is an improvement in bonding and adhesion of the modified materials. This important alteration allows for improved adherence of adhesives, coatings, dyes and inks and enhances the compatibility of dissimilar materials in composite structures. Applications of plasmas are widespread in such industries as packaging, electronics, construction and textiles *(32)*. The variables involved in treatment of both synthetic and cellulosic fibers are discussed below.

Synthetic Fibers. A wide variety of synthetic fibers have been exposed to plasma treatments for reasons ranging from improved adhesion of epoxy-polymer surfaces to enhancement of the anticoagulant properties of synthetic catheters. A discussion of all the variations possible is beyond the scope of this chapter. However, treatments on polyolefins demonstrate the approaches and expected effects.

The effect of the plasma on both synthetic and natural fibers is not only dependent on the nature of the substrate but also on the type of gas employed in the plasma. A large number of different gases have been utilized for surface modification with some of the more common ones listed in Table IV *(36-40)*. As illustrated for polyethylene, the reactions range from oxidation to amination and crosslinking. Oxygen plasmas tend to produce simultaneous wetting and molecular weight changes; while, noble gases appear to alter mainly polymer molecular weight at the surface *(41)*.

Table IV. Plasma Treatments to Improve Adhesion to Polyethylene

Gas	Effect	Reference
Helium	Yellowing, strengthening of weak boundary layer	Schonhorn and Hansen *(36)*
Argon	Carbonyl groups	Yasuda et al. *(37)*
Oxygen	Increased wettability, cellular surface texture	Ladizesky and Ward *(38)*
Air	Increased wettability	Boenig *(39)*
Nitrogen	Nitriles, amines, imides, and/or amides	Yasuda et al. *(37)*
Nitrous oxide	Increased wettability	Boenig *(39)*
Ammonia	Amino groups, aromatic, increase in wettability	Holmes and Schwartz *(40)*

McKelvey *(42)* reviewed the literature for polyethylene surface activation up to about 1960. Most investigators suggested that polar groups were formed at the surfaces through oxidation. Hines *(42, 43)* studied the chemical changes that occurred in a corona discharge by analyzing surface scrapings from treated and untreated polyethylene films. Infrared evidence indicated the presence of hemiformal, formal, polyformaldehyde, vinyliidene, peroxide and other groups.

Kim et al. *(44)* showed by infrared spectroscopy that oxygen plasma treatment of polyethylene produced both carbonyl and carboxyl groups at the polymer surface. ESCA can be utilized to monitor the increasing incorporation of oxygen into the polyethylene surface as the plasma treatment proceeds *(45)*. A greater proportion of singly bonded oxygen is produced in the C1s peak during the initial stages and there is a distinct shift in the O1s peak location. The wettability as measured by contact angle with water was reduced from 90° to 53° with an air or oxygen plasma and to 37° with a nitrogen plasma treatment *(46)*.

Carley and Kitze *(47)* have provided evidence for the presence of fairly stable peroxide structures of the forms RO_2R and RO_3R on polyethylene surfaces after plasma treatment. The peroxides were detected by reaction with diphenyl picryl hydrazyl (DPPH). This compound is capable of detecting as few as 10^{13} peroxide groups per square centimeter. The production of the peroxides was found to be strongly dependent on the energy provided to the film during treatment. The energy is proportional to the quotient of corona current and web speed. Regression analysis demonstrated that air-gap thickness, relative humidity and the number of electrodes were also significant factors; while dielectric constant and corona frequency were not important factors. Carley and Kitze *(47)* felt that the polar component of surface energy, γ_s^p, is the key to understanding the changes in adhesive behavior of films with corona treatment.

Treatment with oxygen plasmas generally can also decrease polymer molecular weight due to chain rupture; while, helium, nitrogen and hydrogen plasmas generally increase the molecular weight through crosslinking *(41)*. Schonhorn and Hansen *(36)* used the term "CASING" (crosslinking by activated species of inert gases) to describe the effect of noble gases on the polymer surfaces. The dramatic improvements in polyethylene-epoxy bond strengths after noble gas plasma treatments were attributed to strengthening of the weak polymer boundary layers through crosslinking, since little or no changes were noted in the wettability after the treatments.

Hall et al. *(48)* have suggested that treatment of polyethylene with activated oxygen also caused crosslinking since they found that the oxygen plasma treated polyethylene gave the same bond strengths as helium plasma treated polyethylene. Although polyethylene is readily crosslinked in noble gas plasmas, these gases do not crosslink polypropylene. Hall et al. *(48)* explained the different effects on polypropylene and polyethylene as due to the greater tendency of polypropylene to yield to chain degradation and quenching of trapped radicals when exposed to oxygen. Polypropylene is also pressure sensitive; for example, a high pressure argon plasma enhances the wettability of polypropylene but at low pressures this plasma has no effect *(48)*.

Holmes and Schwartz *(40)* noted dramatic improvements in the wettability of ultra-high strength polyethylene (UHSPE) fabrics after treatment with ammonia gas plasma. Ammonia plasma treatment is known to implant surface amine groups and these investigators determined the primary amine concentration by dye-ion exchange experiments. Their results, shown in Table V, demonstrate a large decrease in the contact angle with increasing amine concentration. However, the long treatment times (5 minutes) caused a 10% strength loss in single fibers extracted from the fabrics.

Table V. Change in Surface Properties With Plasma Amination

Treatment	Conditions (min/W)	NH_2 Content (%)	Contact Angles (degrees)
Control	0/0	0	101
1	1/100	19.4	54
2	5/100	22.0	0
3	10/100	40.3	0
4	5/50	18.9	0

SOURCE: Adapted from ref. 40

Wesson and Allred *(10, 49)* studied the effects of acidic RF glow discharge plasmas on carbon fibers. They used surface energetics analysis to screen treatment recipes for a variety of reinforcement/matrix combinations. Changes in surface heterogeneity revealed by adsorptive energy distributions from IGC were correlated with wetting data and surface chemical composition deduced from high resolution XPS. Lewis acids and bases were used as probes for IGC and wetting measurements. The carbon fibers were also treated in benzene with 1,2-diphenylguanidine for detection of acid groups and diphenyl phosphate for basic groups. The results obtained by Wesson and Allred *(49)* are summarized in Table VI.

The results in Table VI show that the trends in surface composition, functionality and energetics are in good qualitative agreement. However, each of the techniques is sensitive to different aspects of surface acidity. Both titration and wetting measurements appear to be sensitive to a range of weak acid functionality in addition to high energy sites, since these measurements vary almost quantitatively with surface oxygen concentration. The results from IGC measurements, in contrast, appear to be more closely correlated with carboxyl group concentration as determined by ESCA. Wesson and Allred *(49)* speculated that weak acid functionality is of minor importance to applications of surface treated fibers; while, practical adhesion is more sensitive to small concentrations of sites with very high energies. However, it is difficult to distinguish which of the factors is the most critical to adhesion and all may play a role.

Table VI. Surface Properties of Graphite and Carbon Fibers

Measurement	Graphite	Carbon	Acid Plasma Treated Carbon
ESCA			
% Surface Oxygen	1.6	7.9	13.8
C1s (carboxyl)	--	2.6	6.5
Titration			
Acid sites/nm^2	0.4	1.3	2.1
Wetting			
W$^{a/b}$ (Formamide mN/m)	2.2	15.7	36.8
IGC			
% chemisorption	--	4.7	6.4

SOURCE: Adapted from ref. 49.

Schreiber et al. *(50, 51)* also used acid-based characteristics ascertained from IGC for evaluation of the properties of plasma treated calcium carbonate fillers in polyethylene and polyvinylchloride (PVC) composites. Mechanical properties at large deformation of the filled polymers were shown to depend on surface interactions. Optimum properties of PVC composites were obtained when strong acid-base reactions could take place with the plasma modified fibers.

Cellulosics. Corona discharge treatment has been widely utilized to improve the bonding and adhesion of cellulosic materials *(52)*. Greene *(53)* reported dramatic improvements in the bonding of polyethylene coatings to paper after corona treatments. The peel adhesion of polyethylene to paper without treatment was 45 g/in. but after treatment of the paper the peel strength went up to 360 g/in. Greene found the treatment to be surprisingly persistent, since the corona treated paper samples bonded 24 hours later with polyethylene, still produced peel adhesion strengths of 400 g/in.

Goring *(54)* also noted dramatic strength improvements in wet-bonding of cellulose strips after corona treatment. He reported almost eight-fold increases in bond strength after treatment in air or oxygen plasmas. Kim et al. *(44)* noted synergistic effects of corona treatment in bonding of cellulose to polymer sheets. As shown in Table VII, the effect varied with the type of synthetic polymer bonded to cellulose. In the case of the cellulose-polyethylene composite, corona treatment of the polyethylene produced the greatest effect; while, with the cellulose-PVC composite, treatment of either substrate produced a similar effect on strength. The greatest synergistic effect of the corona treatments was noted for the cellulose-polystyrene composite *(44)*.

Table VII. Bond Strengths of Cellulose to Polymer Sheets After Treatment in an Oxygen Corona Discharge

Polymer Temperature of pressing:	PE 90 °C	PS 110 °C	PVC 110 °C
Untreated control	0.2	3.3	3.6
Cellulose-treated	0.9	7.3	5.4
Polymer-treated	12.2	10.2	5.8
Both treated	13.8	21.8	6.9

PE = polyethylene; PS = polystyrene; PVC = polyvinylchloride; Pressure in Kg/cm^2, SOURCE: Adapted from ref. 44.

Whereas Goring and coworkers *(44, 54)* noted dramatic strength improvements with plasma treatment of cellulose fiber strips after 10-15 minutes of treatments, Tang and Bosisio *(55)* were able to produce significant improvements in wettability of kraft paper strips in only 5 seconds using a microwave oxygen plasma rather than the electrical plasmas used by the other investigators. The duration of the treatment could not exceed 8 seconds; otherwise 50% of the relative rupture strength was lost. If the experiment was continued for more than 10 seconds, the cellulose strips were carbonized in the oxygen plasma. On the other hand, the exposure times could be extended up to two minutes with argon and nitrogen plasmas. However, the best results were obtained in the oxygen plasma.

Other investigators have noted that the type of gas used in the corona can significantly affect the bonding properties as noted previously in the section on synthetic fibers. The bond strengths of cellulose to polyethylene or PVC both treated in an oxygen corona are not much different than the strengths produced for these combinations by an air corona. However, the bonding produced by treatment in a pure nitrogen corona was so high that the cellulose strips failed before the bond released. Covalent bonding may have been promoted by the nitrogen plasma treatment *(44)*.

Several investigators have explored the mechanisms of surface activation of cellulosics by plasma treatment. Goring *(54)* presented infrared evidence for the presence of carboxyl groups on corona treated cellulose films. This was later corroborated by ESCA analysis of corona (air) treated corrugating medium *(56)*. The paperboard showed a decrease in the C1s (I) peak and the appearance of a high binding energy component C1s (IV) assigned to the carboxyl groups in the ESCA spectrum. The oxygen/carbon ratio increased from 0.42 to 0.57 with the corona treatment.

Sawatari et al. *(57)* also used ESCA to characterize corona treated cellulosic type sheets. These investigators analyzed the effect of corona treatment on sheets of filter paper (Whatman no. 1), bleached kraft pulp (BKP) and thermomechanical pulp

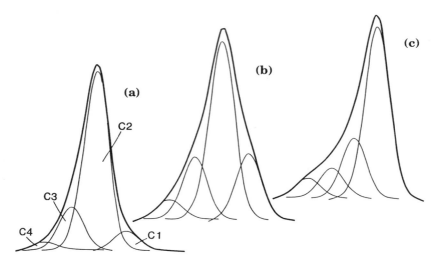

Figure 5. Curve of resolved ESCA C1s peaks of a) untreated; b) one minute; and c) 8 minute hydrogen plasma treated filter paper. (Reproduced from ref. 58, Americal Chemical Society, in press.)

(TMP). After the discharge treatment the samples were derivatized with pentafluorophenylhydrazine (PFPH) and the F1s and C1s peaks analyzed by ESCA. While the O1s/C1s ratio increased only slightly with time of treatment, the relative intensity of the F1s/C1s ratio increased considerably for the derivatized cellulose samples with corona treatment time. Thus Sawarati et al. (58) concluded that a preponderance of carbonyl groups were present at the surface of the corona treated cellulose (filter paper and BKP) samples.

When TMP sheets were corona treated, the oxygen/carbon ratio increased to a much greater extent as compared to the results for the pure cellulose samples (57). Extraction with ethanol/benzene demonstrated that the extractives existing at the surface of the TMP sheets were preferentially oxidized by the corona treatment. Based on their results, Sawatari et al. (57) proposed that the tendency toward surface activation in a corona discharge followed the sequence: extractives > lignin > cellulose.

Carlsson and Ström (58, 59) recently used ESCA to illustrate the different effects of hydrogen and oxygen plasmas on cellulose filter paper, as shown in Figures 5 and 6. The hydrogen plasma treatment reduced the hydroxyl groups on the cellulose; while, the oxygen plasma treatment both oxidized and reduced the surface. Carlsson and Ström postulated that low molecular weight, low polarity materials were formed at the surface of the filter paper when treated with the hydrogen plasma. This was based on the observation that the C1 peak, which increased during plasma treatment, was reduced and the C2 peak increased after solvent extraction.

Grease-proof paper, which contains large amounts of resin, was also analyzed with ESCA by Carlsson and Ström (59). After longer plasma exposure times the effect on the grease-proof paper was similar to that observed for the filter paper. The oxygen/carbon ratio increased from 0.46 to 0.60 when the grease-proof paper was

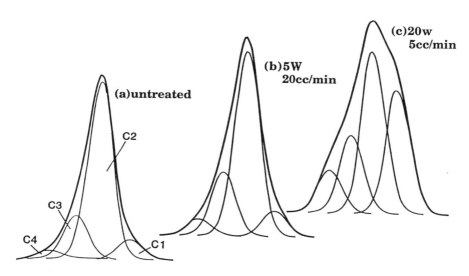

Figure 6. Curve of ESCA C1s peak of untreated filter paper and 2 minutes in oxygen plasma at 2 different power levels and rates of flow (58). (Reproduced from ref. 58, American Chemical Society, in press.)

treated in an oxygen plasma. This increase was, in part, due to the formation of hydroperoxide groups which were detected by sulphur dioxide tagging and analysis of the ESCA S2p signal. Carlsson and Ström estimated the content of hydroperoxide groups at about three per 100 glucopyranose units.

Although the water wettability of filter paper was reduced ($\theta = 115°$) by the hydrogen plasma treatment, the wettability of the grease-proof paper improved; the water contact angle was reduced from $\theta = 94°$ to $\theta = 40°$ with increasing time of plasma treatment. The dispersion component of the surface free energy, as measured with diiodomethane, remained unchanged with the treatment (58,59).

Brown and Swanson (60) noted significant increases in polarity of cellophane surfaces treated with corona discharge. The surface free energy was substantially increased by the corona treatment and this change was primarily due to the increase in the polar component, γ_s^p. The changes in polarity were most rapid during the early stages of treatment. The dispersion component γ_s^d showed a radical decrease with time of treatment. Water swelling measurements also supported the conclusion that extensive oxidation of the surface had occurred with the corona treatment.

Back (61) evaluated the changes in wettability of wood veneers after both corona and nitric acid treatments. The veneers were initially in a deactivated state due to storage for several months. Both treatments reduced the water contact angle of the veneers; however, the effect of corona treatment on the pine was negligible. The nitric acid treatment may be more effective for removal of extractive type materials which would migrate to the veneer surface on storage.

LITERATURE CITED

1. Fasth, R.; C. H. Eckert, *Chemtech*, July, 1988, 409.

2. Kelley, M. J., *Chemtech*, January, 1987 30; February, 1987, 98; March 1987, 170; April, 1987, 237; May, 1987; 294, August, 1987, 490; October, 1987, 635.
3. National Research Council, *Opportunities in Chemistry*, National Academy of Science Press, Washington, D.C. (1985).
4. Collins, G. E.; *J. Text. Inst.*, 1947, *38*, T73.
5. Young, R. A. In *Cellulose: Structure, Modification and Hydrolysis*, Young, R. A.; Rowell, R. M. Eds., Academic Press, NY, 1985.
6. Young, R. A., *Wood Fiber*, 1976, *8*, 120.
7. Miller, B.; and Young, R. A. *Text. Res. J.* 1975, *45*, 359.
8. Miller, B., In *Surface Characteristics of Fibers and Textiles*, Pt. II, Schick, M. J. Ed., Dekker, NY, 1977.
9. Fowkes, F. M., *J. Phys. Chem.* 1962, *66*, 382; *Ind. Eng. Chem. Prod. Res. Dev.* 1978, *1*.
10. Wesson, S. P.; Allred, R. E., *J. Adhesion Sci. Technol.* 1990, *4*, 277.
11. Wu, S. J. *J. Polym. Sci.*, Pt. C, 1971, *34*, 19.
12. Grozdz, S. J. and H.-P. Weigmann, *J. Appl. Polym. Sci.*, 1984, *29*, 3965.
13. Lloyd, D. R.; Ward, T. C.; and Schreiber, H. P. Eds., *Inverse Gas Chromatography*, Am. Chem. Soc. Symp. Ser. No. 391, Am. Chem. Soc., Washington, D.C., 1989.
14. Aspler, J. S.; and Gray, D. G. *J. Polym. Sci., Polym. Phys. Ed.*, 1983, *21*, 1675.
15. Siegbahn, K.; Nordling, C.; Fahlman, A.; Nordberg, R.; Hamrin, K.; Hedman, J.; Johansson, G.; Bergmark, T.; Karlsson, S.; Lindgren, I.; and Lindberg, B. *ESCA: Atomic Molecular and Solid State Structure Studied by Means of Electron Spectroscopy*, Almquist and Wiksells, Uppsala.
16. Dorris, G. M.; Gray, D. G. *Cellul. Chem. and Technol.* 1978, *12*, 9; 721; 735.
17. Brewis, D. M. *J. Matl. Sci.*, 1968, *3*, 262.
18. Horton, P. F. U.S. Patent 2,668,134.
19. Postema, A. R.; Doornkamp, A. T.; Meyer, J. E.; Vlekkert, H. V. D.; and Pennings, A. *J. Polym. Bull.*, 1986, *16*, 1.
20. Smook, J.; Hamersma, W.; and Pennings, A. J. *J. Matl. Sci.*, 1984, *19*, 1359.
21. Young, R. A.; Krzysik, A.; Fujita, M.; Kelley, S. S.; Rammon, R. M.; River, B. H.; and Gillespie, R. H. In: *Wood Adhesives in 1985: Status and Needs*, Christiansen, A. W. et al., Eds., Forest Prod. Res. Soc., Madison, 1986.
22. Zavarin, E. In: *The Chemistry of Solid Wood*. Rowell, R. M., Ed., Am. Chem. Soc., Washington, D.C. 1984.
23. Young, R. A.; Philippou, J. L.; and Barbutis, J.; *Bonding and Molding of Chemically Modified Whole Wood Fibers*, Paper presented at International Chemical Congress, Pacific Basin Societies, Am. Chem. Soc., Honolulu, Hawaii, December, 1989.
24. Grebeler, E. *Holz als Roh- und Werkstoff*, 1983, *41*, 87.
25. Stenberg, E. L. *Sv. Papperstidn.* 1980, *81*, 49.
26. Hillis, W. E. *Wood Sci. Technol.*, 1984, *18*, 281.
27. Young, R. A., Rammon, R. M.; Kelley, S. S.; and Gillespie, R. H. *Wood Sci.*, 1982. *14*, 110.
28. Rammon, R. M.; Kelley, S. S.; Young, R. A.; and Gillespie, R. H. *J. Adhesion* 1982, *14*, 257.
29. Kelley, S. S.; Young, R. A.; Rammon, R. M.; and Gillespie, R. H. *Forest Prod. J.* 983, *33*, 21.
30. Kelley, S. S.; Young, R. A.; Rammon, R. M. and Gillespie, R. H. *J. Wood Chem. Technol.* 1982, *2*, 317.
31. Young, R. A. and Fujita, M. *Wood Sci. Technol.* 1985, *19*, 363.

32. Hollahan, J. R.; and Bell, A. T. Eds., *Techniques and Applications of Plasma Chemistry*, Wiley-Interscience, NY, 1974.
33. Boenig, H. V. *Plasma Science and Technology*, Cornell Univ. Press, Ithaca, NY, 1982.
34. Simionescu, C. I.; Dines, F.; Macoveanu, M. M.; and Negulescu, I. *Makromol. Chem.*, suppl. 1984, *8*, 17.
35. Yasuda, T. In: Am. Chem. Soc., *Organic Coatings and Appl. Polym. Sci. Proc.*, 1982, *47*, 313.
36. Schonhorn, R. H.; and Hansen, R. H. *J. Appl. Polym. Sci.*, 1967, *11*, 1461.
37. Yasuda, H.; Marsh, H. C.; Brandt, S.; and Reilly, C. N. *J. Polym. Sci., Polym. Chem. Ed.*, 1977, *15*, 991.
38. Ladizesky, N. H.; and Ward, I. M., *J. Matl. Sci.*, 1983, *18*, 533.
39. Boenig, H. V. Ed., *Proceedings of Second Ann. Intl. Conf. on Plasma Chem.*, Technomic Pub., Lancaster, PA, 1986.
40. Holmes, S.; and Schwartz, D. *Comp. Sci. Technol.*, 1990, *38*, 1.
41. Hadis, M. In: *Techniques and Applications of Plasma Chemistry*, Hollahan, J. R.; A. T. Bell, A. T. Eds., Wiley-Interscience, NY, 1974.
42. McKelvey, J. M. *Polymer Processing*, John Wiley & Sons, NY, 1962.
43. Hines, R. A. *An Investigation of Chemical Changes Occurring on the Surface of Polyethylene During Treatment*, Paper presented at 132nd National Am. Chem. Soc. Mtg., NY, September, 1957.
44. Kim, C. Y.; Suranyi, G.; and Goring, D. A. I. *J. Polym. Sci.*, Pt. C, 1970, *30*, 533.
45. Clark, D. T.; Dilks, A. and Shuttleworth, D. In: *Polymer Surfaces*, Clark, D. T.; and Feast, W. J., Eds., Wiley-Interscience 1978.
46. Kim, C. Y.; Evans, J.; and Goring, D. A. I. *J. Appl. Polym. Sci.* 1971, *15*, 1365.
47. Carley, J. F.; and Kitze, P.T. *Polym. Eng. Sci.*, 1978, *18*, 326; 1980, *20*, 330.
48. Hall, J. R.; Westerdahl, C. A. L.; Devine, A. T.; and Bodnar, M. J. *J. Appl. Polym. Sci.*, 1969, *13*, 2085.
49. Wesson, S. P. and Allred, R. E. In: *Inverse Gas Chromatography*, Am. Chem. Soc., Symp. Ser. No. 391, ACS, Washington, D.C., 1989.
50. Schreiber, H. P.; Wertheimer, M. R.; and Lambla, M. *J. Appl. Polym. Sci.* 1982, *27*, 2269.
51. Schreiber, H. P.; and Li, Y., In: *Molecular Characterization of Composite Materials*, Plenum Press, NY, 1985.
52. Pavlath, A. E. In: *Techniques and Applications of Plasma Chemistry*, Hollahan, J. R. and Bell, A. T. Eds., John Wiley & Sons, NY, 1974.
53. Greene, R. E. *Tappi*, 1965, *48*, 80A.
54. Goring, D. A. I. *Pulp Paper Mag. Canada*, 1967, *68*, T-37-2.
55. Tang, T. W. C.; and Bosisio, R. G. *Tappi*, 1980, *63*, 111.
56. Suranyi, G.; Gray, D. G.; and Goring, D. A. I. *Tappi*, 1980, *63*, 153.
57. Sawatari, A.; Hayashi, Y.; and Kurihara, T. In: *Cellulose: Structural and Functional Aspects*, Kennedy, J. F.; Phillips, G. O.; and Williams, P. A. Eds., Ellis-Horwood Ltd., Chichester, 1989.
58. Carlsson, C. M. G.; and Ström, G. *Reduction and Oxidation of Cellulose Surfaces by Means of Cold Plasma*, ACS, Washington, D.C., in press.
59. Carlsson, C. M. G.; and Ström, G. *Adhesion Between Plasma Treated Cellulosic Materials and Polyethylene*, Institute of Surface Chemistry, Stockholm, in press.
60. Brown, P. F.; and Swanson, J. W., 1971, *Tappi*, *54*, 2012.
61. Back, E. L.; and Danielsson, S. *Nordic Pulp Paper Res. J.*, Special Issue, 1987, *2*, 53.

RECEIVED May 21, 1991

Chapter 10

Liquefaction of Lignocellulosics in Organic Solvents and Its Application

Nobuo Shiraishi

Department of Wood Science and Technology, Kyoto University, Sakyo-ku, Kyoto 606, Japan

Liquefaction of lignocellulosics in organic solvents at temperatures of 240-270 °C without catalysts and that at temperatures around 80-150 °C with acidic catalysts has been recently developed. These give very high yields of solvent solubles (around 90-95 % based on the lignocellulosic weight), and are quite different from the conventional liquefaction of lignocellulosics where one to several hours of treatments is required at 300-400 °C with or without catalysts. The latter often uses aqueous and/or organic solvents and usually results in quite low yields of 40-60 % because of the conversion of the lignocellulosics into gaseous compounds. Applications of the newly developed lignocellulosic liquids have also been conducted in the preparation of wood-based reactive adhesives, foams, moldings, fibers, and carbon fibers.

The term "liquefaction of lignocellulosics" has hitherto chiefly meant the procedures for the production of oil from biomass using very severe conditions of conversion (1-3). For example, Appel et al. tried to convert cellulosics to oil using homogeneous Na_2CO_3 catalyst in water and high boiling point solvent mixtures (anthracene oil, cresol, etc.) at pressures of 140-240 atm with synthesis gas, CO/H_2 (3). Treatments of one hour at 300-350 °C result in 40-60 % yield of benzene solubles (oil) and a 95-99 % conversion of the starting materials. Thus, this type of liquefaction can be called oilification of lignocellulosics.

In contrast, this review presents recent progress in lignocellulosic liquefaction done by milder treating conditions; that is, at temperatures of 240-270 °C without catalyst, or at temperatures of 80-150 °C with acidic catalysts. One special group of chemically modified woods can be dissolved in cresols even at room temperature as will be shown later. At any rate, the liquefaction or dissolution of chemically modified wood has been developed (4-8). Even the liquefaction of untreated wood was found to be possible as well (8-11).

LIQUEFACTION OR DISSOLUTION OF WOOD

Chemically Modified Wood

Chemically modified woods have been found to liquefy or dissolve in various neutral aqueous solvents, organic solvents, or organic solutions, depending on the characteristics of the modified wood (12-15). This work has been extended to the preparation of wood-based reactive adhesives, wood-based resins, etc., which will be discussed later. So far, three methods have been found for wood liquefaction.

The first trial of liquefaction of wood was accomplished by using very severe dissolving conditions (16). A series of aliphatic acid esterified wood samples could be liquefied in benzyl ether, styrene oxide, phenol, resorcinol, benzaldehyde, aqueous phenols, chloroform-dioxane mixture, benzene-acetone mixture and so forth after treatment at 200-270 °C for 20-150 min.

Carboxymethylated wood, allylated wood and hydroxyethylated wood have been found to be liquefiable in phenol, resorcinol, or their aqueous solutions, formalin, etc. after standing or stirring at 170 °C for 30-60 min (6).

Another method of liquefaction is to make use of solvolysis during the process (9, 17). By using conditions which allow phenolysis of a part of the lignin, especially in the presence of an appropriate catalyst, the liquefaction of chemically modified wood into phenols could be accomplished under milder conditions (at 80 °C for 30 to 150 min). Allylated wood, methylated wood, ethylated wood, hydroxyethylated wood, acetylated wood and others have been found to dissolve in polyhydric alcohols, such as 1,6-hexanediol, 1,4-butanediol, 1,2-ethanediol, 1,2,3-propane triol (glycerol) and bisphenol A, by use of the liquefaction conditions described above. Each of them caused partial alcoholysis of lignin macromolecules (7). This means that several series of reactive solvents could be used in addition to the phenols to liquefy modified wood.

The liquefaction processes give paste-like solutions with a considerably high concentration of wood solute (70 %). The solutions obtained with high concentration of wood can be used directly to prepare adhesives, foams and other molded products. This has opened a new field for utilizing wood materials.

For the third method of liquefaction or dissolution, postchlorination, has been developed. Sakata and Morita (18) have recently found that when chemically modified woods are chlorinated, their solubility in solvents was enhanced tremendously. For example, at room temperature cyanoethylated wood can be dissolved only 9.25 % in o-cresol. However, once chlorinated, it can dissolve almost completely in the same solvent at room temperature. The chlorinated-cyanoethylated wood can also dissolve in resorcinol, phenol, and a LiCl-dimethylacetamide solution under heating.

Untreated Wood

So far, liquefaction or dissolution of chemically modified woods has been discussed. However, more recently, untreated wood was also found to be liquefied in several organic solvents (9, 10). This phenomenon was discovered during an investigation evaluating the effect of the degree of chemical modification of wood on liquefaction. For example, after treatment at 200-250 °C for 30-180 min, wood chips and wood meal were liquefied in phenols; bisphenols; alcohols such as benzyl alcohol; polyhydric alcohols such as 1,6-hexanediol, 1,4-butanediol; oxyethers such as methyl cellosolve, ethyl cellosolve, diethylene glycol, triethylene glycol, polyethylene glycol; 1,4-dioxane, cyclohexanone, diethyl ketone, ethyl n-propyl ketone and so forth.

Liquefaction of untreated wood can also be achieved at a lower temperature of 150 °C and atmospheric pressure in the presence of acid catalysts (19). As the catalysts, phenolsulfonic acid and sulfuric acid can be used. The liquefied wood obtained is a paste-like solution with a highly concentrated wood solute, as high as 70 %. After liquefaction, wood components were found to be largely degraded, modified, and reactive. Thus, the wood solute can be used to prepare adhesives and other moldings. This also opened a new and practical field for utilizing wood materials.

APPLICATION OF THE LIQUEFACTION OR DISSOLUTION OF WOOD

Chemically Modified Wood

There are many potential applications of the liquefaction or dissolution of chemically modified wood: fractionation of modified wood components (16, 18, 20), preparation of solvent-sensitive and/or reaction-sensitive wood-based adhesives (7, 15, 17, 21), preparation of resinified wood-based moldings such as foam-type moldings (7), and preparation of wood-based fibers and their conversion to carbon fibers (22, 23), and so forth.

Fractionation of modified wood components has been studied by at least three groups of investigators (16, 18, 20). For the fractionation, a dissolution-precipitation technique has been successfully used.

Sakata and Morita showed that quantitative precipitation of the wood components can be made by pouring the cresol solutions of chlorinated-cyanoethylated wood into an excess of ethanol, ethyl ether or other solvent (18). Since the recovered amounts of the precipitates were found to change with the type of the non-solvent, the fractionation of modified wood components was thought to be possible.

Young et al. (20) also described the isolation of the main components of modified wood.

The preparation of adhesives from chemically modified wood has been studied. Phenols, bisphenols, and polyhydric alcohols have been used as solvents for the modified wood. In these cases, the resin should contain meaningful amounts of modified wood (7, 15, 17, 21). Combined use of these reactive solvents with reactive agents, such as cross-linking agents and/or hardeners, if necessary, have given phenol-formaldehyde resins (such as resol resin), polyurethane resins, epoxy resins, etc. The chemically modified woods are designed not only to dissolve and disperse in the final resins, but also to chemically react and bond to the resins. Techniques have been used that make it possible for phenol, to react at the phenyl propane side chain of lignin through carbon-carbon bonds, or for polyhydric alcohols to react with lignin through ether bonds. This can be achieved by liquefaction or dissolution of the chemically modified wood into the reactive solvent using solvolysis techniques. In the case of epoxy resins, it can be also achieved by reacting various alcoholic hydroxyl groups remaining in the modified woods with epichlorohydrin, resulting in introduction of glycidyl groups. Cross-linking within and between wood components, especially between polysaccharide components during the last stage of resinification, by reaction with cross-linking agents can also be used. In order to prepare wood-based resins with meaningful amounts of the wood components, it is very important to liquefy or dissolve the chemically modified wood into the reactive solvents in high concentrations (more than 50 % of the chemically modified wood is preferable).

When hydrophilic chemically modified woods, such as carboxymethylated wood, hydroxyethylated wood, or ethylated wood are used in wood-based adhesives, aqueous resol resin adhesives which maintain their solution state during the preparation are obtainable (15, 17, 21, 24, 25). When a phenol solution with a concentration more than 50 % is sought, the chemically modified wood powder can not be completely immersed in phenol, but can only be partly penetrated by the phenol during the first stage of liquefaction. When the heterogeneous mixture, however, is allowed to stand for about 30 min at 80 °C (without stirring in the presence of appropriate amounts of hydrochloric acid as the catalyst), a homogeneous paste can be obtained. Subsequent stirring of the paste for about 1-1.5 hr enhances the liquefaction.

In this liquefaction process, a certain degree of phenolysis of wood components, especially that of lignin, takes place, which makes it easy to dissolve them in phenol. After neutralizing the paste with aqueous sodium hydroxide, a definite amount of formalin and sodium hydroxide are added and the mixture is resinified in accordance with the conventional procedure to prepare the resol resin adhesives. The appearance of the resin obtained is similar to that of the corresponding commercial phenol resin adhesive. The wood-based resol resin adhesives have superior gluability and workability. The adhesives can be used with fillers, thickeners, and fortifiers such as wheat flour, coconut shell, walnut flour, and polymeric MDI (4,4'-diphenyl methane diisocyanate). The addition of appropriate fortifiers, especially cross-linking agents like polymeric MDI, into the wood-based adhesives enhances their dry-bond and water-proof gluabilities remarkably.

For the preparation of wood-based polyurethane as well as epoxy resin adhesives, the above-mentioned hydrophilic chemically modified woods prepared by conventional methods are liquefied in polyhydric alcohols or bisphenol A in a manner similar to the liquefaction in phenol (7). Concentration of the modified wood is usually more than 50 %. Diluents such as ethanol or methanol are also often added to the liquefaction systems according as required. After liquefaction, the pastes are neutralized and the diluents distilled off. When the pastes are used in combination with suitable polyisocyanate compounds, they become wood-based polyurethane adhesives. When the pastes are further reacted with epichlorohydrin, glycidyl etherified resins are formed. These can be used with hardeners such as amines and acid anhydride, and they become wood-based epoxy resin adhesives. Generally, the wood-based epoxy resins tend to become very viscous or solid, depending on the conditions of preparation, and require dilution or dissolution with solvents such as ethyl acetate, acetone, etc. These resins make satisfactory adhesives which can be used in water-proof gluing.

Molding materials such as foams or shaped moldings can also be obtained from chemically modified wood solutions (7). One of the examples is a wooden polyurethane foam. This can be prepared by adding an adequate amount of water as a foaming agent and a polyisocyanate compound as a hardener, to the 1,6-hexanediol solutions of allylated wood, mixing well and heating. When heated at 100 °C, foaming and resinification of the resins are initiated within 2 minutes and completed within several minutes. If promoters such as triethylamine are added, rapid reactions occur even at room temperature and foams can be obtained within several minutes. The foams thus obtained have low densities of around 0.04 g/cm^2, a substantial strength and elasticity in the compression deformation. In order to elucidate the role of the chemically modified wood within the foams, comparative experiments preparing the foams without the presence of the chemically modified wood have also been conducted. It was found that foaming actually occurs during the resinifying process, but immediately after that, a contraction in volume of the foam occurs, resulting in resin moldings with apparent densities around 0.2 g/cm^3 with little foamed-cell structure remaining. This result reveals that the chemically modified wood plays a positive role in maintaining the shape of the foams during their formation.

One other application of modified wood solutions is the formation of filaments or fibers. Tsujimoto *et al.* (22, 23) have prepared wood-based fibers from acetylated wood. After preparing the phenol solution of the acetylated wood, hexamethylene tetramine is added and the solution heated up to 150 °C to promote addition-condensation for a resinified solution with high spinnability. From the solution, filaments are spun and hardened in a heating oven at a definite heating rate. Maximum temperature for the hardening is 250 °C. By this way, continuous filaments can be easily obtained.

These filaments can be carbonized to give carbon filaments. Carbonization is carried out in an electrically-heated furnace at a maximum temperature of 900 °C with a heating rate of 5.5 °C/min. The strength of the carbon filaments was measured according to Japan Industrial Standard (JIS R7601) and tensile strengths up to 100 kgf/mm^2 have been obtained. This strength is comparable to that of the pitch carbon fibers of general purpose grade. Further improvements of strength may be expected by improving the methods of spinning and carbonization.

Fig. 1. Molding from lignocellulose after liquefaction in phenol, distillation of the free phenol, and curing by hexamethylene tetramine.

Untreated Wood

Almost the same products have been prepared from wood solutions of untreated wood as those from chemically modified wood (26, 27). For example, resol-resin type adhesives could be prepared from five parts of wood chips liquefied in two parts of phenol at 250 °C. The adhesives did not require severe adhesion conditions and were comparable to the corresponding commercial adhesives in their gluabilities. Acceptable water proof adhesion was attained for these adhesives.

Resol-type phenol resin adhesives were also prepared from wood-phenol solutions liquefied at 150 °C with phenolsulfonic acid catalyst, and their gluabilities were examined (19). The results revealed that when these adhesives were used, it was easily possible to realize completely satisfactory water proof adhesion even under hot-press conditions at 120 °C, with a hot-pressing duration of 0.5 min per 1 mm thickness of plywood. The adhesion temperature of 120 °C was at least 15 °C lower than ordinarily used adhesion temperature for resol resin adhesives.

As a second example, foams can be prepared from untreated wood-polyethylene glycol solutions (28). Soft-type and hard-type foams can be prepared by changing the preparation conditions. The prepared foams were of density around 0.04 g/cm^3, having substantial strength and elasticity in the deformations.

As a third example, novolak-resin type moldings can be prepared from untreated wood-phenol solutions (29, 30). After one part of wood meal was liquefied in two parts of phenol, unreacted phenol was distilled under reduced pressure. The wood-phenol liquefied and reacted powders obtained can be cured directly, when wood meal filler and hexamethylene tetramine are added and hot-pressed at 170-200 °C. The flexural strengths of the moldings were found to be comparable with those for commercial novolak moldings. An example of the moldings thus obtained is shown in Fig. 1.

The carbon fibers described previously could also be prepared from untreated wood solutions. Tensile strengths up to 120 kgf/cm^2 have been obtained so far. Further improvements of their physical properties may be also expected.

CONCLUDING REMARKS

The present state of studies on wood liquefaction or dissolution has briefly been reviewed. We believe that this line of study is a new field for the chemical processing of wood with great future potential. To achieve progress in this field, more fundamental and critical studies should be made.

LITERATURE CITED

1. Vanasse, C.; Chornet, E.; Overend, R. P., *Can. J. Chem. Eng.*, 1988, **66**, 112.

2. Appel, H. R.; Wender, I.; Miller, R. D., "Conversion of Urban Refuse to Oil", U. S. Bureau of Mines, Technical Progress Report-25, 1969, 5.

3. Appel, H. R.; Fu, Y. C.; Illig, E. G.; Steffgen, F. W.; Miller, R.D., "Conversion of Cellulosics Wastes to Oil", U. S. Bureau of Mines, RI 8013, 1975, 27.

4. Shiraishi, N., *Kobunshi Kako*, 1982, **31**, 500.

5. Shiraishi, N., Japan Patent, 1988, Sho 63-1992 (Appl. June 6, 1980).

6. Shiraishi, N.; Goda, K., *Mokuzai Kogyo*, 1984, **39**, 329.

7. Shiraishi, N.; Onodera, S.; Ohtani, M.; Masumoto, T., *Mokuzai Gakkaishi*, 1985, **31**, 418.

8. Shiraishi, N., *Tappi Proceedings, 1987 International Dissolving Pulps Conference*, 1987, 95.

9. Shiraishi, N.; Tsujimoto, N.; Pu, S., Japan Pat. (Open), 1986, Sho 61-261358 (Submitted on May 14, 1985).

10. Pu, S.; Shiraishi, N.; Yokota, T., *Abst. Papers Presented at 36th National Meeting, Japan Wood Res. Soc.*, Shizuoka, 1986, 179, 180.

11. Shiraishi, N., *Mokuzai Kogyo*, 1987, **42**, 42.

12. Shiraishi, N., *Sen-i To Kogyo*, 1983, **39**, P-95.

13. Shiraishi, N., "Chemistry of Wood Utilization" (H. Imamura, H. Okamoto, T. Goto, Y. Yasue, T. Yokota and T. Yosimoto Eds.), Kyoritsu Publ. Co., Tokyo, 1983, 294.

14. Shiraishi, N., "Advanced Techniques and Future Approaches in Wood Chemicals", (J. Nakano and T. Haraguchi Eds.), CMC Inc., Tokyo, 1983, 271.

15. Shiraishi, N., *Mokuzai Gakkaishi*, 1986, **32**, 755.

16. Shiraishi, N., Japan Pat. (Open), 1982, Sho 57-2301; Sho 57-2360.

17. Shiraishi, N.; Kishi, H., *J. Appl. Polym. Sci.*, 1986, **32**, 3189.

18. Morita, M.; Shigematsu, K.; Sakata, I., *Abst. Papers Presented at 35th National Meeting, Japan Wood Res. Soc.*, Tokyo, 1985, 215, 216.

19. Tanihara, Y.; Kato, K.; Pu, S.; Saka, S.; Shiraishi, N., *Abst. Papers Presented at 39th National Meeting, Japan Wood Res. Soc.*, Okinawa, 1989, 321.

20. Young, R. A.; Achmadi, S.; Barkalow, D., *Preprints of CELLUCON '84, Cartrefle-Wrexham*, Wales, 1984, 65.

21. Kishi, H.; Shiraishi, N., *Mokuzai Gakkaishi*, 1986, **32**, 520.

22. Tsujimoto, N., *Preprints for 14th Symposium on Chemical Processing of Wood*, Kyoto, 1984, 17.

23. Tsujimoto, N.; Yamakoshi, M.; Fukuchi, R., *Preprints for Poster Presentation, International Symposium on Wood and Pulping Chemistry*, Vancouver, 1985, 19.

24. Shiraishi, N.; Itoh, H.; Lonikar, S. V., *J. Wood Chem. Technol.*, 1987, **7**, 405.

25. Shiraishi, N., "Lignin Properties and Materials", A.C.S. Symp. Series, 397 (W. G. Glasser and S. Sarkanen Eds.), Amer. Chem. Soc., Washington D.C., 1989, 488.

26. Shiraishi, N.; Tamura, Y.; Tsujimoto, N., *Mokuzai Kogyo*, 1987, **42**, 492.

27. Shiraishi, N.; Tamura, Y.; Tsujimoto, N., *Mokuzai Kogyo*, 1988, **43**, 2.

28. Shiraishi, N., "Cellulosic Utilization; Research and Rewards in Cellulosics", Elsevier Appl. Sci., 1989, 97.

29. Kato, K.; Yoshioka, M.; Shiraishi, N., *Abst. Papers Presented at 40th National Meeting Japan Wood Res. Soc.*, Tsukuba, 1990, 26.

30. Shiraishi, N.; Kato, K., Japan Pat., submitted (On May 30, 1989).

RECEIVED May 2, 1991

BIOPOLYMERS:
ALLOYS, DERIVATIVES, AND BLENDS

Chapter 11

Low-Cost Uses of Lignin

Robert A. Northey

Georgia-Pacific Corporation, 1754 Thorne Road, Tacoma, WA 98421

Lignin, an abundant organic polymer produced by vascular terrestrial plants, is isolated in enormous amounts as a by product of pulping processes. These lignin products also contain a variety of impurities such as sugars, sugar acids, extractives, and inorganics. The majority of this material is burned as a fuel source, but a portion is purified for use in low cost polymeric products or as a source of low molecular weight chemicals. Uses for these lignin mixtures are based upon this materials ability to function as a dispersant, binder, emulsifier, and sequestrant. Impurities in the lignin products can have a large effect upon the efficiency of these materials in designated end uses.

The term lignin refers to a class of abundant amorphous polymers which provide strength and rigidity to cell walls in higher plants such as *Pterodophytes* and *Spermatophytes* (1). Enormous quantities of these compounds are available as residues from various pulping processes utilized to produce fiber products from woody substances (2). Although the majority of this lignin is burned as a fuel source, a small fraction is purified for use as low cost polymeric products and as a source of low molecular weight chemicals (3). There are a myriad of uses for these polymeric products most of which are based upon the dispersing, binding, emulsifying, and sequestering ability of lignin. These properties and others can be improved through lignin modification reactions which raise the cost of the final product. In most cases, however, little to no modification is necessary as the low cost of lignin makes it competitive with more efficient but higher cost synthetic compounds.

0097–6156/92/0476–0146$08.50/0
© 1992 American Chemical Society

Lignosulfonates produced in the sulfite pulping process represent the major source of lignin utilized in polymer applications (4). This material consists mainly of sulfonated lignins but also contains a variety of other substances. Kraft lignin produced in the kraft pulping process is of higher purity than lignosulfonates but because of costly isolation procedures is produced in much smaller amounts (5). Although individual impurities in lignosulfonates and kraft lignins represent only a small fraction of the total mixture, their presence can influence the efficiency of the lignin material in various polymer applications as well as affect the properties of the final product.

In this chapter, the low cost uses of lignin will be reviewed. These are applications in which the lignin preparations are utilized with little to no modification. There have been many excellent reviews published on this subject for the various lignin preparations (6,7). What is generally missing in these publications and what will be stressed in this chapter, is the influence of various lignin impurities on these applications. Descriptions of the composition of wood, pulping processes, lignin purification systems, and lignin product composition are included in order to aid in the understanding of the origin, structure, and abundance of the various components.

Composition of Wood

Woody tissue is composed primarily of three major compounds; cellulose, hemicelluloses, and lignin (8). Although present in all woody material, the proportions of each of these chemicals and the compositions of the hemicelluloses and lignins vary according to tree species. Also variable in proportion and composition are the extractive materials and inorganic substances (9). Although only present in small amounts, these compounds contribute much to the physical properties of wood such as color, smell, and decay resistance (10).

Lignin Lignin constitutes the second most abundant substance found in woody material behind cellulose (11). The main physiological function of lignin is to provide rigidity and strength to the plant cell wall (12). Lignin is generally considered to be a random amorphous three-dimensional network polymer comprised of phenyl propane units although a portion may consist of nonrandom two-dimensional structures (13). In gymnosperms, lignin arises from a dehydrogenation polymerization of p-coumaryl and coniferyl alcohol (14). In angiosperms sinapyl alcohol is also involved. An excellent description of the various linkages and functional groups associated with the lignin molecule is presented by Adler

(15). Weight average molecular weight determinations of 20,000 have been obtained for softwood lignin samples isolated through a system which is felt to introduce the least amount of structural modification to native lignin (16). This value, however, is questionable because of the difficulty in obtaining samples without major alterations to the lignin structure.

Cellulose Cellulose is the most abundant chemical found in wood comprising roughly one half of the total weight of material. Wood cellulose is a straight chain polysaccharide composed of approximately 10,000 ß-D-glucopyranose units linked by glycosidic bonds between the one and four atoms of adjacent molecules (17). Unlike hemicelluloses and lignin, the structure of cellulose does not vary between tree species. Because of its chemical and physical properties, cellulose functions as the main structural component of plant cell walls (18).

Hemicellulose Hemicelluloses are low molecular weight polysaccharides which, unlike cellulose, can possess some degree of branching (19). These compounds are composed of a variety of monosaccharides such as glucose, mannose, galactose, xylose, and arabinose occasionally acetylated and/or in combination with uronic acids, pectins, and starch (20). Hemicelluloses are classified according to the sugars present in the molecule such as in galactoglucomannans, glucuronoxylans, arabinogalactans, etc. The amount of hemicelluloses found in wood is usually between 20 and 30 % although this varies according to tree species and location in the tree. The composition of hemicelluloses is also variable particularly between hardwoods and softwoods (21,22).

Extractives Extractives is the name applied to those materials which can be extracted from wood with either polar or non-polar solvents (23). In most wood species, these compounds comprise only a few percent of the total matter (24). In some species of hardwoods, however, much higher extractive contents have been reported (25). Organic compounds in this category include terpenes, lignans, stilbenes, flavanoids, tannins, fats, waxes, fatty acids, fatty alcohols, steroids, and proteins. Also included in this category are the inorganic components of wood which comprise 0.1-1.0% of temperate zone trees and up to 5% of tropical species (26). The inorganics in temperate zone trees consist mainly of calcium, potassium, and magnesium although more than 50 trace elements have been identified (27).

Pulping Processes

Sulfite Pulping The term sulfite pulping refers to a group of wood digestion systems which utilize pH dependent combinations of sulfur dioxide/sulfurous acid, bisulfite ion, or sulfite ion along with a calcium, sodium, magnesium or ammonium base (28). The bulk of commercial lignosulfonates are produced in the acid sulfite pulping process which utilizes a mixture of sulfur dioxide and/or sulfurous acid in combination with either calcium, magnesium, sodium, or ammonium bisulfite at a pH of 1-2. The liquor in the bisulfite process consists of magnesium, sodium, or ammonium bisulfite at a a pH of 3-5. Other processes include the neutral sulfite process at pH 5-7 utilizing ammonium or sodium and the alkaline sulfite process at pH 9-13 utilizing sodium as the cation.

Delignification in these processes is the result, to a limited extent, of bond cleavage but mainly of sulfonation reactions which solubilize the lignin molecule (29). Sulfonation occurs primarily at benzyl alcohol, benzyl aryl ether, and benzyl alkyl ether linkages on the side chain of phenyl propane units (30). Lignin is solubilized in softwoods when the degree of sulfonation reaches one sulfonate group per two methoxyl groups (31). The degree of bond cleavage and sulfonation depend a great deal upon the pH of the cooking liquor (32). An additional reaction which occurs under acid sulfite pulping conditions is condensation which results in the formation of new carbon carbon bonds (33).

Under acidic conditions, glycosidic bonds in polysaccharides are readily cleaved (34). This results in extensive degradation of hemicelluloses under acid sulfite pulping conditions (35). Under bisulfite and neutral sulfite conditions, less cleavage occurs which results in larger polysaccharide fragments in solution. Cleavage of glycosidic bonds in cellulose also occurs but only to a limited extent. Approximately 10-20 % of the monosaccharides released into solution are oxidized further to aldonic acids (36). Additionally, a portion of the monosaccharides are converted to a variety of other chemicals such as trioses, furfural, hydroxymethyl furfural, levulinic acid, formic acid, acetic acid, formaldehyde, and sugar sulfonates (37).

Generally it is necessary in sulfite pulping to utilize woods with low extractive contents to avoid detrimental cross linking reactions which retard delignification (38). The water soluble extractives which are present are released into the pulping liquor quite early in the process. These compounds can undergo further reactions such as dehydrogenation and sulfonation (39).

Acid insoluble compounds such as resin acids, fatty acids and their esters, waxes, steroids, and fatty alcohols are partially polymerized and deposited on the pulp fibers.

Kraft Pulping In the kraft pulping process, delignification occurs through the action of sodium hydroxide and sodium sulfide on ether linkages in the lignin molecule (40). Cleavage of these bonds liberates phenolic hydroxyl groups which increases the hydrophilicity and thus the dissolution of lignin in the alkaline solution (41). Kraft lignin is more degraded and thus possesses a lower molecular weight than lignin produced in sulfite pulping (42). The high alkalinity utilized in kraft pulping causes condensation reactions resulting in the formation of new carbon-carbon bonds (43).

The conditions utilized in kraft pulping are responsible for a considerable amount of carbohydrate dissolution and degradation (44). The majority of hemicelluloses are degraded to hydroxy acids through peeling reactions (45). Only xylans are stable enough under these conditions to remain somewhat polymerized in solution (46). Cellulose is also partially degraded to hydroxy acids such as glucoisosaccharinic acid (47).

Alternative Pulping Processes There have been many other pulping methods developed from which lignin could possibly be recovered including the organosolv (48), steam explosion (49), or hydrolysis processes (50). A great deal of research has been applied towards these methods and some commercial installations are currently operating. The lignin produced in these applications possess unique properties which could prove to be beneficial in many applications. Because of the low volume of production, however, the potential uses of these lignins will not be covered in this chapter.

Lignin Isolation

The composition and physical properties of lignosulfonates depends not only upon the wood source and the cooking conditions, but also upon the procedures utilized in isolation, purification, and modification. Spent sulfite liquor can be subjected to steam stripping, concentrating, fermentation, ultrafiltration, base exchange, as well as a variety of modification reactions. For kraft lignin, isolation is normally achieved through acid precipitation with carbon dioxide and sulfuric acid but is also feasible through ultrafiltration (51). These procedures remove nearly all the impurities leaving a high purity kraft lignin (52). Chemical modification of the kraft lignin is necessary for many applications (53).

The first step in the purification of acid sulfite liquors consists of steam stripping to remove sulfur dioxide (54). This process also removes a portion of the volatile compounds in the liquor such as methanol, acetic acid, formic acid, and furfural (55). The next step is normally the concentration of the liquor in an evaporation step to increase the solids content. Almost all of the remaining volatile materials are removed in this process (56). In bisulfite and neutral sulfite pulping liquors, evaporation steps remove most of the volatiles however the sulfur dioxide remains in the form of sodium, magnesium, or ammonium bisulfite or sulfite. Additional removal of volatiles occurs if the lignin material is converted to a powder through spray drying.

It is often advantageous to remove sugars from lignosulfonates through either fermentation (57) or ultrafiltration. Ethanol is produced from lignosulfonates through the yeast (*Saccharomyces cerevisiae*) fermentation of hexoses (58). Fermentation is also utilized in the conversion of hexose and pentose sugars to protein in the generation of Tortula yeast (*Candita utilis*) (59) or in the Pekilo process (60) by *Paecilomyces varioti*. The removal of sugars by ultrafiltration also results in the elimination of other low molecular weight materials such as sugar acids and extractives. This is often advantageous in applications where these chemicals have deleterious effects. The use of ultrafiltration, however, adds substantial cost to the final product.

It is often necessary to convert lignosulfonates to a different base through a base exchange procedure. With calcium based material, this is achieved through the addition of a soluble sulfate or carbonate salt of the preferred cation followed by removal of the insoluble calcium salts. With sodium, magnesium, or ammonium based material, base exchange is accomplished through the use of a cation exchange resin (61).

Chemical modification of lignin materials by such reactions as oxidation, sulfonation, condensation, sulfomethylation, demethylation, alkoxylation, methyloylation, phenolation, and others is often necessary to improve properties required for certain end use applications (62). With kraft lignin, for example, oxidation or sulfonation is needed in order to convert this material into a water soluble, dispersive chemical (63). Modification reactions, however, can substantially increase the cost of the lignin product. These reactions also affect lignin impurities such as in the oxidation of sugars to sugar acids under alkaline oxidative procedures (64). This can have a large effect on the composition and properties of the final product. In general, the bulk of lignin materials utilized in low cost applications is unmodified.

Composition of Lignin Materials

Published information on the composition of lignosulfonates reflects the great deal of variability between commercial lignin products. Differences in reported compositions are due to variability in manufacturing procedures such as wood source, pulping process, and purification procedures. Differences in analytical techniques also add to this variability.

Lignosulfonates It is difficult to determine the exact amount of sulfonated lignin compounds present in a lignosulfonate mixture. Analytical difficulties in this determination arise from the presence of impurities as well as inaccuracies in the methods. It is reasonable, however, to assume that sulfonated lignins make up 60-65% of the solid matter in steam stripped/evaporated commercial lignosulfonates from the acid sulfite paper grade pulping of softwoods (65). The elemental composition of purified sulfonated lignin obtained from the acid sulfite pulping of Western Hemlock was determined through C13NMR (66) to be:

$$C_6C_{2.93}H_{7.86}O_{2.07}(OCH_3)_{0.88}(PhOH)_{0.26}(SO_3M)_{0.48}(COOH)_{0.03}$$

This data includes functional group information corresponding to 2.0% phenolic hydroxyl, 17.5% sulfonate, 12.5% methoxyl, and 0.6% carboxyl based on the weight of lignosulfonates. A methoxyl content of approximately 18% is obtained for hardwood lignosulfonates due to the presence of syringyl groups (67). Additionally, the number of catechol groups in softwood lignosulfonates corresponds to approximately 0.1 groups per methoxyl group (68). Similar data on functional groups obtained utilizing alternate methods confirms these findings (69). Information on the molecular weight of this material indicates a range of from 1000 to 140,000 for sulfonated lignins from the acid sulfite pulping of softwoods and from 1000 to 77,000 utilizing bisulfite pulping conditions (70). Lower values are obtained for hardwoods (71).

Lignosulfonates from the acid sulfite pulping of softwoods contain monosaccharides produced from the degradation of hemicelluloses and cellulose in amounts of up to 28% based on solids (72). Although the majority of hemicelluloses are degraded during acid sulfite pulping, it has been reported that hemicellulose fragments of 2-3 monosaccharide units in amounts of up to 7% of the total solids have been isolated (73). A much higher percentage of hemicelluloses remain in the form of polymers in liquors obtained from bisulfite and neutral sulfite pulping (74). The value of 28% monosaccharides was based upon total dissolution of hemicelluloses. Allowing for some remaining hemicellulose fragments, a more reasonable value of 20-22% for monosaccharides was obtained from the acid sulfite pulping of Norway spruce (75). The monosaccharide fraction consisted of 4% arabinose, 21% xylose, 51% mannose, 12% galactose, and 12% glucose. Fermentation removes the bulk of these hexoses leaving only the xylose and arabinose.

Aldonic acids produced from the degradation of monosaccharides comprise approximately 4% of the organic solids in softwood lignosulfonates (72). The relative composition of this fraction is 42% xylonic acid, 18% arabonic acid, 19% mannonic acid, 4% gluconic acid, 15% galactonic acid, and 1% rhammonic acid. A small amount (<0.5% of organic matter) of uronic acids such as glucuronic and galacturonic are also present from the dissolution of hemicelluloses. Also produced in small amounts from monosaccharides are stable bisulfite addition compounds of trioses and methylglyoxal (76) as well as sulfonic acids produced by bisulfite oxidation (77).

Softwood lignosulfonates before evaporative concentration or spray drying typically contain 3-4% acetic acid (74), 0.02-0.06% furfural (78), 0.02-0.06% methanol (75), 0.05-0.1% formic acid (79), and trace amounts of formaldehyde, levulinic acid, and hydroxymethyl furfural (28). Additional volatile compounds such as ethanol and acetaldehyde are introduced in the fermentation process (80). The yield of acetic acid is greater in material obtained from hardwood species because of the larger percentage of acetylated hemicelluloses (8). Evaporation or spray drying removes the majority of these volatiles leaving an almost volatile free product.

The composition of extractives in lignosulfonates is dependent mainly upon the wood species utilized but also upon such factors as the percent inclusion of bark in the raw material. Generally, lignosulfonates contain a wide variety of extractives compounds which total approximately 4% of the total solids (74). The bulk of the extractive materials are water soluble compounds such as lignans, flavanoids, condensed tannins, and proteins. The most prominent lignan in softwood lignosulfonates from hemlock or spruce is conidendrin comprising 0.6-0.8% of the total solids (81). Typical flavanoids include catechin and dihydroquercetin alone or as polymers in the form of condensed tannins (82). Nitrogen concentrations of 0.1-0.2% indicate the presence of amino acids/proteins. A portion of these extractive materials will be in a sulfonated form (83).

Inorganic compounds in lignosulfonates originate mainly in the pulping liquor but also to a small extent with the wood supply. When lignosulfonates produced from acid sulfite pulping are ashed, a value of approximately 10% is obtained. This ash consists mainly of the cations and sulfonate groups associated with the lignin molecule but also contains other sulfur compounds in the form of sulfates, thiosulfates, and bisulfite ions. The amount of ash in bisulfite and neutral sulfite lignosulfonates is much higher due to residual bisulfite and sulfite ions which, unlike sulfur dioxide in acid sulfite pulping, are not removed from the spent sulfite liquor.

Kraft Lignin The purification procedures utilized in the isolation of kraft lignin remove the majority of impurities leaving only a residual amount of inorganics, hydroxy acids, carbohydrates, and extractives associated with the lignin (52,84,85). Highly purified kraft lignins have been shown to contain up to 1.5% carbohydrates and 0.02-0.12% nitrogen which corresponds to 0.1-0.6% amino acids/protein (86). The elemental composition of a commercial softwood kraft lignin isolated through acid precipitation was reported as (69):

$$C_9H_{6.52}O_2(H_2O)_{0.79}(OCH_3)_{0.79}S_{0.03}$$

Kraft lignins possess a large number of free phenolic hydroxyl groups (4.0%) and low molecular weight (M_N 2000) due to the high degree of lignin degradation during pulping (87). According to the elemental analysis, methoxyl groups correspond to approximately 13% and organically bound sulfur 0.5% of the phenyl propane units. The figure for sulfur is rather low as percentages in the 1.5-2.5% range have been reported (88).

Utilization of Lignin

The first practical use of isolated lignins occurred only four years after the construction of the first sulfite mill in Sweden in 1874 (89). A patent was issued in 1878 for the use of lignosulfonates as an extender for vegetable tannins in the process of leather tanning (90). A great deal of research was subsequently applied towards new uses for lignosulfonates. The same is true for alkali lignins which were first isolated in 1910 (91). Although a great many uses for lignosulfonates and kraft lignins had been developed by the late 1950's, the majority of these materials were still either sewered (lignosulfonates) or burned (kraft lignin). Of the approximately 2,800,000 tons of lignosulfonate solids produced in the United States in 1956, only 69,000 tons were sold and the majority of this was as unconcentrated road dust binder (92). Environmental constraints finally forced the sulfite pulping industry to either develop recovery systems or to increase lignin sales in order to eliminate the polluting discharge. The majority of mills opted to burn lignin. The remaining lignosulfonate producers continued to be very active in developing new lignin markets as now the production of pulp depended upon the sale of lignin. Currently, the majority of lignosulfonates and kraft lignin are burned with only a small portion utilized as polymeric products.

The uses of lignin materials are based upon these chemicals ability to function as dispersants, binders, emulsifiers/emulsifier stabilizers, and sequesterants. Often in a specific application, lignins are utilized for more than one of these functions. The efficiency of lignin materials in these areas is often enhanced or diminished by the presence of impurities such as sugars, sugar acids and extractives. In the following sections, the uses of lignin materials will be reviewed. Included will be information regarding the relative amount of lignin utilized in these applications (Table I) as well as the mechanisms involved. Special emphasis will be placed upon the effects of lignin impurities on these applications when data is available.

Table I. Non-Fuel Use of Lignin in the United States

Large (100-200 million lbs. solids/year)
  Concrete Admixtures
  Animal Feed Molasses Additive
  Animal Feed Binder
  Road Dust Control
  Vanillin Manufacture

Medium (20-100 million lbs. solids/year)
  Gypsum Wallboard
  Oil Well Drilling Muds Additive
  Pesticides Dispersant
  Dye Dispersant
  Binder for Molding Applications
  Adhesives

Small (1-20 million lbs. solids/year)
  Cement Manufacture - Grinding Aid
  Carbon Black and Pigments Dispersant
  Emulsifier/Emulsion Stabilizer
  Water Treatment and Cleaning Applications
  Micronutrients
  Battery Expander
  Leather Tanning
  Rubber Additive

Dispersion Lignosulfonates or modified kraft lignins are utilized as surface active agents to prevent the formation of aggregates of finely divided insoluble particles in suspension (93). Agglomeration is the result of Van der Waals forces which cause an attraction between particles (94). This attractive force is counteracted somewhat by electrostatic repulsive forces due to the charged outer layers on the outside of the particles (95). This layer is formed by ions which are attracted to the charged surface of the particles.

Lignosulfonates and modified kraft lignins consist of hydrophobic molecules containing hydrophilic groups such as sulfonate or carboxylate groups. The hydrophobic portion of lignin is absorbed on the surface of particles leaving the charged groups facing the water phase. These hydrophilic groups increase the repulsive nature of the outer layer and thus help to prevent agglomeration. This results in the formation of a more stable dispersion of reduced viscosity and sedimentation.

The dispersing efficiency of lignin molecules is influenced by several factors including molecular weight and number of hydrophilic groups. In dispersing applications, lignins of molecular weight of 10,000-40,000 are most efficient (96). Improved dispersing ability is obtained through purification steps which remove the very high and low molecular weight materials as well as by oxidative procedures which lower the overall molecular weight (97). Increases in the degree of sulfonation (98) or in the amount of carboxyl groups (99) also increases dispersing ability.

Binding Lignosulfonates function as binders in a variety of extrusion, compaction, granulation, briquetting, and coating applications (100). Lignosulfonates are naturally tacky when moist and thus possess excellent adhesive properties for the agglomeration of particles. Impurities, especially sugars, are responsible for a considerable amount of this tacky nature (101). The exact nature of adhesion is not fully understood although there are several theories regarding this process (102). Adhesion is the phenomena by which surfaces are held together by interfacial forces (103). The more intimate the contact between particles, the better the adhesion. The application of liquid lignosulfonates to a particle system results in the dispersion of the particles. When the system dries, this results in more intimate contact between particles and thus better adhesion. Lignosulfonates also function as surface wetting agents which is beneficial for adhesion (104).

Emulsions Lignosulfonates and kraft lignin are utilized in emulsion technology as emulsifying agents or emulsion stabilizers. An emulsion is a stable mixture of mutually insoluble liquids in which one is dispersed as fine droplets throughout the other (105). In the formation of emulsions, a mixture of two liquids is normally subjected to mechanical agitation which results in the formation of droplets. The breaking up or disruption of these droplets into finer droplets is the mechanism of emulsification. Droplet disruption is dependent upon interfacial tension between the liquids. The surfactant characteristics of emulsifying agents such as kraft lignin are utilized to lower the interfacial tension and thus increase droplet disruption (106)

There are five different classifications for the breakdown processes that occur in emulsion systems. These include sedimentation or creaming, flocculation, coalescence, Ostwald ripening, and emulsion inversion (107). Although each of these processes are unique, the basic mechanisms involve either the separation of the fine droplets from the solution or the formation of larger particles. Emulsion stabilizers, such as lignosulfonates, are added to prevent these mechanisms by either charge or steric stabilization. The process of charge stabilization by lignosulfonates is similar to the dispersion of particles discussed in a previous section. The principles of steric stabilization involves the adsorption of macromolecules on the surface of the droplets. These molecules extend into solution in the form of "loops and tails" forming an adsorbed layer which functions to repulse other droplets (108).

Sequestering The sequestering ability of lignosulfonates is utilized in applications in which the solubilization of metal atoms is desired. Sequesterants are materials which combine with metal ions to form soluble compounds of sufficient stability to withstand conditions which would normally cause the metal ions to precipitate. The sequestrant molecule contains donor atoms which combine through coordinate bonding with a single metal atom to form a cyclic structure called a chelate (109). On the lignin molecule, catechol, sulfonate, and carboxyl groups possess limited chelating ability. Similar groups on the impurities in lignosulfonates add to the sequestering ability of the mixture.

Fuel Source The lowest cost and highest volume use of lignins is as a fuel source. Recovery furnace systems are utilized in kraft pulping (110) and for sulfite pulping systems which utilize either sodium, magnesium, or ammonium bases (54). Calcium based spent sulfite liquor can be burned, however chemical recovery is difficult. The fuel value of lignin (12,700 Btu/lb solids) (111) is diluted by the presence of impurities to 5800-6600 Btu/lb of kraft liquor solids (112) and 8000 Btu/lb of lignosulfonate solids (5).

Gypsum Wallboard In the manufacture of gypsum wallboard, stucco (calcium sulfate hemihydrate) is combined with water and a variety of additives to form a flowable slurry. Lignosulfonates are utilized in this system as dispersants to reduce the slurry viscosity and thus lower the water requirements (113). This results in reduced energy costs and faster throughput in the drying stage. Certain physical properties of the board are also improved such as compressive strength, density, and impact strength (114). Lignosulfonates are typically added at levels of 1-3 pounds per thousand square feet of one-half inch board.

The speed at which gypsum board can be produced is influenced by the setting or hardening rate of the stucco slurry. Lignosulfonates can retard the set and thus slow the production rate (115). To overcome this problem, accelerators such as potassium sulfate or finely ground landplaster (calcium sulfate dihydrate) are added to the mixture (116). Another method utilized to reduce the retarding nature of lignosulfonate is ultrafiltration to remove low molecular weight impurities.

A small amount (3-25%) of lignosulfonates are utilized in the grinding of landplaster for use as an accelerator (117). The use of lignosulfonates in the grinding stage results in the production of finer sized particles which are more efficient accelerators. The lignosulfonates also protect the landplaster from dehydrating upon grinding and storage.

In lignosulfonates, sugars provide no stucco dispersing ability and are weakly set retarding. In the grinding of landplaster, however, the presence of sugars is beneficial (118). Sugar acids and their lactones are dispersants and are also moderately set retarding (119). Of the extractive compounds, hydrolyzable tannins are strongly set retarding while simple phenols, flavanoids, and condensed tannins are weak to moderately retarding. The degree of set retardation is also affected by the base of the lignosulfonate mixture (120).

Concrete Admixtures Lignosulfonates are utilized in concrete admixtures as dispersing agents to reduce the amount of water needed to form a plastic and workable mixture (121). Because of the lower water requirements, the resulting concrete is of increased density, better uniformity, higher compressive strength, and better durability (122). Lignosulfonates are also utilized as set retarding agents in applications where concrete needs to remain fluid over extended periods of time (123) and as air entraining agents.

The water reducing ability of lignosulfonates in concrete is mainly due to sulfonated lignin compounds. At application levels above 0.2-0.4%, these chemicals also contribute to set retardation (124). Sugars are not concrete dispersing so their removal improves the water reducing ability of the mixture. These compounds are set retarding, however, so their presence adds to this phenomena (125). Sugar acids impart some dispersing ability to the mixture but mainly serve as set retarders. Gluconic acid is approximately ten times more retarding than glucose (126). Set retardation is also affected by extractives such as simple phenols, flavanoids, and tannins as well as by the cations associated with the lignosulfonates (127,128).

Cement Manufacture Lignosulfonates and sulfonated kraft lignin are utilized in the manufacture of Portland cement both as dispersants in the initial grinding if done by the wet process and as grinding aids in the grinding of clinker (129). In the wet process of cement manufacture, the ingredients to be ground are mixed with water to form a flowable slurry which is sent through the milling process (130). The addition of sulfonated lignin materials as dispersants allows for a reduction in water usage which leads to fuel savings, greater throughput, and lower production costs. In the grinding of cement clinker, the addition of lignosulfonates enables the mill to grind the clinker to smaller size, reduces caking problems, reduces pack set, and lowers energy consumption (131). Typical addition levels of lignosulfonates amount to 0.02-0.2% on dry cement.

Sugars in lignosulfonates assist in the grinding of cement clinker. In the wet grinding process, sugars dilute the dispersing efficiency of lignosulfonates because of their lack of dispersing ability. Alternatively, sugar acids increase this dispersion efficiency. The use of lignosulfonates in these grinding processes also effects properties of the final product such as set retardation in the poured cement (132).

Oil Well Drilling Muds Oil well drilling muds consist of a mixture of fine clay, water, and additives. These mixtures are used to cool and lubricate the drill, transport cuttings to the surface, hold cuttings in suspension during delays, seal and support the walls of the hole, and form a hydrostatic head to control high pressure gas, oil, and water flows (133). The primary use for lignosulfonates in oil well drilling muds is as a thinning agent (134) although these materials can also function as clay conditioners, dispersants, viscosity control agents, and fluid loss additives. Lignosulfonates are especially useful in fresh water muds when gypsum salts are present (135). The efficiency of lignosulfonates increases through the formation of complexes with either chromium or a mixture of chromium and iron (136).

The bulk of the mud thinning ability of lignosulfonates is due to sulfonated lignin compounds. Sugars reduce the thinning ability of the mud and decrease its quality through gel formation (137) and through fermentation reactions which can lead to odor problems (138). Sugar acids (139) and tannins (140) are excellent dispersants in this system especially in the form of chrome salts.

Pesticide Dispersant Certain pesticides are sold as dry powders to be mixed with water to form dispersions or emulsions for use in spray applications (141). It is important that these pesticide particles remain suspended in solution to prevent pluggage of the spray nozzle or uneven distribution of the pesticide. Lignosulfonates function as effective dispersants in these wettable powders thereby eliminating sedimentation (142). Lignosulfonates are also used as dispersants and binders in applications in which the pesticide is granulated with an inert carrier to form a dry flowable product (143). Typical addition levels of lignosulfonates in these products is in the 2-10% range.

Kraft lignins may be utilized as dispersants in these applications but only if modified through sulfonation or oxidation (144). Sugars do not function as dispersants so their presence dilutes the dispersing ability of the mixture. The humectant property of the sugars is also undesirable as it leads to poor storage characteristics (138). The dispersing properties of sugar acids is beneficial in these application.

Dye Manufacture Lignosulfonates and sulfonated kraft lignins are utilized in the manufacture of disperse and vat dyes as primary dispersants, leveling agents, extenders, protective colloids, and grinding aids (7). The primary function of these compounds is to increase the stability of the dyestuff towards settling, recoagulation, and excessive thickening and to impart the characteristic of ready redispersion (145). An ideal lignin dye dispersant should posses excellent heat stability, low azo dye reducing ability, minimal foaming tendency, low fiber staining, good grinding ability, and good dye paste dispersion properties (146). Unfortunately, while both lignosulfonates and sulfonated kraft lignins are excellent dispersants and grinding aids, these materials function poorly in the other areas such as azo dye reduction and fiber staining so that modification is necessary to obtain a satisfactory product (147).

Much of the azo dye reducing and staining properties of lignosulfonates has been attributed to the catecholic and other phenolic hydroxyl groups on the lignin molecule (148). Sugars are also reducing substances and their presence may contribute to the azo dye reducing problem (149). Also contributing are any metals present in the lignosulfonates which will catalyze both azo dye reducing reactions and lignin color formation reactions which lead to fiber staining (148). The presence of sugars reduces the dispersing ability of the mixture while sugar acids have been shown to be excellent dye dispersants (145).

Carbon Black And Pigments The dispersing ability of lignosulfonates is utilized in a wide range of pigment applications to inhibit settling and to decrease solution viscosity (100). Lignosulfonates are added to dark pigment systems used to color textile fibers, paper, emulsion paints, coatings, and inks. Lignosulfonates are utilized in carbon black applications as a grinding aid in the initial grinding of the carbon black particles, as a binder in the pelletizing process, and as a dispersant in the addition of the particles to rubber (150). Typical addition rates for lignosulfonates in carbon black dispersion are 3-8%.

The dispersing ability of lignosulfonates in these systems is improved dramatically by oxidation reactions. Unfortunately, these reactions can lead to darkening of the lignosulfonates which may cause problems in lighter pigment applications. Sugars in the lignosulfonates are not pigment dispersing so their presence dilutes the dispersing ability of the mixture. Alternatively, sugar acids increase the dispersing ability of the mixture.

Animal Feed Molasses Additive Lignosulfonates are added to animal feed molasses to reduce solution viscosity (151) which decreases the water required to produce a pumpable product. This reduction in water also increases the solution's resistance to fermentation and spoilage. These processes produce obnoxious flavors and odors which reduce the palatability of the feed (152). The U.S. Food and Drug Administration has approved the use of lignosulfonates in animal feed molasses in levels of up to 11%.

Lignosulfonates possess little nutritional value so their presence reduces the overall food energy of the molasses. What little metabolizable energy lignosulfonates possess comes from sugars in the mixture. Sugar acids assist in the viscosity reducing process and act as dispersants for suspended solids in the molasses.

Animal Feed Binder Large volumes of lignosulfonates are utilized as binders and lubricants in the manufacture of animal feed pellets (153) particularly for consumption by dairy cattle. Lignosulfonates are nearly devoid of nutritional value so cannot be use in applications where any dilution of nutritional value of the feed is unacceptable such as in chicken feed (153). The incorporation of lignosulfonates results in the formation of stronger, more durable pellets which resist abrasion and the resultant dust formation. In the pelletizing process, the lubricating ability of lignosulfonates increases production rates, reduces wear in the pelletizing equipment, and reduces energy consumption. Lignosulfonates are typically added in amounts of 1-4%.

The presence of sugars in lignosulfonates represents the only source of metabolizable energy among the lignosulfonate components (155). The pentose sugars, such as xylose, increase the efficiency of protein utilization by ruminants (156) which results in better milk production by dairy cattle (157). Sugars also add to the binding ability of the lignosulfonate mixture. The hydrophilic nature of the sugars, however, can be detrimental as this can lead to poor storage characteristics (158).

Road Dust Control The application of lignosulfonates on secondary dirt roads is an excellent method of dust control and road stabilization (159). The binding ability of lignosulfonates reduces the dust while increasing the load-bearing strength of the roadbed. Unpaved roads can thus withstand heavy traffic and will resist erosion and frost-heave in cold weather (160). The use of lignosulfonates on roads also creates a stabilized base for subsequent application of asphalt or other road materials.

Lignosulfonates are naturally tacky when moist and thus dust adheres to the mixture when it is applied. As the material dries, it loses some of its ability to attract dust but re-wetting or rain rejuvenates this ability. The humectant ability of the sugars in the mixture (101) attracts moisture and thus keeps the material somewhat tacky between water applications. Sugars are responsible for a great deal of the dust binding ability of the material upon application but because of biological degradation do little for long term dust control.

Binder For Molding Applications Lignosulfonates are utilized in molding applications such as ceramics, refractories, building bricks, and foundry cores (100). In these applications, the lignosulfonates function as a water reducer, a dispersant, a clay conditioner, and an extrusion lubricant. The products produced possess greater green, dried and fired strengths as well as suffering less shrinkage during drying and firing. This results in less product breakage and lower energy costs.

In the manufacture of refractory bricks, lignosulfonates function as binders to prevent the disintegration of the moist brick upon removal from the mold (161) and to improve the dry fired strength. In refractory plastic mixes, lignosulfonates increase the plasticity of the mixes allowing for the reduction of plastic clay (162). This results in heightened refractoriness and decreased shrinkage. Lignosulfonates are typically added in amounts of 4-10% on refractory solids.

In the manufacture of building bricks, the dispersing ability of lignosulfonates allows for a reduction in the use of water during brick formation (163). This produces a brick of greater green strength which shrinks less on drying and firing. The lignosulfonates also lubricate the pug mill which reduces power consumption. Typically, lignosulfonates are added at levels of 0.5-2.0% on green clay.

In the manufacture of ceramic materials, lignosulfonates are utilized to improve the characteristics of slips for casting, for spray drying, and for plastic bodies (164). In extrusion processes, lignosulfonates reduce the ceramic viscosity which results in water reduction, increased throughput, and lower energy costs. The products produced utilizing lignosulfonates are of increased green and fired strengths. Applications levels of 0.05-0.4% are standard.

Lignosulfonates can be utilized as a binder in foundry applications to hold together sand molds and cores (165). It is often necessary, however, to either modify the lignin (166) or to combine it with a phenolic resin to achieve enough strength and smoothness in the mold (167). The lignosulfonates also improve the flowability of the sand.

In molding applications, sugars in the lignosulfonates assist in the binding process (168). Mold and bacteria growth and water absorption due to these sugars are not a problem in these applications. Sugar acids in the lignosulfonates add to the clay dispersing ability of the mixture. In some applications, such as in ceramics utilized in electronics, it is necessary that the lignosulfonates contain only very low levels of metals.

Emulsions Lignins find use as emulsifiers, extenders, and emulsion stabilizing agents in slow-setting, cold mix cationic or anionic asphalt emulsions utilized in road construction and maintenance materials (169). These emulsions consist of asphalt, water, and a cationic or anionic emulsifier system. Lignins can serve as a primary emulsifiers but in most cases are combined with a more effective agent. In cationic emulsions, lignin amines prepared from kraft lignin (170) serve as emulsion stabilizers when combined with an emulsifier such as fatty amines or quatenary amines. In anionic systems, unmodified kraft lignins or partially sulfonated kraft lignins function as emulsifying agents or stabilizers. Unmodified lignosulfonates can also function as anionic stabilizing agents although improved stabilization ability is imparted through partial desulfonation (171).

Lignosulfonates also function as efficient stabilizers for a wide range of oil and wax emulsions (172) such as wax emulsion sizing for use in the manufacture of insulation and particle board. These emulsions are made from a combination of wax or oil, water, an emulsifying agent, and lignosulfonates. Once formed, these emulsions are stable over a wide range of oil to water ratios, pH ranges, and temperatures. Typical addition levels for lignosulfonates are 2.5-3.5% on oil/wax and water.

Sugars have little to no effect on the emulsifying ability of lignosulfonates. Their presence, however, can lead to increased solution viscosity (173) and to the growth of unpleasant molds and bacteria (174). The oxidation procedures which are utilized to partially desulfonate lignosulfonates also oxidize sugars to acids. Sugar acids have been shown to possess emulsion stabilization abilities (175).

Water Treatment and Cleaning Applications Water utilized in boilers and in cooling tower applications is treated with lignosulfonates to prevent scale deposits (176). The lignosulfonates are able to sequester hard water salts and thus prevent their deposition on metal surfaces. The dispersing ability of lignosulfonates is utilized to prevent the precipitation of certain insoluble heat-coagulable particles (177). Addition levels of lignosulfonates are between 1-1000 ppm.

Lignosulfonates (178) and kraft lignins (179) are added to commercial cleaning compositions as dirt dispersants and suspending agents. The largest utilization is in industrial acid and alkaline metal cleaning solutions where lignosulfonates function to disperse dirt particles allowing for easier rinsing of the cleaned surface. Their presence also reduces the amount of wetting agent needed. Normal addition levels of lignosulfonates are 0.05-2.0% in water.

Although sugars dilute the dispersing ability of the lignosulfonates, the sequestering ability of these compounds assist in prevention of scale formation (180).

Sugar acids are good dispersants and excellent sequestering agents so their presence is quite beneficial in these applications (181). Other excellent sequestering agents in the lignosulfonates include catechols, flavanoids, and tannins (182).

Micronutrients Iron, copper, zinc, boron, molybdenum, or manganese either alone or in combinations are chelated by lignosulfonates to provide micronutrients to plants growing in soils deficient in these metals (183). In this form, the metals can be applied directly to the foliage and are readily absorbed by the plant (184). In soil treatments, the lignosulfonates maintain these metals in a plant available state much longer than if the metals were applied alone (185). Lignosulfonates are compatible with most insecticides, fungicides and herbicides but as calcium salts will cause precipitation problems with liquid phosphate fertilizers.

Lignosulfonates are intermediate strength complexing agents in comparison to other compounds utilized in the manufacture of micronutrient fertilizers. Sugars are weak complexing agents at best so they add little to the sequestering ability of the mixture. Sugar acids, flavanoids, and tannins are good complexing agents and impart beneficial attributes to the mixture (186).

Battery Expander Lignosulfonates are utilized as an expander in the manufacture of lead-acid batteries. These batteries consist of lead-antimony alloy grids containing a paste of primarily lead oxide (187). Lignosulfonates are added in amounts of less than 0.3% to the paste to prevent the contraction of the spongy lead upon solidification which would lead to a loss of capacity and would reduce the life of the battery (188). The addition of lignosulfonates also serves to reduce polarization of the negative electrodes.

The presence of sugars, phenols, and conidendrin reduces the expander ability of lignosulfonates (189). It has been shown that this reduction in expander efficiency is not due to a chemical reaction between these compounds and sulfonated lignins. Rather, because these compounds posses no expander capabilities, their physical presence interferes with the action of the sulfonated lignins.

Leather Tanning Lignosulfonates are utilized in the "re-tan" process of tanning leather (190) which employs an initial chrome tannage followed by a vegetable tannage. This process produces leather which possesses the beneficial characteristics imparted by each process. Addition of lignosulfonates in the initial stage slows the subsequent tannage reactions and thus allows for up to a 65% reduction in the use of vegetable tannins (191).

Purified lignosulfonates function as tanning agents but are not as efficient as natural vegetable tannins. Sugars are not tanning agents and their presence reduces the tanning efficiency of the lignosulfonates. Removal of hexose sugars and oxidation of pentose sugars produces an improved tanning material (192). This is due to the formation of sugar acids which improve the tanning ability of the mixture and impart beneficial characteristics to the leather (193). Tannins present in the lignosulfonates also assist in the tanning process.

Adhesives There has been a great deal of effort applied towards increasing the amount of lignin utilized in adhesive applications (194). Currently only limited amounts of lignosulfonates and kraft lignin are used as extenders or co-reactants in phenol-formaldehyde and urea-formaldehyde resins for use in fiberglass applications, particleboard, and plywood (195). Typically, lignosulfonates are able to replace 10-15 % of the resin solids. Larger volumes of lignins find use in linoleum pastes and in temporary adhesives for rewettable gum tapes and palletized containers.

Carbohydrates are added to synthetic resins to modify the resin or to partially replace resin constituents (196). The addition of monosaccharides is especially efficient in the presence of nitrogen compounds such as urea. Tannins and flavanoids in the lignosulfonates are also effective as resin co-reactants (197).

Rubber Additive Kraft lignin can be utilized as an extender, modifier, and reinforcing pigment in rubber compounding (198). It is a favorable replacement for carbon black in styrene-butadiene rubber because of its lower density and opacifying diameter. Addition of 5 parts of dry kraft lignin to rubber for automobile tires increases the bond strength between the tire elements without negatively affecting the other physical and mechanical properties of the tires. The lignin must be added to the rubber latex before coagulation as dry mix addition eliminates the reinforcing ability of the lignin (199).

Vanillin Vanillin is the only low molecular weight chemical produced commercially from lignosulfonates. This process consists of an alkaline air oxidation reaction which converts approximately 7% of the lignosulfonate solids to vanillin. The residual lignosulfonates are of reduced molecular weight and degree of sulfonation and function as efficient dispersing agents (200). Vanillin is the major flavoring ingredient in artificial vanilla flavoring and functions as a precursor in the production of pharmaceuticals.

Since vanillin is produced from the aromatic units in lignin, impurities in the lignosulfonate mixture reduce the yield of vanillin. Higher yields, therefore, can be obtained from ultrafiltered lignosulfonate materials (201). Sugars present in the mixture are also easily oxidized to acids which increases the consumption of base and thus production costs.

Miscellaneous Applications Lignosulfonates will precipitate proteins from aqueous solutions under slightly acidic conditions and are therefore useful for treatment of food processing wastewaters (202). In mining operations, lignosulfonates are utilized as flotation aids to recover fines and as slime depressants in the treatment of effluents. In the processing of ores, lignosulfonates are utilized as flotation and wetting aids as well as sequestering agents (135). The sequestering ability of lignosulfonates is also used in deicing applications to reduce the corrosive nature of salt solutions (203). Lignosulfonates are also utilized as flotation aids for post harvest fruit flotation and as solar evaporation enhansors.

High Value Uses of Lignin Unmodified lignosulfonates function quite well in the majority of the previously described applications. The use of modification reactions and/or further purification to achieve certain improved properties in lignosulfonates drastically increases the product cost. With kraft lignin, costly isolation procedures automatically make this product costlier than lignosulfonates. The degree to which the previously described applications can be categorized as high value depends upon the amount of purification or modification required in each end use.

In is not within the scope of this chapter to discuss all of the possible modification reactions utilized with lignin or the uses for these improved products. A tremendous amount of research has been applied in this area specifically towards the development of specialty polymers from lignin. An excellent source of information on this topic is the ACS symposium series on Lignin edited by Glasser and Sarkanen (205).

A unique high value use of lignin involves the production of a magnetic fluid from lignosulfonates (206) which can be utilized in such applications as seed cleaning (207), corrosion removal (208) and ore separation (209). Highly purified lignins have also been shown to process unique medicinal values (210).

Future of Lignin Utilization

The future of lignin utilization will depend upon the ability of lignin to compete with more efficient but costlier materials as well as continued research into new applications. The low cost of lignin has kept this material in markets where more efficient materials are available. The enormous volume of lignin available has spurred a tremendous amount of research into new applications. Many of these, such as enhanced oil recovery (204) or adhesive applications, have the potential to utilize an extremely large amount of these products. For lignin to remain competitive in current markets and to be utilized in new markets, continued improvements in lignin properties must be obtained through modification procedures. These include both changes to the lignin molecule as well as modifications to lignin impurities. Additional help for lignin will come from environmental constraints placed upon the competing synthetic compounds. As more restrictions are placed upon these materials, the non-toxic, biodegradable aspects of lignins will increase the attractiveness of these materials.

Literature Cited

1. Panshin, A. J.; Zeeuw, C. D. *Textbook of Wood Technology*; McGraw-Hill: New York, NY, 1970, 71.
2. Falkehag, I.; In *1989 International Symposium on Wood and Pulping Chemistry, Raleigh*, Tappi Proceeding, Tappi Press: Atlanta, GA, 1989, Vol 1, 107.
3. Pearl, I. A. *Tappi*. 1982, 65(5), 68.
4. Nakano, J. In *Wood and Cellulosics*; Kennedy, J.F., Ed.; J. Wiley: New York, NY, 1987, 499-512.
5. Wiley, P.R.; *Tappi*. 1961, 44(11), 22A.
6. Hoyt, C.H.; Goheen, D.W. In *Lignins: Occurrence, Formation, Stucture and Reactions*; Sarkanen, K. V.; Ludwig, C. H., Eds.; J. Wiley: New York, NY, 1971, 833-865.
7. Lin, S. Y. *Progress in Biomass Conversion*; Academic Press: Orlando, FL, 1983, Vol 4, 31-78.
8. Fengel, D.; Wegener, G. *Wood: Chemistry, Ultrastructure, Reactions*; Walter de Gruyter: Berlin, 1984.
9. *Wood Extractives and Their Significance to the Pulp and Paper Industry*; Hillis, W. E., Ed.; Academic Press, New York, NY, 1962.
10. Buchanan, M. A. In *The Chemistry of Wood*; Browning, B. L., Ed; Intersci. Publ.: New York, NY, 1963, 313-367.
11. Veeramani, H.; Wani, G. A. *Chem. Ind. Dev.*, Incorp. CP&E 1977, Dec., 13.
12. Glasser, W. G. In *Pulp and Paper: Chemistry and Chemical Technology*, Casey, J.P., Ed.; John Wiley &

Sons, New York, NY, 1980, Third Edition, Vol I, 39-111.
13. Goring, D. A. I. In *Lignin: Properties and Materials*; Glasser, W. G.; Sarkanen, S., Eds.; ACS Symposium Series 397; American Chemical Society: Washington DC, 1989, 2-10.
14. Sarkanen, K. V. In *Lignins: Ocurrence, Formation, Structure and Reactions*; Sarkanen, K. V.; Ludwig, C. H., Eds.; J. Wiley: New York, NY, 1971, 95-163.
15. Adler, E.; *Wood Sci. Technol.* 1977, 11, 169.
16. Obiaga, T. I. Ph.D. Thesis, University of Toronto, 1972.
17. Guthrie, R. D. *Introduction to Carbohydrate Chemistry*; Clarendon Press: Oxford, 1974, Fourth Ed.
18. Lewis, H. F.; Ritter, G. J. In *Cellulose*; Ott, E.; Spurlin, H. M.; Grafflin, M. W. Eds.; High Polymers: Vol V; Interscience Pub.: New York, NY, 1954, Part I, 443-509.
19. Whistler, R. L.; Richards, E. L. In *The Carbohydrates*; Pigman, W.; Horton, D., Eds.; Academic Press: New York, NY, 1970, 2nd ed., Vol 2A, 447-469.
20. Timell, T. E. *Wood Sci. Technol.* 1967, 1, 45.
21. Timell, T. E. *Svensk Papperstid.* 1969, 72, 173.
22. Côté, W. A.; Simson, B. W.; Timell, T. E. *Svensk Papperstid.* 1966, 69, 547.
23. Wenzl, H. F. J. *The Chemical Technology of Wood*; Academic Press: New York, NY, 1970, 141-158.
24. Assarsson, A.; and Akerlund, G. *Svensk Papperstid.* 1966, 69, 517.
25. Lai, Y. Z.; Butnaru, R.; Simionescu, C. *Cell. Chem. Technol.* 1977, 11, 59.
26. Cutter, B. E.; McGinnes, E. A.; McKown, D. H. *Wood Fiber* 1980, 12, 72.
27. Young, H. E.; Quinn, V. P. *Tappi* 1966, 49, 190.
28. Rydholm, S. A. *Pulping Processes*; J. Wiley: New York, NY, 1965.
29. Hägglund, E. *Chemistry of Wood* 1951; Academic Press: New York, NY.
30. Gellerstedt, G.; Gierer, J. *Svensk Papperstid.* 1971, 74, 117.
31. Erdtman, H. *Svensk Papperstid.* 1945, 48(4), 75.
32. Glennie, D. W. In *Lignins: Occurence, Formation, Structure, and Reactions*; Sarkenen, K. V.; Ludwig, C. H. Eds.; J. Wiley: New York, NY, 1971, 597-637.
33. Gierer, J. *Wood Sci. Technol.* 1985, 19, 289.
34. Hamilton, J. K., *Proc. Wood Chemistry Symp.*, (Applied Chem. Div., Intl. Union of Pure and Applied Chem.), London, England, 1962, 197-217.
35. Janson, J.; Sjöström, E. *Svensk Papperstitn.* 1964, 67(19), 764.
36. Samuelson, O.; Ljungqvist, K. J.; et al. *Svensk Papperstidn.* 1958, 24, 1043.
37. Slavik, I. *Svensk Papperstidn.* 1961, 64, 427.

38. Pew, J. C. *Tappi* 1949, 32(1), 39.
39. Herrick, F. W.; Hergert, H. L. In *Adv.Phytochem.*, (Proc. 16th Ann. Mtg. Phytochem. Soc. N. Am. Avg., 1976, Vancouver); Vol 11, 1977, 443-515.
40. Marton, J. In *Lignins: Occurence, Formation, Structure, and Reactions*; Sarkenen; K. V.; Ludwig, C. H. Eds.; J. Wiley: New York, NY, 1971, 639-694.
41. Gierer, J. *Svensk Papperstidn.* 1970, 73, 571.
42. Forss, K. G.; Stenlund, B. G.; Sagfors, P. E. *Appl. Polymer. Symp.* 1976, 28, 1185.
43. Gierer, J.; Imsgard, F.; Pettersson, I. *Appl. Polymer. Symp.* 1976, 28, 1195.
44. Sjöström, E.; Enström, B. *Tappi* 1977, 60(9), 151.
45. Malinen, R.; Sjöström, E. *Pap. Puu* 1975, 57, 728.
46. Saarnio, J.; Gustafsson, C. *Paperi Puu* 1953, 35(3), 65.
47. Corbet, W. M.; Richards, G. N. *Svensk Papperstidn.* 1957, 60(21), 791.
48. Lora, J. H.; Wu, C. F.; et al In *Lignin: Properties and Materials*; Glasser, W.; Sarkanen, S. Eds.; ACS Symposium Series 397, American Chemical Society: Washington DC, 1989, 312-323.
49. Lora, J. H.; Wayman, M. *Tappi* 1978, 61(6), 47.
50. Goldstein, I. S. *Tappi* 1980, 63(9), 141.
51. Uloth. V. C.; Wearing, J. T. *Pulp & Paper Can.* 1989, 90(9), 67.
52. Kim, H.; Hill, M. K.; et al. *Tappi* 1987, December, 112.
53. Öster, R.; Kringstad, K. P. *Nor. Pulp Pap. Res. J.* 1988, 2, 68.
54. Hart, J. S. In *The Pulping of Wood*; MacDonald, R. G.; Franklin, J. N. Eds.; McGraw-Hill: New York, NY, 1969, Vol 1, 277-346.
55. Edge, D.; U.S. Pat. 4,155,804, 1979.
56. Puurunen, J.; U.S. Pat. 4,088,660, 1978.
57. Watson, C. A. *Forest Prod. J.* 1959, 9(3), 25.
58. Christensen, L. M.; U.S. Pat. 2,362,451, 1941.
59. Anderson, R.; Weisbaum, R. B.; et al. *Food Process* 1974, 35(7), 58.
60. Romanyschuk, H.; Lehtomaki, M. *Process Biochem.* 1978, 13(6) 16.
61. Swenson, L. K.; U.S. Pat. 2,916,355, 1959
62. Glasser, W. G., In *Encyclopedia of Polymer Science and Engineering*, J. Wiley: New York, NY, 1987, 795-851.
63. Öster, R. Nor. *Pulp Pap. Res. J.* 1988, 2, 82.
64. Theander, O. In *Chemistry of Delignification with Oxygen, Ozone, and Peroxides*; Symposium Proceedings, Raleigh NC, May 27-29, 1975; Uni Pub.: Tokyo, 1975, 43-59.
65. Forss, K. *Paperi ja Puu* 1974, 3, 174.
66. Ludwig, C. H.; Zdybak, W. T. Presented at the 185th National Meeting, Cellulose, Paper and Textile Division, March 23, 1983.

67. Björkman, A.; Person, B. *Svensk Papperstidn.* 1957, 60, 158.
68. Hayashi, A.; Goring D. A. I. *Pulp Pap. Mag. Can.* 1965, 66(C), T-154. 1984.
69. Fengel, D.; Wegener, G.; et al. *Holforschung* 1981, 35, 111.
70. Yean, W. Q.; Goring, D. A. I. *Svensk Papperstidn.* 1952, 55, 563.
71. Sjöström, E.; Hagglund, P.; et al. *Svensk Papperstidn.* 1962, 65, 855.
72. Pfister, K.; Sjöström, E. *Paperi ja Puu - Papper och Trä* 1977, 11, 711.
73. Shaw, A. C. *Pulp Pap. Mag. Can.* 1956, 57(1), 95.
74. Sjöström, E. *Wood Chemistry: Fundametals and Applications;* Academic Press: New York, NY, 1981, 122.
75. Forss, K. Ph.D. Thesis, Abo Akademi, 1961
76. Adler, E. *Svensk Papperstidn.* 1947, 50(261), 9.
77. Cordingly, R. H. *Tappi* 1959, 42(8), 654.
78. Gadd, G. O. *Pappers och Trävarutidskr.* Finland 1946, 28(7A), 61.
79. Ahlén, L.; Samuelson, O. *Svensk Papperstidn.* 1955, 11, 421.
80. Reed, G. In *Kirk-Othmer: Encyclopedia of Chemical Technology;* Grayson, M. Ed.; J. Wiley: New York, NY, Third Ed., Vol 24, 771-806.
81. Lackey, H. B.; Moyer, W. W.; et al. *Tappi* 1949, 32(10), 469.
82. Hergert, H. L. *Forest Prod. J.* 1960, 10(11), 610.
83. Sears, K. D. *J. Org. Chem.* 1972, 37, 3546.
84. Fengel, V. D.; Wegener, G.; et al *Holforschung* 1981, 35, 51.
85. Iverson, T.; Westmark, U. *Cellulose. Chem. Technol.* 1985, 19, 531.
86. Gellerstedt, G.; Lindfors, E. *Holzforshung* 1984, 38, 151.
87. Glasser, W. G. *Forest Prod. J.* 1981, 31(3), 24.
88. Marton, J. *Tappi* 1964, 47, 713.
89. Bryce, J. R. In *Pulp and Paper: Chemistry and Chemical Technology;* Casey, J. P. Ed.; J. Wiley: New York, NY, 1980, Third Edition, Vol 1, 291-376.
90. Mitscherlich, A.; Ger. Pat. 4178, 1878.
91. Hough, W. J.; U.S. Pat. 949,324, 1910.
92. Hearon, W. M. *J. Chem. Ed.* 1958, 35, 498.
93. Baum, M. A., Harris, E. E., et al. *Tappi* 1962, 45(1), 195A.
94. Maron, S. H.; Lando, J. B. *Fundamentals of Physical Chemistry;* Macmillan: New York, NY, 1974, 218-220.
95. Goodman, R. M. In *Kirk-Othmer: Encyclopedia of Chemical Technology;* Grayson, M. Ed.; J. Wiley: New York, NY, 1979, Third Ed., Vol 5, 833-848.
96. Goring, D. A. I. In *Lignins: Occurrence, Formation, Structure, and Reactions;* Sarkanen, K. V.; Ludwig, C. H. Eds.; J. Wiley: New York, NY, 1971, 695-768.

97. Jiang, J. E.; Chen, C. L.; et al; *1989 International Symposium on Wood and Pulping Chemistry*; Tappi Proceedings, Tappi Press: Atlanta, GA, 1989, 197-202.
98. Oita, O.; Nakano, J.; et al. *Mokuzai Gakkaishi* 1966, 12(5), 239.
99. Detroit, W. J.; U.S. Pat. 3,726,850, 1973.
100. Anon., Lignin Chemicals : Presented by Borregaard, Technical Report 600E.
101. Shallenberger, R. S.; Birch, G. G. *Sugar Chemistry*; Avi: Westport, CT, 1975, 158-159.
102. Keimel, F. A. In *Kirk-Othmer: Encyclopedia of Chemical Technology*; Grayson, M. Ed.; J. Wiley: New York, NY, 1979, Third Ed., Vol 1, 488-510.
103. Rumpf, H. In *Agglomeration*; Knepper, W. A. Ed.; J. Wiley: New York, NY, 1962, 379-414.
104. Capes, C. E. In *Kirk-Othmer: Encyclopedia of Chemical Technology*; Grayson, M. Ed.; J. Wiley: New York, NY, 1979, Third Ed., Vol 21, 77-104.
105. Becher, P. *Emulsions: Theory and Practice*; Reinhold Pub.: New York, NY, 1965.
106. Walstra, P. In *Encyclopedia of Emulsion Technology*; Becher, P., Ed.; Marcel Dekker: New York, NY, 1983, Vol 1; 57-128.
107. Tadros, T. F.; Vincent, B. In *Encyclopedia of Emulsion Technology*; Becher, P., Ed.; Marcel Dekker, New York: NY, 1983, Vol. 1; 129-285.
108. Heller, W.; Pugh, T. L. *J. Chem. Phys.* 1954, 22, 1778.
109. McCrary, A. L. In *Kirk-Othmer: Encycolpedia of Chemical Technology*; Grayson, M. Ed.; J. Wiley: New York, NY, 1979, Third Ed., Vol 5, 339-368.
110. Britt, K. W. In *Handbook of Pulp and Paper Technology*; Britt, K. W. Ed.; Van Nostrand Reinhold: New York, NY, 1970, Second Ed., 239-246.
111. Falkehag, S. I. *J. Appl. Polymer Sci.* 1975, 28, 247.
112. Hough, G. W. In *Chemical Recovery in the Alkaline Pulping Process*; Hough, G. W. Ed.; Tappi Press: Atlanta, GA, 1985, 10.
113. Kirk, G. B.; U.S. Pat. 2,856,304, 1958.
114. Khater, A. M.; Khater, E. H.; et al. *Pakistan J. Sci. Ind. Res.* 1987, 30(4), 275.
115. Anon. The Use of Lignosite in the Manufacture of Gypsum Board; Georgia Pacific Technical Bulletin, GPLIG 3/70.
116. Karmazsin, E.; Murat, M.; et al. *Bull. Soc. Fr. Ceram.* 1979, 124, 3.
117. DeRooy, F. J.; Danial, T. D.; et al.; U.S. Pat. 4,059,456, 1977.
118. Kinkade, W.; O'Neill, E. E.; U.S. Pat. 3,573,947, 1971.
119. Simatupang, M. H.; Xian, L. X. *Holz Roh-und Werkstoff* 1985, 43, 325.

120. Combe, E. C.; Smith, D. C. *J. Appl. Chem.* 1964, 14, 544.
121. King, E. G.; Adolphsen, C.; U.S. Pat. 3,126,291, 1964.
122. Collepardi, M. In *Concrete Admixtures Handbook: Properties, Science, and Technology*; Ramachandran, V. S. Ed.; Noyes Pub.: Park Ridge, NJ, 1984, 116-210.
123. Dodson, V.H.; Farkas, E.; U.S. Pat. 3,351,478, 1967.
124. Forss, B.; U.S. Pat. 4,450,106, 1984.
125. Milestone, N. B. *J. Am. Ceram. Soc.* 1979, 62(7), 321.
126. Milestone, N. B. *Cem. Concr. Res.* 1977, 7, 45.
127. Singh, N.B.; Singh, S. P.; et al. *Adv. Cem. Res.* 1989, 2(6), 43.
128. Yasuda, S.; Hirano, J. *J. Wood Chem. Tech.* 1989, 9(1), 123.
129. Kennedy, H. L.; Mark, J. G.; U.S. Pat. 2,141,571, 1938.
130. Helmuth, R. A.; Miller, F. M.; et al. In *Kirk-Othmer: Encyclopedia of Chemical Technology*; Grayson, M. Ed.; J. Wiley: New York, NY, 1979 Third Ed., Vol 5, 163-193.
131. Adams, A. B.; Farkas, E.; U.S. Pat. 3,094,425, 1963.
132. Moorer, H. H.; Anderegg, C. M.; U.S. Pat. 4,204,877, 1980.
133. Hoyt, C. H.; U.S. Pat. 3,035,042, 1962.
134. Lauzon, R. V. *Oil Gas J.* 1982, April 19, 93.
135. Anon.; Lignosite: Lignin Chemicals; Georgia Pacific Technical Bulletin.
136. Detroit, W. J.; U.S. Pat. 4,447,339, 1984.
137. Detroit, W. J.; U.S. Pat. 4,505,825, 1985.
138. Steinberg, J. C.; Gray, K. R.; U.S. Pat. 3,505,243, 1970.
139. Erasmus, A.; U.S. Pat. 3,720,610, 1973.
140. Floyd, J. C.; Hoepner, J.J.; U.S. Pat. 3,479,287, 1969.
141. Metcalf, R. L. In *Kirk-Othmer: Encyclopedia of Chemical Technology*; Grayson, M. Ed.; J. Wiley: New York, NY, 1979, Third Ed., Vol 13, 413-484.
142. King, E. G.; U.S. Pat. 3,639,617, 1972.
143. Pearce, D. A.; U.S. Pat. 3,137,618, 1964.
144. Moorer, H. H., Sandefur, C. W.; U.S. pat. 3,986,979, 1976.
145. Blaisdell, L. A.; U.S. Pat. 3,156,520, 1964.
146. Lin, S. Y. *J. Am. Assoc. Textile Chem. Colorists* 1981, 13(11), 261.
147. Dilling, P.; U.S. Pat. 4,131,564, 1978.
148. Dilling, P. *J. Am. Assoc. Textile Chem. Colorists* 1986, 18(2), 17.
149. Linhart, K.; Scholl, W.; et al.; Ger. Offen. 2,913,719, 1979.

150. Dannenberg, E. M. In *Kirk-Othmer: Encyclopedia of Chemical Technology*; Grayson, M. Ed.; J. Wiley: New York, NY, 1979, Third Ed., Vol 4, 631-666.
151. Anon. Lignosite: Additive M Information Sheet, Georgia Pacific Corp., Oct., 1989.
152. Tribble, T. B. *Feed Flavor and Animal Nutrition*; Agriaids Inc.: Chicago, IL, 1962.
153. Knodt, C. B.; U. S. Pat. 3,035,920, 1962.
154. Winowiski, T. S.; Lin, S. Y.; U.S. Pat. 4,952,415, 1990.
155. Scott, R. W.; Millett, M. A.; et al. *F. Prod. J.* 1969, 19(4), 1969.
156. Winowiski, T. S.; U.S. Pat. 4,957,748, 1990.
157. Larsen, H. J.; U.S. Pat. 4,377,595, 1983.
158. Morrison, D. G., U.S. Pat. 4,698,225, 1987.
159. Sherman, W. A. *Paper Trade J.* 1950, October 12, 19.
160. Anthone, R.; Parks, M. P.; U.S. Pat. 4,001,033, 1977.
161. Anon.; Lignosite: The Use of Lignosite in Refractory Clay Products; Georgia Pacific Technical Bulletin, 5/89
162. Hamme, J. V. In *Kirk-Othmer: Encyclopedia of Chemical Technology*; Grayson, M. Ed.; J. Wiley: New York, NY, 1979, Third Ed., Vol 5, 241.
163. King, E. G.; Adolphson, C.; U.S. Pat. 3,109,742, 1963.
164. Burgess, D. G. *Cer. Bul.* 1957, 36(5), 168.
165. Wallace, B. P.; Romney, V. D.; et al.; U.S. Pat. 2,863,781, 1958.
166. Cowan, J. C.; Beasley, E. B.; et al.; U.S. Pat. 3,403,037, 1968.
167. Iyer, R.; U.S. Pat. 4,336,179, 1982.
168. Cooper, R. H.; U.S. Pat. 3,307,959, 1967.
169. Johnson, J. C. *Emulsions and Emulsifying Techniques*; Noyes Data Corp.: Park Ridge, NJ, 1979, 321-364.
170. Doughty, J. B.; U.S. Pat. 3,871,893, 1975.
171. Detroit, W. J.; U.S. Pat. 4,293,459, 1981.
172. Stazdins, E.; U.S. Pat. 3,709,708, 1973.
173. Roualt, E.; French Pat. 748,886, 1933.
174. Wiley, A. J.; Kummer, M. F.; et al. *Tappi* 1951, 34(12), 556.
175. Tolstoguzov, V. B.; Braudo, E. E.; Mikheeva, N. V.; U.S.S.R. Pat. 303,819, 1971.
176. Bonewitz, P. W.; Fults, E. H.; Hockett, S. W.; U.S. Pat. 2,826,552, 1958.
177. Kaye, S.; U.S. Pat. 3,317,431, 1967.
178. Von Pless, J. A.; U.S Pat. 3,247,120, 1966.
179. Dimitri, M. S.; U.S. Pat. 3,803,041, 1974.
180. Angyal, S. J. *Pure Appl. Chem.* 1973, 35(2), 131.
181. Hamada, M. Japan Kokai 76,139,807, 1976.
182. Robertson, R. S., U.S. Pat. 3,375,200, 1968.
183. Anon.; Lignosite: KE-MIN Micronutrient Products, Georgia Pacific Technical Bulletin, 5/89.

184. Adolphson, C.; Simmons, R. W.; U.S. Pat. 3,244,505, 1966.
185. Wallace, A.; Ashcroft, R. T. *Soil Sci.* 1956, 82(3), 233.
186. Fisher, F. L. *Solutions* 1984, Jan., 52.
187. Doe, J. B. In *Kirk-Othmer: Encyclopedia of Chemical Technology*; Grayson, M. Ed.; J. Wiley: New York, NY, Third Ed., Vol 3, 640-663.
188. Shilling, P.; U.S. Pat. 3,446,670, 1969.
189. Limbert, J. L.; Proctor, H. G.; Poe, D. T., U.S. Pat. 3,523,041, 1970.
190. Simmons, R. W.; U.S. Pat. 3,447,889, 1969.
191. Buchanan, M. A.; Lollar, R. M. *J. Am. Leather Chem. Assoc.* 1947, 42, 232.
192. Wallace, F. J.; U.S. Pat. 2,244,410, 1941.
193. Wallace, F. J.; U.S. Pat. 2,142,739, 1939.
194. Lewis, N. G.; Lantzy, T. R. In *Adhesives from Renewable Resources*; Hemmingway, R. W.; Conner, A. H.; Branham, S. J. Eds.; ACS Symposium Series 385, Am. Che. Soc.: Washington DC, 1987, 1-26.
195. Nimz, H. In *Wood Adhesives: Chemistry and Technology*; Pizzi, A. Ed.; Marcel Dekker Inc.: New York, NY, 1983 247-288.
196. Conner, A. H. In *Adhesives form Renewable Resources*; Hemmingway, R. W.; Conner, A. H.; Branham, S. J. Eds.; ACS Symposium Series 385, American Chemical Society: Washington DC, 1989, 271-288.
197. Pizzi, A. In *Wood Adhesives: Chemistry and Technology*; Pizzi, A. Ed.; Marcel Dekker Inc.: New York, NY, 1983, 177-246.
198. Rao, G. V.; Jagannadhan, V.; et al. *Ind. Pulp Paper* 1978, June-July, 11.
199. Falkehag, S. I.; Braddon, D. V.; et al.; In *ACS Symp. Renewable Resources For Plastics* (Philadelphia, PA.); 1975, 68-72.
200. Bryan, C. C.; U.S. Pat. 2,506,540, 1950.
201. Evju, H.; U.S. Pat. 4,151,207, 1979.
202. Sherman, R. *J. Food Tech.* 1979, 33(6), 50.
203. Neal, J.; U.S. Pat. 4,668,416, 1986.
204. Babu, D. R.; Neale, G.; et al. *Cellulose Cem. Technol.* 1986, 20, 603.
205. *Lignin - Properties and Materials*; Glasser, W. G. and Sarkanen S. Eda, ACS Symposium Series 397, American Chemical Society, Washington DC, 1989.
206. Briggs, W. S., Kjargaard, N. J.; U.S. Pat. 4,019,995, 1977.
207. Beriage, A. G., Krishmann, P, et al; U.S. Pat 4,765,486, 1986.
208. Hulford, K. F., Price, J. W.; U.S. Pat. 4,445,940, 1984.
209. Andres, UY. T., Devernoe, A. L., Walker, M. S., U.S. Pat. 4,961,841, 1990.
210. Samson, R. H., Hollis, J. W.; European Pat. 0 303 236 A2, 1988.

RECEIVED July 16, 1991

Chapter 12

New Developments in Cellulosic Derivatives and Copolymers

David N.-S. Hon

Department of Forest Resources, Wood Chemistry Laboratory, Clemson University, Clemson, SC 29634-1003

Cellulose is a wonder material. Many cellulose derivatives and copolymers have been developed to meet certain areas of applications. Cellulose chemistry and new developments in derivatives and copolymers since 1980s were reviewed with emphasis on present technologies for making oyxcelluloses, esters, ethers, deoxycelluloses, and block- and graft-copolymers.

Cellulose as a material is a marvel of nature which has been and is of immense importance to mankind (1). Although it plays an important role since early civilization and in the development of polymer science in the 1930s, the establishment of petroleum chemistry and the ease of processing oil into useful products have stymied cellulose research. In the mid-1970s, ironically, with the growing shortage of raw materials as well as the uncertainties about oil prices and supplies, and environmental consciousness, active cellulose research has been revived by the chemical industries.

Cellulose can be considered as one of the oldest polymers and has established a long record of industrial applications. In its native form, it is distributed throughout the plant kingdom, starting from the tree to shrubs and grass. It is a principal product of photosynthesis and is the major organic component in biomass. Its annual production amounts to over 10^{11} tons (1), and more importantly, it is a renewable resource. Today, cellulose, in its various forms, constitutes about half of all of the polymers consumed industrially in the world. Its importance is indisputable.

Although cellulose possesses many intrinsic properties which are unique in many ways, its fundamental research and industrial applications are sometimes unfortunately hindered by its natural structures, e.g., high crystallinity or less accessibility of its reactive sites. These properties oftentimes limit its dissolving characteristics in common organic solvents, hence, limit its accessibility for chemical reactions and biological and microbial treatments (2). Nonetheless, many modified products have been championed since the chemical separation of cellulose from plants by Payen in 1839 (3). Distinctly, the first man-made plastic, cellulose nitrate, was a derivative of cellulose. Chemical modification of polymers is a very wide domain of polymer science. Many chemical modifications of polymers were actually spin-offs of cellulose modification. Ward (4) once wrote that "no discussion of polymer modification could even hope to approach completeness without mentioning modification of cellulose."

The products of chemical modification generally enhance the usefulness of cellulose because more desirable properties are incorporated. Despite the fact that various families of cellulose derivatives have been made (5,6), the restless and probing minds of chemists and the needs and desires of an ever-expanding technology have continually sought to prepare novel derivatives having unique and useful properties. A quick search of cellulose modification research between 1966 and 1990 revealed that approximately 16,500 articles were published in the areas of esterification, etherification, crosslinking and graft copolymerization reactions (2,7). Many publications dealing with these topics are available (5,6,8-11). During the past ten years, cellulose and its derivatives have progressed from its "controversial" physical structures to products of considerable industrial importance. In this chapter only the new cellulose derivatives that were reported since 1980 will be discussed.

History and Background

Cellulose is a polydisperse linear syndiotactic natural polymer. The basic monomeric unit of cellulose is D-glucose, which links successively through a glycosidic bond in the beta configuration between carbon 1 and carbon 4 of adjacent units to form long-chain 1,4-β-glucans (see Figure 1). Each β-D-glucopyranose unit within a cellulose chain has a primary and two secondary hydroxyl groups. These hydroxyl groups frequently form intra- and inter-molecular hydrogen bondings within and among molecules, respectively. These bondings, combine with other secondary forces, such as Van der Waals attraction, aggregate portions of the molecular chains into various degrees of lateral order ranging from perfect geometrical packing of the crystal lattice (the so-called crystalline region) to random conditions (amorphous region). This topochemistry actually controls the chemical reactivity of cellulose. Essentially, the hydroxyl groups located in the amorphous regions, being in a highly accessible environment, react readily in many chemical reactions. However, in the crystalline regions, where there is a close packing and strong interchain bonding, these groups are not readily accessible to reactant molecules and are occasionally completely inaccessible to some. In order to enable a significant portion of cellulose molecules to participate in a reaction, crystalline regions must be made accessible to reactants.

Accessibility of the cellulose molecules in the fiber is frequently determined by fiber reactivity, which obviously also depends on the nature of the reactants as well as on such conditions as time, temperature, pressure and solvents. Due to the strong hydrogen bondings, cellulose is not readily dissolved in common organic solvents. Thus, most of the reactions on cellulose are heterogeneous in nature. As with other semi-crystalline polymers, cellulose undergoes two-phase crystalline-amorphous reaction, in which four reaction patterns are possible (12): (a) surface reaction, (b) macroheterogenous reaction, (c) microheterogenous reaction, and (d) permutoid reaction. Surface reactions involve only the cellulose molecules at the microscopic surface, or, more precisely, their segments located at the surface. The macroheterogeneous reaction starts at the surface but proceeds through the fiber from layer to layer as the reacted cellulose dissolves or swells in the surrounding solvent. The microheterogeneous reaction occurs when the cellulose is swollen by the reaction medium, but the crystalline, ordered, or intrafibrillar regions are not accessible to the reagent. Permutoid reactions (or so-called intrafibrillar or intracrystalline reactions) extend into the highly ordered and crystalline regions without dissolving them and usually lead to the transformation of the lattice. Microheterogeneous and permutoid reactions are also often distinguished as intermicellar and intramicellar reactions, respectively. While photochemical and weathering reactions of cellulose generally are considered a surface reaction, the nitration of cellulose is considered to be permutoid. Reactions (c) and (d) are determined by the solvent power. The acetylation of cellulose in a solvent for cellulose acetate that does not swell cellulose is a macroheterogeneous reaction. If a partial acetylation of cellulose is carried out by applying the acetic anhydride from a swelling agent for cellulose, then it is a microheterogeneous reaction.

Many activation treatments, such as swelling, solvent exchange, inclusion of structure-loosening additives or mechanical action, can be used to increase reactivity. Once the original hydrogen bonds have been broken and intramicellar swelling achieved, the cellulose hydroxyls are capable of reacting like an ordinary aliphatic hydroxyl group. Recently, many co-solvents have been developed for cellulose which consequently improve cellulose accessibility and reactivity. These co-solvents are dimethyl-sulfoxide/paraformaldehyde, dimethylacetamide/paraformaldehyde, lithium chloride/dimethylacetamide, dinitrogen tetroxide/dimethylformamide and sulfur dioxide/nitrosyl chloride (8,13). With these co-solvents, cellulose modification can be conducted in a homogeneous system. For example, a high degree of substitution, organosoluble trimethyl silyl cellulose has been prepared in DMAC/LiCl system (14). Several cellulose sulfonates have been prepared in the DMF/Chloral solvent system (15,16). The number of hydroxyl groups available for reactions can vary from as few as 10-15% in highly crystalline cellulose materials to as much as 85-95% in decrystallized cellulose (17). Even higher accessibility (i.e., 98-100%) can be achieved from a regenerated non-crystalline cellulose material (18).

The history of cellulose modification began around 1830 when Brocannot first described the nitration of cellulose (19). Regeneration of cellulose fiber from cellulose nitrate through denitration, dissolution, spinning and regeneration was detailed by Count Hilaire de Chardonnett in 1885 (20). After 1890, additional methods were found to solubilize cellulose by acetylation, xanthation and cuproxyammoniation, to spin the resulting solutions by coagulating them into the form of a filament or fiber. This period was considered to be the beginning of cellulose chemical modification.

Although modifications of cellulose can be performed under either homogenous or heterogeneous conditions, currently the application of heterogeneous processes still prevails in industry, particularly for high volume polymers. Hence, non-uniform distribution of the substituent groups on the cellulose matrix is a major concerned, because they have profound effects on mechanical, physical and biological properties of modified products. These properties include solubility, viscosity and other solution characteristics, gel strength, strength-resilience balance, and susceptibility to enzymic degradation.

Although working under the constraint of heterogeneous system, chemical modification of cellulose has been successful in esterification, etherification, crosslinking and graft copolymerization (21). All these modification methods are used in industry for the manufacture of a wide variety of useful products.

Present Technology

Oxycellulose and Polyalcohol from Cellulose

One of the most selective processes of cellulose modification is the oxidation of cellulose by periodic acid and its salts to form a dialdehyde cellulose, which can further be oxidized to dicarboxyl cellulose, tricarboxyl cellulose, or reduced into an acyclic, stereoregular polymer of [(2r,4s,5r)-2,4,5-tris(hydroxymethyl)-1,3-dioxopentamethylene] (22), as shown in Figure 2. A recent review article on this subject is available (23). It has been noted that under suitable conditions, periodate oxidation of cellulose can yield products containing high levels of carboxyl or acidic enediol function (24), and methyl ester derivatives (25). The high level of 2,3-dicarboxycellulose produced by oxidation with $HClO_2$ was completely soluble in water and took up various metallic ions other than alkali metals to form precipitates (26). Homogeneous periodate oxidation of cellulose was attempted by using methylol cellulose, in which a uniform cleavage of C_2-C_3 bonds by the periodate ion was achieved (27). Examination of the thermal deformation and tensile properties revealed that no notable cellulose degradation occurred during the reaction.

The impregnation of crotonized crotonaldehyde cellulose with ε-aminocaproic acid and $CaCl_2$, $(CH_3CO_2)_2Cu$, $CuSO_4$, or $AgNO_3$ leads to the formation of bioactive gauzes with hemostatic and antimicrobic action (28). Moreover, the oxidation of unprotected (regenerated) cellulose with DMSO-Ac_2O or DCC/DMSO/pyridine/ trifluoroacetic acid affords a mixture of 2-oxy-, 3-oxy-, and 2,3-dioxycelluloses (29). Recently, it was observed that oxidation of unprotected cellulose with DMSO-Ac_2O in the DMSO/paraformaldehyde solvent system affords exclusively 3-oxycellulose, due to the reversible formation of O-6 and O-2 hydroxymethyl and poly(oxymethylene)ol side chains (30).

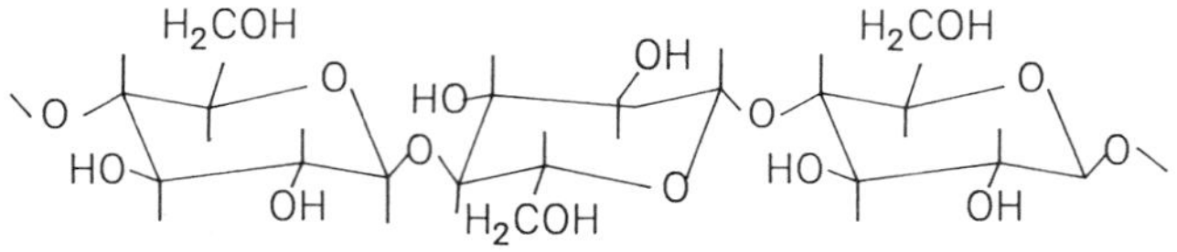

Figure 1. Partial structures of cellulose.

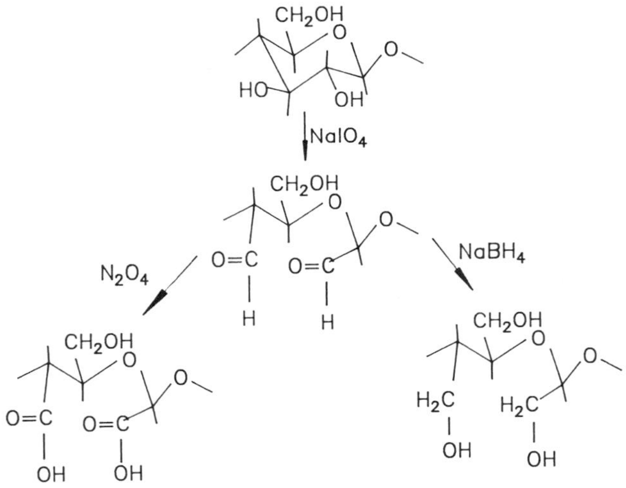

Figure 2. Preparation of oxycelluloses.

Esterification

Because it is a polyhydroxyl alcohol, cellulose can be esterified in strong acid mediums, anhydrides, and catalysts, and requires the absence of water for completion, as it is a reversible reaction. Normally, the reaction of the mixture proceeds rapidly and is permitted to continue until the three hydroxyl groups on each anhydroglucose unit have been replaced with the acyl group of the organic acid or mixture of acids.

Many cellulose esters such as cellulose nitrate, cellulose acetate and mixed esters of cellulose acetate butyrate have enjoyed their success in the commercial scale production. Many new esters continue to appear in the market. Traditionally, esterification is conducted on a heterogeneous system, however, homogeneous system by using mixed organic solvents have been developed recently. For example, Ikeda et al. (31) demonstrated that homogeneous esterification and acetalization of cellulose in LiCl/dimethyl-acetamide (DMAC) can be achieved.

Cellulose Sulfates: Cellulose can be sulfated by the use of sulfating agents such as sulfuric acid in organic solvents, chlorosulfonic acid in the presence of amines, and liquid or gaseous sulfurtrioxide (32). When prepared to the appropriate degree of sulfate ester substitution, cellulose sulfate esters are water soluble and of interest as detergents, antistatic coatings for photographic film, viscosity modifiers for enhanced oil recovery, as thickening agent for food, cosmetics, and pharmaceutical, and low-calorie food additives (33-38).

Although cellulose sulfates have been known since 1819, new processes for making this inorganic cellulose ester continue to appear. High molecular weight cellulose sulfate esters with a high degree of sulfate ester substitution, and an excellent thermal stability have been synthesized (39,40). This method used preformed dialkylamide sulfur trioxide complexes as a sulfating reagent in the corresponding dialkylamide solvent. The reaction is heterogeneous, and the cellulose remains fibrous throughout the sulfation. Completely water-soluble, highly viscous sodium cellulose sulfate semiesters are obtained in homogeneous systems by the reaction of cellulose nitrite (41). The intermediate, cellulose nitrite, that is formed and dissolved is obtained in the N_2O_4/dimethylformamide (DMF) system and is at the same time transesterified by the SO_3/DMF complex (41). Such transesterified products can be cross-linked by metal ions to form highly effective thickening agents in aqueous media (42). This process has also been developed to produce cellulose sulfate ester with interesting rheological and gel forming properties (43).

Being ionic compounds, cellulose sulfates have ion-exchange properties. They have been recommended for use as cation exchangers (39,44,45). Sodium cellulose sulfates are also known to have blood anticoagulant activity. The correlation of molecular characteristics of this derivative with its anticoagulant activity has been investigated (46). However, it has been reported that cellulose sulfate exhibits certain toxicity (47).

Cellulose Carbamates: Cellulose reacts with isocyanates in anhydrous pyridine or with urea and substituted ureas at relatively high temperature to yield carbamates. The optimum carbamation reaction of microcrystalline cellulose with urea in a dry solid mixture has been studied (48). In addition, a preferentially C-6 modified cellulose carbamate derivative has been obtained (49). Heating of cellulose with thiourea at 180°C yielded cellulose thiocarbamate (50). Heat treatment of cellulose isocyanate products has been utilized for the production of urethanes (51). Treatment of cellulose with urea at temperatures at or above the latter's melting point (where urea decomposes into isocyanic acid and ammonia) has been employed for the production of cellulose carbamates fibers (52). The advantages and disadvantages of using urea as an intermediate for production of fiber have been discussed (53).

Metal chelating amino acid derivatives of cellulose have recently been obtained via modification of cellulose with 2,4-toluene diisocyanate, followed by treatment with amino acid ester derivatives (54,55). Diisocyanates are able to crosslink cellulose chains and/or to yield reactive cellulose isocyanate, depending on the reaction conditions. Sato and his co-workers (56) examined the optimum conditions for the reaction between cellulose and 2,4-toluene-diisocyanate and succeeded in introducing 0.30 mol of free isocyanate group per glucose unit. Cellulose isocyanate was further converted into isothiocyanate (57). This derivative has also been synthesized by condensation of cellulose with 2,4-diisocyanototoluene, followed by hydrolysis and thiophosgene treatment (57).

Cellulose carbamate and its derivatives are able to immobilize enzymes easily with the help of dialdehydes such as dialdehyde starch, glutaraldehyde, and glyoxal (58). Since cellulose triphenylcarbamate or tricarbanilate prepared without degradation showed good solubility in many organic solvents, it was used for determining the DP and DP distribution of cellulose by gel permeation chromatographic analysis (59,60).

Cellulose Acetate: Cellulose acetate is the most important organic ester because of its broad application in fibers and plastics. Cellulose acetate can also be further modified to improve its properties. For example, cellulose diacetate can be perfluoroacylated with straight-chain perfluoroalkanoyl chlorides and with oligo(hexafluoropropene oxides) in the presence of amines, to generate mixed esters having oil and water repellency properties (61). Some mixed cellulose esters, such as cellulose acetate propionate and cellulose acetate butyrate, can be prepared in one step from cellulose with the corresponding acid and acetic anhydride in the presence of sulfuric acid. These mixed cellulose esters find applications as lacquers, plastics, and hot melt coatings (62). Graft copolymerization of cellulose acetate to improve its properties has gained popularity recently (63). Details are discussed in a subsequent section.

Cellulose Phosphate: Cellulose phosphate esters are of considerable interest because of their inherent flame resistance and ion exchange capability. Cellulose phosphates with a low phosphorous content are obtained by reacting cellulose or linters with phosphoric acid in an urea melt (64). Higher phosphorous contents and a lower degradation rate of the cellulose may be obtained with excess urea at higher temperature (130-150°C) for 15 min. The Ban-Flame process (65), one of the first commercially feasible flameproofing procedures for cotton fabric, was based on this method. Water soluble cellulose phosphate with a high degree of substitution may be obtained from a mixture of phosphoric acid and phosphorus pentoxide in an alcoholic medium.

Others: A relatively novel class of derivatives is obtained by the covalent incorporation of organometallic moieties into cellulose. For example, cellulose ferrocenyl derivatives have, for instance, been prepared by esterification of cellulose with an intermediate derived from ferrocene carboxylic acid and triphenyl phosphite in the presence of pyridine (66). An enzymatically-cleavable cellulose ester has been developed (67), and prodrugs have been coupled to the hydroxyl or carboxyl functions of C-terminal aromatic amino acids of cellulose peptide derivatives for controlled release applications (68).

Etherification

In contrast to esterification, etherification is carried out in an alkaline medium, and the etherifying agents are alkyl halides. The general reaction is termed aliphatic, nucleophilic substitution and, employed under normal conditions, it is of the bimolecular type.

Cellulose ethers have gained their positions on the market due to their multifunctional effects. They are either soluble in water or organic solvents, function as thickeners, flow-control agents, suspending aids, protective colloids, water binders, liquid crystals, film formers, or thermoplastics. With these properties, they are used in such diverse industries as food, paint, oil recovery, paper, cosmetics, pharmaceutical, adhesives, printing, agriculture, ceramics, textiles, and building materials (69).

Classical cellulose ethers, such as carboxymethylcellulose (CMC), alkyl cellulose, hydroxyalkyl cellulose will not be discussed here. Readers should refer to classical publications (5,6,8,69).

The reactivity of cellulose towards tri-(p-toluenesulfonyl) methane chloride has been recently examined (70). The tosyl reagent is more reactive than trityl chloride, and the primary hydroxyl position exhibited a 43 times higher reactivity than the secondary hydroxyl groups. The products were used as intermediates in the synthesis of selectively modified cellulose derivatives (70). As mentioned earlier, a high DS, organosoluble trimethylsilyl cellulose has been prepared in DMAc/LiCl (16). The condensation of polysaccharides with triphenylmethyl (trityl) chloride proceeds generally with preference for the primary hydroxyl positions. The tritylation of cellulose occurs initially 58 times faster at the hydroxyl group at C_6 than at either C_2 or C_3 (71).

Cellulose can be modified with organostannane chlorides, such as dibutyl or triphenyl derivatives (72,73), or with organotin halides in the presence of bis(ethylene-diamine) copper(II)-hydroxide (74). Epoxy-activated cellulose was prepared by reacting cellulose acetate fibers with sodium methoxide, followed by reacting it with epichlorohydrin in DMSO. This epoxy-activated cellulose has proved to be a useful intermediate to react with substances containing active hydrogen, such as amine, amino acid or carboxylic acids (75), as shown in Figure 3. Epoxidized cellulose has also been converted into a thiol derivative via reduction of a thiosulphate intermediate (76), and sulfoethyl cellulose has been obtained from sodium chloroethanesulfonate (77).

Modified Cellulose Ethers: Isogai and coworkers (78) recently prepared a series of tri-O-alkyl cellulose ethers using a technique that was originally developed for permethylations, and involves the use of alkyl halides, powdered sodium hydroxide and non-aqueous solvents. Water soluble phosphonomethyl cellulose products have been produced by modification of cellulose ethers with chloromethane phosphonic acid derivatives (79). Low levels of hydrocarbon residues can be incorporated into cellulose ethers, such as hydroxyethylcellulose, to yield high viscosity, water soluble products, which display non-Newtonian behavior at low shear rates (80,81).

Chemical modification of hydroxyethylcellulose or hydroxypropylcellulose with long-chain hydrocarbon alkylating reagents, such as C8-C24 epoxides or halides, has been reported to yield novel water-soluble compositions exhibiting enhanced low shear-rate solution viscosities and polymeric surfactant properties (82,83). Patents have also issued for water-soluble phosphonomethylcellulose and phosphonomethylhydroxyethyl cellulose (84,85).

The preparation of predominantly O-6-substituted carboxymethyl cellulose can be achieved in a homogeneous solution, using the N-methylmorpholine-N-oxide/ DMSO solvent system (86).

The incorporation of mercury into cellulose has been accomplished via treatment of cellulose aniline ether derivatives with mercuric acetate (87). Arsenic-containing cellulose derivatives have been obtained from sodium arsenate and diazotized cellulose precursors (88). Platinum-containing polysaccharide derivatives have also been reported (89).

Treatment of cyanoethylcellulose with borane-dimethyl sulfide or borane-tetrahydrofuran complexes in tetrahydrofuran has resulted in the quantitative conversion into 3-aminopropyl cellulose. Such aminopropyl cellulose derivative have also been employed as intermediates for acetamido- or aryluredo-products, and in grafting reactions (90).

Deoxycellulose

Due to the practical importance of deoxycellulose, the halogenation of cellulose has been the subject of many studies which were reviewed by Gal'braikh and Rogovin (22), Vigo (91) and Ishizu (92).

Deoxycelluloses denote anhydroglucose units in which the three hydroxyl groups are partially or completely replaced by other functional groups which do not contain an oxygen atom attached to the ring carbon atom. The possibility of synthesizing deoxycellulose was first demonstrated by Shorygin and Makarov-Zemlyanskaya during an investigation of the cleavage of cellulose ethers by solutions of metallic sodium in liquid ammonia (22). Today, many reactions have been used to prepare deoxycelluloses. Of these, halogenation is the most popular one.

Halodeoxycellulose: Chlorodeoxycellulose has been the most widely studied halodeoxycellulose. This derivative has useful properties such as resistance to flame and rotting. Slight fluorination increases oil resistance and lowers the soiling potential of cellulose fibers. It has also been employed as an intermediate in the preparation of many cellulose derivatives, as shown in Figure 4. The most widely used method for preparing halodeoxycellulose is the nucleophilic displacement of good leaving groups by halides in various solvents. The most frequently employed leaving groups are tosylate and mesylate. Other leaving groups, such as nitrate, and N,N-dimethylformamide from a formiminium salt intermediate, have also been used. Recently, sulfuryl chloride (SO_2Cl_2)-pyridine system was also used (93). Chlorination with SO_2Cl_2 and pyridine proceeded in parallel at C_6 and C_3 in a heterogeneous system. Homogeneous LiCl-dimethyl-acetamide system was also attempted by Ishizu (92).

Carboxyl-substituted aminodeoxycellulose (Cell-CH_2NH-R-CO_2H) prepared by the reaction of amino acids with chlorodeoxycellulose adsorbed various heavy metal ions with high efficiency (94). Cellulose isocyanate reacted with amino acids or their esters in DMSO at low temperature to yield cellulose derivatives containing amino acid residues (56). These derivatives adsorbed various kinds of metal ions in their free acid state with relatively high adsorption values for Cu(II) and Fe(III) (95,96). The derivative containing cysteine was especially effective for the adsorption of Hg(II) ions (96).

Iododeoxycellulose: Since iodide ion is a strong nucleophile, the action of iodides on cellulose tosylate, mesylate and nitrate in ketones, and chlorodeoxycellulose in DMF or 2,5-hexanedione can generate iododeoxycellulose. In use of chlorodeoxycellulose, the chlorine substituents were almost completely replaced by iodide in 2,5-hexanedione. The iododeoxycellulose thus prepared was 6-dexoy-6-iodocellulose. Iododeoxycelluloses having iodo-substituents at C_2 and C_3 were also prepared by treating various sulfonates having a DS higher than 1.0 or 6-O-trityl-2(3)-p-nitrobenzenesulfonate of cellulose with sodium iodide in DMF. Ishii (97,98) succeeded in the almost quantitative preparation of 5,6-cellulosene acetate by

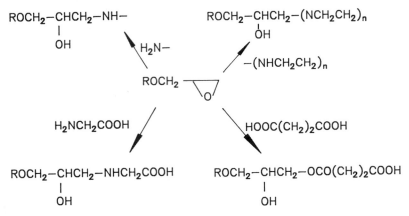

Figure 3. Reactions of epoxy-activated cellulose. R denotes cellulose.

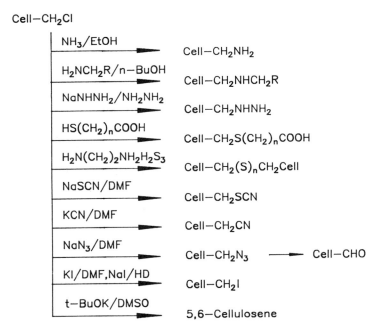

Figure 4. Reactions of chlorodeoxycellulose. Cell denotes cellulose.

the treatment of acetylated 6-deoxy-6-iodocellulose (DS 0.8) with 1,8-diazobicyclo[5,4,0]undec-7-ene in DMF.

Novel cellulose derivatives containing viologen were prepared from a halogenated cellulose or tosylated cellulose. These products are expected to be a good polymeric electron carrier because of forming a stable hydrophilic membrane as compared with other polymers (99). Deoxy(thiocyanate)cellulose fabrics exhibited moderate antibacterial activity (100).

Block and Graft Copolymers

Whereas block copolymers are linear chains formed by introducing active sites in the terminal units, graft copolymers are their branched equivalents where the active site is included on an internal monomer unit.

Two block Copolymers of triemthylcellulose-[b-poly(oxyethyl-ene)] have been reported (101). The trimethylcellulose-blocks containing one α - chloro ether end were treated with silver hexafluoroantimonate, $AgSbF_6$, in THF solution between -10°c and +23°C to facilitate a living cationic THF polymerization of poly(oxytetramethylene) blocks, as shown in Figure 5.

Monofunctional 1-hydroxy cellulose triesters, such as tributyrate and propionate acetate derivatives, have been coupled to bis-(4-isocyanotophenyl)-disulfide to obtain macroinitiators for the radical syntheses of three block copolymers of the type Cellulose-Initiator-Initiator-Cellulose (102).

A large number of graft copolymers of cellulose have been prepared, and although there has been little commercial exploitation, there is considerable interest from the view of modification of cellulose. Three methods have been commonly used for graft copolymerization, namely, radical polymerization, ionic polymerization, and condensation and ring opening copolymerization. Of these, radical modification is the most popular one. However, problems of using this method frequently occur. There are some formidable challenges facing those engaged in the synthetic aspects of cellulose grafting if viable commercial processes are to be developed. Stannett (103) pointed out that the following factors must be considered:

1. The elimination, or at least minimization, of concurrent homopolymer formation.
2. The involvement of all or most of the cellulose molecules in the grafting process.
3. Better control of the molecular weights and molecular weight distribution of the grafted side chains.
4. Better reproducibility of the grafting yields, properties and other features of the graft copolymers.

In addition, problems of using radical polymerization approaches have been addressed (104):

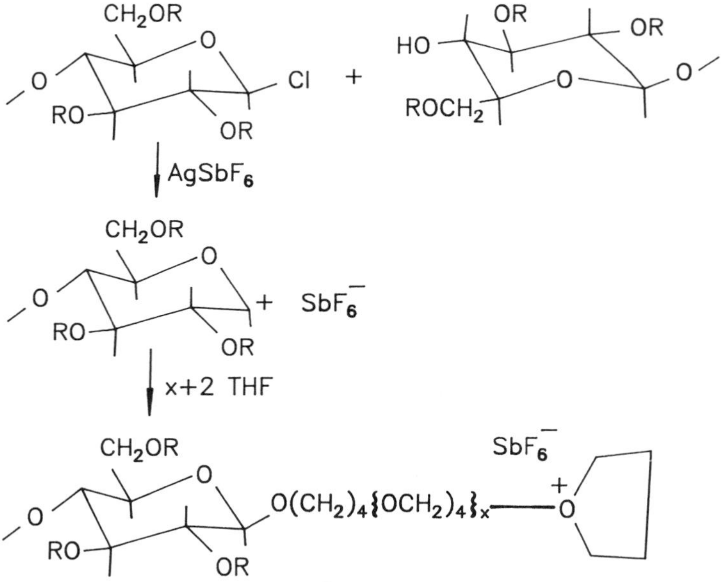

Figure 5. Synthesis of trimethylcellulose-[b-poly(oxyethylene)] block copolymer starting with an α-chloroether end of the trimethylcellulose.

1. It is difficult to control or change the molecular weight of the side chain grafts.
2. The molecular weight of the grafts are very high and very polydisperse.
3. Grafting of only a few high molecular weight chains occurs.
4. Considerable amounts of homopolymer are sometimes formed, some of which may get embedded in the cellulose matrix and are difficult to remove.
5. No knowledge exists of the nature of the linkage between the cellulose backbone and the graft, nor the ability to control it.
6. Reproducibility and control of the grafting yields, properties and other features of the graft copolymers are not realized.

In spite of these problems, many graft copolymers of cellulose are continued to appear using the radical initiation method. Recently, grafted products of cellulose exhibited thermoplastic properties have also been reported (105).

Graft polymerization of 2-methyl-5-vinyltetrazole onto cellulose fiber decreased the temperature for initial oxidative thermal degradation and increased the activation energy of degradation compared with that for the initial cellulose (106). It was reported that the oxidative thermal degradation of the grafted fiber depended on the amount of grafted polymer and on the structure of the cellulose substrate (cotton cellulose, rayon, etc.).

UV or γ radiation-induced grafting of cellulose films with styrene in methanol solution was significantly enhanced in the presence of H_2SO_4, lithium salts, urea and other organic compounds and polyfunctional monomers (107). In the graft polymerization of methacryloyllupinine onto cotton cellulose in the presence of $Ce(NH_4)_2(NO_3)_6$, the induction period of methacryloyllupinine polymerization increased with decreasing temperature and was higher in air than in a vacuum.

Vinyl acetylene copolymers, derived from grafting of dimethylvinylethyl carbinol onto cellulose, afford metal containing polymer derivatives on treatment with copper or silver salts. The preparation of cation-exchange cellulose was examined by grafting of acrylic acid (108) and itaconic acid (109). Wood cellulose grafted with poly(acrylic acid) or 2-arcylamido-2-methylpropane-sulfonic acid were reacted with hexadecyltrimethylammonium bromide to improve its hydrophobicity in order to utilize the graft copolymers as an oil absorbent. This system was developed by Fanta and his coworkers (110).

Scoured cotton was modified chemically to dialdehyde cellulose and hydrazinated dialdehyde cellulose by treating cellulose successively with aq. $NaIO_4$, N_2H_4 and H_2O. Vinyl grafted cellulose was subsequently prepared by graft copolymerization of methyl methacrylate on hydrazinated dialdehyde cellulose using $K_2S_2O_3$ as the catalysts in a limited aq. system. SEM studies of cellulose, dialdehyde cellulose and the vinyl grafted fibers were made to get an idea of the changes brought about in the surface morphology of cellulose by different types and degrees of chemical modifications. The orientational patterns of the fibrils of cellulose get somewhat blurred on oxidation and the fiber surface suffers some damage on hydrazination. Vinyl grafting causes deposition of vinyl polymer on the fiber surface thus leading to further damage and masking of the fibrils and their orientation pattern (111).

Modification of cellulose derivatives via graft copolymerization reaction has gradually gained popularity. An excellent source covering grafting on chemically modified celluloses up to 1980 was reviewed by Hebeish and Guthrie (64). Several new graft copolymers of cellulose derivatives have been developed since then.

It was reported that grafting levels of acrylonitrile on allylic-modified bleached cellulosic material were higher than graft polymer of acrylonitrile on the cellulosic material (112).

Poly(glutamate) and poly(glutamine) have been grafted onto cellulose or its acetate to afford blood compatible polymers. Single amino acid residues have been grafted onto cellulose acetate (121). Vinyl acetate have been grafted onto cellulose acetate and ethylene oxide-modified wood cellulose (63). Dextran has been grafted onto carboxymethyl cellulose (113).

UV graft polymerization of arylamide onto cellulose acetate reverse osmosis membranes yielded grafted membranes with higher salt retention and lower water flux compared with pristine cellulose acetate (114). Acid-catalyzed grafting of styrene on cellulose acetate reverse-osmosis membranes imparted higher salt-rejection rate (92.4%) to the membrane compared with those of ungrafted membranes (80.8%) and heat-shrinked membranes (90.2%) (115).

Cellulose acetate butyrate and hydrolyzable silanes were reacted at 80°C for 10 hr to give a silyl group modified CAB for coating application (116).

Kinetic data for the graft polymerization of acrylamide with hydroxyethyl cellulose in the presence of Co(III)-cyclohexanol redox initiating system are available (117).

Several graft copolymers were prepared based on anionic polymerization method to overcome some of the major problems encountered in radical-initiated grafting. This method is used to prepare a living synthetic polymer with mono- or di-carbanions to react with modified cellulosic substrates under homogenous conditions. For example, polyacrylonitrile carbanion was prepared to react with cellulose acetate to generate a cellulose acetate-polyacrylonitrile graft polymer (104).

Well-defined tailored cellulose-styrene graft copolymers were also prepared by anionic polymerization. Preliminary bonding studies showed that these graft copolymers could function effectively as compatibilizers or interfacial agents to bond hydrophobic polystyrene to wood, evolving into a new class of composites (118). Likewise, polystyrene was grafted onto cellulose with precise control over molecular weights and narrow molecular weight distribution. Crosslinked cellulose grafted copolymers grafted with exactly defined polymer chain segments between crosslink points have been synthesized (104).

Another unconventional technique of grafting with success is to graft a polymer with terminating functional groups which are capable of reacting with hydroxyl groups in cellulose. For example, reaction of anhydride-terminated polyisobutylene with sodium cellulosate gives polyisobutylene-grafted cellulose products in high graft yields (110%) and grafting efficiency (119). Likewise, cellulose is reacted with polybutadiene bearing succinic anhydride groups in dimethylformamide and N,N-dimethylbenzylamine between 90° and 150°C through esterification. The grafted copolymer possesses high tack which is suitable for surgical adhesive and skin barrier applications (120).

Future Trends

For the past ninety years, efforts were made to produce industrial products of cellulose with specific properties by means of chemical modifications. Today this activity is still continued to receive worldwide attention as evidenced by many papers presented in various conferences, voluminous patents and journal literature.

The advent of energy crisis, shortage of raw materials and the concern of environment have certainly drawn more attention to cellulose research and development. On the other hand, with the chemical industries coming under federal and state regulations on discharges of air, water and solid wastes, much attention has to be directed to improve the effectiveness and efficiency of existing technologies, and to produce cellulose derivatives which are environmentally acceptable. They must be low in cost, and, to a greater extent, reusable and recyclable.

It is suspected that more cellulose derivatives will be produced in solution or homogeneous systems. The success of this system has to be depended heavily on non-polluting, solvents-recyclable processes. An area of future expansion of research and development should be in the area of biomedical applications of cellulose derivatives.

The future of cellulose looks good. The years ahead offer great challenges but also great rewards for chemists and material scientists with the vision to take advantage of cellulose greatest asset -- their renewability, versatility and adaptability.

Literature Cited

1. Hon, D.N.-S. *POLYMERNEWS* **1988**, *13*, 134.
2. Hon, D.N.-S. *POLYMERNEWS* **1991**, in press.
3. Payen, A. *Compt. Rend.* **1839**, *8*, 51.
4. Ward, K., Jr.; Morak, A. J. *Reaction of Cellulose* in *Chemical Reactions of Polymers*; Fettes, S.M., Ed.; Wiley-Interscience: New York, **1964**, p 321.

5. *Cellulose and Cellulose Derivatives*; Ott, E.; Spurlin, H.M.; Grafflin, M. W., Eds.; Interscience: New York, **1954**; Parts I - III.
6. *Cellulose and Cellulose Derivatives*; Bikales, N.M.; Segal, L., Eds.; Wiley-Interscience: New York, **1971**, Vol. V; Part V, Chapter XVII.
7. Hon, D.N.-S. *Chemical Modifications of Lignocellulosic Materials: An Overview*; Paper presented at The 1989 International Chemical Congress of Pacific Basin Societies, Honolulu, Hawaii, **December 17-22, 1989**.
8. *Cellulose Chemistry and Its Applications*; Nevell, J.P.; Zeronian, S.H. Eds.; Ellis Horwood: Chichester, **1985**.
9. *Cellulose and Its Derivatives*; Kennedy, J.F.; Phillips, G.O.; Wedlock, D.J.; Williams, P.A. Eds.; Ellis Horwood, Chichester, **1985**.
10. Fengel, D.; Wegener, G. *Wood: Chemistry, Ultrastructure, Reactions*; Walter de Gruyter, Berlin, NY, **1984**; pp 482-525.
11. *Wood and Cellulosic Chemistry*; Hon, D.N.-S.; Shiraishi, N. Eds.; Marcel Dekker: New York, **1990**.
12. Heuser, E. *Textile Res. J.* **1950**, *20*, 828.
13. Hudson, S.M.; Cuculo, J.A. *J. Macromol. Sci. Rev. Macromol. Chem.* **1980**, *C18*, 1.
14. Schempp, W.; Krause, Th.; Seilfried, U.; Koura, A. *Papier* **1984**, *38*, 607.
15. Ishii, T.; Ishizu, A.; Nakano, J. *J. Carbohydr. Res.* **1977**, *21*, 2085.
16. Nakao, O.; Yamazaki, S.; Amano, T. *Japan Patent* **1972**, 39,951.
17. Bose, J.L.; Roberts, E.J.; Rowland, S.P. *J. Appl. Polym. Sci.* **1971**, *15*, 2999-3007.
18. Schroeder, L.R.; Gentile, V.M.; Atalla, R.H. *J. Wood Chem. Technol.* **1986**, *6*, 1-14.
19. Braconnot, H. *Ann.* **1833**, *1*, 242,245.
20. Mark, H.F. *A Century of Polymer Science and Technology* in *Applied Polymer Science* Tess, R.W.; Poehlein, G.W., Eds., ACS Symposium Series 285, Washington, D.C., **1985**, p 4.
21. Hon, D.N.-S. *Cellulose: Chemistry and Technology* in *Concise Encyclopedia of Wood & Wood-Based Materials* Schniewind, A.P. Ed.; Pergamon Press: **1989**, pp 39-44.
22. Leoni, R.; Baldini, A. *Carbohydr. Polym.* **1982**, *2*, 298.
23. Gal'braikh, L.S.; Z.A. Rogovin, *Derivatives with Unusual Functional Groups* in *Cellulose and Cellulose Derivatives*; Wiley-Interscience: NY, **1971**; pp 894-901.
24. Perlin, A.S. *The Carbohydrates, Chemistry and Biochemistry*; Vol. 1b, Pigman, W. and Horton, D., Eds.; **1980**, pp 1167-1215.
25. Mehltretter, C.L. *Starch/Stärke* **1983**, *37*, 294-297.
26. Maekawa, E.; Koshimima, T. *J. Appl. Polym. Sci.* **1984**, *29*, 2289.
27. Morooka, T.; Norimoto, M.; Yamada, T. *Some Physical Properties of Cellulose Derivatives Prepared by Homogeneous Periodate Oxidation* in *Cellulose and Wood: Chemistry and Technology*; Schuerch, C. Ed.; John Wiley & Sons: New York, **1988**, p 1103.

28. Dimitrov, D.G.; Dimitrov, C.D.; Tsankova, C.D. *Cellulose Chem. Technol.* **1982**, *16*, 19.
29. Bredereck, K.; *Tetrahedron Lett.* **1967**, pp 695-698.
30. Baize, C.; Defaye, J.; Gadelle, A.; Wong, C.C.; Pederson, C. *J. Chem. Soc. Perkin Trans. I* **1982**, pp 1579-1585.
31. Ikeda, I.; Kurata S.; Suzuki, K. *Cellulose: Structural and Functional Aspects*; Kennedy, J.F.; Phillips, G.O.; Williams, P.A. Eds.; Ellis Horwood: **1990**, p 219.
32. Phillip, B.; Wegenknecht, W. *Cellulose Chem. Technol.* **1974**, *14*, 13.
33. Diehl, F. *U.S. Patent* **1974**, 3,794,605.
34. Dodwell, G.M. *Ger. Offen.* **1974**, 2,348,409.
35. Schweiger, R.G. *U.S. Patent* **1973**, 3,726,796.
36. Schweiger, R.G. *U.S. Patent* **1972**, 3,637,520.
37. Bischoff, K.H.; Dantzenberg, H. *Ger. Offen.* 112,456.
38. Turbak, A.F.; Burke, N.I. *U.S. Patent* **1974**, 3,886,295.
39. Schweiger, R.G.; *Carbohydr. Res.* **1972**, *21*, 219.
40. Schweiger, R.G.; *U.S. Patent* **1972**, 3,639,655.
41. Akelah, A.; Sherrington, D.C. *J. Appl. Polym. Sci.* **1981**, *26*, 3377.
42. Kamel, M.; Hebeish, A.; Allam, M.; Al-Aref, A. *J. Appl. Polym. Sci.* **1973**, *17*, 2725.
43. Zhdanova, Y.P.; Rogovin, Z.A.; Sletkina, L.S.; Baibikov, A.; Kashkin, A.V. *Cellulose Chem. Technol.* **1980**, *14*, 623.
44. Isogai, A.; Ishizu, A.; Nakano, J. *J. Appl. Polym. Sci.* **1984**, *29*, 3873.
45. Traskman, B.; Tammela, V. *J. Appl. Polym. Sci.* **1986**, *31*, 2043.
46. Kamide, K.; Okajima, K.; Matsui, T.; Ohnishi, M.; Kobayashi, H. *Polym. J.* **1983**, *15*, 309.
47. Kamide, K.; Okajima, K.; Matsui, T.; Ohnishi, M.; Kobayashi, H. *Polym. J.* **1983**, *15*, 309.
48. Nozawa, Y.; Higashide, F.; *J. Appl. Polym. Sci.* **1981**, *26*, 2103.
49. Ekman, K.; Eklund, V.; Fors, J.; Huttunen, J.I.; Selin, J.-F.; Turunen, O.T. in *Cellulose, Structure, Modification and Hydrolysis*; Young, R.A.; Rowell, R.M. Eds.; Wiley Interscience: NY, **1986**, pp 131-148.
50. Nagieb, A.Z.; El-Gammel, A.A. *J. Appl. Polym. Sci.* **1986**, *31*, 179.
51. Wadeson, F.P. *United Kingdom Patent* **1983**, 2,112,791.
52. Selin, J.-F.; Huttunen, J.; Turunen, O.; Fors, J.; Eklund, V.; Ekman, K.U. *U.S. Patent* **1985**, 4,530,999.
53. Struszczyk, H. *Wlokna Chem.* **1988**, *14*, 367.
54. Sato, T.; Karatsu, K. *Sen-i-Gakkaishi* **1983**, *39*, T519-524.
55. Sato, T.; Motomura, S.; Karatsu, K. *Sen-i-Gakkaishi* **1985**, *41*, T235-T240.
56. Sato, T.; Karatsu, K.; Kitamura, H.; Ohno,Y. *Kobunshi Ronbunshu* **1982**, *39*, 699.
57. Gemeiner, P.; Augustin, J.; Drobnica, L. *Carbohydr. Res.* **1977**, *53*, 217.
58. Nozawa, Y.; Hasegawa, M.; Higashide, F. *Kobunshi Ronbunshu* **1981**, *38*, 39.

59. Danbelka, J.; Kossler, I.; *J. Polym. Sci. Polym. Chem. Ed.* **1976**, *14*, 287.
60. Schroeder, L.R.; Haigh, F.C. *Tappi* **1979**, *62*, 103.
61. Ishikawa, N.; Matsuhisa, H. *Nippon Kagaku kaishi* **1985**, 1247.
62. Bogan, R.T.; Brewer, R.J. *Encyclopedia of Polymer Science and Engineering*; Mark, H.; Bikales, M., Eds; John Wiley: **1985**, Vol. 3; 2nd ed., pp 164-175.
63. Hebeish, A.; Guthrie, J.T. *The Chemistry and Technology of Cellulosic Copolymers*; Springer-Verlag: Berlin, Heidelberg, NY, **1981**; p 351.
64. Rogovin, Z. A. *Chem. Abstract* **1966**, *65*, 15646c.
65. *British Patent* **1948**, 604,197.
66. Simionescu, C.; Lixandru, T.; Tatara, L.; Mazilu, I.; Vata, M.; Sentaru, D. in *Metal-containing Polymeric Systems*; Sheats, J.E.; Carraher, C.E.; Pittman, C.U. Eds., Plenum Press: NY, **1983**, pp 69-81.
67. Klemm, D.; Geschwend, G.; Hartman, M. *Angew. Makromol. Chem.*, **1984**, *126*, 59-65.
68. Lapicque, F.; Dellacherie, E. *J. Contr. Release* **1986**, *4*, 39-45.
69. Greminger, G.K., Jr.; Krumel, K.L. *Alkyl and Hydroalkylalkylcellulose* in *Handbook of Water-Soluble Gums and Resins*; Davison, R.L. Ed.; McGraw-Hill: New York, **1980**, Chapter 3.
70. Kaifu, K.; Nishi, N.; Komai, T. *J. Polym. Chem., Polym. Chem. Ed.* **1981**, *19*, 2361-2363.
71. Green, J.W.; *Methods Carbohydr. Chem.* **1963**, *3*, 327-331.
72. Carraher, C.E.; Gehrke, T.; Giron, D.; Cerutis, D.; Molloy, H.M.J. *Macromol. Sci.-Chem.* **1983**, *A19*, 1121.
73. Carraher, C.E.; Gehrke, T.J. *Modification of Polymers*; Carraher, C.E.; Moore, J.A., Eds.; Plenum Press: New York, **1981**, pp 229-245.
74. Naoshima, Y.; Carraher, C.E.; Hess, G.G.; Kurokawa, M.; Hirorno, S. *Okayama Riga Daigaku Kiyo*, A **1984**, *20*, 33-41, 1985.
75. Ikeda, I.; Tomita, H.; Suzuki, K. *Sen-i Gakkaishi*, **1990**, *46*, 63.
76. Gemeiner, P.; Benes, M. *Coll. Czech. Chem. Commun.* **1983**, *48*, 267-278.
77. Vaughanm, C.L.P. *U.S. Patent* **1952**, 2,591,748.
78. Isogai, A.; Ishizu, A.; Nakano, J. *J. Appl. Polym. Sci.*, **1986**, *31*, 341-352.
79. Brandt, L.; Holst, A. A., *Canad. Patent* **1986**, 1,174,234; Brandt, L.; Holst, A.U. *U.S. Patent* **1983**, 4,396,433.
80. Landoll, L.M. *U.S. Patent* **1981**, 4,243,802.
81. Gelman, R.A.; Barth, H.G. *ACS Symp. Ser.*, **1986**, *235*, 101-110.
82. Landoll, L.M. *U.S. Patent* **1981**, 4,243,802.
83. Landoll, L.M. *J. Polym. Sci. Polym. Chem. Ed.* **1982**, *20*, 443.
84. Brandt, L.; Holst, A. *U.S. Patent* **1983**, 4,379,918.
85. Brandt, L.; Holst, A. *U.S. Patent* **1983**, 4,396,433.
86. Nicholson, M.D.; Johnson, D.C.; Haigh, F.C. *Appl. Polym. Symp.* **1976**, *28*, 931.
87. Rogovin, Z.A. *Vysokomol. Soed.*, **1965**, *7*, 1314.

88. Achwal, W.B. *Colourage,* **September 23, 1971,** 32-34.
89. Ceskoslovenska Akademie VED, *GB Patent,* **1986,** 2,168,063A.
90. Daly, W.H.; Caldwell, J.D.; Kien, V.P.; Tang, R. *Polym. Prepr.* **1982,** *23,* 145-146.
91. Vigo, T. *Cellulose Derivatives* in *Encyclopedia of Polymer Science and Engineering* John Wiley & Sons: NY, **1985,** 2nd edition.
92. Ishizu, A. *Chemical Modification of Cellulose* in *Wood and Cellulosic Chemistry;* Hon, D.N.-S; Shiraishi, N. Eds.; Mercer Dekker: **1990,** Chapter 16, pp 757-800.
93. Krylova, R.G.; Usov, A.I.; Shashkov, A.S. *Biorgan. Khim.,* **1981,** *7,* 86.
94. Sato, T.; Ohno, Y. *Proc. Int. Symp. Fiber Sci. Technol.,* **1985,** 287.
95. Sato, T.; Karatsu, K.; Kitarmura, H.; Ohno, Y. *Kobunshi Ronbunshin* **1983,** *39,* T-519.
96. Sato, T.; Motomura, S.; Ohno,Y. *Kobinnshi Ronbunshu,* **1985,** *41,* T-235.
97. Ishii, T. *Carbohydr. Res.* **1986,** *154,* 63.
98. Ishii, T. *Proc. Int. Symp. Fiber Sci. Technol.* Hakone, Japan Aug. 20-24, 285, **1985** (ABIPC, 57, 75, 1986). Ishii,T. *Carbohydr. Res.* **1986,** *154,* 63-70.
99. Sato,T.; Nambu, Y.; Endo, T. *Preparation and Redox Behavior of Cellulose Derivatives Having Viologen Moiety* in *Cellulose and Wood: Chemistry and Technology;* Schuerch, C., Ed.; John Wiley & Sons: NY, **1988,** p 1119.
100. Vigo, T.L.; Danna, G.F.; Welch, C.M. *Carbohydr. Res.,* **1975,** *44,* 45.
101. Feger C.; Cantow, H.-J. *Polym. Bull.,* **1980,** *3,* 407.
102. Mezger, T.; Cantow,H.-J. *Angew. Makromol. Chem.* **1983,** *116,* 13.
103. Stannett, V. *Some Challenges in Grafting to Cellulose and Cellulose Derivatives* in *Graft Copolymerization of Lignocellulosic Materials;* Hon, D.N.-S., Ed.; ACS Symposium Series; **1982,** *187,* 1-20.
104. Narayan, R. *Synthesis of Controlled Cellulose - Synthetic Polymer Graft Copolymer Structures* in *Cellulose and Wood: Chemistry and Technology;* Schuerch, C., Ed.; John Wiley & Sons: NY, **1988,** p 945.
105. Shiraishi, N. *Chemtech* **June 1983,** 366.
106. Vysotskaya, E.P.; Gal'braikh, J.S.; Andreeva, I.N.; Kuznetsova, S. Yu; Fronchek, E.V. *Khim. Drev.* **1989,** *3,* 24.
107. Dworjanyn, P.A.; Fields, B.; Garnett, J.L. *ACS Symp. Ser.* **1989,** *381,* 112.
108. Simionescu, C.I.; Brandisteanu, S.; Rumega-Stoianovici, D.G. *Cellulose Chem. Technol.* **1983,** *17,* 687.
109. Dimov, K.; Dimitrov, D.; Terlemezian, E.; Semkova, M. Bandova, M. *Cellulose Chem. Technol.* **1980,** *14,* 665.
110. Fanta, G.F.; Burr, R.C.; Doane, W.M. *Oil Absorbency of Graft Copolymers from Softwood Pulp* in *Renewable-Resource Materials: New Polymer Sources;* Carraher, C.E., Jr. Sperling, L.H., Eds.; Plenum Press: NY, **1986,** p 107.

111. Ghosh, P.; Dalal, J.C. *Indian J. Technol.*, **1989**, *27*, 189.
112. Agbonlahor, F.O.; Ejike, E.N.; Gbinije, A.; Okieimen, F.E.; Otaigbe, J.U. *Acta Polym.* **1989**, *40*, 723.
113. Sikkhema, D.J.; *J. Appl. Polym. Sci.*; **1985**, *30*, 3523.
114. Yan, W.; Yang P.; Wang, Y. *Shuichuli Jishu* **1988**, *12*, 213.
115. Yan, W.; Yang, P.; Zhang, X.; Zhang, H. *Mo Xexue Yu Jishu* **1987**, *7*, 38.
116. Yamakado, N.; Kawakami, S.; Hata, H. *Japan Kokai Tokkyo Koho JP* 63,254,101.
117. Nemchinov, I.A.; Molotkov, V.A.; Kurlyankina, Vysokomol, V.I. *Soedin, Ser. A.* **1989**, *31*, 123.
118. Narayan, R.; Biermann, C.J.; Hunt, M.O.; Horn, D.P. *ACS Symp. Ser.* **1989**, 385,337.
119. Coleman-Kammula, S.; Hulskers, H. *The Preparation and Characterization of Polyisobutylene-grafted Cellulose Pulp* in *Cellulose and Cellulosics*; Kennedy, J.F.; Phillips G.O.; Williams, P.A., Eds.; Ellis Horwood: Chichester, England, **1987**, pp 195-202.
120. Hon, D.N.-S.; Xing, L.-M. *Cellulose-Polybutadiene Copolymer: Properties and Applications* Paper presented at the 200th ACS National Meeting, Cellulose, Paper and Textile Division, Washington, D.C., **1990**. (Abstracts CELL-15).

RECEIVED May 29, 1991

Chapter 13

Emerging Polymeric Materials Based on Starch

William M. Doane, Charles L. Swanson, and George F. Fanta

Plant Polymer Research, National Center for Agricultural Utilization Research, U.S. Department of Agriculture, Agricultural Research Service, 1815 North University Street, Peoria, IL 61604

Interest in natural products as annually renewable raw materials for industry has greatly intensified, especially during the last fifteen years. Although much of this interest can be attributed to the oil embargo of the early 1970s, the increased abundance of agricultural production beyond available markets has generated an oversupply of many commodities and with it an increased interest in such commodities as raw materials for industry to develop new and expanded markets.

Plant and animal materials have long served beyond food and feed needs in special industrial markets. With the advent of the petrochemical industry during the last fifty years, some of the traditional markets served by materials of agricultural origin were replaced by petrochemical synthetics. This situation continues today due largely to the vast array of synthetic materials that can be produced with excellent properties and satisfactory economics.

Now there is a perception in many countries that greater utilization of products from plants, animals and microbes from land and sea can and should play a more significant role in meeting society's needs for a broad range of industrial materials. Perceptions include improved economies through increased processing of domestically produced commodities, reduction of imported oil, and products from natural sources that are environmentally more acceptable.

Starch is one of the natural materials that is receiving considerable attention in this renewable resources scenario. In answer to the question: why is starch of major interest as a renewable material, one might answer that starch is one of the most abundant materials produced in nature, is easily recovered from plant organs holding it, is relatively low in cost and is readily converted chemically, physically and biologically into useful low molecular weight compounds or high molecular weight polymerics.

The mention of firm names or trade products does not imply that they are endorsed or recommended by the U.S. Department of Agriculture over other firms or similar products not mentioned.

This chapter not subject to U.S. copyright
Published 1992 American Chemical Society

Starch: Occurrence, Composition, Properties, Uses

What is starch? Where is it found? What are its properties? How is it/can it be used as an industrial material? These and other related questions have been addressed quite thoroughly in several publications (1-4). For the purpose of this report, which is to consider starch as a source of new polymeric materials, we will review only briefly some of the fundamental information on starch.

Starch is the name given to the major food reserve polysaccharide produced by photosynthetic plants. It occurs in various plant tissues as discrete granules with a size and shape characteristic of the source. The granules may vary in size, depending on the source, from a few microns to fifty or more microns. Insolubility of the starch granule in cold water facilitates its recovery from plant tissue in rather pure form. Although starch occurs in many plant tissues, commercially it mostly is recovered from seeds, roots and tubers.

In the United States, cereal grains, predominately corn, provide the major source of starch. The average corn crop contains in excess of 300 billion pounds of starch with only about 15% of the crop being processed to separate the starch or starch-protein (flour) matrix from the corn kernels. The corn processing industry is expanding, doubling the amount processed during the last decade, and has both the interest and capability to further expand as market opportunities increase.

Starch granules do not contain starch as a well-defined homogeneous polymer. Rather, the granules contain starch most often as a mixture of two polysaccharides differing in structure and molecular dispersity. Starch is characterized as a mixture of a predominately linear α-(1——>4)-glucan, termed amylose, and a highly branched α-(1——>4)-glucan with branch points occurring through α-(1——>6) linkages, termed amylopectin (Figure 1). The amylose molecules have a molecular weight of approximately 1 million, whereas the molecular weight of the amylopectin molecules may be on the order of 10 million or more. Depending on the source, the two components are present in quite varied ratios. Some sources of starch contain almost none of the linear component, while others contain only small amounts of the branched polysaccharide. Typically, starches contain 20-30% of the linear amylose fraction.

Although starch is highly hydrophilic, the granules do not dissolve in ambient temperature water due to the ordered arrangement of molecules within the granule. Segments of molecules are so arranged as to give rise to apparent crystallinity within the granule. This insolubility not only allows for facile recovery from plant organs, it also allows for chemical modification of the starch molecules without disruption of the granule, facilitating recovery of the modified product. Recovery of product becomes more difficult when granules are disrupted and starch molecules become more soluble.

When granules are heated in water, they swell and lose their ordered arrangement, a process known as gelatinization. Rupturing of the granules releases the individual amylose and amylopectin molecules, which can become completely soluble at a temperature of 130-150°C. Gelatinization of starch can be carried out at low

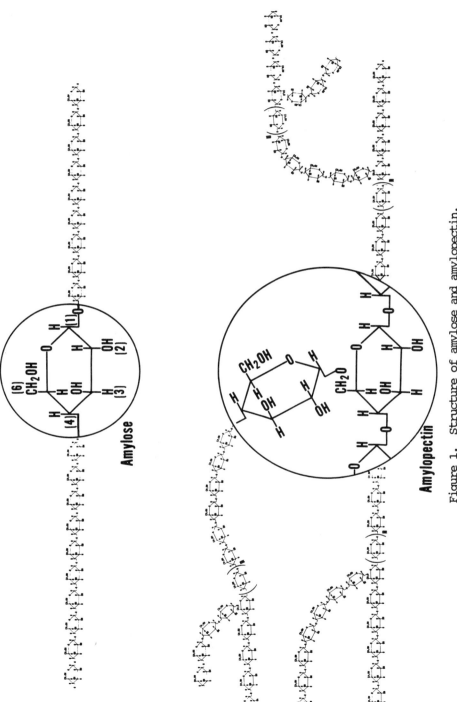

Figure 1. Structure of amylose and amylopectin.

temperature by treatment with alkali or other reagents to disrupt the hydrogen bonds that give rise to the crystallinity. Solubilization of the starch molecules allows for access of the entire molecule to chemical or enzymatic conversion. Also, it allows for separation of the amylose from the amylopectin, most often accomplished through complexing of amylose with butanol or thymol, which causes precipitation of the complex.

In the U.S., about 5 billion pounds of starch or flour are provided for industrial (non-food) uses. This does not include the starch in nearly 365 million bushels of corn converted into ethanol in 1989. Although a variety of industrial markets are served by starch due to its inherent adhesive and film forming properties, the dominant use for starch is in paper making applications. About three-and-a-half billion pounds are used in the paper, paperboard, and related industries, where starch serves a variety of adhesive functions (Table I).

Table I. Starch Adhesive Applications

Paper	
Surface Sizing (size press 1.2×10^9) (wet end 0.4×10^9)	1.6×10^9 lb
Pigment Bonding	0.6×10^9 lb
Corrugating Board	0.9×10^9 lb
Textiles	0.15×10^9 lb
Miscellaneous (bags, cartons, labels, envelopes, briquettes)	0.3×10^9 lb

Smaller but significant amounts are used in other operations, such as the textile, oil recovery and mining industries. In these applications, starch is used in native form or after partial acid or enzyme hydrolysis, oxidation, esterification, etherification or cross-linking. Appropriate selection of chemical reagent allows for introduction of anionic or cationic charge into the starch molecules. The modification of starch, mostly by classical methods, and the properties and uses of such modified starches have been recently reviewed (5).

Starch: Role In Biodegradable Plastics

In about the mid-1980s there began to appear many articles and commentaries, especially in the popular press, on the need to develop biodegradable polymers to replace plastics. It was perceived (often stated) that use of such biodegradables would greatly lessen, if not solve, the solid waste disposal problem in landfills. These various reports were followed by considerable legislative activity from local to national fronts and resulted in a variety of laws specifying biodegradability requirements for certain polymeric materials. Plastics, the major non-energy product of petroleum chemicals, are considered to be nonbiodegradable, or at best only slowly degradable over many years. This, coupled with the amount of plastics produced and ending up as litter or in landfills, is primarily responsible for the activity towards plastics from natural materials that would biodegrade. In the U.S., about 58 billion pounds of petroleum derived plastics were produced in 1989 (6). Municipal solid waste contains about 7% by weight (7) and 17-25% by volume (8,9) of plastic, largely from packaging materials. Traditional plastics can be altered to enable facile chemical degradation, but the toxicity of the residues is as yet undefined. Some chemically-degraded petroleum based polymer residues may biodegrade to further increase global CO_2 levels.

Degradable plastic materials based on annually renewable biological products, such as starch, are seen by many as a solution to these problems. Röper and Koch recently reviewed the role of starch in thermoplastic materials (10). Replacement of petrochemically based plastics by biologically derived plastics, where feasible, would reduce petroleum usage. It would also slow introduction of fossil fuel derived CO_2 into the atmosphere since incineration or biological digestion of annually renewable-biomass derived polymers simply recycles CO_2 to maintain the ambient level. Litter from such plastics would disappear into its surroundings to leave only normal biological residues. Integrated waste management practices that include off-landfill composting of biodegradable wastes, incineration, source reduction of packaging materials, barring of toxic colorants, and recycling may bring waste disposal under control.

Due largely to independent studies conducted in the 1970s by a scientist in the United States and one in the United Kingdom, many of the articles appearing in the press beginning in the mid-1980s and many of the legislative proposals singled out starch as an additive to plastics to impart biodegradability.

Otey, in the U.S., was studying starch-synthetic polymer films for use as biodegradable agricultural mulch, while Griffin, in the U.K., was developing polyethylene films containing granular starch as a filler to impart better hand feel and printability to polyethylene shopping bags. More detail of the studies by these researchers will be discussed later in this paper.

What has been generated by the legislative actions and numerous articles on biodegradable plastics, in addition to much continuing debate, is a heightened awareness of the need for more scientific data addressing this biodegradability issue. Scientists in industry, academia and the public sector are responding to this need, and today many scientific articles are appearing in the literature and are being presented at scientific meetings at the local, state, national and international levels.

Probable markets for these biodegradable materials include many single use items such as agricultural mulch films, garbage bags, shopping and produce bags, diaper linings, bottles, drums, sanitary applicators, and fast food service items. Penetration into these markets requires matching of physical properties and costs of the plastics to the needs of the applications. Costs much above those of conventional non-degradable materials may be resisted by consumers unless legislative fiat requires use of more expensive materials.

In this presentation we will first discuss the research on starch leading to plastic materials that are more, if not totally, biodegradable as replacements for the current plastics derived from petroleum. Following this, we will discuss other starch-based polymers for areas of application other than plastics.

Biopolymer Plastics via Starch Fermentation

Fermentation of starch or starch-derived sugars has long been practiced to produce a variety of alcohols, polyols, aldehydes, ketones and acids. One of the acids, lactic, has received considerable attention as the basis of biodegradable thermoplastic polymers with a host of potential industrial applications. Polymerization of lactic acid to poly(lactic acid) was first studied about 50 years ago and continues to be a topic of research today. Although lactic acid can be directly polymerized by condensation polymerization, polymerization is more efficient and the polymer has better properties if the lactic acid is first converted to the lactide, the dilactone of lactic acid (Figure 2). To improve properties of poly(lactic acid), copolymerization with glycolic acid or epsilon caprolactone has received considerable attention. A wide range of properties result on varying the ratio of comonomers in the mixture, as reported by Sinclair (11). Commercial use of these polymers has been restricted mostly to the medical field, where they function as biocompatible, biodegradable, reabsorbing sutures and prosthetic devices.

Workers at Battelle, Columbus, Ohio have done considerable research and development of lactic acid based polymers during the last two decades. A review article by Lipinsky and Sinclair (12) discusses the properties and market opportunities for these polymers and problems that need to be overcome. They project the potential of multihundred million pound markets in commodity plastics and controlled release agrochemical formulations. In such applications, the environmentally benign poly(lactic acid), or copolymers with glycolic acid or caprolactone, would biodegrade in the environment to natural products. The wide range of physical properties achievable on processing these copolymers of various compositions confirms the excellent potential for their use in place of many of the current commercial thermoplastics.

While properties of the polymers are excellent, full realization of their potential will depend on improved preparation and recovery of the basic fermentation chemicals. Improved biotechnology that leads to higher solids fermentation to produce lactic acid in pure state is needed. Enhanced recovery technology is required to readily recover lactic acid from the fermentation broth. Direct fermentation of starch, rather than conversion first to glucose, has been reported and could assist in improving the economics of lactic acid production (13).

Although fermentation of sugars to produce polyhydroxybutyrate (PHB) has been known for decades, several shortcomings of the polymer have prevented its use on a significant commercial scale. Copolymers of hydroxybutyrate with hydroxyvalerate (PHBV) overcome many of these shortcomings, and such copolymers are now in commercial use (Figure 3). The bacterium *Alcaligenes eutrophus* ferments sugars to PHBV. The British Company ICI has developed a range of PHBVs with up to 30% hydroxyvalerate(HV). Whereas PHB is a rather brittle polymer, PHBV, with 25% hydroxyvalerate, is quite flexible. ICI now offers PHB and PHBV with up to 25% HV for a variety of applications. A 1988 product bulletin of ICI Americas Inc. lists several applications for PHBV with varying HV content in such areas as medical implants, injection molding, extrusion/injection blow molding for packaging materials, and slow release delivery of medicinals.

The biodegradability of PHB and PHBV polymers has drawn attention to these natural polymers, as interest in replacements for nonbiodegradable polymers has grown. The polymers, which have good shelf stability, undergo microbial degradation when buried in soil. ICI has announced that its Biopol (PHBV) is now being used in Europe in blow molded shampoo bottles, and ICI is actively seeking other markets for its biodegradable polymers. The price of the polymers is expected to drop from the initial $15 per pound to around $2 to $4 per pound by the mid-1990s.

Granular Starch As Filler In Plastics

Plastic materials can be made from starch in numerous ways. Starch hydroxyl groups can be esterified or etherified with hydrophobic groups to reduce hydrophilicity and retrogradation. Derivatization of a majority of the hydroxyl groups produces thermoplastic materials (14). However, due primarily to the cost of such derivatives, little commercial interest has been shown in them. Rather, a major interest has evolved in plastics in which granular starch is incorporated as a filler.

Granular starch has been used as a filler in numerous composite plastic compositions having utility as packaging films and containers. Griffin (15,16) demonstrated compounding of natural granular starch with PE (8-23% starch, w/w) to produce films that became embrittled in less than three months in moist compost and thus were said to be biodegradable. Embrittlement of the petroleum-based polymers was assisted by incorporation of about 5% unsaturated fatty acid or fatty acid ester in the formulations. Soil-born transition metal salts catalyzed oxidation of the unsaturated additives to peroxides, the active degradative agents for the synthetic polymers. Later formulations included metal salts (17). Starch was predried to less than 1% moisture to avoid steam-generated defects in the plastics. Alternatively, desiccants such as CaO (18) in the formulations trapped moisture from the starch. Esterification of the hydrophilic hydroxyl groups at the surfaces of the starch granules with alkyl siliconate (19) or

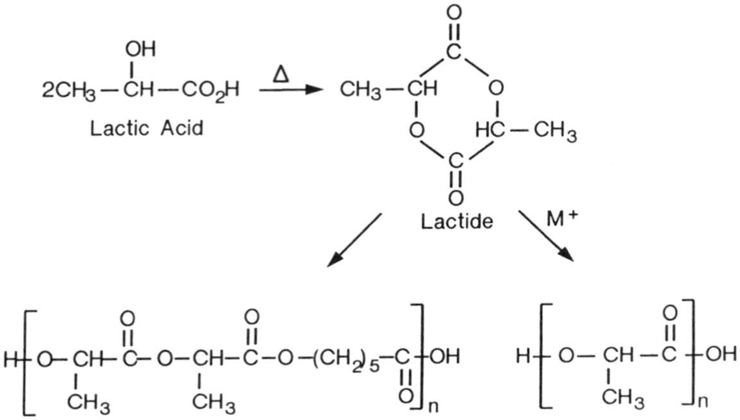

Figure 2. Polymers from lactic acid.

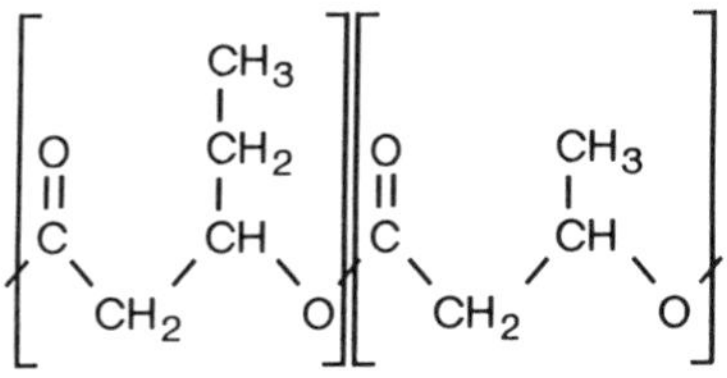

Figure 3. Poly(hydroxybutyrate-co-hydroxyvalerate)

octenyl-succinate (20) groups improved adhesion to the surrounding hydrophobic plastic matrix. For some applications, mixing of the dried starch with paraffin wax was sufficient surface treatment (21).

Jane et al. (22) produced sub-granular sized starch particles (average diameters of 2.2, 3.8, and 8 microns) for plastic filler applications in thin films, while Griffin proposed use of small wheat starch granules, and Maddever (17) suggested use of rice starch. Jane et al. (23) demonstrated the use of oxidized PE as a bridging agent between granular starch and PE. Duan et al. (24) showed that the physico- mechanical properties of starch-PE films were improved by added graft copolymer or polymer emulsion compatibilizers. Emulsions of soft polymers, such as poly(butyl acrylate-co-ethylene-co-vinyl acetate), improved elongation at the expense of tensile strength, while emulsions of hard polymers, such as poly(acrylonitrile-co-styrene), gave higher tensile strengths with lower elongation. PE films containing granular starch are now being produced directly by, or under license from, Coloroll Ltd in England, St. Lawrence Starch in Canada, and Archer Daniel Midland Company in the U.S., for use as garbage bags, grocery shopping bags, and over-wraps for mail. Other markets are being developed.

Common plastics other than PE that have been compounded with granular starch include: PS, Poly(vinylidene chloride-co-vinylidene acetate) (PVDC/A), and PVC. All of these composites support microbial growth and lose strength when buried near the surface. PVDC/A spray dried with 90% granular starch or PVC compounded on rubber rolls with 50% granular starch produced resins suitable for molding meat display trays and transplanting pots (19). Smoke production of burning PVC was reduced by incorporating 30 percent granular starch that had been permeated with 12% ammonium molybdate and spray dried (21). Griffin (16) produced molding sheets of PS containing 9-50% granular starch that were suitable for thermoforming into thin walled containers and drinking cups. Compatibility between starch and various synthetic polymers was improved by admixture of copolymers of the synthetic polymer with monomers that contained carboxyl groups (25).

Various techniques were investigated by Westhoff et al. (26) for incorporating large amounts of starch as a filler in PVC plastics. Starch-PVC films were prepared, and their properties were measured in Weather-Ometer and outdoor exposure tests. By varying the composition, films were obtained that lasted from 40 to 900 hr. in the Weather-Ometer and from 30 to more than 120 days in the soil. All samples tested under standard conditions with common soil microorganisms showed microbial growth, with the greatest amount of growth recorded for samples containing the highest amount of starch.

Otey et al. (27) used cornstarch, wheat starch, dialdehyde cornstarch, cornstarch graft copolymers, and dialdehyde cornstarch graft copolymers at levels of 15 to 55% as inert fillers in PVC. The starch or starch graft copolymers were dry blended with PVC and dioctyl phthalate on a rubber mill and were then pressed into films. Tensile strength of many of the molded starch-PVC products, even with 50% starch in the plastic were on the order of 3000 psi

(20.7 MPa) or higher; however elongation decreased rapidly as the starch level increased. Clarity of the plastics was good, except for plastics made by dry blending of unmodified starch. Plastics made from dialdehyde starch grafted with a mixture of polyacrylonitrile and poly(methyl methacrylate) had outstanding tensile strength.

Gelatinized Starch As A Component In Plastics

Considerable work has been reported where starch is gelatinized and thus may form a continuous phase with the synthetic polymer rather than merely being present as a particulate filler. Starch-PVA films, investigated by Westhoff et al. (28) represent such a continuous phase system and may have application as a degradable agricultural mulch. A composition containing 60 to 65% starch, 16% PVA, 16-22% glycerol, 1 to 3% formaldehyde, and 2% ammonium chloride was combined with water to give a mixture containing 13% solids and was then heated at 95°C for 1 hr. The hot mixture was then cast and dried at 130°C to form a clear film. Films were passed through a solution of PVC or Saran to give the film a water-repellent coating, since uncoated films had poor wet strength. Coated films retained good strength even after water soaking, and Weather-Ometer tests suggested that film with 15 to 20% coating (by weight) might last 3 to 4 months in outdoor exposure. Nwufo and Griffin (29) found poor adhesion between granular starch and PVA in calendered film.

Uncoated starch-PVA films are used commercially to produce a water-soluble laundry bag for use by hospitals to store soiled or contaminated clothing prior to washing. The bag and its contents are placed directly into the washing machine, where the bag dissolves. To provide enhanced solubility, a slightly derivatized starch is used for this application. Such water-soluble bags are also being suggested for packaging agricultural chemical pesticides to improve safety during handling.

Otey et al. (30,31) prepared films from various combinations of starch and EAA that have potential application in biodegradable mulch, packaging, and other products. EAA contains about 20% copolymerized acrylic acid and is dispersible in aqueous ammonium hydroxide. Films were prepared by solution casting and oven drying heated aqueous dispersions of starch and EAA or by fluxing dry mixtures of starch and EAA on a rubber mill. Cast films containing 30, 40, 50, 70, and 90% starch were exposed to outdoor soil contact, with the ends buried in the soil, to observe their resistance to sunshine, rain, and soil microorganisms. Films with more than 40% starch deteriorated within 7 days, but those containing 30 to 40% starch remained flexible and provided mulch protection for at least 70 days. Other tests revealed that starch-EAA films have sufficient strength, flexibility, water resistance, and heat sealability for a variety of mulch and packaging applications.

In further research on blending starch with synthetic polymers, Otey developed a semi-dry process to extrude thermoplastic starch-EAA mixtures containing up to 60% starch and 2-10% moisture (3.3-18.3% on a starch basis at the 60% starch level). Films were blown at a die temperature of 125-145°C (32), and film properties are shown in Table II. Otey also produced transparent starch-EAA films with semipermeability properties (33,34) by using aqueous sodium hydroxide to gelatinize the starch in the extruder at temperatures near 100°C. Alternatively, ammonium hydroxide-urea solution was used to destructure the starch granules 35), and films prepared in this manner were blown on a commercial film blowing apparatus. LDPE was incorporated as a partial replacement for EAA, which further reduced film cost and in (some instances improved properties. Swanson et al. (36) showed that shelf life of these films can be extended by low level

Table II. Starch-EAA Film Properties

Composition of Thermoplastic						Extrusion Temperature (°C)	Tensile Strength (MN/m^2)	Elongation (%)
Starch (%)	EAA (%)	LDPE (%)	Urea (%)	Water (pph)	Base[a] (pph)			
60	40	0	0	5-10	7.5N	110	25.0	7
34	51	0	15	5-8	8.5A	125-145	6.2	120
40	45	0	15	5-8	8.5A	125-145	11.0	90
40	50	10	0	5-8	3.6A	125-145	25.8	80
40	40	10	10	<14	8.5A	125	9.6	80

SOURCE: Adapted from ref. 32, 34 and 35.
[a] N=NaOH; A=Ammonium hydroxide

Table III. Physical Properties of LDPE-Starch-EAA[a-c]

Starch (%)	0	20	40
LDPE (%)	100	65	40
Tear Strength (KN/m)	53	17(17)	6(2)
UTS (MN/m^2)	17	13(13)	11(13)
Elongation (%)	254	211(231)	30(13)

SOURCE: Adapted from ref. 36.
[a] Samples were equilibrated at 50% RH and 23°C.
[b] Properties were measured at 1 week and 4 weeks (parentheses).
[c] Starch/EAA = 4:1 (w/w), water/starch = 0.15:1, NH_4OH/starch = 0.1:1, 10% urea in films with starch (based on total solids). Processed in Brabender 7:1 twin screw extruder.

hydroxypropylation of the starch and inclusion of urea in the formulations. Jasberg et al. (37) demonstrated that starch-EAA-PE formulations could be compounded in a single pass through a twin-screw extruder. The products were easily granulated and blown into film (See Table III).

Press releases from the Ferruzzi Group in Italy and Warner-Lambert in the USA and Europe forecast the production of plastics composed of 40-100% starch. Chamroux (38), of the Ferruzzi Group, reports preparation of films, pots, bowls, and plates containing more than 50% destructured starch. Synthetic low molecular weight hydrophilic polymers in the formulation contribute fluidity and rheological behavior similar to that of low density PE. Durkin (39) reports that films of these materials biodegrade under anaerobic and aerobic conditions. Starch decomposes quickly while the synthetic additives degrade more slowly. The films reportedly possess excellent printability, high resistance to oil and organic solvents, good gas barrier properties, and natural antistatic properties.

Warner-Lambert Company has been assigned a number of patents, typified by Wittwer (40) and Lay (41), which describe injection molding of starch at 5-30% moisture with a large number of materials known to have compatibility with starch or to modify its physical properties. These patents are based on work by Stepto and Tomka (42) which describes the plastic behavior of high solids starch and gelatin gels. They pointed out that higher temperatures and pressures enable plasticization of starch with lower water levels than are required for production of gels and pastes at 100°C and one atm. Starch granules are completely destructured under pressures of 60-300 MN/m^2 and temperatures of 80-240°C in an extruder barrel to produce extremely thick pastes. The pastes become solid on cooling, as in an injection mold. The apparent T_g of starch with 20% moisture is 25°C while that of starch with 15% moisture is 62°C, so these materials become brittle as moisture is lost. As yet, little has been released about the physical properties of the products, but they are likely to depend greatly on the particular co-plasticizers and polymeric additives present in the formulations. Product examples in the patents are often capsules.

An expanded packaging material suitable as a biodegradable replacement for styrofoam has been described by Lacourse (43). Hydroxypropyl cornstarch with at least 45% amylose content (70% amylose, 10% hydroxypropyl preferred) and containing 21% or less water is extruded at a temperature of 150-250°C to produce an open cell foam with resilience and compressibility values similar to styrofoam.

Not only can starch be mixed with synthetic polymers and exhibit utility as a filler, extender, or reinforcing agent, it can also become an integral part of such polymers through chemical bonds. (review by Doane, 44). Urethane is an example of a system where starch has been chemically bonded to a resin. To produce relatively low cost rigid urethane plastics that might have application in solvent-resistant floor tile, a system was developed based on 10 to 60% starch, castor oil, the reactive products of castor oil and starch-derived glycol glycosides, and polymeric diisocyanates. The addition of starch to the isocyanate resins substantially reduced chemical costs and improved solvent resistance and strength properties. Evidence showed that the starch chemically combined with the resin molecules. The degree of reactivity between starch and isocyanates was greatly enhanced by modifying starch with nonpolar groups, such as fatty esters, before the isocyanate reaction. Maximum reactivity of the modified starches was achieved when the degree of substitution was about 0.7.

Griffin prepared starch filled polyester plastics with good starch-polyester compatibility by ester interchange on the surface of starch granules pretreated with alkali or alkaline earth metal ions (45). Elastomers have been prepared where starch was a filler and cross-linking agent for diisocyanate modified polyesters. Also, starch can be incorporated into urethane systems to produce shock-resistant foams. Up to 40% starch or dextrin can be incorporated into rigid urethane foam, and such foams are more flame resistant and more readily attacked by microorganisms.

Thermoplastic Starch Graft Polymers

The structure of a starch graft copolymer is shown schematically in Figure 4, where AGU represents a glucopyranosyl (or anhydroglucose) unit, and M is the repeating unit of the monomer used in the polymerization reaction. Reviews on starch graft copolymers have been published (46,47) and these articles present a comprehensive picture of the numerous monomer and initiating systems that have been used to synthesize these polymers. In this paper, we will limit our discussion to starch graft copolymers having the greatest potential for commercial application.

Graft copolymers are prepared by first generating free radicals on starch and then allowing these free radicals to serve as macroinitiators for the vinyl or acrylic monomer. Generally, free radical initiated graft copolymers have high molecular weight branches that are infrequently spaced along the starch backbone. A number of free radical initiating systems have been used to prepare graft copolymers, and these may be divided into two broad categories: chemical initiation (usually with ceric salts) and initiation by high energy irradiation, such as cobalt-60. The choice depends on the particular monomer to be polymerized. Graft polymerizations are usually run in water and can be carried out with either granular, unswollen starch or with starch that has been gelatinized by heating. Copolymers with hydrophobic grafts cannot be dispersed in water, but remain as insoluble solids, even after prolonged heating. Copolymers with water-soluble grafts, however, swell in water at room temperature and either dissolve or disperse to give smooth pastes when heated.

Starch graft copolymers having thermoplastic grafted branches, e.g., poly(methyl acrylate) (PMA) or polystyrene, are an important group of polymers that have potential for commercialization. Starch-g-PMA is prepared by ceric-initiated graft polymerization of methyl acrylate onto either granular or gelatinized starch (48). The resulting graft copolymer typically contains about 50-60% PMA having a molecular weight of about 500,000; and conversion of monomer to polymer approaches 95%.

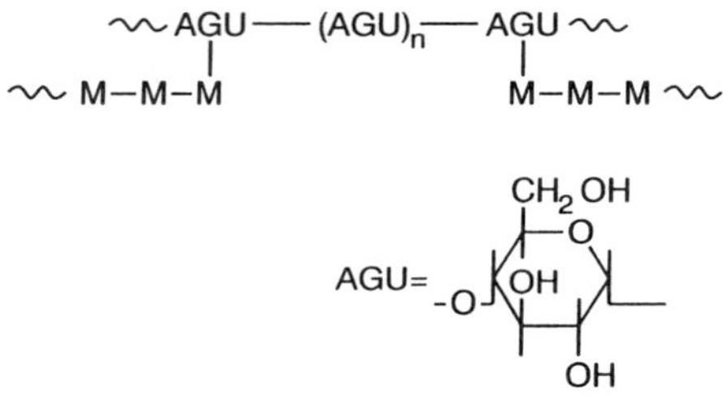

Figure 4. Schematic representation of a starch copolymer. (Reproduced from ref. 47.)

Since styrene does not graft copolymerize with starch in the presence of ceric ion, two alternative methods were used to prepare starch-g-polystyrene. In the first method (49), a small quantity of water is added to a mixture composed of equal weights of granular corn starch and styrene, and the resulting paste is then irradiated with cobalt-60 to initiate polymerization. Polymerization does not take place if water is omitted from the reaction mixture. In the second method (50), potassium persulfate is used to initiate polymerization in a semi-solid mixture of starch, styrene, and water. Since weight % polystyrene in the graft copolymer prepared by this method was too low (28%) for processing into a continuous plastic, the copolymer was heated in dilute hydrochloric acid to hydrolyze and thus remove part of the starch component. The final product contained 58% polystyrene and could be easily processed.

Graft polymers were processed into plastic ribbons by extruding the powdery products through a laboratory model Brabender extruder (48). Table IV shows some extrusion conditions used for starch-g-polystyrene prepared by cobalt-60 initiation. In the first entry of the table ungrafted polystyrene was removed before

Table IV. Extrusion of Starch-g-polystyrene

Material	Extrusion Temp., °C	Die Swell	Properties
Graft copolymer (no homopolmer)	175	None	Excellent, smooth, glossy, strong but brittle
Blend (90:10 graft copolymer:polystyrene)	175	None	Poor, inhomogeneous
Blend (20:80 graft copolymer:polystyrene)	150	Large,	Poor, inhomogeneous variable
Graft copolymer (homopolymer not removed)	175	None	Tensile strength 51.7-62.7 MPa (7500-9100 psi)

SOURCE: Adapted from ref. 48.

extrusion by extraction of the graft copolymer with benzene; whereas in the last entry, the sample contained about 10% homopolymer which was not removed. Extruded ribbons from these two samples were similar in appearance, and were excellent looking plastic products, although they were quite brittle. These samples exhibited little or no die swell (51) which is an important property of these materials, since the presence of die swell complicates the design of dies made to extrude precise shapes. It is apparent from the second and third entries in the table that physical mixtures of graft copolymer and ungrafted polystyrene, prepared by blending the polymers in the dry state, gave poor, non-homogeneous extrudates.

Polystyrene, therefore, must be grafted onto and within the starch granules to give acceptable plastic products on extrusion. A small amount of homopolymer is not detrimental as long as it too is formed onto and within the individual starch granules.

Table V shows some extrusion conditions used for starch-g-PMA copolymers. Since PMA has a lower Tg than polystyrene (8°C vs 100°C), lower extrusion temperatures could be used. Also, graft copolymers prepared from gelatinized starch could be processed at lower temperatures than comparable products made from granular starch. In the last entry of the table, addition of 20% water to gelatinized starch-g-PMA allowed the extrusion temperature to be lowered to 95°C. Extruded ribbons were tough, leathery, and

Table V. Extrusion of Starch-g-poly(methyl acrylate)[a]

Granule state of starch	Water in sample	Barrel Temp., °C	Die Temp, °C	Properties
Granular	None	150	160	Smooth, translucent, leathery, tensile strength: 20.7 MPa (3000 psi)
Granular	None	125	125	Poorly formed
Gelatinized	None	125	140	Smooth, translucent, leathery, tensile strength: 17.2 MPa (2500 psi)
Gelatinized	None	100	100	Poorly formed
Gelatinized	20%	90	95	Tough, translucent, leathery

SOURCE: Adapted from ref. 48.
[a] Homopolymer removed by extraction.

translucent, and ultimate tensile strengths were on the order of 2500-3000 psi (17.2-20.7 MPa). Patel et al. (52) have substituted acrylonitrile for part of the methyl acrylate in the graft polymerization reaction and have examined the effect of copolymerized acrylonitrile on tensile properties.

Dennenberg and coworkers (53) studied the effects of water on extruded ribbons of starch-g-PMA, and their results are shown in Table VI. Although the graft copolymer maintained its integrity and did not disintegrate after it had stood for 70 hr in water at room temperature, the sample whitened and lost about 80% of its original strength. The elongation exhibited by the sample before breaking also increased dramatically. When placed in acetone or benzene, which are good solvents for PMA, the extruded graft copolymer

Table VI. Properties of Starch-g-poly(methyl acrylate) Extrudates After Immersion in Different Solvents

Sample	Tensile strength MPa (psi)	% Elongation
Low density polyethylene film	13.8 (2000)	500
Starch-g-PMA[a]	22.4 (3250)	255
Starch-g-PMA 70 hr in water	4.6 (670)	535
Starch-g-PMA 30 min in acetone or benezene	Disintegrates to white powder	

SOURCE: Adapted from ref. 53 and 54.
[a] Contains 65% poly(methyl acrylate) (abreviated as PMA); extrudate thickness: 0.5 mm.

disintegrated to a white powder within a few minutes (54), confirming the theory that bonding between graft copolymer particles is due to fusion between heat-softened PMA branches. Dennenberg and coworkers (53) also established that starch-g-PMA supported the growth of some common microorganisms, and these authors measured the accompanying weight losses due to digestion of the starch component. Henderson and Rudin (55) have made a detailed study of the effects of water soaking on tensile properties of starch-g-PMA and starch-g-polystyrene and confirm the loss in tensile strength and increase in percent elongation observed by Dennenberg et al. Henderson and Rudin also report dynamic mechanical properties of the two graft copolymers.

Starch-g-PMA copolymers can also be extrusion blown to form continuous films (56). Copolymers used for this technology are prepared by the same method described earlier, except that starch is dissolved in hot water before the grafting reaction rather than merely being pasted or gelatinized. Since unmodified corn starch is difficult to dissolve and requires special techniques to attain solution and to avoid retrogradation, either derivatized or partially depolymerized starches have been used to prepare graft copolymers used for extrusion blowing. Soluble starch is necessary because grafting of PMA onto insoluble starch fragments leads to chunks of graft copolymer which are large enough to disrupt the integrity of a thin extrusion-blown film of polymer.

One derivatized starch that has worked well in both the graft polymerization and in the subsequent extrusion-blowing process is a commercial cationic waxy corn starch with a D.S. of about 0.035. Table VII shows some tensile properties of these extrusion-blown films. In most experiments, a solution of 4.7 g of urea in 15 ml of water was mixed into 90 g of air-dried graft copolymer just before passing it through the extruder. Urea and water act as plasticizers for starch, and urea also functions as a humectant to help keep water in the system at the temperature used for extrusion (about

95°C). If the polymer formulation becomes too dry, starch becomes too rigid for blowing, and a poor quality film is obtained. Film samples collected immediately after leaving the die contained about 10% water. Good quality films can also be obtained in the absence of urea; however, urea facilitates the blowing operation. Commercial acid-modified corn starches are also hot water-soluble, and these starches similarly yield high tensile strength films (Table VII, No. 2).

The importance of starch solubility is apparent from Numbers 3 and 4 of Table VII. Unmodified cornstarch does not completely dissolve, even after heating in water at 95°C; and as a result, a poor quality film is obtained. Similarly, if cationic waxy cornstarch is not dissolved prior to graft polymerization but is simply slurried in water at room temperature and then graft polymerized as whole granules, the resulting blown film is also one of inferior quality.

Table VII. Extrusion-Blown Films from Starch-g-poly(methyl acrylate) copolymers[a]

No.	Starch	wt % PMA[b] in product	Urea, g per 90 g polymer	Tensile strength MPA (psi)	% Elongation
1	Cationic waxy (d.s.=0.035)	58	0	25.5 (3700)	78
			4.7	22.1 (3200)	67
2	Acid modified	59	4.7	26.2 (3800)	74
3	Unmodified	59	4.7	Poor quality film	
4	Cationic waxy (not dissolved)	59	4.7	Poor quality film	
5	Cationic waxy	58 (PMA not grafted)	4.7	Could not be extrusion-blown	

SOURCE: Adapted from ref. 56.
[a] 100 g starch in 2000 ml water, heated for 30 min at 95°C, and graft polymerized with 150 g methyl acrylate (except for No. 4, where starch was merely slurried in water at room temperature before grafting).
[b] PMA = poly(methyl acrylate)

The last entry of Table VII shows the importance of chemical bonding between PMA and starch, as opposed to merely having a physical blend of the two components. To obtain an intimate physical mixture, polymerization was carried out in the same manner as No. 1 except that the initiating system (ammonium persulfate-sodium metabisulfite) was chosen to produce largely homopolymer with little accompanying graft polymerization. The resulting mixture of starch and PMA did not form a continuous film when extrusion blowing was attempted.

A useful property of starch-g-PMA blown films that could have commercial importance is their tendency to shrink to a fraction of their original size when they are allowed to stand at room temperature in an atmosphere of high relative humidity. Table VIII shows that shrinkage only takes place at relative humidities of 75% or higher. When placed around irregularly shaped objects, these films conform on shrinking to the shape of the object without forming stress cracks.

Shrinkable films are used extensively by the packaging industry and are currently prepared from polymers that are totally synthetic and contain no starch. The ability to shrink in size comes about from an elastic memory imparted to the plastic film during manufacture. Films are first oriented by stretching at elevated temperatures, and the oriented films are then either cooled or lightly crosslinked to lock in the oriented structure. Shrinkage takes place when heat is applied, since heat causes the film to soften and thus allows it to revert back to its original unoriented

Table VIII. Shrinking of Starch-g-poly (methyl acrylate) Blown Films

Relative Humidity, %	% Shrinkage	
	In width	In length[a]
50	0	0
65	1	3
75	40	21
100[b]	62	39

SOURCE: Adapted from ref. 56.
[a] Machine direction
[b] At this relative humidity, the moisture content of the film increased to 15.6%, and the film thickness increased about 320%.

state. Films that shrink at room temperature by simply exposing them to high relative humidity have not been previously reported; and such films could find use for packaging heat-sensitive objects. Since these films contain starch, they are also water sensitive and can thus be easily removed by soaking the coated object in water.

A logical explanation for the shrinking behavior of starch-g-PMA can be proposed. On extrusion blowing, most of the water is driven out of the film by the high temperature encountered in the extruder die. Since water serves as a plasticizer for starch, the starch portion reverts back to a rigid polymer, and the resulting film thus remains dimensionally stable as long as it is stored under conditions of low to moderate relative humidity. However, when relative humidities approach 100%, the starch portion can absorb enough water from the atmosphere to plasticize it and thus permit it to flow. Since the PMA portion of the film is already a soft, flowable polymer at room temperature (Tg = 8°C), the entire graft copolymer is now capable of flowing, and the film shrinks because of relaxation of high molecular weight polymer chains that had been stretched and extended during the blowing operation.

Gugliemelli and coworkers have also carried out graft polymerizations onto water-soluble cationic starch and have used this technique to prepare starch graft copolymer latexes with particle sizes small enough to preclude settling, even after prolonged standing. Acrylonitrile (57,58), acrylonitrile/isoprene mixtures (59), and chloroprene (60) were used as monomers, and cast films of good quality could be prepared by simply allowing latex dispersions to air-dry. Weaver (61-63) has found that latexes prepared from starch-g-PMA and starch-g-poly(butyl acrylate) will stabilize soil to water erosion, and has studied the effect of a number of graft polymerization parameters on soil stabilization.

Water Dispersible Starch Graft Polymers

Perhaps the most important hydrophilic starch graft polymer from a commercial point of view is one prepared from starch and acrylonitrile in which the nitrile substituents of grafted polyacrylonitrile (PAN) have been hydrolyzed with aqueous alkali. This polymer (commonly abbreviated as HSPAN) has the ability to absorb an amount of water equal to hundreds of times the dry weight of polymer (64). The discovery of these highly absorbent starch graft copolymers in the mid 1970's did much to spark the tremendous interest in superabsorbents that we have seen in recent years.

The synthesis of HSPAN is outlined in Figure 5. Graft polymerizations may be carried out with either granular, unswollen starch (65) or with starch that has been gelatinized by heating an aqueous water slurry to about 85-95°C before the room temperature graft polymerization with acrylonitrile (66). Gelatinization of starch before graft polymerization has a profound effect on the structure of starch-g-PAN, as seen in Table IX (67). When starch is graft polymerized as unswollen granules to a PAN content of about 50% by weight, the molecular weight of grafted PAN is about 100,000; whereas the grafting frequency, expressed as the average number of AGU per grafted branch, is about 600. However, when starch is gelatinized before graft polymerization, the M_v of PAN grafts is about 800,000, and the grafting frequency is about 4,000 AGU per graft.

When starch-g-PAN is treated with either sodium hydroxide or potassium hydroxide solutions at a temperature near 100°C (68) nitrile substituents of PAN are converted to a mixture of carboxamide and alkali metal carboxylate, and ammonia is evolved (Figure 5). Carboxamide to carboxylate ratios will vary depending on saponification conditions, but are typically on the order of 1:2 (69). Complete saponification to give poly(sodium acrylate) does not occur.

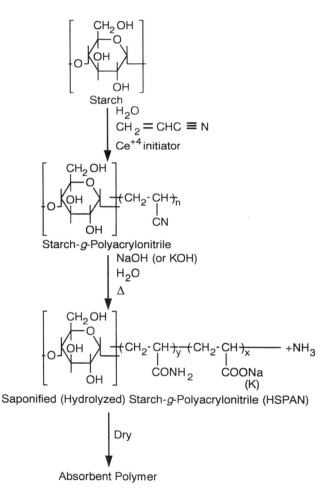

Figure 5. Preparation of saponified starch-g-polyacrylonitrile absorbent polymer. (Reproduced from ref. 47.)

Table IX. Ceric Ammonium Nitrate Initiated Graft Polymerization of Acrylonitrile—Influence of Swelling and Disruption of Starch Granules[a]

Starch pretreatment[b] (°C)	Conversion of acrylonitrile (%)		Graft Copolymer		
	To grafted polyacrylonitrile	To ungrafted polyacrylonitrile	Polyacrylonitrile content (%)	Mol wt of graft	Anhydroglucose units/graft	Percent of total reaction product[c]

Starch pretreatment[b] (°C)	To grafted polyacrylonitrile	To ungrafted polyacrylonitrile	Polyacrylonitrile content (%)	Mol wt of graft	Anhydroglucose units/graft	Percent of total reaction product[c]
25	79	4.4	52.8	116,000	640	93
60	83	7.2	55.6	566,000	2,770	85
85	75	6.6	56.4	810,000	3,880	78

SOURCE: Reprinted with permission from ref. 46.
[a] Wheat starch, 0.135 mol; acrylonitrile, 0.6 mol; water, 507 ml; ceric ammonium nitrate, 1.5×10^{-3} mol/l. Reaction time, 3 hr at 25°C.
[b] Aqueous starch slurries stirred for 1 hr at specified temperature before room temperature reaction with monomer and initiator.
[c] The fraction remaining after room temperature extraction with water, dimethylformamide, and dimethyl sulfoxide.

HSPAN remains largely in the form of a highly swollen but insoluble gel after saponification. Taylor and Bagley (70) correctly conclude that the thickening action of HSPAN in water (71) is due to nearly complete absorption of water by the gel to give a system consisting of highly swollen, deformable gel particles that are closely packed and in intimate contact. Neither the minor amounts of graft copolymer in solution (15 to 20% for HSPAN prepared from gelatinized starch) nor the size of the gel particles exerts a large influence on rheological properties. The excellent thickening properties of HSPAN for aqueous systems have been utilized in firefighting fluids and in electrolyte solutions for alkaline batteries (72-73). Rodehed and Rånby (74) have used differential scanning calorimetry to characterize the absorbed water in HSPAN with moisture contents ranging from 0-1.05 g per g of dry polymer. Free crystallizable water was detected only when the water content exceeded 0.44 g/g. Both free and bound water increased linearly with larger water amounts.

Precipitation of HSPAN by addition of a water miscible non-solvent, such as methanol, is a commonly used method of isolation. Rodehed and Rånby (75) studied the effect of methanol and ethanol concentrations in water on the amount of aqueous alcohol absorbed by HSPAN and have determined the alcohol concentrations needed to give total collapse of the HSPAN gel. Since excess alkali and inorganic salts can be removed by washing, this method gives a purified HSPAN, which will absorb about 1000 times its weight of deionized water, when gelatinized starch is used as the substrate for graft polymerization. HSPAN derived from granular starch has an absorbency of about 200 to 300 g/g. Since HSPAN polymers are polyelectrolytes, ion-containing fluids are absorbed to a lesser extent. The swelling behavior of HSPAN in the presence of inorganic salts has been studied by Castel et al. (76). Alcohol-precipitated HSPAN copolymers, which exhibit fast wicking and reduced gel blocking, are prepared either by subjecting the polymers to an ion exchange reaction with a small amount of a high molecular weight quaternary ammonium chloride (77) or by adding an aluminum salt (78).

If maximum purity and absorbency are not needed (e.g., for agricultural applications), the viscous reaction mass from saponification can be simply dried on heated drums to give HSPAN in the form of coarse flakes. This process is inexpensive (68) and produces no by-products other than steam from the product isolation step. Although the deionized water absorbency of drum-dried HSPAN from gelatinized starch is only about 300 g/g, an important advantage of drum-dried HSPAN flakes is their rapid rate of liquid absorption and the absence of gel blocking.

The basic HSPAN technology described in this section has been modified in several ways to provide new additions to this family of absorbents. In one method, flour is simply substituted for starch in the ceric-initiated graft polymerization with acrylonitrile, and the saponification is carried out in water in the same manner used for starch-g-PAN (79). In a second HSPAN modification, either granular or gelatinized starch is graft polymerized with a monomer system in which a minor amount (up to 10 mol %) of the acrylonitrile is replaced with a comonomer (79). Of the various co-monomers studied, 2-acrylamido-2-methylpropanesulfonic acid imparted the highest absorbency to the final saponified polymers. Finally, it has been found that HSPAN may be blended with certain acid-modified starches (dextrins) without adversely affecting absorbency properties (80).

In addition to the HSPAN family of products, starch graft copolymer absorbents have also been prepared by grafting water-soluble monomers directly onto starch (as opposed to the two-step process in which acrylonitrile is first graft polymerized and the resulting hydrophobic PAN grafts are then saponified). These products are also used commercially. One such polymer is prepared by graft polymerization of acrylic acid and sodium acrylate onto starch (81). A crosslinking agent such as N,N'-methylene-bisacrylamide is included in the monomer mixture to help reduce water solubility, which is an inherent property of starch graft copolymers prepared from water-soluble monomers.

A major use for starch graft copolymer absorbents is for the absorption of body fluids, and these polymers have been added to disposable soft goods such as diapers, adult incontinent pads, feminine napkins, and hospital underpads to increase absorbency. HSPAN has also been marketed for addition to ostomy and urinal bags to prevent leaks and odors. When HSPAN is partially hydrated, the polymer will readily asborb blood, serum, and pus from a wound and will thus promote wound healing. Results of successful animal trials have been published (82,83) and studies with human patients suffering from decubitus ulcers also gave good results (84). HSPAN has also been used as a perspiration absorbent in body powder formulations (85).

In addition to the medical and personal care markets, there are numerous agricultural applications for starch graft copolymer absorbents. Use of HSPAN as a seed coating seems particularly promising, since studies have shown that the polymer absorbs water and holds it at the seed surface, thus increasing both rate of germination and percentage of the total number of planted seeds that germinate (86,87). Also, HSPAN can be applied as a gel slurry to the root zone of plants before transplanting to prevent roots from drying, to reduce wilting and transplant shock, and to improve plant survival (87,88). Use of superabsorbents as soil additives is another area of application. Shrader and Mostejeran (89) report the effect of HSPAN addition on water-holding capacity for a variety of different soil types. HSPAN not only increased the amount of water held by sandy soils, but the water was held in a form that was readily available to plant roots. The authors concluded that a sandy soil treated with 0.2 wt % HSPAN had about the same water-holding capacity as the best corn belt soil. HSPAN is also used as a component in hydroseeding or hydromulching, where a water slurry of either wood fibers, straw, or a similar material is blown onto an area, such as a construction site, to provide erosion control and to promote plant establishment (90). This same slurry may also contain seed, fertilizer, and various polymers or polymer mixtures (tackifiers) that are designed to hold the fiber mulch in place, to hold moisture, and to promote seed germination.

HSPAN is also used for removal of suspended water from organic solvents. Successful laboratory tests have been carried out (91) using HSPAN as a dehydrating agent for ethanol-gasoline mixtures (gasohol) to avoid the azeotropic distillation step necessary to remove final traces of water from ethanol. Although the polymer will not remove water from ethanol, its addition to a cloudy, two-phase mixture of wet ethanol and gasoline affords a clear gasohol solution with a water content of only about 0.4 to 0.5%. Gasohol with an even lower water content may be obtained by passing the two-phase mixture through a column packed with the absorbent polymer. HSPAN selectively absorbs water and allows the ethanol component to totally dissolve into the gasoline phase. Fuel filters containing HSPAN for water removal are now being manufactured. Also, another company is marketing porous bags containing HSPAN which can be lowered to the bottom of fuel storage tanks to absorb condensed water.

In addition to the superabsorbents, there are other hydrophilic starch graft copolymers that have potential for commercial use. For example, graft polymerization of acrylamide, acrylic acid, and cationic amino-containing monomers (Figure 6) onto starch has yielded products useful as flocculating agents, thickening agents, and retention aids for papermaking.

Use of starch graft copolymers for flocculation and flotation of suspended solids has been reviewed by Burr et al. (92). In laboratory tests, graft copolymers prepared from cationic monomers, either alone or in combination with acrylamide, functioned as flocculants for diatomaceous silica (93-95). Also, a graft copolymer containing 15% poly(2-hydroxy-3-methacryloyloxypropyl-trimethyl ammonium chloride) (HMAC) and one containing 30% poly(HMAC-co-acrylamide) were selective depressants for silica in the flotation-beneficiation of Florida pebble phosphate ore (96). In the flocculation of bauxite red mud suspensions, Jones and Elmquist (97) report that starch-g-poly(acrylic acid) copolymers were more effective than either cationic products containing poly(N,N,N-trimethylaminoethyl methacrylate methyl sulfate) (TMAEMA·MS) or graft copolymers with poly(acrylamide-co-acrylic acid) grafts. Performance was proportional to poly(acrylic acid) content. Restaino and Reed (98) also found that starch-g-poly(acrylic acid) and starch-g-poly(methacrylic acid) are good flocculants for this application. Finally, starch-g-polyacrylamide copolymers have been examined as flocculants for clays (99-101) and for coal refuse slurries (102).

Graft copolymers prepared from cationic amine-containing monomers are effective retention aids in the manufacture of mineral-filled paper. Copolymers from the reaction of cobalt-60 preirradiated starch with mixtures of acrylamide and TMAEMA·MS function particularly well for this application, and their performance has been evaluated on a 32-in. pilot Fourdrinier machine. In a study by Heath and co-workers (103), excellent suspended solids reduction in the white water was obtained with a starch-g-poly(acrylamide-co-TMAEMA·MS) copolymer with 46% add-on and an acrylamide:cationic monomer ratio of 34.9:11.1, by weight. The graft copolymer was used without removal of homopolymer, since its removal would be uneconomical in any commercial polymerization

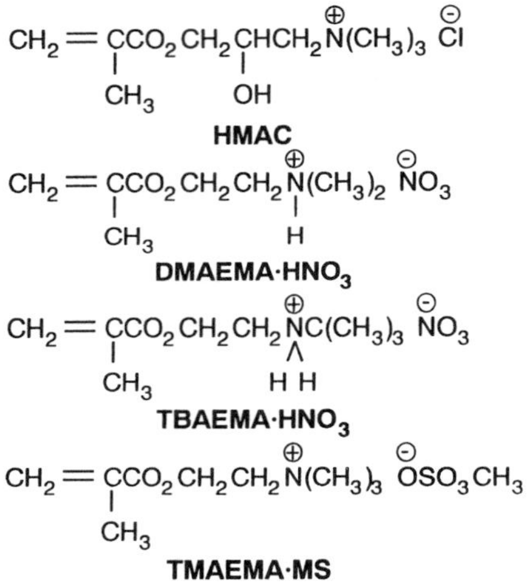

$$CH_2 = \underset{CH_3}{\underset{|}{C}}CO_2CH_2\underset{OH}{\underset{|}{CH}}CH_2\overset{\oplus}{N}(CH_3)_3 \; \overset{\ominus}{Cl}$$

HMAC

$$CH_2 = \underset{CH_3}{\underset{|}{C}}CO_2CH_2CH_2\underset{H}{\underset{|}{\overset{\oplus}{N}}}(CH_3)_2 \; \overset{\ominus}{NO_3}$$

DMAEMA·HNO₃

$$CH_2 = \underset{CH_3}{\underset{|}{C}}CO_2CH_2CH_2\underset{H\;H}{\underset{\wedge}{\overset{\oplus}{N}C}}(CH_3)_3 \; \overset{\ominus}{NO_3}$$

TBAEMA·HNO₃

$$CH_2 = \underset{CH_3}{\underset{|}{C}}CO_2CH_2CH_2\overset{\oplus}{N}(CH_3)_3 \; \overset{\ominus}{OSO_3}CH_3$$

TMAEMA·MS

Figure 6. Structures of some cationic monomers. HMAC = 2-hydroxy-3-methacryloyloxypropyltrimethylammonium chloride: DMAEMA·HNO, = nitric acid salt of N,N,dimethylaminoethyl methacrylate: TBAEMA·HNO, = nitric acid salt of N,tertbutylaminoethyl methacrylate: TMAEMA·MS = N,N,N-trimethylaminoethyl methacrylate methyl sulfate. (Reproduced from ref. 47.)

process. Percent reduction of white water-suspended solids from the control was determined at different levels of polymer addition (Table X) for the starch graft copolymer and a commercial cationic polyacrylamide. The graft copolymer was comparable to cationic polyacrylamide at the higher levels of addition. Moreover, starch-g-poly(acrylamide-co-TMAEMA·MS) improved drainage rate and also increased dry tensile strength of paper handsheets prepared without clay filler.

Heath and co-workers (104) also used 1:10, 1:20, and 1:30 blends of the starch graft copolymer with unmodified corn starch to increase paper strength as well as to efficiently incorporate clay filler in the paper sheet. Commercial cationic starches, used for comparison, gave paper with higher bursting strengths than equivalent amounts of starch blended with cationic graft copolymer; however, 1:10 blends gave higher suspended solids reduction. Also, starch in the blends was better retained by the paper.

Selected graft copolymers have been examined as thickeners. For example, a starch-g-polyacrylamide with 39% add-on and an HSPAN containing 58% synthetic polymer were found to be effective print paste thickeners if combined with one another or with a commercial thickener, such as sodium alginate (105). Starch-g-polyacrylamide copolymers have been evaluated as agents to increase the viscosity of aqueous fluids used for enhanced oil recovery (106,107) and there is evidence that a fewer number of long-

Table X. Reduction of White Water-Suspended Solids by Starch Graft Copolymer and Commercial Retention Aids

Retention aid	Clay (%)	pH	Reduction of suspended solids from control at different levels of addition[a] (%)			
			0.025	0.05	0.1	0.4
Graft copolymer[b]	10	5.0	31	53	67	84
	10	6.5	41	56	70	87
	20	6.5	18	55	74	93
Cationic polyacrylamide	20	6.5	47	66	74	..
Cationic starch	10	5.5	..	5	22	28

SOURCE: Adapted from ref. 103.
[a] Based on dry weight of pulp: headbox addition.
[b] Graft copolymer contains 34.9% acrylamide. 11.1% N,N,N-trimethylaminoethyl methacrylate methyl sulfate (TMAEMA·MS) and 54% starch.

chain polyacrylamide grafts results in larger hydrodynamic dimensions for the graft copolymer than a large number of short-chain grafts. Finally, Meister et al. (108) have graft polymerized starch with mixtures of acrylamide and 2-acrylamido-2-methylpropanesulfonic acid and have studied the rheology of the resulting polymers.

Conclusion

Starch, due to its great abundance in nature, ease of recovery from plant organs, relative low cost and ease of chemical and biological modification, is playing an increasing role as a renewable resource for polymeric materials. Worldwide interest in biodegradable polymers that are more environmentally benign than current synthetic polymers has seen expanded activities in scores of laboratories on the development of new polymeric materials based on starch. Biopolymers produced by fermentation of starch or starch-derived sugars, although known for many years, now are entering the market place as biodegradable containers and the like. Polyolefin films and bottles containing starch as a biodegradable component, along with other additives to cause chemical degradation, have found special niches and are in early stages of what is predicted to become a substantial market volume. Plastic compositions containing starch as a filler (up to 20% starch) or as a part of the continuous phase (from 40% to nearly 100% starch) are being developed. Starch polymeric materials with covalently bonded thermoplastic or hydrophilic side chains offer much promise commercially, due not only to unique and useful properties but also to the relative ease with which structures and properties can be controlled. The decade of the 1990s should see many new starch-based polymers and new applications as we seek petroleum-sparing, more environmentally benign materials.

Literature Cited

1. *Starch Chemistry and Technology.* Whistler, R. L.; BeMiller, J. N.; Paschall, E. F. Eds.; 2nd ed., Academic Press Inc.; 1984.
2. *Starch: Properties and Potential.* Galliard, T. Ed.; Critical Reports of Applied Chemistry. John Wiley and Sons; 1987; Vol. 13.
3. Koch, H. and Röper, H. New Industrial Products From Starch. *Starch/Stärke*, **1988**, *40*(4):121-131.
4. Doane, W. M. New Industrial Markets For Starch. *Proceedings of Corn Utilization Conference II*, 1988; pp 1-19.
5. *Modified Starches: Properties and Uses.* Wurzburg, O. B., Ed.; CRC Press; 1986.
6. Modern Plastics Staff. Resin Report. *Modern Plastics*, **1990**, *67*(1):53.
7. Franklin Associates, Ltd. Characterization of Municipal Solid Waste in the United States, 1960-2000 (Update 1988). Prepared for the U.S. Environmental Protection Agency. Contract No. 68-01-7310.
8. Studt, T. Degradable Plastics. *Research and Development*, **1990**, *32*(4)50.

9. Raloff, J. Helping Plastics Waste Away. *Science News*, **1989**, 135 (May):282.
10. Röper, H.; Koch, H. The Role Of Starch In Thermoplastic Materials. *Starch/Stärke*, **1990**, 42(4):123-130.
11. Sinclair, R. Lactic Acid Polymers- Controlled Release Applications for Biomedical Use and Pesticide Delivery. *Proceedings Of Corn Utilization Conference I*, 1987; pp 221-236.
12. Lipinsky, E. S.; Sinclair, R. G. Is Lactic Acid A Commodity Chemical? *Chemical Engineering Progress*, August, **1986**, pp 26-32.
13. Mueller, R., Jannotti, G., Bojpai, R., Cheng, P., and Jaeger, S. Lactic acid production from liquified corn starch using *Lactobacillus amylovoris* and newly isolated *Lactobacilli*. *Proceedings of Corn Utilization Conference III*, 1990; pp. 15-18.
14. Jarowenko, W. Starch. *Encyclopedia of Polymer Science and Technology*. Mark, H., Ed.; Interscience, New York, 1970; pp 787-862.
15. Griffin, G. J. L. Biodegradable fillers in thermoplastics. *Am. Chem. Soc. Div. Org. Coat. Plast. Chem.*, **1973**, 33(2):88-96.
16. Griffin, G. J. L. U.S. Patent 4 016 117, 1977.
17. Maddever, W. J. Current status of starch based degradable plastics. *Proceedings of Corn Utilization Conference III*, 1990; pp 143-165.
18. Griffin, G. J. L. U.S. Patent 4 021 388, 1977.
19. Griffin, G. J. L. U.S. Patent 4 125 495, 1978.
20. Jane, J.; Gelina, R. J.; Nikolov, Z.; Evangelista, R. L. Degradable plastics from octenylsuccinate starch. U.S. Patent application 407 294, 1989.
21. Griffin, G. J. L. U.S. Patent 4 218 350, 1980.
22. Jane, J. L. U.S. Patent Application 382 491, 1989.
23. Jane, J.; Evangelista, R.; Wang, L. Use of modified starches in degradable plastics. *Proceedings of Corn Utilization Conference III*, 1990; pp 167-171.
24. Duan, M.; Yu, J.; Chen, Y.; Tian, R.; Xu, Z.; Xu, X. Studies on the compatibilizing agent for the polyethylene and starch blend system. *Gaofenzi Cailiao Kexue Yu Gongcheng*, **1988**, 4(4):47-52.
25. Osterreichische Agrar-Industrie Gesellschaft M. B. H. Austrian Patent 365 619, 1982.
26. Westhoff, R. P.; Otey, F. H.; Mehltretter, C. L.; Russell, C. R. Starch-filled polyvinylchloride plastics--preparation and evaluation, *Ind. Eng. Chem., Prod. Res. Dev.*, **1974**, 13(2):123-125.
27. Otey, F. H.; Westhoff, R. P.; Russell, C. R. Starch graft copolymers--degradable fillers for poly(vinyl chloride) plastics. *Ind. Eng. Chem., Prod. Res. Dev.*, **1976**, 15(2):139-142.
28. Westhoff, R. P.; Kwolek, W. F.; Otey, F. H. Starch-polyvinyl alcohol films--effect of various plasticizers. *Starch/Stärke*, **1979**, 31(5):163-165.

29. Nwufo, B. T.; Griffin, G. J. L. Microscopic appearance of fractured surfaces and mechanical properties of starch-extended poly(vinyl alcohol). *J. Polym. Sci., Polym. Chem. Ed.*, **1983**, *23*(7):2023-2031.
30. Otey, F. H.; Westhoff, R. P.; Russell, C. R. Biodegradable films from starch and ethylene-acrylic acid copolymer. *Ind. Eng. Chem., Prod. Res. Dev.*, **1977**, *16*(4):305-308.
31. Otey, F. H.; Westhoff, R. P.; Doane, W. M. Starch-based blown films. *Ind. Eng. Chem., Prod. Res. Dev.*, **1980**, *19*(4):592-595.
32. Otey, F. H.; Westhoff, R. P. U.S. Patent 4 337 181, 1982.
33. Otey, F. H.; Westhoff, R. P. U.S. Patent 4 454 268, 1984.
34. Otey, F. H.; Westhoff, R. P. Starch-based films. Preliminary diffusion evaluation. *Ind. Eng. Chem., Prod. Res. Dev.*, **1984**, *23*(2):284-287.
35. Otey, F. H.; Westhoff, R. P.; Doane, W. M. Starch-based blown films. 2. *Ind. Eng. Chem., Prod. Res. Dev.*, **1987**, *26*(8):1659-1663.
36. Swanson, C. L.; Westhoff, R. P.; Doane, W. M. Modified starch in plastic films. *Proceedings of Corn Utilization Conf. II*, 1988; pp 499-521.
37. Jasberg, B. K.; Swanson, C. L.; Nelson, T. C.; Doane, W. M. Twin-screw mixing of polyethylene/starch formulations for blown films. *Proceedings of Corn Utilization Conference III*, 1990; pp 566-569.
38. Chamroux, I. Demain, des plastiques d'origine agricole. Cultivar No. 251, 1989, pp 31-32.
39. Durkin, H. L. Thermoplastic starch materials. *Proceedings of Corn Utilization Conference III*, 1990; pp 489-492.
40. Wittwer, F.; Tomka, I. U.S. Patent 4 673 438, 1987.
41. Lay, G.; Rohm, J.; Stepto, R. F. T.; Thoma, M. Polymeric materials made from destructurized starch and at least one synthetic thermoplastic polymeric material. European Patent Application Publication No. 0 327 505 A 2, 1988.
42. Stepto, R. F. T.; Tomka, I. Injection molding of natural hydrophobic polymers in the presence of water. *Chimia*, **1987**, *41*(3):76-81.
43. Lacourse, N. L.; Altieri, P. A. U.S. Patent 4 863 655, 1989.
44. Doane, W. M. Starch: Renewable raw material for the chemical industry. *J. Coat. Technol.*, **1978**, *50*(636):88-98.
45. Griffin, G. J. L. U.S. Patent 4 420 576, 1983.
46. Fanta, G. F.; Bagley, E. B. Starch, graft copolymers. *Encycl. Polym. Sci. Technol.*, 1977, Suppl. 2, 665-699.
47. Fanta, G. F.; Doane, W. M. Grafted Starches. *Modified Starches: Properties and Uses*, Wurzburg, O. B., Ed.; CRC Press, 1986; pp 149-178.
48. Bagley, E. B.; Fanta, G. F.; Burr, R. C.; Doane, W. M.; Russell, C. R. Graft copolymers of polysaccharides with thermoplastic polymers. A new type of filled plastic. *Polym. Eng. Sci.*, **1977**, *17*:311-316.
49. Fanta, G. F.; Burr, R. C.; Doane, W. M.; Russell, C. R. Graft polymerization of styrene onto starch by simultaneous cobalt-60 irradiation. *J. Appl. Polym. Sci.*, **1977**, *21*:425-433.

50. Fanta, G. F.; Swanson, C. L.; Burr, R. C.; Doane, W. M. Polysaccharide-g-polystyrene copolymers by persulfate initiation: preparation and properties. *J. Appl. Polym. Sci.*, **1983**, *28*:2455-2461.
51. Westover, R. F. Melt extrusion. *Encycl. Polym. Sci. Technol.*, **1968**, *8*:573.
52. Patel, A. R.; Patel, M. R.; Patel, K. C.; Patel, R. D. Processibility of starch-graft-polyacrylonitrile copolymers. *Angew. Makromol. Chemie*, **1985**, *136*:135-145.
53. Dennenberg, R. J.; Bothast, R. J.; Abbott, T. P. A new biodegradable plastic made from starch graft poly(methylacrylate) copolymer. *J. Appl. Polym. Sci.*, **1978**, *22*:459-465.
54. Swanson, C. L.; Fanta, G. F.; Fecht, R. G.; Burr, R. C. Starch-g-poly(methyl acrylate)--Effects of starch level and molecular weight on tensile strength. C. E. Carraher, Jr.; Sperling, Eds.; *Polymer Applications of Renewable-Resource Materials*, Plenum; 1983; pp 59-71.
55. Henderson, A. M.; Rudin, A. Effects of water on starch-g-poly-styrene and starch-g-poly(methyl acrylate) extrudates. *J. Appl. Polym. Sci.*, **1982**, *27*:4115-4135.
56. Fanta, G. F.; Otey, F. H. U.S. Patent 4 839 450, 1989.
57. Gugliemelli, L. A.; Swanson, C. L.; Baker, F. L.; Doane, W. M.; Russell, C. R. Cationic starch-polyacrylonitrile graft copolymer latexes. *J. Polym. Sci., Polym. Chem. Ed.*, **1974**, *12*:2638-2692.
58. Gugliemelli, L. A.; Swanson, C. L.; Doane, W. M.; Russell, C. R. Latexes of starch-based graft copolymers containing polymerized acrylonitrile. *J. Appl. Polym. Sci.*, **1976**, *20*:3175-3183.
59. Gugliemelli, L. A.; Doane, W. M.; Russell, C. R. Preparation of soapless latexes by sonification of starch-based poly-(isoprene-co-acrylonitrile) graft reaction mixtures. *J. Appl. Polym. Sci.*, **1979**, *23*:635-644.
60. Gugliemelli, L. A.; Swanson, C. L.; Doane, W. M.; Russell, C. R. Cationic starch graft polychloroprene latexes. *J. Polym. Sci., Polym. Letters Ed.*, **1976**, *14*:215-218.
61. Weaver, M. O.; Fanta, G. F. Cationic starch graft copolymer latexes as soil stabilizers. *J. Polym. Mater.*, **1987**, *4*:51-65.
62. Weaver, M. O. Starch-g-poly(methyl acrylate) latexes for stabilizing soil to water erosion: extending the range of polymer add-on. *Stärke*, **1989**, *41*:106-110.
63. Weaver, M. O. Starch-g-poly(methyl acrylate) latexes for stabilizing soil to water erosion: varying the starch moiety. *J. Polym. Mater.*, **1989**, *6*:271-276.
64. Stannett, V. T.; Fanta, G. F.; Doane, W. M. Polymer grafted cellulose and starch. Chatterjee, P. K., Ed.; *Absorbency, Textile Science and Technology 8*, Elsevier; 1985; pp 257-279.
65. Smith, T. U.S. Patent 3 661 815, 1972.
66. Weaver, M. O.; Bagley, E. B.; Fanta, G. F.; Doane, W. M. U. S. Patent 3 997 484, 1976.

67. Burr, R. C.; Fanta, G. F.; Russell, C. R.; Rist, C. E. Influence of swelling and disruption of the starch granule on the composition of the starch-polyacrylonitrile copolymer. *J. Macromol Sci.*, **1967**, Chem. A1:1381-1385.
68. Weaver, M. O.; Montgomery, R. R.; Miller, L. D.; Sohns, V. E.; Fanta, G. F.; Doane, W. M. A practical process for preparation of Super Slurper, a starch-based polymer with a large capacity to absorb water. *Stärke*, **1977**, *29*:413-422.
69. Weaver, M. O.; Gugliemelli, L. A.; Doane, W. M.; Russell, C. R. Hydrolyzed Starch-polyacrylonitrile graft copolymers: Effect of structure on properties. *J. Appl. Polym. Sci.*, **1971**, *15*:3015-3024.
70. Taylor, N. W.; Bagley, E. B. Dispersions or solutions? A mechanism for certain thickening agents. *J. Appl. Polym. Sci.*, **1974**, *18*:2747-2761.
71. Gugliemelli, L. A.; Weaver, M. O.; Russell, C. R.; Rist, C. E. Base hydrolyzed starch-polyacrylonitrile (S-PAN) graft copolymer. S-PAN-1:1, PAN M. W. 794,000. *J. Appl. Polym. Sci.*, **1969**, *13*:2007-2017.
72. Goodman, J. T.; Graham, T. O. Belgian Patent 886 842, 1981.
73. Graham, T. O.; Parsen, F. E. Belgian Patent 886 843, 1981.
74. Rodehed, C.; Rånby, B. Characterization of sorbed water in saponified starch-g-polyacrylonitrile with differential scanning calorimetry. *J. Appl. Polym Sci.*, **1986**, *32*:3309-3315.
75. Rodehed, C.; Rånby, B. Gel collapse in the saponified starch-g-polyacrylonitrile/water-alcohol system. *Polymer*, **1986**, *27*:313-316.
76. Castel, D.; Richard, A.; Audebert, R. Swelling of anionic and cationic starch-based superabsorbents in water and saline solution. *J. Appl. Polym. Sci.*, **1990**, *39*:11-29.
77. Jones, D. A.; Elmquist, L. F. U.S. Patent 4 159 260, 1979.
78. Elmquist, L. F. U.S. Patent 4 302 369, 1981.
79. Fanta, G. F.; Burr, R. C.; Doane, W. M.; Russell, C. R. Absorbent polymers from starch and flour through graft polymerization of acrylonitrile and comonomer mixtures. *Stärke*, **1978**, *30*:237-242.
80. Fanta, G. F.; Doane, W. M.; Stout, E. I. U.S. Patent 4 483 950, 1984.
81. Masuda, F.; Nishida, K.; Nakamura, A. U.S. Patent 4 076 663, 1978.
82. Valdez, H. A hydrogel preparation for cleansing and protecting equine wounds. *Equine Pract.*, **1980**, *2*:33-36.
83. Geronemus, R. G.; Robins, P. The effect of two new dressings on epidermal wound healing. *J. Dermatol. Surg. Oncol.*, **1982**, *8*:850-852.
84. Spence, W. R. U.S. Patent 4 226 232, 1980.
85. Spence, W. R. U.S. Patent 4 272 514, 1981.
86. Tanzy, K. Is there a Super Slurper in your future? *Soybean Dig.*, **1981**, *41*(1):306-308.
87. Deterling, D. "Super Slurper" gets your crop moving earlier. *Progressive Farmer*, **1981**, *96*(2):56k, 56p.
88. Hamilton, J. L.; Lowe, R. H. Use of a water absorbent polymer in tobacco seedling production and transplanting. *Tobacco Sci.*, **1982**, *26*:17-20.

89. Shrader, W. D.; Mostejeran, A. Potentials of a polyacrylonitrile-starch polymer, Super slurper, to modify water holding properties of soils. *Coatings Plastics Prepr. Am. Chem. Soc. Div. Org. Coatings and Plastics Chem.*, **1977**, *37*(1):683-687.
90. Kay, B. L. Mulch choices for erosion control and plant establishment. *Weeds Trees Turf.*, **1980**, *19*(8):16-24.
91. Fanta, G. F.; Burr, R. C.; Orton, W. L.; Doane, W. M. Liquid phase dehydration of aqueous ethanol-gasoline mixtures. *Science*, **1980**, *210*:646-647.
92. Burr, R. C.; Fanta, G. F.; Doane, W. M.; Russell, C. R. Starch graft copolymers for water treatment. *Stärke*, **1975**, *27*:155-159.
93. Fanta, G. F.; Burr, R. C.; Russell, C. R.; Rist, C. E. Graft copolymers of starch and poly(2-hydroxy-3-methacryloyloxypropyltrimethyl-ammonium chloride). Preparation and testing as flocculating agents. *J. Appl. Polym. Sci.*, **1970**, *14*:2601-2609.
94. Fanta, G. F.; Burr, R. C.; Russell, C. R.; Rist, C. E. Graft copolymers of starch with mineral acid salts of dimethylaminoethyl methacrylate. Preparation and testing as flocculating agents. *J. Appl. Polym. Sci.*, **1971**, *15*:1889-1902.
95. Fanta, G. F.; Burr, R. C.; Doane, W. M.; Russell, C. R. Graft copolymers of starch with mixtures of acrylamide and the nitric acid salt of dimethylaminoethyl methacrylate. *J. Appl. Polym. Sci.*, **1972**, *16*:2835-2845.
96. Jones, D. A.; Jordan, W. A. Starch graft polymers. I. Graft co- and terpolymers of starch with 2-hydroxyl-3-methacryloyloxypropyltrimethyl-ammonium chloride: preparation and evaluation as silica depressants. *J. Appl. Poly. Sci.*, **1971**, *15*:2461-2469.
97. Jones, D. A.; Elmquist, L. F. Starch graft copolymers. III. Preparation of graft copolymers containing acrylamide, acrylic acid and β-methacryloyloxyethyltrimethylammonium monomethyl sulfate and evaluation as flocculants for bauxite ore red mud suspensions. *Stärke*, **1973**, *25*:83-89.
98. Restaino, A. J.; Reed, W. N. U.S. Patent 3 561 933, 1971.
99. Restaino, A. J.; Reed, W. N. U.S. Patent 3 635 857, 1972.
100. Inano, M. Reactions of Starch IV. Flocculation of kaolin suspensions by starch type polymer flocculants. *Kogyo Kagaku Zasshi*, **1969**, *72*:2298-2303.
101. Hirano, T.; Hasegawa, T. Preparation and applications of starch grafted copolymers. Flocculation of kaolinite suspensions by starch-polyacylamide graft copolymers. *Kogyo Kagaku Zasshi*, **1971**, *74*:91-96.
102. Hirano, T.; Hasegawa, T. The flocculation of powdered coal suspension by starch-polyacrylamide. *Kogyo Kagaku Zasshi*, **1971**, *74*:456-460.
103. Heath, H. D.; Hofreiter, B. T.; Ernst, A. J.; Doane, W. M.; Hamerstrand, G. E.; Shulte, M. E. Cationic and ionic starch graft polymers for filler retention. *Stärke*, **1975**, *27*:76-82.

104. Heath, H. D.; Ernst, A. J.; Hofreiter, B. T.; Phillips, B. S.; Russell, C. R. Flocculating agent-starch blends for interfiber bonding and filler retention: comparative performance with cationic starches. *Tappi*, **1974**, *57*:109-111.
105. Jones, D. A.; Elmquist, L. F. Starch graft polymers. II. Preparation of graft polymers containing acrylamide and acrylic acid and evaluation as textile print paste thickeners. *Stärke*, **1972**, *24*:23-27.
106. Pledger, H., Jr.; Meister, J. J.; Hogen-Esch, T. E.; Butler, G. B. Starch-acrylamide graft copolymers for use in enhanced oil recovery. *Polym. Prepr.*, **1981**, *22*(2):72-73.
107. Meister, J. J. Rheology of Starch-acrylamide graft copolymer solutions. *J. Rheol.*, **1981**, *25*:487-506.
108. Meister, J. J.; Patil, D. R.; Jewell, M. C.; Krohn, K. Synthesis, characterization, and properties in aqueous solution of poly(starch-g-[1-amido-ethylene)-co-(sodium-1-(2-methylprop-2N-yl-1-sulfonate) amidoethylene)]). *J. Appl. Polym. Sci.*, **1987**, *33*:1887-1907.

RECEIVED May 2, 1991

Chapter 14

Monomers and Polymers Based on Mono-, Di-, and Oligosaccharides

Stoil K. Dirlikov

Coatings Research Institute, 122 Sill Hall, Eastern Michigan University, Ypsilanti, MI 48197

> The application of mono-, di-, and oligosaccharides as monomers in polymer chemistry has been reviewed. In addition, author's research on preparation, characterization, and potential applications of different saccharides and their derivatives has been summarized. Several diols: isosorbide, isomannide, 1,4-lactone of 3,6-anhydrogluconic acid, and 1,4:3,6-dilactone of mannosaccharic acid have been prepared. Their polymers exhibit unique solubility properties. Linear soluble polyurethanes and polyesters based on cellobiose, glucosamine, trehalose, and sucrose have been obtained. 1,4:2,5:3,6-Trianhydromannitol undergoes easy ring opening and appears to be an attractive volume expansion monomer. D-glucose methacrylate was prepared by direct addition of glucose to methacroyl chloride. 1,2:5,6-Dianhydrosorbitols produce good epoxy resins. Finally, polymer blends with cyclodextrin form films which are attractive for membrane separation. In conclusion, mono-, di-, and oligosaccharides appear to be a very attractive renewable resource for preparation of a broad variety of unique monomers and polymers.

Polymers containing mono- or disaccharides in the main chain or as pendant groups are receiving increasing attention. They possess a number of attractive, useful, and unique properties such as high hydrophilicity, optical activity (chirality), etc. In addition, some of these polymers have biological activity. Others are anticipated to be tractable and combine useful properties of both natural polysaccharides and synthetic polymers. Finally, polymers based on saccharide monomers are expected to undergo easier biodegradation than the synthetic polymers. Consequently, this class of biodegradable polymers has potential utility in reducing "plastic" pollution which is a major problem in our modern world. This is probably the most attractive and challenging application of saccharides in polymer chemistry. Many "saccharide monomers" such as glucose, sucrose, etc. or their derivatives, sorbitol, gluconic acid, and others, are industrially produced in large scale from sugar waste, starch, etc. and are available at a low price normally less than one dollar per pound. Other mono- and disaccharides are potentially available from different renewable resources (biomass), again at a low cost. Glucosamine is available from chitin and cellobiose, which is the cellulose dimer, can be obtained from cellulose by enzymatic degradation.

This chapter describes the preparation, characterization, and potential applications of different mono-, di-, and oligosaccharides and their derivatives as monomers in polymer chemistry. It is based on our results, part of which have been published (1), and reviews related research by other authors. In a sense, the chapter is a review of the potential of saccharides and their derivatives for preparation of a new class of polymers with many attractive (industrial) applications. The goal, however, is only to illustrate the potential of saccharides in polymer chemistry and is not a complete review of this field. The saccharides described here have been selected either because their polymers have unique properties or they are attractive inexpensive industrial products. Ring-opening (cationic) polymerization of anhydrosugar derivatives has been purposely omitted because there are several excellent reviews in this field published recently (2-6).

Isosorbide

Sorbitol (I) is an attractive and inexpensive raw material available from renewable resources. It is industrially produced by hydrogenation of glucose (II) from sugar waste and is available at about 50 cents/pound. Alternative sorbitol can be produced from cellulose.

Sorbitol easily dehydrates into its 1,4:3,6-dianhydroderivative, isosorbide (III), in the presence of an acid catalyst such as sulfuric, hydrochloric, or toluenesulfonic acid or a cation exchange resin in a one-step preparation in high yield (75-85%)(7):

```
    CHO              CH₂OH             CH₂⎤
     |                |                 |  |
    CHOH             CHOH              CHOH|
     |                |                 |  |
    HOCH             HOCH         ⎡────CH  O
     |        →       |        →  |     |  |
    CHOH             CHOH         |    CH──┘
     |                |           |     |
    CHOH             CHOH         O    CHOH
     |                |           |     |
    CH₂OH            CH₂OH        └────CH₂

     II               I                III
```

Isosorbide is a rigid diol with two non-equivalent hydroxyl groups (endo-5 and exo-2) and two tetrahydrofuran rings:

Its polymers, therefore might have good complexation ability.

A literature search shows that isosorbide has been used in the preparation of polyesters (8,9) and polycarbonates (10,11).

We have prepared and characterized three linear isosorbide containing polyurethanes with toluene diisocyanate (TDI), 4,4'-diphenylmethane diisocyanate (MDI), and 1,6-hexamethylene diisocyanate (HMDI): P(I-TDI), P(I-MDI), and P(I-HMDI). These polyurethanes have been synthesized as described in the experimental section by solution polymerization of isosorbide with the corresponding diisocyanate in dimethylacetamide using dibutyltin dilaurate as a catalyst at 75°C for 24 hours. All polymers have been isolated in quantitative yield by precipitation in methanol or water (12).

All polyurethanes have been obtained by reacting 1:1 molar ratios of isosorbide and diisocyanate. A slight excess of about 5-10% of the diisocyanate or of the diol resulted in the generation of soluble polyurethanes with lower molecular weight and possessing either functional isocyanate or hydroxyl groups, respectively.

All three polymers possess high molecular weights in the range of 20-25,000 by gel permeation chromatography.

```
      CH₂ ┐
       |  |
      CHOH |
       |  |
   ┌──CH   O
   |   |  |
   |  -CH──┘
   |   |
   |  CHOH
   O   |
   |  CH₂
   └────┘
```
                        +   OCN.R.NCO ─────>

```
      CH₂ ┐
       |  |
      CHO──┐         ┐
       |   |         |
   ┌──CH   O    ─────|─────
   |   |  |          |
   |  CH──┘          |
   |   |             |
   O  CHO────OC.NH.R.NH.CO──|──
   |   |                    | n
   └──CH₂
```

They are soluble inpolar solvents such as dimethylacetamide, dimethyl sulfoxide, and dimethylformamide, and insoluble in non-polar solvents such as chloroform and carbon tetrachloride. The other properties of these polymers depend on the diisocyanate used in their preparation.

The isosorbide polyurethane based on the aliphatic diisocyanate P(I-HMDI) has flexible molecules. It is a thermoplastic with a glass transition temperature of $110°C$ and softening temperature of $190°C$. Both transitions are well below its degradation onset which occurs at approximately $260°C$. It forms good films by evaporation from its solutions and colorless transparent compression moldings.

The isosorbide polyurethanes based on the aromatic diisocyanates: P(I-TDI) and P(I-MDI), possess more rigid structures with both polymers forming brittle films and brittle compression moldings. Their glass transition temperatures are above their decomposition temperature of $260°C$. The thermostability of isosorbide polyurethanes correspond to that of conventional polyurethanes with similar structure based on 1,4-cyclohexanedimethanol for which degradation temperature of $260°C$ has been determined.

The limiting oxygen index (LOI) corresponds to the minimum oxygen content in air which supports still combustion of the polymer. LOI of P(I-MDI) is 25 and it is not flammable since atmospheric air has only 21% oxygen. In contrast, the conventional polyurethane with a similar structure based on MDI and 1,4-cyclohexane-dimethanol has LOI of 20 and it is flammable in air. It indicates that isosorbide polyurethanes require higher oxygen content to support combustion and they have lower flammability than the corresponding pure "synthetic" polymers.

Isosorbide polyurethanes, especially those based on aliphatic isocyanates, may be useful in the same applications as conventional polyurethanes i.e. thermoplastics, coatings, and foams. In fact, excellent rigid foams have been obtained from P(I-MDI)(12). Isosorbide has a low melting point of 61°C and it is suitable for use in reactive injection molding processes alone or in the form of a mixture with other conventional diols. In addition, its polymers may also find specific applications due to the anticipated high complexation ability of the two tetrahydrofuran rings in their isosorbide units.

Isomannide

Isomannide or 1,4:3,6-dianhydromannitol (**IV**) is an isomer of isosorbide. It has been prepared by dehydration of mannitol (**V**) according to a procedure similar to that for the preparation of isosorbide (13) which has been described previously.

Isomannide is a rigid diol with a low melting point of 85°C. It has two adjacent tetrahydrofuran rings in a "clam" configuration and two equivalent hydroxyl groups, both in the endo-configuration, suitably situated for complexation via their oxygen atoms. One could speculate that isomannide will have better complexation ability than isosorbide since its four oxygen atoms are oriented in the same direction.

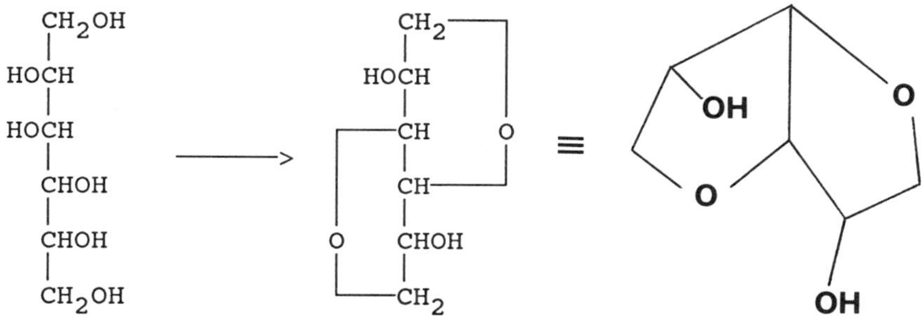

$$\begin{array}{c}CH_2OH\\|\\HOCH\\|\\HOCH\\|\\CHOH\\|\\CHOH\\|\\CH_2OH\end{array} \longrightarrow \quad \text{IV} \quad \equiv$$

V   IV

There is a growing interest in polymers which contain tetrahydrofuran rings in their main chains because of their good/excellent ability to complex with different cations. Their applications, however, are limited by difficult multi-step preparations. Isomannide polymers appear to be attractive candidates and easy alternatives in such applications. Isomannide polyurethanes have been prepared by the same procedure described for isosorbide polyurethanes in the experimental section. Their properties are under investigation.

Dimethyl Isosorbide

Other molecules of the same type, which are expected to have good complexation ability, are the methyl or ethyl ethers of isosorbide, isomannide, and isoidide. They appear to have the potential as solvents with high solvation power and low reactivity. All these molecules possess rigid structures with two adjacent tetrahydrofuran rings in "clam-like" configuration and four oxygen atoms with glyme distribution, $-CH_2OCH_2CH_2OCH_2CH_2OCH_2-$. They are expected, therefore, to exhibit stronger solvating power and complexation ability in comparison to that of tetrahydrofuran and other glymes. Dimethyl isosorbide (**VI**) and dimethyl isomannide (**VII**) illustrate the structures of these materials.

Dimethyl isosorbide (**VI**) has been prepared in quantitative yield from isosorbide and dimethyl sulphate according to known procedure (7). It is a liquid possessing a low vapor pressure at room temperature and boiling point of 95°C at 0.1 mm Hg. In addition, it also possesses optical activity. It is expected to be a relatively inexpensive solvent (for an optical active solvent with 100% purity of the optical isomer) in comparison to other "solvating" (non-optical active) solvents such as tetrahydrofuran (one dollar per liter), dioxane (one dollar per liter), and pyridine (two dollars and 50 cents per liter).

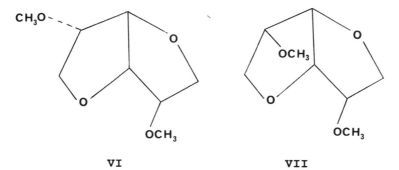

VI　　　　　　　　　VII

Dimethyl isomannide (**VII**) has both methoxy groups in endo-position and might even have better complexation than dimethyl isosorbide (as discussed above for isomannide). However, it does not have optical activity.

Dimethyl isoidide is the third possible isomer with two exo groups.

1,4:2,5:3,6-Trianhydromannitol

1,4:2,5:3,6-Trianhydromannitol (**VIII**) is another attractive monomer. It is obtained from isosorbide in a two-step synthesis with about 50% overall yield as outlined below (14,15):

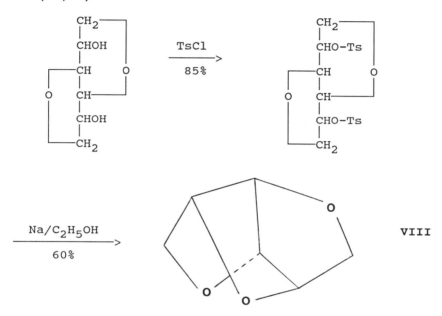

1,4:2,5:3,6-Trianhydromannitol has a low melting point of 64°C. It is a highly strained molecule with three adjacent tetrahydrofuran rings. In our studies, we have found that it polymerizes in melt or in solution (in methylene dichloride) at normal or elevated temperature (70°C) in the presence of $BF_3 \cdot O(C_2H)_2$ as a catalyst with the formation of an insoluble polymer. Thermo-gravimetric analysis (TGA) of these polymers shows an onset of thermodegradation in the range of 135-150°C. The polymerization evidently proceeds with opening of one or two of the three tetrahydrofuran rings of the monomer. As a result, we believe it has a potential as a volume-expansion monomer and this property is under investigation. Another possibility for its utilization is for preparation of glyme-like cyclic oligomers by opening of only one of its tetrahydrofuran rings by polymerization or copolymerization of the monomer in solution at milder conditions and lower temperatures.

1,4-Lactone of 3,6-Anhydrogluconic Acid (LAGA)

LAGA (**IX**) is available from gluconic acid (**X**) according to the following scheme. Gluconic acid (**X**) is an inexpensive (70 cents/lb.), commercially available raw material produced by oxidation of glucose (**II**) from sugar waste. Cellulose could be used as an alternative resource. Gluconic acid exists in its "acidic" form only in aqueous solutions. Upon evaporation, it forms its crystalline 1,4- (**XI**) or 1,5-lactones (**XII**) depending on reaction conditions.

LAGA is a known compound. Its preparation from glucose involves a long, multi-step blocking and deblocking procedure which proceeds in low overall yield (16-20).

We have been able to develop a one-step procedure (21) for the preparation of LAGA by acid catalyzed dehydration directly from gluconic acid or from its 1,4- or 1,5-lactones:

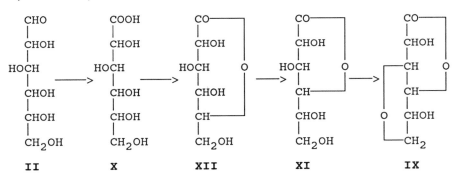

It was found that the 1,5-lactone of gluconic acid rearranges into 1,4-lactone of LAGA in the course of the reaction. The formation of the 3,6-anhydro ring proceeds easily in the presence of DOWEX 50WX4 ion exchange resin as a catalyst at $110^{\circ}C$ with the removal of water as an water/toluene azeotrope. The process was carried out as described for the preparation of isosorbide from sorbitol. The procedure, however, has not been optimized and the yield of LAGA was only 37%. The optimization of the procedure will certainly increase the yield of LAGA.

The 1,4-lactones of other 3,6-anhydro acids are available in a similar manner.

LAGA is a rigid diol with a melting point of $117^{\circ}C$ (large prisms from ethyl acetate). It has two non-equivalent hydroxyl groups in endo-5 and exo-2 positions.

LAGA is a monocarbonyl derivative of isosorbide. In contrast to isosorbide, however it is not hydroscopic and easily opens its lactone ring with the formation of potential cross-linking sites.

14. DIRLIKOV Mono-, Di-, and Oligosaccharides

```
    CO
    |
    CHOH
    |
   -CH     O
    |
    CH-
    |
 O  CHOH
    |
   -CH₂
```
≡
[structure of LAGA showing HO, O, =O, OH]

A literature search does not indicate any publications concerning the use of LAGA in polymer applications.

We have prepared two linear soluble polyurethanes: P(LAGA-MDI) and P(LAGA-HMDI) by solution polymerization of LAGA with MDI and HMDI respectively in dimethylacetamide as a solvent using dibutyltin dilaurate as a catalyst at 75°C over 24 hours according to the procedure used in the preparation of the isosorbide polyurethanes as described in the experimental section. Both polymers have been isolated in quantitative yield in their "<u>lactone</u>" form by precipitation in chloroform (22).

As obtained, P(LAGA-MDI) and P(LAGA-HMDI) are soluble in polar solvents such as dimethylacetamide, dimethyl sulfoxide, and dimethylformamide, and insoluble in nonpolar solvents as chloroform and carbon tetrachloride. These polymers possess high molecular weight in the range of 20-25,000.

The properties of the polyurethanes depend on the nature of the diisocyanate used in their preparation.

P(LAGA-MDI) has rigid macromolecules possessing a glass transition temperature above the onset of degradation which is observed at about 240°C. It forms brittle films and brittle compression moldings. In contrast, the aliphatic polyurethane, P(LAGA-HMDI) has more flexible macromolecules. It is a thermoplastic with a glass transition temperature of 85°C and a softening temperature of 150°C. Its degradation onset is observed around 240°C. This polyurethane forms good films from solution and colourless transparent compression moldings.

```
    CO
    |
    CHOH
    |
   -CH     O
    |
    CH-
    |
 O  CHOH
    |
   -CH₂
```
    +   OCN.R.NCO   ——→

```
      CO─────┐
      │      │
      CHO────┼──────    ──┼──
      │      │
   ┌──CH     O
   │  │      │
   │  CH─────┘
   │  │
   O  CHO────────CO.NH.R.NH.CO──┼──────
   │  │                         │n
   └──CH₂
```

These results show that in general, LAGA polyurethanes in their <u>lactone</u> form have practically the same transitions and properties as the corresponding isosorbide polyurethanes (12,22).

In contrast to isosorbide polymers precipitation of the initially formed LAGA polymers in water (instead of chloroform) results in slow opening of the lactone rings with formation of hydroxyl and carboxyl groups. This hydrolysis results in a dramatic change in solubility, TGA, and IR, etc.. Although the rate of the lactone ring opening is very slow at the beginning the formation of the carboxyl groups lowers the pH of the aqueous media and accelerates the hydrolysis of the remaining lactone rings. As a result, the solubility of the polymer in water sharply increases with hydrolysis.

```
      CO─────┐                          COOH
      │      │                          │
      CHO────┼──────                    CHO────────
      │      │                          │
   ┌──CH     O           ⇌           ┌──CH
   │  │      │                       │  │
   │  CH─────┘                       │  CHOH
   │  │                              │  │
   O  CH──O─────                     O  CH──O─────
   │  │                              │  │
   └──CH₂                            └──CH₂
```

The infrared spectra of the hydrolyzed polymer show a strong broad absorption band in the range of 2500-3500 cm^{-1} which is not observed in the spectra of the initial non-hydrolyzed polymer. This absorption band corresponds to the O-H vibrations of the free hydroxyl and carboxyl groups. Its intensity increases with the degree of hydrolysis. The broad character of this absorption without distinguished separate band indicates that the free hydroxyl and carboxyl groups form intra- and intermolecular hydrogen bondings (22).

The TGA curves of the initial P(LAGA-HMDI) polymer in the lactone form show a weight loss onset at approximately 240°C. In contrast, the hydrolyzed polymer in the so-called acidic form is very hydrophilic, absorbs large amounts of water from the air at room temperature and decreases the temperature at which weight loss occurs. The weight loss which occurs below 180°C, probably corresponds entirely to the release of absorbed water. The glass transition temperature of the hydrolyzed polymers is not observed in DSC due to the intra- and intermolecular hydrogen bonding between the hydroxyl and carboxyl groups.

These polymers, which slowly dissolve in water, might be suitable for control release and preparation of biodegradable polymers. LAGA homo- and copolymers open the lactone rings of their LAGA monomeric units with formation of hydroxyl and carboxyl groups and thus, increase the polymer solubility, hydrophilicity and the rate of their biodegradation. In this sense LAGA might offer a new approach for polymer controlled biodegradation.

The partially hydrolyzed LAGA polymers have free hydroxyl and carboxyl groups which act as sites for cross-linking. Their cross-linked polymers have been obtained in the presence of additional amount of diisocyanate (22).

1,4:3,6-Dilactone of Mannosaccharic Acid

Another group of diol monomers of a similar type is based on the dilactones of the saccharic acids (DLSA) (**XIII**) which exist in two forms. Their acidic form (**XIV**) is formed only in aqueous solutions. Upon the evaporation of water, the saccharic acids (**XIV**) close their two lactone rings producing the corresponding stable diols (**XIII**).

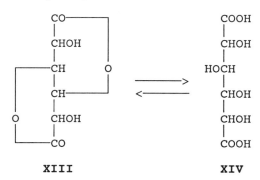

XIII   XIV

DLSA are rigid diols. They have the same skeleton as isosorbide and LAGA. For instance, 1,4:3,6-dilactone of mannosaccharic acid is a dicarbonyl derivative of isomannide and monocarbonyl derivative of the lactone of the anhydromannonic acid. Therefore, DLSA could find applications as diols and their polymers are expected to have similar properties to that of isosorbide and LAGA polymers.

There are two possible routes for DLSA preparation: 1). They are available by oxidation of the corresponding monosaccharides with nitric acid (23). This reaction generates relatively low yields in the range of 40%.

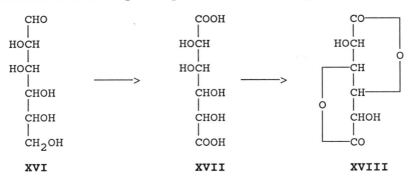

2). An alternative method for their preparation is by oxidation of uronic acids (**XV**) which are available from alginic acids (from kelp).

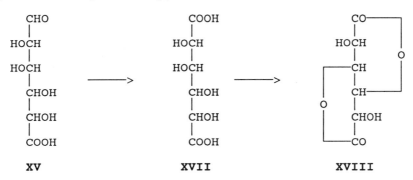

Our experience indicates that DLSA are less available than isosorbide and LAGA. We have prepared mannosaccharic acid (**XVII**) and its 1,4:3,6-dilactone (**XVIII**) by oxidation of mannose (**XVI**) with nitric acid according to (16). Pure dilactone with a melting point of 187°C has been isolated in 26% yield.

Its polyurethanes are under investigation.

1,4:3,6-Dilactone of D-Glucaric Acid

Recently, Hashimoto et al. (24) reported an improved procedure for 1,4:3,6-dilactone of D-glucaric acid (**XIII**) and preparation of its polyamide, poly(p-xylylene-D-glucaramide) directly from **XIII** and p-xylylenediamine:

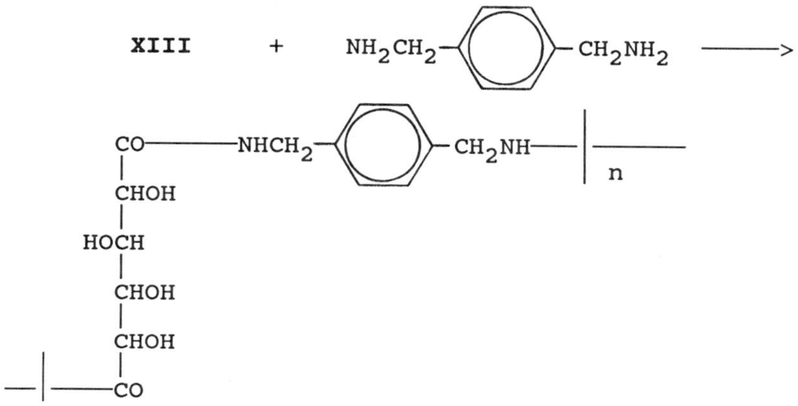

$$\textbf{XIII} + NH_2CH_2-\langle C_6H_4\rangle-CH_2NH_2 \longrightarrow$$

$$-\left[\begin{array}{l}CO-NHCH_2-\langle C_6H_4\rangle-CH_2NH-\\ | \\ CHOH\\ | \\ HOCH\\ | \\ CHOH\\ | \\ CHOH\\ | \\ CO\end{array}\right]_n$$

The reactivity of the lactone rings of **XIII** is enhanced by the hydroxyl groups adjacent to the lactone carbonyl groups and the ring-opening polyaddition proceeds in DMF or $(CH_3)_2SO$ solution even at 25°C without any catalyst. The polyamide is obtained with no protection of the hydroxyl groups of **XIII**.

Similar polyamides have been reported by polycondensation of tetra-O-acetyl-D-galactaric dichloride with diamines followed by deacetylation (25).

Nylons Based on Saccharides

Two types of high viscosity nylons have been prepared by interfacial polycondensation from (a) 1,6-diamino-1,6-dideoxy-di-O-methylenehexitol and acid (sebacoyl or adipoyl) dichloride, and (b) from a di-O-methylenehexaroyl dichloride and hexa- or decamethylenediamine, followed by a deblocking procedure (26). The final "sugar" nylons have high melt temperature (over 270°C), high water absorption, but are not soluble in water.

Epoxy Resins: 1,2:5,6-Dianhydro-3,4-O-Isopropylidene-D-Mannitol

One would also think that saccharides are more attractive starting raw materials for direct preparation of epoxy resins than hydrocarbons since they already possess oxygen atoms needed for the epoxy rings. Such diepoxy derivatives of hexitols are known compounds, for instance 1,2:5,6-dianhydro-3,4-O-isopropylidene-D-mannitol (**XIX**). Their preparation, however is difficult and requires 15 to 18 steps which proceed in very low overall yield (27-29).

As a first step in this direction, we have been able to prepare 1,2:5,6-dianhydro-3,4-O-isopropylidene-D-mannitolas a model diepoxy hexitol derivative in four steps with overall 50% yield as follows:

```
   CH2OH              CH2O     CH3           CH2OH
   |                  |     \C/              |
   HOCH               CHO    /\CH3           HOCH
   |                  |                      |
   HOCH               CHO    CH3             CHO     CH3
   |          →       |     \C/        →     |     \C/
   CHOH               CHO    /\CH3           CHO    /\CH3
   |                  |                      |
   CHOH               CHO    CH3             CHOH
   |                  |     \C/              |
   CH2OH              CH2O   /\CH3           CH2OH

     V                    XX                    XXI
```

```
   CH2Cl                     CH2\
   |                         |   \O
   HOCH                      CH  /
   |                         |
   CHO     CH3               CHO    CH3
   |     \C/          →      |     \C/
   CHO    /\CH3              CHO    /\CH3
   |                         |
   CHOH                      CH  \
   |                         |   \O
   CH2Cl                     CH2 /

    XXII                        XIX
```

The first step, which proceeds in 75% yield to form (**XX**), involves the blocking of mannitol (**V**) with acetone at room temperature in the presence of 1% concentrated sulfuric acid as a catalyst (30).

The second step involves unblocking of the primary hydroxyl groups in 70% acetic acid at 40°C for 1.5 hours (31). By this procedure, 3,4-O-isopropylidene-D-mannitol (**XXI**) has been obtained in 80% yield.

Then, selective halogenation has been carried out at mild conditions under which only the halogenation of the two primary hydroxyl groups occurs. The secondary hydroxyl groups remain intact under these conditions. The reaction proceeds smoothly with carbon tetrachloride and triphenylphosphine in anhydrous pyridine at 5°C for 18 hours (32). 1,6-Dichloro-1,6-dideoxy-3,4-O-isopropylidene-D-mannitol (**XXII**) has been obtained as described in the experimental part in quantitative yield according to the NMR spectrum of the reaction product (33). Its purification, however, is difficult and requires tedious chromatographic separation on a silica gel column. The purification conditions, however, have not been optimized and this results in a lower yield.

Finally, epoxy ring formation has been carried out in methanol at room temperature with sodium methoxide as a catalyst (34). 1,2:5,6-Dianhydro-3,4-O-isopropylidene-D-mannitol (**XIX**) has been obtained in near quantitative yield.

The NMR and IR spectra confirm the structure of 1,6-dichloro-1,6-dideoxy-3,4-O-isopropylidene-D-mannitol (**XXII**). The resonance signals of the CH_3 protons of the isopropylidene residues of both compounds **XXII** and **XIX** appear at about 8,60 ppm. The signal for the CHOH protons of **XXII** at 5,45 ppm is not observed in the NMR spectra of **XIX**. Instead, two new multiplets centered at 6,93 and 7,30 ppm for the methine and the methylene protons of the epoxy groups of **XIX** appear. At the same time, the relative intensity of the >CH-O and -CH_2-Cl protons of **XXII** at 6,15 ppm decreases with six protons in the spectrum of **XIX**. The infrared spectrum of **XXII** shows an absorption band at 3500 cm^{-1} which corresponds to its hydroxyl groups. This band is not observed in the spectrum of **XIX**.

This diepoxy mannitol derivative (**XIX**) cures with amines at room or elevated temperatures. When reacted with Versamid 140 polyamide resin (Henkel Corporation) with an amine value of 370-400, commonly used for curing of commercial epoxy resins, at epoxy/amine molar ratio of 1:1, it produces epoxy resins with good "physico-mechanical" properties and excellent adhesion to glass. The homogeneous mixture is cured as a thin layer between two (glass) microslides. The curing is carried out either at room temperature for 24 hours and then at 150°C for 2 hours, or at room temperature for 7 days. In both cases, transparent cured epoxy resins are obtained. Attempts to separate the two microslides always result in breaking the microslide which indicates strong epoxy resin/glass adhesion. The curing process is nearly complete since extraction with different solvents does not give any extractables.

Other renewable resources (sorbitol (**I**), cellobiose (**XXVII**), etc.) can be used in similar ways as starting materials in the preparation of epoxy resins.

An attractive direct two-step preparation of diepoxy-galactitol derivative has been reported in low yield by direct halogenation of the two primary hydroxyl

```
    CH2OH              CH2Br              CH2
    |                  |                  |   \O
    CHOH               CHOH               CH  /
    |                  |                  |
    HOCH               HOCH               HOCH
    |      ———>        |      ———>        |
    HOCH               HOCH               HOCH
    |                  |                  |
    CHOH               CHOH               CH  \
    |                  |                  |    O
    CH2OH              CH2Br              CH2 /

      I                XXIII              XXIV
```

groups with the formation of 1,6-dibromo-1,6-dideoxy-galactitol, (**XXIII**), followed by epoxy ring formation 1,2:5,6-dianhydrogalactitol (**XXIV**) (17,19,35).
There is no doubt that the yield will improve under more selective conditions. Diepoxy derivatives of mannitol and sorbitol have been prepared in a similar manner from the corresponding 1,6-di-O-methanesulphonoxy- or 1,6-dibromo derivatives (17,36). It is rather surprising that in all these reactions the three-membered epoxy ring is prefered over the formation of four-membered ring of the corresponding 1,3:4,6-dianhydrohexitols. It is confirmed by the quantitative formation of 2,3:4,5-dianhydroiditol (**XXV**) from the 3,4-di-O-p-tosylsulphonyl-mannitol (**XXVI**) with barium methoxide in methanol (37):

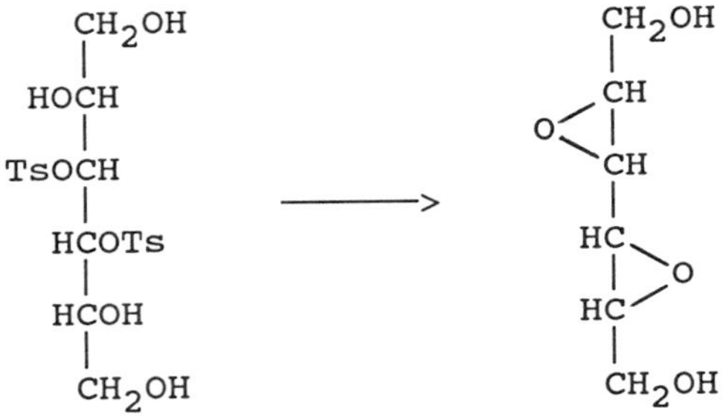

$$\begin{array}{c} CH_2OH \\ | \\ HOCH \\ | \\ TsOCH \\ | \\ HCOTs \\ | \\ HCOH \\ | \\ CH_2OH \end{array} \longrightarrow \begin{array}{c} CH_2OH \\ | \\ CH \\ O< | \\ CH \\ | \\ HC \\ | \quad >O \\ HC \\ | \\ CH_2OH \end{array}$$

XXVI   **XXV**

All these diepoxy hexitol derivatives are stable compounds.

Cellobiose

Cellobiose (**XXVII**), the basic unit of cellulose, is a potentially inexpensive monomer obtained in 90-95% yield by bacterial hydrolysis of cellulose. It is another attractive disaccharide with many potential applications in polymer chemistry.
The two primary hydroxyl groups of cellobiose and other mono- and disaccharides have higher reactivity than the remaining six secondary hydroxyl groups (38). Kurita and co-workers have recently reported the preparation of soluble polyurethanes by direct polyaddition of cellobiose to different diisocyanates: MDI, TDI, HMDI, etc., without blocking the excess hydroxyl groups of cellobiose by using the difference in reactivity between its primary and secondary hydroxyl groups (39-41).
The only other example of such polymers found in the literature are block-type copolymers derived from diisocyanates and depolymerized low molecular weight cellulose or amylose triacetate having hydroxyl end groups (42-44). The blocking acetyl groups in the resulting polyure-

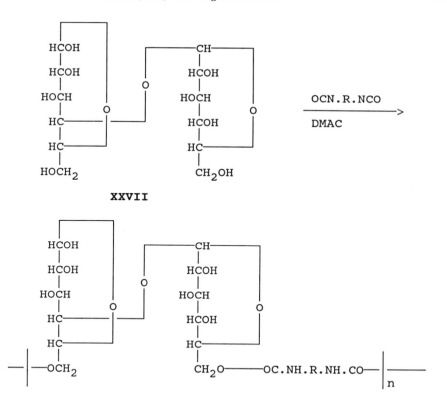

thanes have to be removed thereupon to regenerate the native saccharide structures.

We have carried out the polymerization of cellobiose with MDI in dimethylacetamide at 17°C with stirring for 24 hours as described by Kurita (39,41). The polymerization proceeds without gel formation. The polymer has been isolated in quantitative yield by precipitation in chloroform.

These cellobiose polyurethanes are soluble in polar solvents as dimethylacetamide, dimethyl sulfoxide, and dimethylformamide, etc., in marked contrast to the insolubility of cellulose. As expected, these polymers are insoluble in non-polar solvents as chloroform and carbon tetrachloride. They are very hydrophilic polymers and rapidly absorb water from air. MDI polymer contains about 20% water at 67% relative humidity at 0°C. TGA analysis shows that it releases absorbed water up to approximately 200°C with a 12% decrease in weight. Our gel permeation chromatography (GPC) shows an unexpectedly high average molecular weight in the range of 100,000 which is in agreement with the high polymer inherent viscosities up to 1.08 dl.g^{-1}. Polymers prepared from several different polymerization runs have the same molecular weight. This high molecular weight indicates that some branching probably occurs during the polymerization by the participation of small amount of the secondary hydroxyl groups of cellobiose in addition to the main type of polymerization of its two primary hydroxyl groups.

The pendant hydroxyl groups of cellobiose have been confirmed as useful for crosslinking of its polymers in the presence of additional amount of diisocyanate. Cross-linked insoluble films have been obtained by casting a polymer solution in dimethylacetamide containing 7% of an additional amount of MDI. The pendant hydroxyl groups undergo acetylation as well. However, the acetylated polyurethanes characterize with much lower water absorption and are amorphous.

The polymer limiting oxygen index (LOI) value is 26,5 which corresponds to that of the isosorbide polyurethanes. It is again higher than that of conventional MDI polyurethanes with similar structure based on 1,4-cyclohexanedimethanol (LOI = 20) and indicates lower flammability of cellobiose polymers.

TGA analysis shows that polymer degradation starts at about 235°C which corresponds to the temperature of decomposition of the cellobiose monomer (m.p. 239°C with decom.). Torsion Braid analysis and differential scanning calorimetry measurements show that this polymer is very rigid and does not exhibit any transition in the range of -100 to +250°C, e.g. the polymer decomposition occurs below any transition temperature. This result is expected since both of the monomers, cellobiose and MDI, have rigid molecules and because cellobiose units of the polymer form intra- and intermolecular hydrogen bondings.

Cellobiose polyurethanes based on aliphatic diisocyanates, e.g. HMDI, are expected to have more flexible macromolecules.

Cellobiose polyesters have been obtained directly from cellobiose under similar conditions with carboxylic (adipoyl, terephthaloyl, and isophthaloyl) acid dichlorides in the presence of pyridine (acid acceptor) (45):

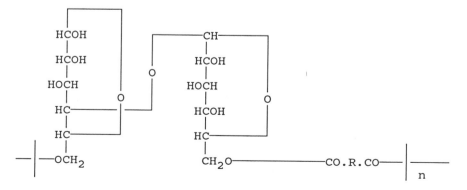

The resulting soluble polyesters have similar properties: Solubility, water absorption, thermostability, etc. to those of the polyurethanes.
All cellobiose polyurethanes and polyesters are crystalline. The cellobiose monomeric unit, obviously renders symmetry, rigidity, and intramolecular hydrogen bondings to its polymers, including cellulose and they all are crystalline.

D-Glucosamine

Kurita et al. (46-48) have also prepared soluble polymers by direct addition polymerization of D-glucosamine (**XXVIII**):

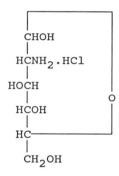

XXVIII

D-Glucosamine, the basic unit of chitin and chitosan, is a pentafunctional compound. Its primary hydroxyl and amino groups have higher reactivity than the three secondary hydroxyl groups and it behaves as a difunctional (aminoalcohol) monomer.
D-Glucosamine is unstable in a free-base form especially in solution, thus its hydrochloride is used as a monomer. The hydrochloride is subjected to additional polymerization in aprotic solvents with different diisocyanates (DMI, TDI, HMDI, etc.) in the presence of triethylamine as an acid acceptor without using the troublesome blocking and deblocking procedure of the excess hydroxyl groups of glucosamine (46,47):
The polymerization proceeds without gel formation. However, the possibile reaction of the secondary hydroxyl groups can not be ruled out completely and a small extent of branching probably proceeds as well.
Both natural glucosamine polymers, chitin and chitosan, are completely insoluble and intractable. Their full potential has not been explored because of problems related to their fabrication. In contrast, glucosamine poly(urea-urethane)s are soluble in polar solvents (DMAc, N-methyl-2-pyrrolidone (NMP), DMSO, etc.)

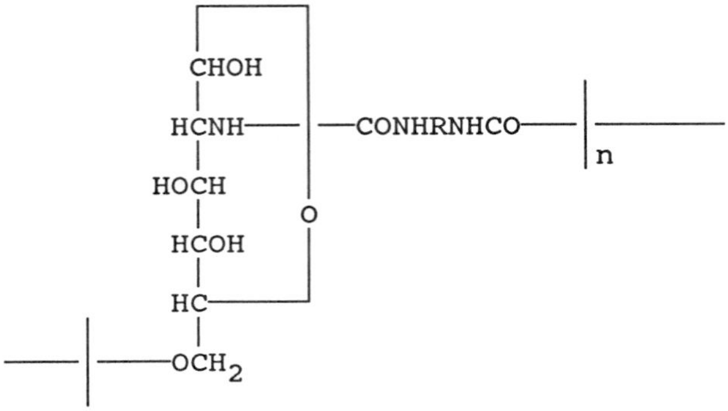

```
         ┌─────────────────┐
         CHOH
         |
    HCNH─┤   ─CONHRNHCO─ ┬─
         |               │n
         HOCH            │
         |         O
         HCOH            │
         |               │
         HC──────────────┘
         |
 ─┬──OCH₂
  |
```

and have high inherent viscosity of up to 1.27 d.g^{-1}. They form tough transparent films. The three pendant hydroxyl groups in each unit exhibit good reactivity (toward acetylation) and render high water absorption (18%). Polymer degradation starts at about 160°C due to the lower thermostability of glucosamine ring.

Glucosamine poly(amide-ester)s have been obtained under similar conditions with dicarboxylic (adipoyl, terephthaloyl, and isophthaloyl) acid dichlorides (48):

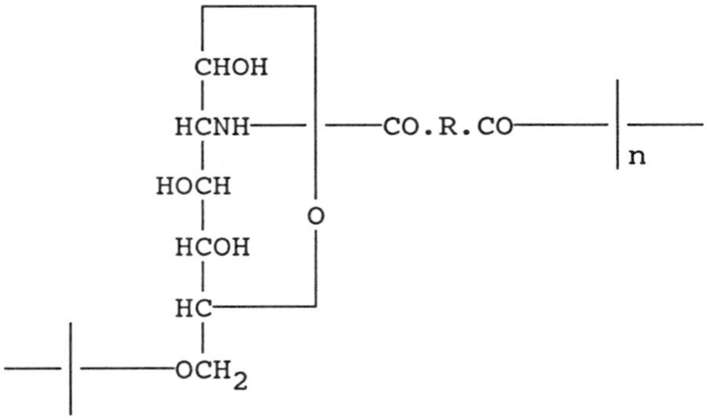

```
         ┌─────────────────┐
         CHOH
         |
    HCNH─┤   ─CO.R.CO─  ┬─
         |              │n
         HOCH           │
         |        O
         HCOH           │
         |              │
         HC─────────────┘
         |
 ─┬──OCH₂
  |
```

They have similar properties: good solubility and relatively low thermostability (160°C). In contrast to poly(urea-urethane)s, however they all are crystalline.

This is another example of Kurita's concept for a new type of tractable (half-natural and half-synthetic) polymers having sugar residues obtained from intractable polysaccharides such as chitin, cellulose, etc.

2,6-Diamino-2,6-dideoxyglucose based polyurethanes have been prepared in similar manner (49).

Trehalose

Alpha, alpha-Trehalose (**XXIX**) has eight hydroxyl

groups with two of them primary and structure similar to that of cellobiose:

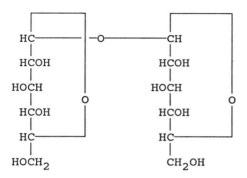

XXIX

In contrast to cellobiose and glucosamine, however it does not have readily oxidizable hemiacetal hydroxyl group and produce polymers with appreciable stability.

Kurita et al. (50) have described the preparation of soluble trehalose polyurethanes by solution polymerization directly from trehalose and different aromatic or aliphatic diisocyanates: MDI, TDI, HMDI, etc., as described above for cellobiose.

Trehalose polyurethanes have good solubility in polar solvents: DMAc, NMP, DMSO, etc., like the other polymers containing sugar residues. The MDI (and other) polyurethanes characterize with high water absorption: 19% at 67% relative humidity at 0°C. In contrast, the fully acetylated polymer has only 2% water absorption under the same conditions. TGA shows good polymer thermo-oxidative stability to about 300°C.

Sucrose

Sucrose (**XXX**) is the organic compound produced in the largest volume at the present moment:

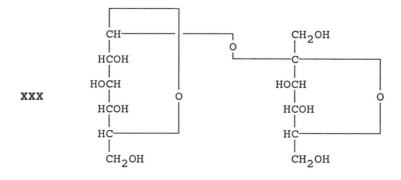

Sucrose is widely used for preparation of rigid cross-linked polyurethane (foams). Synthesis of linear sucrose polymers, however is difficult. Kurita's approach (39-41, 50) is applicable for multifunctional monomers with only two groups with higher reactivity. Sucrose has three primary (higher reactivity) and five secondary hydroxyl groups.

Recently, Patil (51) reported linear polyesters by direct polycondensation of sucrose with bis(2,2,2-trifluoroethyl) adipate in pyridine at 45°C in the presence of different enzymes, from which proleather has the highest activity.

Poly(sucrose adipate) is soluble in water and most polar solvents. It, however, has a relative lower molecular weight of Mw=2100 by GPC. ^{13}C-NMR spectra of the polymer show sucrose substitution at C6 and C1' positions. Sucrose C6' hydroxyl group does not react under these conditions:

XXX + $CF_3CH_2OCO(CH_2)_4COOCH_2CF_3$ ⟶

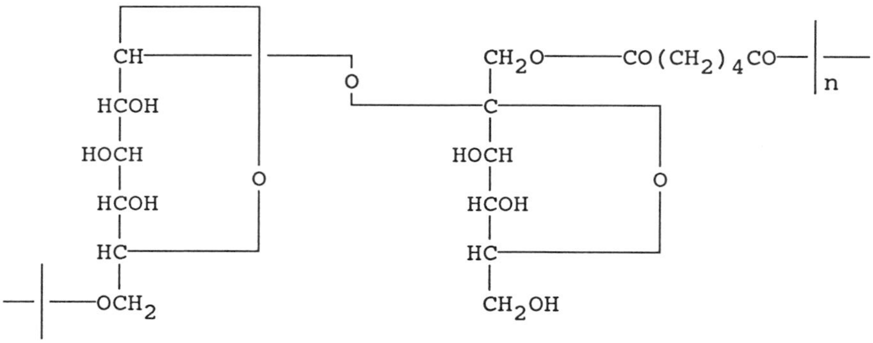

$$\begin{array}{c}\text{structure of poly(sucrose adipate)}\end{array}$$

D-Glucose Methacrylate

D-Glucose methacrylate (**XXXI**) is another monomer. Its polymers have many potential applications in medicine: soft contact lenses, surgical implants, external prothesis, orthopaedic appliances, secondary oil recovery, membrane separation (reverse osmosis), etc. (52,53). High molecular weight polymers M_n=715,000 have been prepared by emulsion polymerization (54).

The industrial production and applications of D-glucose methacrylate are limited by its difficult multi-step preparation with blocking and deblocking procedures of the secondary hydroxyl groups of glucose according to the procedures described in the literature (53).

1,2:5,6-Di-O-isopropylideneglucofuranose (55,56), (**XXXII**) and 1,2:3,4-di-O-isopropylidenegalactopyranose methacrylates (57,58), (**XXXIII**) etc. are common starting monomers:

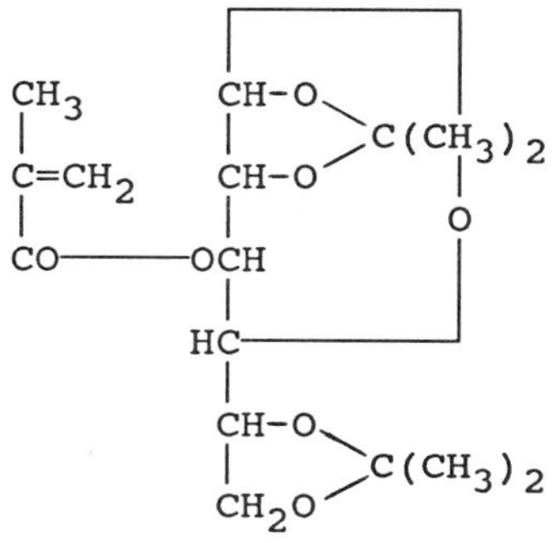

XXXII polymers are obtained by radical (or cationic) polymerization and subsequent acid hydrolysis of the protecting acetone groups. An interesting feature of the hydrolysis is that is accompanied by a change in optical rotation and by a furanose ⟶ pyranose ring conversion:

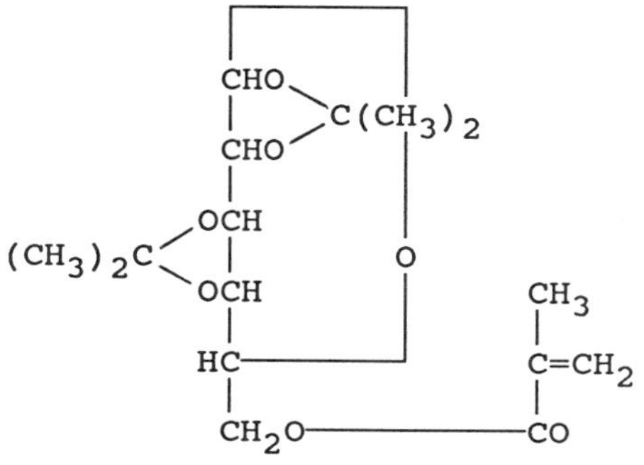

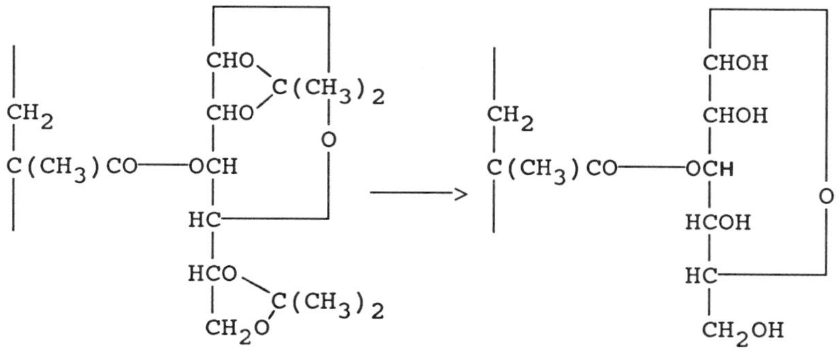

This is a completely new type of glucose polymer for it has a free reducing group at C-1 position on every glucose unit, while in all natural glucose polysaccharides, such as starch, cellulose, dextran, etc., the hemi-acetal hydroxyl group is combined in the glucosidic linkage (except in the one reducing unit at the end of each polysaccharide molecule). The reactions at the reducing group of these synthetic polymers open new applications not possible with natural polysaccharides (53). In addition, water-soluble sugar methacrylates may be cross-linked simply by heating at 60°C in vacuum over phosphoric oxide by a mechanism analogous to the polycondensation of reducing sugars when they are heated above their melting point under vacuum (57). This condensation occurs via the anomeric hydroxyl group and a secondary hydroxyl group on a neighboring polymer chain.

Sugar methacrylates may be a valuable cross-linking agent for many polymer systems.

The primary hydroxyl group of glucose has higher reactivity than the remaining four secondary hydroxyl groups. Our initial results indicate that direct preparation of glucose methacrylate is possible from glucose (II) (dried over P_2O_5) and freshly distilled (in vacuum) methacroyl chloride in 1:1 molar ratio in anhydrous dimethylacetamide and pyridine (in molar ratio as a hydrochloride scavenger) at room temperature under nitrogen with stirring for 20 hours, by a procedure similar to that described for the preparation of cellobiose polyurethanes, without blocking/deblocking procedure of the excess secondary hydroxyl groups, just by using the difference in reactivity between glucose primary and secondary hydroxyl groups.

In agreement with the results of other authors on other sugar methacrylates (57) we observe a spontaneous polymerization of glucose methacrylate during the monomer preparation. This polymerization proceeds without gel formation and a very hydrophilic water soluble polymer with high viscosity is obtained by precipitation in acetone. Our procedure requires obviously an inhibitor and further improvement.

```
HOCH                                HOCH
 |              CH3                   |
CHOH      +    CH2=C-COCl            CHOH
 |                                    |
HOCH           ────────>             HOCH
 |      O                             |      O
CHOH                                 CHOH
 |                                    |           CH3
CH──┘                                CH──┘        |
 |                                    |
CH2OH                                CH2-O.CO-C=CH2

  II                                     XXXI
```

Similar procedure has been reported for preparation of di- and trisaccharide methacrylates by controlled reaction of unprotected sugar with methacrylic acid anhydride in pyridine, transfer into an aqueous solution and subsequent polymerization (53).

Determination of the reactivity ratios (58,59) in the radical copolymerization of sugar (meth)acrylates with other (meth)acrylates, styrene, acrylonitrite, etc. shows that these complex, carbohydrate monomers behave as a simple synthetic monomer. They are valuable intermediates for the introduction of hydrophilic properties into hydrophobic polymers.

Saccharide - Carrying Styrene

Similar polymers, based on saccharide-carrying styrenes have been prepared by Kobayashi et al. (60-62). The reducing terminal of glucose, maltose, lactose, and maltotriose (trisaccharide) is oxidized to the corresponding saccharide lactone, followed by coupling the lactone with p-vinylbenzylamine to form an amine linkage. The reaction scheme is illustrated with glucose:

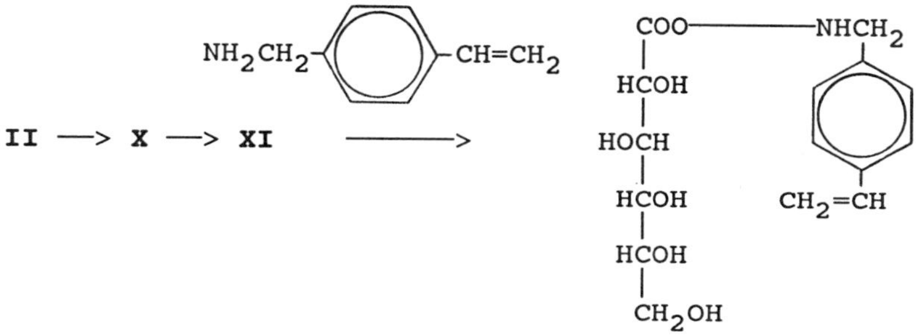

II ⟶ X ⟶ XI ⟶

Another saccharide-carrying styrene monomer (**XXXIV**) has been prepared from chloromethylstyrene and 1,2:5,6-di-O-isopropylidene-alpha-glucofuranose (63,64):

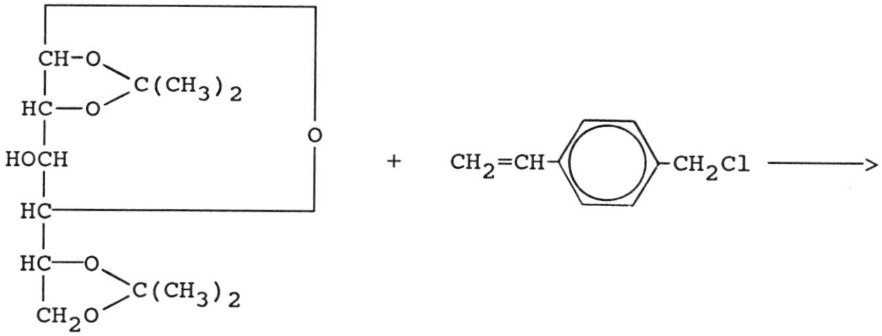

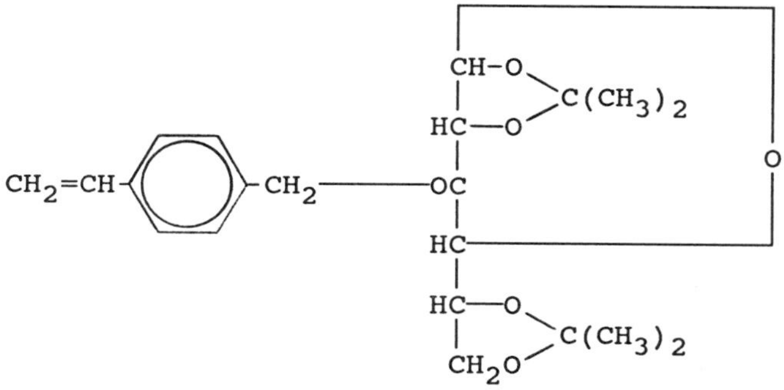

XXXIV

Oligosaccharide-substituted polystyrenes are obtained by radical polymerization with azodiisobutyronitrile as an initiator in $(CH_3)_2SO$ at 60°C followed by deblocking procedure.

The sugar moeity of these polymers has affinity for water whereas the vinylbenzyl residue avoids water. They are water soluble but the polymer sequence resists spreading out into water to result in a tightly - coiled micellar conformation. The intramolecular aggregation of the vinylbenzyl residues forms the hydrophobic regions which are enclosed in hydrophilic surroundings of water-solvated sugar residues. These polymers act as a neutral nonionic soap and have strong binding properties in water to different organic solutes. They also tend to be adsorbed on different surfaces and the resulting polymer-coated dishes have enhanced specific cell adhesion properties very useful for growth of cell cultures.

Poly(Vinyl Saccharide)s

Two poly(vinyl saccharide)s have been reported by cationic polymerization of 6-O-vinyl ethers of 1,2:3,4-di-O-isopropylidenegalactopyranose, **XXXV** (65,66) and of 1,2:5,6-di-O-isopropylideneglucofuranose (67), followed by deblocking procedure:

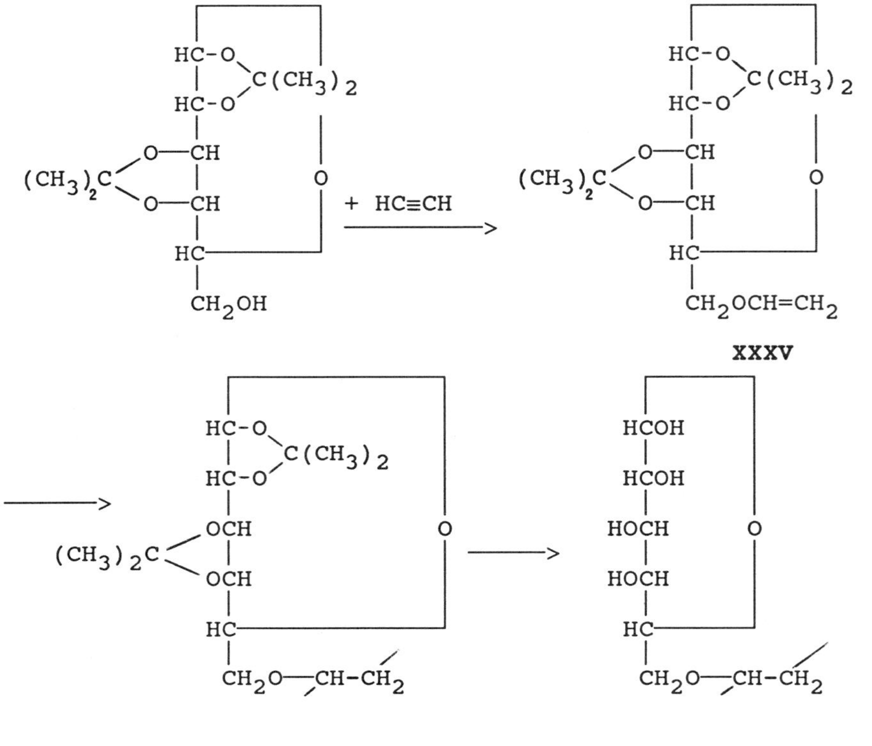

Cyclodexrin

Cyclodextrins (**XXXVI**) are cyclic oligomers of amylose, which are produced by enzymatic degradation of starch:

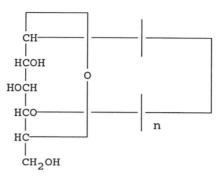

XXXVI

Several types of cyclodextrins are known, three of which are produced industrially and available in pure form: alpha, beta, and gamma. They correspond to the cyclohexa- (n=6), cyclohepta- (n=7) and cyclooctaamylose (n=8) respectively. Beta-cyclodextrin is the least expensive ($8.00 per pound).

The cyclodextrins are cylinder-shaped molecules with axial void cavities or pores. The diameters of the alpha, beta, and gamma cyclodextrin cavities are 5.7, 7.8, and 9.5 A respectively. The lengths of each cavity are 6.7, 7.0, and 7.8 A for alpha-, beta-, and gamma-cyclodextrin respectively:

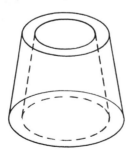

Our objective is the preparation and characterization of a new type of "porous" polymer membranes, which contain small cylindrical pores with equal dimensions in the range of several A, by introducing cyclic hollow molecules of cyclodextrins (or their derivatives) into the polymer matrix (68).

The principle advantage of these membranes is the

exact and equal size of their pores which is determined by the cyclodextrin rigid structure. All other known porous membranes characterize with a broad pore size distribution.

Membrane technology and separation is often ruled by a compromise between selectivity and flux rate. If selectivity is improved the flux rate is decreased and vice versa, if the flux rate is increased the selectivity is sacrificed. Cyclodextrin membranes are expected to have two distinct permeability rates. Molecules smaller than cyclodextrin cavities are expected to have much higher permeability than the molecules with a larger diameter. They might be able to simultaneously improve selectivity and increase flux rate for the smaller molecules.

The cavity of the cyclodextrin molecules is very small, below 10 Å, and their potential applications are obviously for separation of smaller molecules, for example: carbon dioxide/methane, oxygen/nitrogen, water desalination, bio-products concentration or purification, etc.

We have used two types of cyclodextrin derivatives for membrane preparation: permethylated-beta-cyclodextrin and the corresponding peracetylated derivative. Both are easily available.

Any polymer, mixable with cyclodextrins, is suitable for membrane preparation. Acetylcellulose with 39.8 percent acetyl content appears very attractive for initial characterization of cyclodextrin membranes since it is often used in membrane applications. Other cellulose derivatives, however acetate butyrate, propionate, and triacetate produce similar membranes. In addition, cyclodextrins polymerize with bi- or polyfunctional isocyanates, epoxies, etc. to oligomers, long-chain polymers, and to cross-linked polymer networks with similar permeability properties.

Cyclodextrin membranes were easily prepared from 10% solution of acetylcellulose and cyclodextrin derivative in MEK or acetone by a usual casting procedure. Both the acetyl- and the methyl derivative of cyclodextrin at 50% weight level in acetyl cellulose produce excellent homogeneous transparent amorphous membranes. They have excellent physico-mechanical properties and we do not observe any phase separation for more than six months. At higher concentration (70 weight percent) the methylated cyclodextrin, phase separates partially in the acetylcellulose and produces slightly turbid membranes. The peracetylated beta-cyclodextrin, however, has better compatibility with acetylcellulose at 70 weight percent and produces still excellent membranes, perhaps slightly brittle.

The volume percentage of pores in the membranes depends on the amount and the type of cyclodextrin derivative used for their preparation. At 50 percent weight level it corresponds to 7.3% and 10.3% volume percent of pores with diameter of 7.8 A and height of 7.8 A, for the acetylated and methylated beta-cyclodextrins respectively. At 70 weight percent, it is 10.2% and 14.5% respectively. Higher volume percentage of pores (above 15%) is achieved by using mixtures of acetyl and methyl derivative of beta-cyclodextrins, which do not co-crystallize and do not phase separate from acetylcellulose membrane, or by using gamma-cyclodextrin derivatives which have a larger pore.

The cavity diameter varies obviously by using different cyclodextrins.

The membrane glass transition temperature (Tg) depends on the amount of cyclodextrin in the acetylcellulose used. The Tg of the pure acetylcellulose is in the range of 204 - 211°C. The Tg of the membranes, based on acetylcellulose and peracetylated beta-cyclodextrin in 1:1 weight ratio, is about 153°C.

The permeability and selectivity properties of these membranes are under evaluation.

Another interesting application of the cyclodextrins in the field of free radical polymerization was recently reported. Partially and permethylated beta-cyclodextrins can promote the two-phase (water/organic) free radical polymerization of water-soluble vinyl monomers, such as acrylamide, sodium p-styrenesulfonate, acrylic acid, etc. initiated by water-insoluble free radical initiators such as azo or peroxide compounds: 2,2'-azobis(isobutyron-itrile), etc. (69). On the basis of the high lipo-philic character of the methylated beta-cyclodextrins as well as their ability to form a stable inclusion complex with organic compounds (70), it was suggested that, in this polymerization, the free radical initiator (or a radical species generated from the initiator) was entrapped by the lipophilic beta-cyclodextrin in the organic phase and then efficiently transferred to the aqueous phase to initiate the polymerization of the water-soluble vinyl monomers. It is also found that the two-phase polymerization of water-insoluble vinyl monomers, which have an aryl group (benzyl methacrylate, etc.), using water-soluble free radical initiators (ammonium peroxydisulfate, etc.) is also accelerated by the addition of cyclodextrins (71).

Experimental

All monomers and intermediates have been prepared according to the references given in the text. This experimental part contains only procedures which have been developed in our laboratory. All solvents and reagents are from Aldrich Chemicals unless otherwise indicated.

1,4-Lactone of 3,6-Anhydrogluconic Acid (LAGA) (IX).
50.0g of 1,5-lactone of gluconic acid (Sigma), 5.0g of DOWEX (cation exchange) resin 50WX4 and 500 ml of toluene were loaded into a flask equipped with a reflux condensor and a Dean-Stark adapter for water separation. The mixture was refluxed for 5 hours and cooled. Toluene was separated by decantation and the residue dissolved in 500 ml of methanol. The solution was filtered several times for separation of DOWEX resin and methanol distilled on a Rotavapor. 15.7 g of pure LAGA was isolated by vacuum distillation using a Kugelrohr distillation apparatus (Aldrich) at 160-170°C/0.2 mm Hg (temperature of the heating bath) (37% yield).

1,6-Dichloro-1,6-Dideoxy-3,4-O-Isopropylidene-D-Mannitol (XXII). 100 ml of carbon tetrachloride was gradually added to a solution of 11.1 g (5 mmole) of 3,4-O-isopropylidene-D-mannitol (XXI) and 26.3 g (10 mmole) of triphenylphosphine in 500 ml of anhydrous pyridine at 0°C. After holding the resulting solution at 5°C for 18 hours, methanol was added, and the mixture was evaporated to a crystalline residue which was chromatographically separated on a silica gel column. Elution first with chloroform and then with 6/1 v/v chloroform/acetone gave XXII fraction. An additional recrystallization from toluene/heptane produced 6.50 g of pure XXII (50% of yield) with m.p. 75-76°C (lit. 24: m.p. 76°C).

Poly(Isosorbide-Hexamethylene Diisocyanate). HMDI (Polysciences) was purified prior to use by vacuum distillation through a short Vigreux column (10 cm), b.p. 127°/10 mm. Isosorbide was prepared according to method (1). It was purified by vacuum distillation using a Kugelrohr Distillation Apparatus (Aldrich) at 115-130°C (temperature of the heating bath) and 0.2 mm vacuum, followed by vacuum drying at 50°C over phosphorus pentoxide. N,N-Dimethylacetamide (DMAc) (Aldrich, 99+%, GOLD label) was stored over molecular sieve 3Å and used without further purification. Dibutyltin dilaurate (Polysciences) was used without purification. The polymerization was carried out in a nitrogen atmosphere in 250 ml four-neck flask equipped with a dropping funnel, reflux condenser, mechanical stirrer and nitrogen inlet tube.
16.82 g (0.1 mole) of HMDI was added dropwise at room temperature from the dropping funnel into the rapidly stirred solution of 14.62 g (0.1 mole) of isosorbide in 100 ml of DMAc. After the addition was completed, 0.2 ml of dibutyltin dilaurate as a catalyst was added and the polymerization was carried out at 75°C for 24 hours. A very viscous completely transparent solution was obtained. It was cooled to room temperature, diluted with 500 ml of DMAc and precipitated by dropwise addition

into 5 L of methanol. The suspension was left overnight, filtered, washed with methanol and dried in vacuum at 25°C. A white powder of P(I-HMDI) was obtained in 100% yield.

All other polyurethanes have been prepared in a similar manner.

Cyclodextrin (XXXVI). The peracetylated beta-cyclodextrin was prepared by Dr. E. Compere, Eastern Michigan University, in quantitative yield from beta-cyclodextrin and boiling acetic anhydride for half an hour in the presence of sodium acetate as a catalyst followed by precipitation in ice water and subsequent recrystallization from toluene (72). Permethylated cyclodextrin was available from Aldrich.

Measurements. The melting points of the monomers and the glass transition temperatures and softening points of the polymers have been determined on capillary melting point apparatus, DuPont differential scanning calorimeter (with 10°C/min.), and Dennis thermal bar, respectively. Temperature of decomposition is measured as first onset of weight loss on the thermogravimetric analysis curve in air at 10°C/min.. Limiting Oxygen Index is measured on a home-made instrument. Molecular weight of the polymers is determined by gel permeation chromatography with a DuPont Zorbax Bimodal PSM 1000s column and dimethylsulfoxide as a mobile phase. Infrared (in KBr) and NMR spectra (in deuterochloroform) are taken on Perkin-Elmer infrared and 60 MHz Varian NMR spectrometers, respectively.

Conclusion

Mono-, di-, and oligosaccharides appear to be very attractive renewable resources for preparation of broad variety of unique monomers and polymers. Most of the research is at an exploratory stage and requires further effort to mature and reach industrial production and applications. The most challenging, just emerging research area is application of polymers with saccharide units in their main chains for controlled (bio)degradation. Such high molecular weight polymers are expected to undergo easy hydrolysis and enzymatic degradation into lower molecular weight biodegradable oligomers. In general, saccharide polymers are expected to undergo easier (bio)degradation than the pure synthetic polymers. They have a potential for reducing the "plastic" pollution, a major problem in our modern world.

Literature Cited

(1) Dirlikov, S. In _Agricultural and Synthetic Polymers, Biodegradability and Utilization_; Glass,

(2) Sumitomo, H.; Okada, M. Adv. Polym. Sci. 1978, 28, 47.
(3) Schuerch, C. Adv. Carbohydr. Chem. Biochem. 1981, 39, 157.
(4) Sumitomo, H.; Okada, M. In Ring-Opening Polymerization; Ivin, K.J.; Saegusa, T., Eds.; Elsevier Applied Science Publishers: New York, 1984, Vol. 1; pp. 299.
(5) Schuerch, C. In Cationic Polymerization and Related Processes; Goethals, E.J., Ed.; Academic Press: London, 1984; pp. 204-217.
(6) Sumitomo, H.; Okada, M. In Current Topics in Polymer Science; Ottenbrite, R.M.; Utracki, L.A.; Inone, S., Eds.; Carl Hanser Verlag: Munich, 1987, Vol 1; pp. 15.
(7) Montgomery, R.; Wiggins, L.F. J. Chem. Soc. 1946, 390.
(8) Courtaulds Ltd. Neth. Appl. 6,405,497, 1964; Chem. Abstr. 1965, 62, 10588.
(9) Thiem J.; Lüders, H. Starch/Stärke 1984, 36, 170.
(10) Medem, H.; Schreckenberg, M.; Dhein, R.; Nouvertne, W.; Rudolph, H. Ger. Offen 3,002,762, 1981; Chem. Abstr. 1981, 95, 151439n.
(11) Medem, H.; Schreckenberg, M.; Dhein, R.; Nouvertne, W.; Rudolph, H. Ger. Offen. 2,938,464, 1981; Chem. Abstr. 1981, 95, 44118k.
(12) Dirlikov, S.; Schneider, C. U.S. Pat. 4,443,563, 1984; Chem. Abstr. 1984, 101, 24146g.
(13) Montgomery, R.; Wiggins, L.F. J. Chem. Soc. (London) 1948, 2204.
(14) Montgomery, R.; Wiggins, L.F. J. Chem. Soc. (London) 1946, 393.
(15) Cope, A.C.; Shen, T.Y. J. Amer. Chem. Soc. 1956, 78, 3177, 5912, and 5916.
(16) Chle, H.; Dickhauser, E. Chem. Ber. 1925, 58, 2593.
(17) Chle, H.; Von Vargha, L.; Erlbach, H. Chem. Ber. 1928, 61, 1203.
(18) Haworth, W.; Owen, L.; Smith, F. J. Chem. Soc. (London) 1941, 88.
(19) Fischer, E.; Zach, K. Chem. Ber. 1912, 45, 456.
(20) Fischer, E.; Zach, K. Chem. Ber. 1912, 45, 2068.
(21) Dirlikov, S.; Schneider, C. U.S. Pat. 4,581,465, 1986; Chem. Abstr. 1986, 105, 60909z.
(22) Dirlikov, S.; Schneider, C. U.S. Pat. 4,438,226, 1984; Chem. Abstr. 1984, 100, 192955t.
(23) Hayworth, W.; Heslop, D.; Salt, E.; Smith, F. J. Chem. Soc. (London) 1944, 217.
(24) Hashimoto, K.; Okada, M.; Honjou, N. Makromol. Chem., Rapid Commun. 1990, 11, 393.

(25) Mansour, E.M.E.; Kandil, S.H.; Hassan, H.H.A.M.; Skaban, M.A.E. Eur. Polym. J. **1990**, 26, 267.
(26) (a) Bird, T.P.; Black, W.A.P.; Dewar, E.T.; Hare, J.B. J. Chem. Soc. **1963**, 1208; (b) Ibid. **1963**, 3389.
(27) Jarman, M.; Ross, W. Carbohydr. Res. **1969**, 9, 139.
(28) Kuszmann, J. Carbohydr. Res. **1979**, 71, 123.
(29) Institoris, L.; Horvath, I.P.; Csanyi, E. Arzneimittel-Forsch. **1967**, 17, 145.
(30) Fischer, E. Chem. Ber. **1895**, 28, 1167.
(31) Wiggins, L.F. J. Chem. Soc. (London) **1946**, 13.
(32) Anisuzzaman, A.; Whistler, R. Carbohydr. Res. **1978**, 61, 511.
(33) Dirlikov, S.; Schneider, C. U.S. Pat. 4,709,059, 1987; Chem Abstr. **1987**, 108, 187805e.
(34) Wiggins, L.F. J. Chem. Soc. (London) **1946**, 384.
(35) Horvath, I.P.; Institoris, L. Arzneimittel-Forsch. **1967**, 17, 149.
(36) Jarman, M.; Ross, W.C.J. Chem. Ind. (London) **1967**, 1789.
(37) Tipson, R.S.; Cohen, A. Carbohydr. Res. **1968**, 7, 232.
(38) (a) Sugihara, J.M. Adv. Carbohydr. Chem. **1953**, 8, 1; (b) Wolfrom, M.L.; Szarek, W.A. In The Carbohydrates; Pigman, W.; Horton, D., Eds.; Academic Press: New York, 1972, Vol. 1A; pp. 217; (c) Hanessian, P.; Lavalle, P. Carbohydr. Res. **1973**, 28, 303.
(39) Kurita, K.; Hirakawa, N.; Iwakura, Y. Polym. Prepr. **1979**, 20(1), 496.
(40) Kurita, K.; Hirakawa, N.; Iwakura, Y. Makromol. Chem. **1979**, 180, 855.
(41) Kurita, K.; Hirakawa, N.; Iwakura, Y. Makromol. Chem. **1980**, 181, 1861.
(42) (a) Steinman, H.W. U.S. Pat. 3,386,932, 1968; Chem. Abstr. **1968**, 69, 28723; (b) Polym. Prepr., Am. Chem. Soc., Div. Polym. Chem. **1970**, 11 (1), 285.
(43) Kim, S.; Stannet, V.T.; Gilbert, R.D. J. Polym. Sci., Polym. Lett. **1973**, 11, 731.
(44) Lynn, M.M.; Stannet, V.T.; Gilbert, R.D. Polym. Prepr., Am. Chem. Soc., Div. Polym. Chem. **1978**, 19(2), 106.
(45) Kurita, K.; Hirakawa, N.; Iwakura, Y. J. Polym. Sci., Polym. Chem. Ed. **1980**, 18, 365.
(46) Kurita, K.; Hirakawa, N.; Iwakura, Y. Makromol. Chem. **1977**, 178, 2939.
(47) Kurita, K.; Hirakawa, N.; Iwakura, Y. Makromol. Chem. **1979**, 180, 2331.
(48) Kurita, K.; Miyajima, K.; Sannan, T.; Iwakura, Y. J. Polym. Sci., Polym. Chem. Ed. **1980**, 18, 359.
(49) Kurita, K.; Murakami, K.; Kobayashi, K.; Takahashi, M.; Koyama, Y. Makromol. Chem. **1986**, 187, 1359.

(50) Kurita, K.; Hirakawa, N.; Morinaga, H.; Iwakura, Y. Makromol. Chem. **1979**, *180*, 2769.
(51) Patil, D.; Dordick, J.; Rathwisch, D. Biotech. Bioeng. **1991**, *37*, 639.
(52) Klein, J.; Kulicke, W.M. Polymer Prepr. **1981**, *22(2)*, 88.
(53) Colquhoun, J.A.; Dewar, E.T. Process Biochem. **1968**, 31.
(54) Klein, J.; Herzog, D.; Hajibegli, A. Makromol. Chem., Rapid Commun. **1985**, *6*, 675.
(55) Kimura, S.; Hirai, K.; Imoto, M. Makromol. Chem. **1961**, *50*, 155.
(56) Imoto, M.; Kimura, S. Makromol. Chem. **1962**, *53*, 210.
(57) Black, W.A.P.; Colquhoun, J.A.; Dewar, E.T. Makromol. Chem. **1968**, *117*, 210.
(58) Black, W.A.P.; Colquhoun, J.A.; Dewar, E.T. Makromol.Chem. **1969**, *122*, 244.
(59) Black, W.A.P.; Colquhoun, J.A.; Dewar, E.T. Makromol. Chem. **1967**, *102*, 266.
(60) Kobayashi, K.; Sumitomo, Y.; Ina, Y. Polym. J. **1983**, *15*, 667.
(61) Kobayashi, K.; Sumitomo, Y.; Ina, Y. Polym. J. **1985**, *17*, 567.
(62) Kobayashi, A.; Akaike, T.; Kobayashi, K.; Sumitomo, H. Makromol. Chem., Rapid Commun. **1986**, *7*, 645.
(63) Kobayashi, K.; Sumitomo, H. Macromolecules, **1980**, *13*, 234.
(64) Kobayashi, K.; Sumitomo, H. Polym. J. **1981**, *13*, 517.
(65) Klein, J.; Blumenberg, K. Makromol. Chem., Rapid Commun. **1986**, *7*, 621.
(66) Whistler, R.L.; Panzer, H.P.; Goatley, J.L. J. Org. Chem. **1962**, *27*, 2961.
(67) (a) Black, W.A.P.; Dewar, E.T.; Rutherford, D. J. Chem. Soc. **1963**, 4433; (b) Chem. Ind. (London) **1962**, 1624.
(68) Dirlikov, S.; Compere, E. Polym. Prepr. **1989**, *30(1)*, 20.
(69) (a) Kunieda, N.; Taguchi, H.; Shiode, S.; Kinoshita, M. Makromol. Chem., Rapid Commun. **1982**, *3*, 395; (b) Taguchi, H.; Kunieda, N.; Kinoshita, M. Ibid. **1982**, *3*, 495; (c) Taguchi, H.; Kunieda, N.; Kinoshita, M. Makromol. Chem. **1983**, *148*, 925.
(70) (a) Czugler, M.; Eckle, E.; Stezowski, J.J. J. Chem. Soc., Chem. Commun. **1981**, 1291; (b) Casu, B.; Reggiani, M.; Standerson, G.R. Carbohydr. Res. **1979**, *76*, 59.
(71) Kunieda, N.; Shiode, S.; Ryoshi, H.; Taguchi, H.; Kinoshita, M. Makromol. Chem., Rapid Commun. **1984**, *5*, 137.
(72) French, D.; Levine, M.L.; Pagur, J.H.; Norberg, E. J. Amer. Chem. Soc. **1949**, *71*, 357.

Received May 8, 1991

Chapter 15

Cellulose Ethers

Self-Cross-linking Mixed Ether Silyl Derivatives

Arjun C. Sau and Thomas G. Majewicz

Research and Development Center, Aqualon Company, Little Falls Centre One, 2711 Centerville Road, P.O. Box 15417, Wilmington, DE 19850–5417

Functionalization of polysaccharides with silanolate groups, $-Si-O^-$, leads to the formation of a new class of silated polysaccharide derivatives. These polymers are alkali-soluble and self-crosslink upon drying to form water-resistant films. They also exhibit good adhesion to hydroxylic surfaces and crosslink in solution with polyvalent metal ions to form viscoelastic gels. The synthesis and properties of silated hydroxyethylcellulose (SIL-HEC) are described in this article.

Cellulose is the most abundant and renewable natural material. Structurally, it is a syndiotactic polymer of anhydroglucose units which are connected by (1→4) ß-glycosidic linkages (Figure 1). It is characterized by a

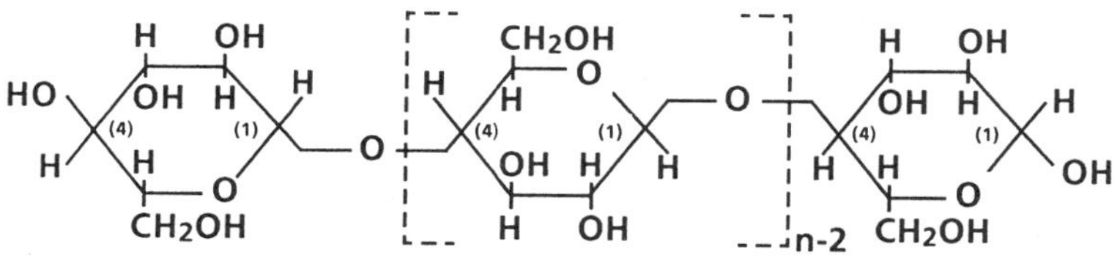

Figure 1. The structure of cellulose.

high degree of crystallinity and is water-insoluble – the result of inter- and intra-molecular hydrogen bonding. Derivatization of the cellulose hydroxyl groups under appropriate conditions disrupts the hydrogen bonding network and leads to water-solubility. Typically, water-soluble derivatives result from etherification of cellulose in the presence of an alkali *(1-2)*. Examples of etherifying agents used to manufacture several commercially available water-soluble cellulose ethers are shown below.

0097–6156/92/0476–0265$06.00/0

Cellulose Ether	Etherifying Agent	Substituent Group
Methylcellulose (MC)	CH_3Cl	$-CH_3$
Carboxymethylcellulose (CMC)	$ClCH_2CO_2^-\ Na^+$	$-CH_2CO_2^-\ Na^+$
Hydroxyethylcellulose (HEC)	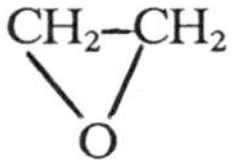 $\underset{O}{CH_2\!-\!CH_2}$	$-(CH_2\text{-}CH_2\text{-}O)_n\text{-}H$

Hydroxypropylcellulose (HPC)	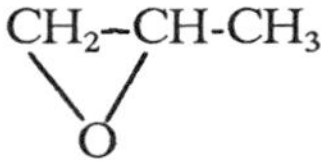 $\underset{O}{CH_2\!-\!CH\text{-}CH_3}$	$-(CH_2\text{-}CH\text{-}O)_n\text{-}H$
		        |
		       CH_3

The water-soluble cellulose ethers (CE's) shown above are widely used as thickeners to control the rheology of water-based formulations, such as latex paints, drilling muds, cosmetics, pharmaceuticals and building materials.

Further modification of mono-substituted derivatives with another functional group can enhance or produce novel properties. For example, modification of methylcellulose with low levels of hydroxypropyl or hydroxyethyl groups increases its gelation and flocculation temperatures in water *(1)*. Carboxymethylation of hydroxyethylcellulose produces a product which exhibits excellent tolerance to mono- and di-valent metal ions in solution, but readily crosslinks with tri- or tetra-valent ions to give highly viscoelastic gels *(1)*. Cellulose ethers containing more than one type of functional group are commonly referred to as mixed cellulose ethers. Examples of commercial mixed cellulose ethers include carboxymethylhydroxyethylcellulose (CMHEC), methylhydroxypropylcellulose (MHPC), methylhydroxyethylcellulose (MHEC) and cationic hydroxyethylcellulose *(1-2)*.

A more recently commercialized mixed cellulose ether is hydrophobically modified hydroxyethylcellulose (HM-HEC) *(3-4)* – a polymeric surfactant. HM-HEC is made by modifying HEC with a low level of a hydrophobic group, such as a long hydrocarbon alkyl chain. The hydrophobic moiety imparts associative properties to hydroxyethylcellulose. HM-HECs exhibit enhanced low-shear solution viscosity and surface activity. Detailed synthesis and solution/rheological properties of HM-HEC have recently been reported *(5-6)*. Its primary application is in latex paints where it functions as an associative thickener *(4)*.

It is known that silicon plays a central role in determining the properties of polymers containing silyl functionality *(7)*. Several polysaccharides including cellulose *(8-9)* have been modified with trimethylsilyl ($-SiMe_3$) substituents to achieve organosolubility. We now have discovered a new class of silated polysaccharides which are soluble in dilute aqueous alkali and self-crosslink on drying to afford water-resistant films *(10-11)*. This paper describes the synthesis and properties of silated HEC (SIL-HEC) – a new mixed cellulose ether containing reactive silyl substituents.

Experimental

Preparation of SIL-HEC. The SIL-HEC samples were prepared by reacting HEC (hydroxyethyl molar substitution (HE MS) ~ 2.5-3.3) with various amounts of (3-glycidoxypropyl)trimethoxysilane (GPTMS) (Aldrich) in the presence of an alkali at 95-115°C *(11)*.

Results and Discussion

1. Preparation of SIL-HEC. HEC reacts with (3-glycidoxypropyl)trimethoxysilane (GPTMS) (1) in the presence of an alkali to form a silanolate derivative (2) as shown below *(10)*.

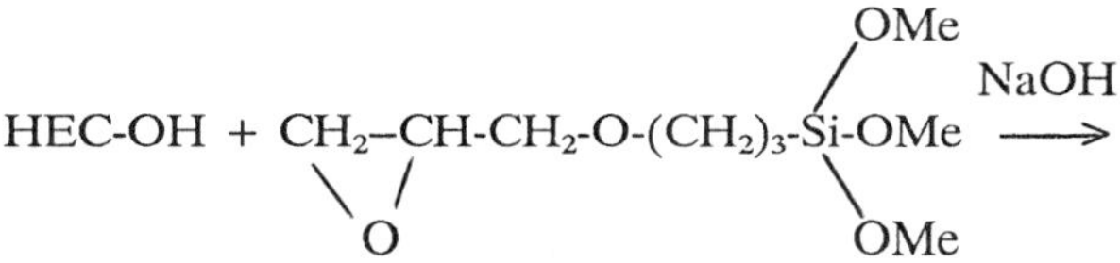

$$\text{HEC-OH} + \underset{(\underline{1})}{\text{CH}_2\text{-CH-CH}_2\text{-O-(CH}_2)_3\text{-Si(OMe)}_3} \xrightarrow{\text{NaOH}}$$

$$\underset{(\underline{2})}{\text{HEC-O-CH}_2\text{-CH(OH)-CH}_2\text{-O-(CH}_2)_3\text{-Si(O}^-\text{Na}^+)_3}$$

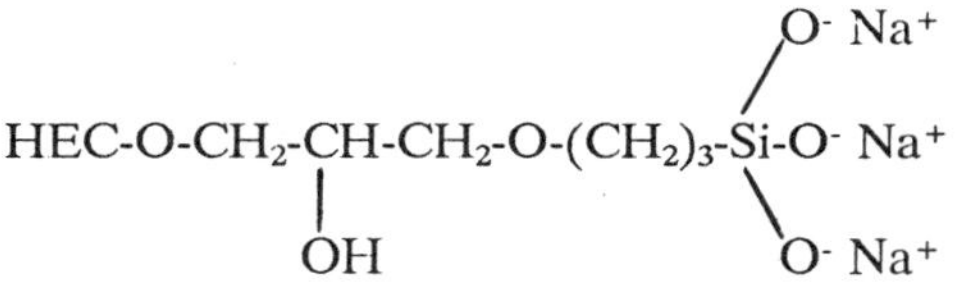

The silanolate groups in (2) results from the alkaline hydrolysis of methoxysilyl groups (-Si-OMe) of (1).

$$\text{-Si-OMe} + \text{NaOH} \longrightarrow \text{-Si-O}^- \text{Na}^+ + \text{MeOH}$$

Alkoxysilyl moieties are known to react with alkali metal hydroxides to form alkali-metal silanolates *(12-14)*. We speculate that the glycidoxysilanolate (3), generated *in situ*, reacts with HEC to form the SIL-HEC (2).

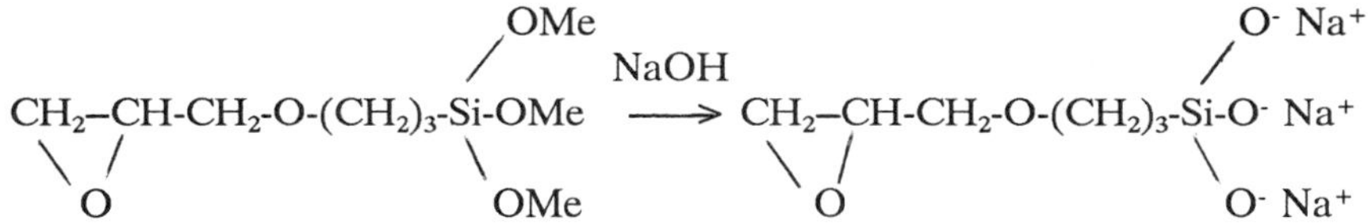

$$\underset{(\underline{1})}{\text{CH}_2\text{-CH-CH}_2\text{-O-(CH}_2)_3\text{-Si(OMe)}_3} \xrightarrow{\text{NaOH}} \underset{(\underline{3})}{\text{CH}_2\text{-CH-CH}_2\text{-O-(CH}_2)_3\text{-Si(O}^-\text{Na}^+)_3}$$

2. Properties of SIL-HEC

a) Solution Properties. SIL-HEC, isolated in a neutralized state, is water-insoluble due to crosslinking of the silanol groups with each other and with the hydroxyls of the HEC (see discussion below on the crosslinking mechanism). These crosslinks, however, are labile in the presence of an alkali or ammonia; hence, SIL-HEC readily dissolves at solution pHs > ~11.

Once in solution, SIL-HEC does not precipitate out of solution upon acidification: however, viscosity enhancement or gelation may occur. This behavior is controlled by the composition of the SIL-HEC (molecular weight (MW) and silyl MS), solution concentration, and solution pH. For example, at moderately high concentration (~10%), the viscosity of a SIL-HEC with MW ~80,000 and silyl MS ~0.1 remains unchanged over a pH range from 7-12. However, a product with MW ~300,000 and silyl MS ~0.04 exhibits progressive viscosity enhancement below pH 11.2 and eventually gels at pH 9.5 at 2% polymer concentration (see Figure 2). This process is reversible, i.e., the gel can be broken down by raising its pH. The viscosity enhancement or gelation of a SIL-HEC solution at lower pHs is due to intermolecular crosslinking. As the solution pH is lowered, the silanolate groups (-Si-O$^-$) are converted into silanol groups (-Si-OH) (eqn. 1) which condense and/or hydrogen bond with other silanol or carbinol groups to form a three-dimensional network (see discussion below).

$$\text{HEC} \sim \text{Si-O}^- \text{Na}^+ \underset{\text{NaOH}}{\overset{\text{H}_3\text{O}^+}{\rightleftarrows}} \text{HEC} \sim \text{Si-OH} \quad \ldots(1)$$

b) Self-Crosslinking Properties. Besides being able to crosslink in solution in the presence of an acid, SIL-HEC undergoes self-crosslinking when an aqueous solution (alkaline) is air-dried. This behavior is a unique property of SIL-HEC. Solution-cast films of SIL-HECs hydrate but remain insoluble in water. The hydrated films are transparent and somewhat elastic. These crosslinked films, however, reversibly dissolve in alkalies.

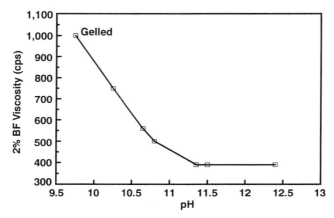

Figure 2. pH Versus 2% Brookfield (BF) viscosity of SIL-HEC (HE MS ~ 3.2; silyl MS ~ 0.04; MW ~ 300,000).

Mechanism of Self-Crosslinking. Sodium methyltrisilanolate (4), which has the same silanolate functionality of SIL-HEC (2), crosslinks when its alkaline

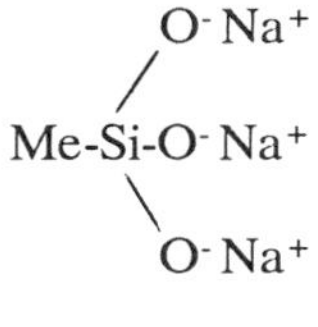

$$\text{Me-Si}\begin{array}{c}\text{O}^-\text{Na}^+\\ \text{O}^-\text{Na}^+\\ \text{O}^-\text{Na}^+\end{array}$$

(4)

solution is air dried. It has been reported *(15-16)* that the silanolate groups of (4) are neutralized with atmospheric CO_2 to form the silanetriol intermediate, $MeSi(OH)_3$ (5), which in turn condenses to form methylpolysiloxane (6) as shown below.

$$MeSi(O^-Na^+)_3 + CO_2 + H_2O \longrightarrow [MeSi(OH)_3] + Na_2CO_3$$

(4)   (5)

$$\underline{n}[MeSi(OH)_3] \longrightarrow (MeSiO_{3/2})_n + 3/2\underline{n}H_2O$$

(5)   (6)

A similar mechanism is proposed to explain crosslinking of the SIL-HEC. In the presence of atmospheric CO_2, silanolate groups of the SIL-HEC are converted into silanol groups. Crosslinking occurs via condensation of silanol groups and/or condensation of silanol groups with carbinol (C-OH) groups. These reactions are shown below.

$$Si\text{-}OH + HO\text{-}Si \longrightarrow Si\text{-}O\text{-}Si + H_2O \quad......(2)$$

$$Si\text{-}OH + HO\text{-}C \longrightarrow Si\text{-}O\text{-}C + H_2O \quad......(3)$$

The crosslinked structure of a room temperature cured SIL-HEC (HE MS ~3.2; silyl MS ~0.15; MW ~300,000) is schematically shown in Figure 3. This proposed structure is consistent with the three silicon signals discernible in the ^{29}Si NMR spectrum of the crosslinked SIL-HEC (Figure 4). Based on the ^{29}Si NMR study by Clayden and Palasz *(17)* of the crosslinked product resulting from silyl condensation of trimethoxysilane functionalized isocyanurate, we assign these signals to the following silicon environments (I-III).

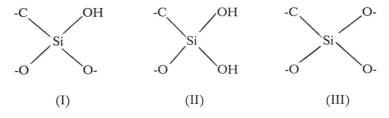

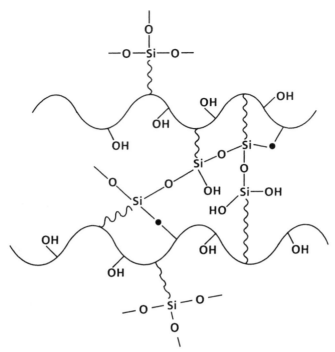

Figure 3. Schematic diagram showing the proposed structure for a room temperature cured SIL-HEC. (∩∧∩ = HEC backbone; -OH = hydroxyl group of the HEC; ∼∼∼ = $-CH_2CH(OH)CH_2O(CH_2)_3-$ and Si-O- = Si-O-C bond formed by the reaction of the Si-OH with the hydroxyl (C-OH) of the SIL-HEC).

3. Other Properties of SIL-HEC. Besides self-crosslinking, SIL-HEC can also crosslink other hydroxyl containing water-soluble polysaccharides to form water- and acid-resistant composite films. Silanol functions are known to react with hydroxylic polymers *(18)* to form Si-O-C linkages; however, their reactivity depends on the nature of the hydroxyl group. It has been reported *(19)* that primary hydroxyls are ten times more reactive than secondary hydroxyls to silanols.

In solution, SIL-HEC crosslinks with polyvalent metal ions, such as Ti^{+4}, to form viscoelastic gels.

SIL-HECs, in general, exhibit good adhesive properties. Their aqueous solutions can be used as an adhesive to glue various substrates containing surface hydroxyls or other active hydrogens.

Conclusions

Functionalization of water-soluble cellulose ethers, e.g., HEC, with silanolate functionality ($Si-O^-$) affords a new class of alkali-soluble polymers. They are self-crosslinkable and exhibit adhesion to hydroxylic surfaces. Their solution behavior (viscosity enhancement, gelation, etc.) is dictated by their composition (MW and silyl MS), solution concentration, and solution pH.

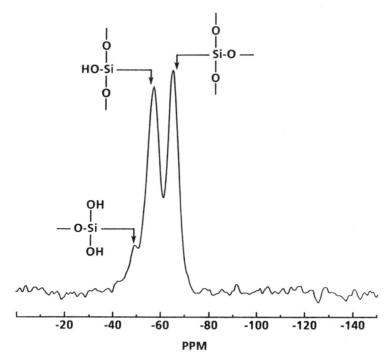

Figure 4. 39 MHz ^{29}Si CP MAS NMR spectrum of a crosslinked SIL-HEC (silyl MS ~ 0.15).

The unique properties of SIL-HEC could be exploited in a number of industrial applications that rely on:
a) water- and acid-resistant coatings;
b) metal-crosslinked gels;
c) adhesion to various substrates.

The silation chemistry described here can be extended to prepare analogous derivatives from other polysaccharides, such as guar, starch and their ether derivatives.

Literature Cited

1. Just, E. K.; Majewicz, T. G. In *Encyclopedia of Polymer Science and Engineering*; John Wiley & Sons, Inc., Second Edition, 1985, Vol. 3; p. 226.
2. Donges, R. *Brit. Polym. J.* 1990, *23*, 315.
3. Shaw, K. G.; Leipold, D. P. *J. Coatings Technol.* 1985, 57(727), 63.
4. *NATROSOL PLUS - Modified Hydroxyethylcellulose*, The Aqualon Company (a Hercules Incorporated Company), Wilmington, Delaware, 1988.
5. Sau, A. C.; Landoll, L. M. *Polymers in Aqueous Media: Performance Through Association*; Glass, J. E. Ed.; Advances in Chemistry Series, American Chemical Society, Washington, D.C., 1989, 223, p. 343.

6. Goodwin, J. W.; Hughes, R. W.; Lam, C. K.; Miles. J. A.; Warren, B. C. H. In *Polymers in Aqueous Media: Performance Through Association*; Glass, J. E. Ed.; Advances in Chemistry Series, American Chemical Society, Washington, D.C., 1989, 223, p. 365.
7. *Silcon-Based Polymer Science: A Comprehensive Resource*; Zeigler, J. M.; Gordon Fearson, F. W. Eds.; Advances in Chemistry Series, American Chemical Society, D.C., 1990, Vol. 224.
8. Herman, R. E.; De, K. K.; Gupta, S. K. *Carbohydr. Res.* 1973, 31, 407 and references cited therein.
9. Klebe, J. F.; Finkbeiner, H. L. *J. Polym. Sci.* 1969, 7, 1947.
10. Sau, A. C. *Polymer Preprints*, ACS Meeting in Boston, April, 1990, Vol. 31(1), 636.
11. Sau, A. C. *U.S. Patent* 4,992,538, February 12, 1991.
12. Noll, W. *Chemistry and Technology of Silicones*, Academic Press, New York, 1968, p. 86.
13. Hyde, J. F.; Johannson, O. K.; Daudt, W. H.; Fleming, R. F.; Laudenslager H. B.; Roche, M. P. *J. Amer. Chem. Soc.* 1953, 75, 5615.
14. Plueddemann, E. P. *Silane Coupling Agents*, Plenum Press, New York, 1982, p. 67.
15. Ref. 11, p. 607.
16. Ref. 13, p. 209.
17. Clayden, N. J.; Palasz, P. *J. Chem. Research (S)* 1990, 68.
18. West, R.; Burton, T. J. *J. Chem. Educ.* 1980, 57, 165.
19. Brown, L. H. *Film Forming Compositions*, Myers, R. R.; Long, J. S. Eds.; Marcel Dekker, New York, 1972; Vol. 1, Part III, Chapter 13, p. 548.

RECEIVED June 8, 1991

Chapter 16

Polymeric Materials from Agricultural Commodities

S. F. Thames and P. W. Poole

Department of Polymer Science, University of Southern Mississippi, Hattiesburg, MS 39406-0076

The guayule shrub (*Parthenium Argentatum Grey*) is processed into the following five major components of potential utility as polymeric materials: (1) high molecular weight natural rubber, (2) low molecular weight natural rubber, (3) organic soluble resins, (4) water soluble extracts, and (5) bagasse. This research has emphasized the isolation, characterization, and derivation of guayule coproducts.

The entire natural rubber demands of the United States are currently met via importation. It is important, therefore that a domestic supply of natural rubber, a strategic material, be available. The current demands for natural rubber are being met via imported Hevea rubber to the extent approximating 800,000 metric tons per year. The demands for the 1990's are projected to reach 920,000 metric tons per year at an estimated cost of one dollar per pound (*1*).

With no anticipated decline in demand and the ever present increasing conversion of rubber plantations into more profitable crops (i.e., coffee or coconut), short supply and accompanying price escalations for natural rubber are apparently inevitable. It is clear, therefore, that the factors of long term projected demands, market availability, and lack of a domestic source strongly supports the development of a U.S. natural rubber industry.

Guayule as a Source

Among rubber producing plants, Guayule provides high molecular weight natural rubber of high quality, with similar, if not identical properties, to those of Hevea (Malaysian rubber) (*2,3*). In contrast, however, guayule rubber (GR) is one of five fractions obtained via a selective solvent extraction process. Currently, the value and quantity of the high molecular weight rubber is insufficient to offset the costs of planting, cultivating, harvesting, and processing.

Therefore, if Guayule is to become the basis of a domestic, natural rubber industry, value added products must be developed from guayule co-products.

Rubber Extraction Process

Strands of guayule generally found in semi-arid regions such as Texas, Arizona, California and northern Mexico provide a promising source of natural rubber. Approximately five to twenty percent of the shrub's dry weight contains high molecular weight rubber. Low molecular weight natural rubber, guayule resin, a water soluble component, and bagasse make up the remainder of the fractions. Though the molecular structure of guayule is very similar to Hevea, the process for rubber isolation is quite different. For instance, guayule rubber is extracted from the walls of individual parenchyma cells requiring the destruction of the entire shrub (4,5,6,7).

While our efforts center about the use of guayule coproducts, cost effective processing methods are necessary. These efforts are currently under investigation by Dr. John Wagner and co-workers at Texas A&M University. In particular, they have conducted research in sequential and simultaneous solvent extraction processes.

Sequential Solvent Extraction (7). The sequential solvent extraction process involves two primary steps. First, polar solvent with acetone and a variety of alcohols are used to de-resinate the material, after which it is re-extracted with non-polar solvents from which the natural rubber is obtained. Flash evaporation of solvent with steam injection is necessary, with the water being removed from the rubber slurry via a combination of dewatering operations of thermal or mechanical origin. The authors do not considered this process economically favorable due to the high cost of water and solvent removal.

Simultaneous Solvent Extraction (7). Simultaneous extraction is a single step process and separates resins and rubber, thereby minimizing processing time. Filtration and/or centrifugation are used to remove particulate matter, while deresination and rubber recovery are affected via the addition of polar solvent(s). The natural rubber precipitates while the resins remain in solution. The polar solvent(s) is removed to insure high quality rubber.

Separation of Low Molecular Weight Rubber. The process used to isolate high molecular weight guayule rubber results in the formation of a resinous byproduct containing, among other components low molecular weight guayule rubber (LMWGR). The guayule resin is dissolved in acetone and upon its removal provides a sticky, black residue. Further treatment with 90% ethanol results in precipitation of LMWGR as a solid mass, while the solvent, containing the remaining resinous material(s), is removed by decantation. ^{1}H and ^{13}C NMR analysis of the rubber fraction has served as structural identification for LMWGR or poly(cis-isoprene) (8). An alternative separation process involves the use of xylene, with subsequent rubber precipitation by addition of ethanol. This process insures the isolation of rubber of high purity, grey color and essentially no tackiness for which gel permeation chromatography (GPC) has confirmed a molecular weight of 40 to 50 thousand.

Other Constituents of the Extraction Process. The organic soluble resins remaining after extraction and isolation of LMWGR are separated into several fractions. Moreover, the water soluble resinous portion and bagasse, a woody pulp containing lignins and cellulosics, are fractions of potential utility.

The extraction process provides materials useful for derivation, and thereby allows for the development of value added materials. The following sections will emphasize derivation and formulation of guayule coproducts into useful and economically important products.

High Molecular Weight Guayule Rubber (HMWGR)

The HMWGR from the guayule shrub is high quality poly(cis-isoprene) of approximately 1,000,000 molecular weight with properties equal to those of Hevea. Vulcanized HMWGR, utilized in the formation of high quality rubber products, possesses properties equal or superior to those of Hevea products.

Low Molecular Weight Guayule Rubber (LMWGR)

LMWGR does not possess mechanical properties equal to the HMWGR and thus, functional derivatives have been synthesized for application in a variety of uses. Moreover, since the traditional mastication or molecular weight reduction process used with Hevea rubber is not necessary with LMWGR, impetus is given to the development of products which otherwise require mastication. The alkene character of LMWGR affords the opportunity for chlorination, epoxidation, maleinization, cyclization, and hydrogenation. Consequently, we have employed this functionality in the synthesis of a number of derivatives and have shown that they can be utilized in surface coating formulations and in crosslinking reactions.

Characterization of LMWGR. LMWGR can be characterized by several methods; among them are NMR, IR, and GPC. ^{1}H and ^{13}C NMR spectra are shown in Figure 1.

LMWGR typically contains wax and other hydrocarbons and must be purified before derivation. Purification can be monitored by ^{1}H spectroscopy with the peak assignments in Figure 1 used to document purity (8):

       1.67 ppm   (cis double bond methyl protons)
       2.00 ppm   (methylene protons)
       5.12 ppm   (vinyl proton)

The small peak at 1.25 ppm is characteristic of an impurity that can be removed by additional precipitation from an ethanolic solution (8). The assignments are in excellent agreement to literature values for Hevea as is the infrared spectrum (Figure 2).

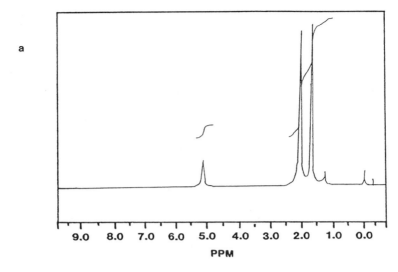
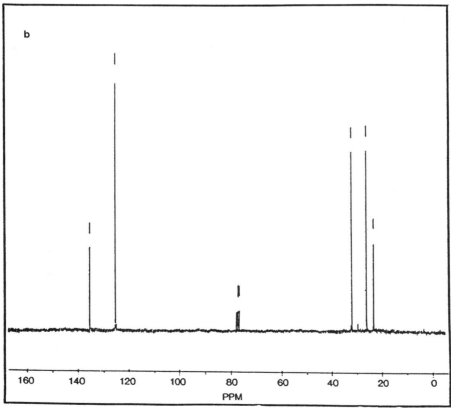

Figure 1. (a) ^{1}H NMR spectrum of LMWGR, (b) ^{13}C NMR spectrum of LMWGR.

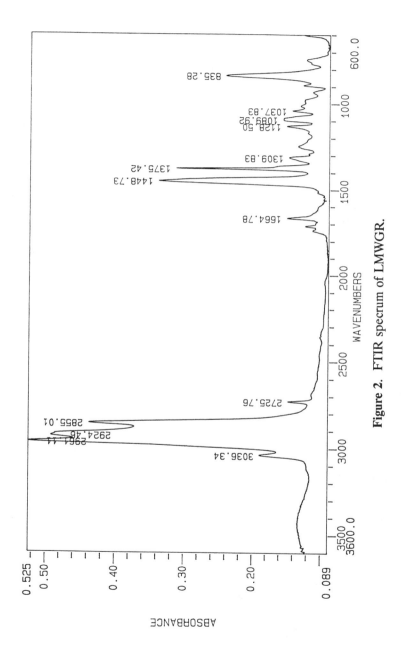

Figure 2. FTIR spectrum of LMWGR.

In summary, low molecular weight GR of high purity can be readily isolated and avoids the mastication process required with Hevea rubber.

Chlorination of LMWGR. Hevea rubber has been successfully chlorinated (9) and yields a product useful as a result of its chemical resistance and increased adhesion due to its polar nature. It was therefore of interest to determine the efficacy of chlorinating unmasticated LMWGR. The reaction of chlorine with LMWGR produces a fine white powder recoverable by steam distillation of the inert reaction solvent, carbon tetrachloride(Figure 3)(10).

The empirical formula for chlorinated natural rubber is $C_5H_8Cl_{3.5}$, suggesting more than one isoprene unit reacts (Figure 4) (11) to form a soluble, cyclized product, rather than one of a high crosslink density (Figure 5) (10).

Azo-bis-isobutyronitrile (AIBN) and LMWGR produces grades of chlorinated rubber lower in viscosity than that from Hevea rubber (Table I). It is significant that more than two pounds of chlorinated product is produced from one pound of LMWGR, a result of the addition of chlorine. The higher market value of chlorinated- in comparison to natural-rubber clearly establishes chlorinated LMWGR as a value added *coproduct* for the emerging domestic Guayule industry. Product yields of chlorinated guayule rubber, using AIBN as a catalyst, have approached the literature cited theoretical chlorine contents of 64.7% (8).

Experimental (8). A dilute solution (5-10%) of LMWGR in carbon tetrachloride is placed in a three neck round bottomed flask equipped with a condenser, gas dispersion tube, and adaptor. A chlorine cylinder is fitted with two gas traps and connected to the reaction flask with Teflon tubing that is changed periodically for safety purposes. Ice-cooled traps of 2 N sodium hydroxide solution are used to collect any residual chlorine and/or hydrogen chloride. Oxygen is purged from the reaction vessel via nitrogen at the outset of the reaction and is maintained as a continuous nitrogen blanket for the duration of the experiment. The solution is allowed to reflux at 79°C with stirring. The addition of ethanol provides chlorinated rubber in the form of a fine, white precipitate.

Characterization of the Chlorinated Guayule Rubber (CR) (8). ^{1}H NMR and ^{13}C NMR spectra of chlorinated LMWGR were obtained on a 300 Mhz Bruker fourier transform spectrometer. The spectra were prepared in $CDCl_3$ and tetramethylsilane was used as an internal standard. Major ^{13}C NMR chemical shifts of LMWGR chlorinated rubber agree with literature values (Table II)(Figure 6). Fourier transform infrared spectroscopy (FTIR)(Figure 7) is likewise consistent with the NMR data.

Gel permeation chromatography (GPC) was performed on a Waters Associates, Inc. GPC equipped with a refractive index detector. Operating conditions were: mobile phase, THF; flow rate, 1 ml/min; columns 10^6, 10^4, 500 ,100 E. Calibration curves were obtained from polystyrene standards. Figures 8 and 9 show relative elution times and volumes as compared to commercially available chlorinated rubber products.

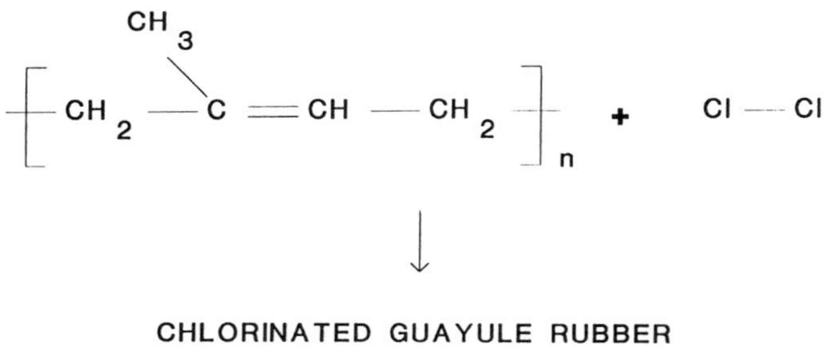

Figure 3. Reaction of LMWGR with chlorine gas.

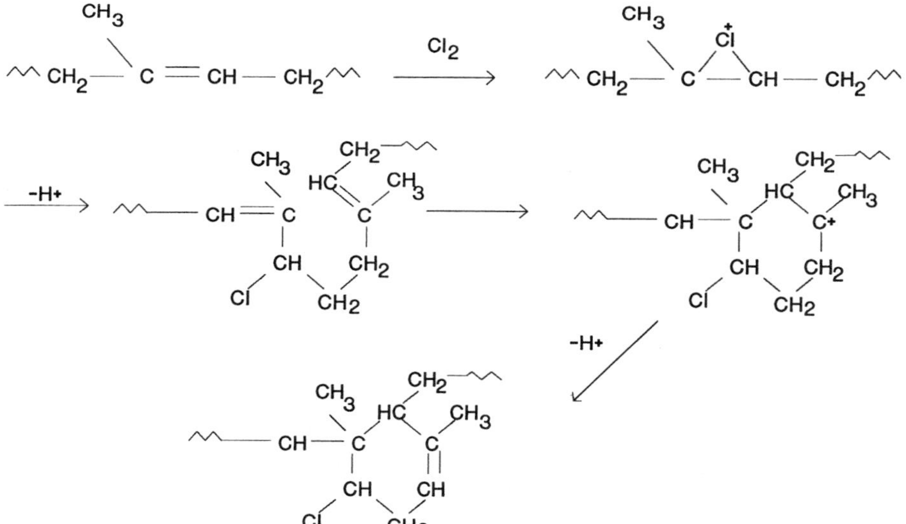

Figure 4. Mechanism of the chlorination of LMWGR.

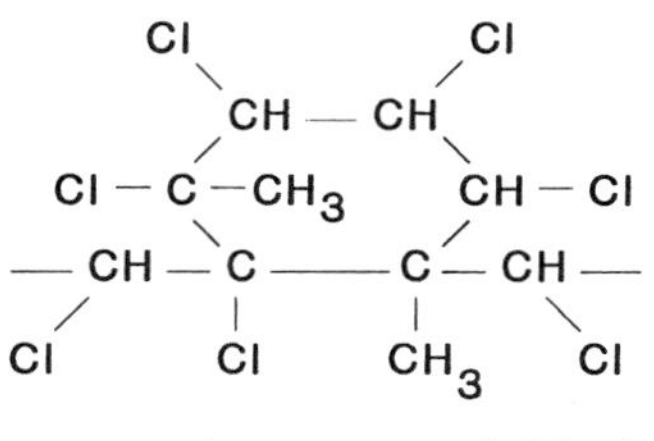

Figure 5. Final proposed structure of chlorinated LMWGR.

Table I
Experimental Conditions Used in the Synthesis of Chlorinated Guayule Rubber

Batch No.	Amount Guayule Used, grams	Amount of Chlorinated Rubber Isolated, grams
1[a]	3.20	6.00
2[b]	5.00	11.80
3[c]	5.00	13.14

[a]CCl_4, 200 ml
[b]CCl_4, 100 ml
[c]CCl_4, 100 ml and 37.5 mg of AIBN

Table II
Characteristic Shifts of Guayule Chlorinated Rubber and Commerical CR

Guayule CR, ppm	Literature value, ppm	Assignment
21.5, 28.7, 34.7, 37.6	21.5, 28, 34.7, 37.4 21.5, 38, 37.4	CH_2, CH_3
45.4, 48.01	45.4, 48	$-CH_2-Cl$
62.25, 64.37	62-64	$=CHCl$
75.1, 77.18	74-77	$=CCl$

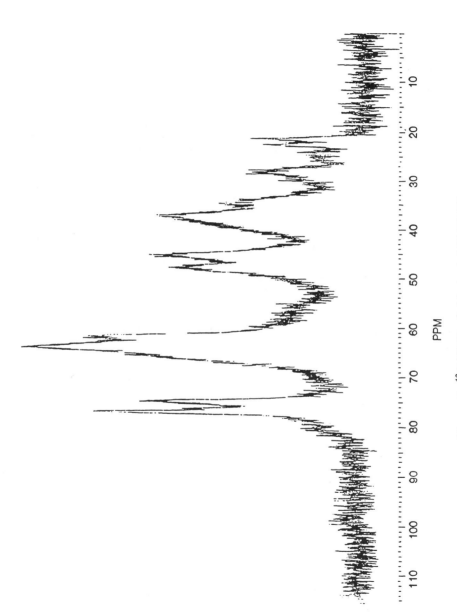

Figure 6. ^{13}C NMR of chlorinated LMWGR.

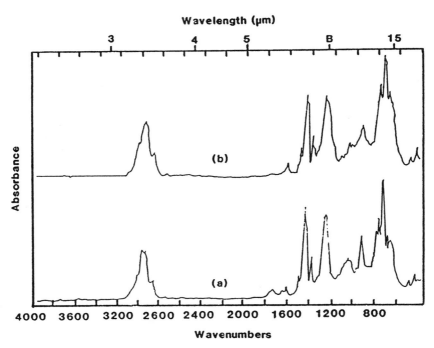

Figure 7. FTIR spectra: (a) guayule chlorinated rubber, (b) commercial chlorinated rubber.

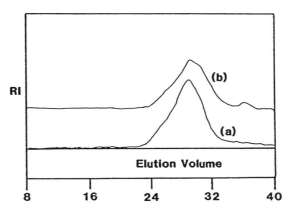

Figure 8. Gel permeation chromatograms: (a) guayule CR, (b) commercial CR.

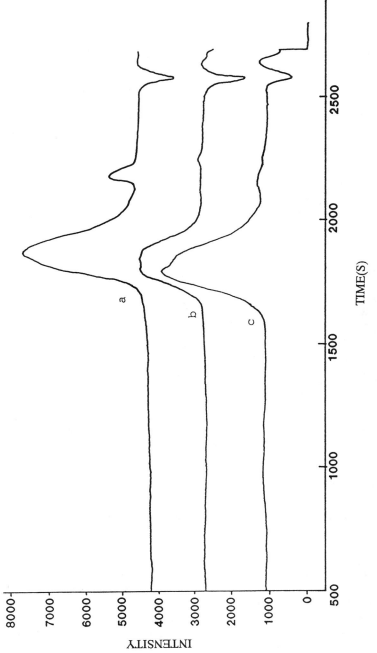

Figure 9. Gel permeation chromatograms: (a) guayule CR with AIBN and excess chlorine, (b) guayule CR with AIBN (64% Cl), (c) commercial grade CR-5 (64% Cl).

Thermal analyses were performed on a Dupont Model 9900 thermal analyzer in a nitrogen atmosphere. The lower transitions for guayule CR are likely a result of the presence of trace amounts of waxy materials acting as plasticizers. The glass transition temperature (Tg) of the chlorinated rubber of Figure 10.

Utility of Guayule Chlorinated Rubber (CR). Chlorinated rubber is typically employed in applications where chemical resistance and adhesion are important, i.e., marine paints, floor finishes, pool paints, and clear coatings. Accordingly, we have shown that guayule CR is equal or superior to commercial grades of chlorinated Hevea natural rubber (Tables III and IV).

In all cases, paints prepared with chlorinated GR were comparable in performance to those containing Hevea chlorinated rubber or Alloprene. The clear, non-pigmented coating contained only CR dissolved in toluene. It is noteworthy that the clear coat derived from guayule chlorinated rubber showed higher impact resistance than the Alloprene clear coat; indicating that chlorinated GR provides a superior combination of flexibility and adhesion.

In summary, LMWGR can be successfully chlorinated to obtain coating grade CR. Elemental analysis and other characterization techniques confirm that the chemical structure of the material is equivalent to that of commercial grades. The use of AIBN lowers the apparent molecular weight and allows the formation of lower viscosity grades of CR.

The Epoxidation of LMWGR. Oxirane rings are formed by the reaction of olefinic bonds with peracids, a reaction commonly known as the "Prilezhaev reaction" (*12*). Reactions of this type utilize a variety of peracids such as meta-chloroperbenzoic acid, 3,5 dinitroperbenzoic acid (*13*) and trifluoroperacetic acid (*14*). The generally accepted mechanism of epoxidation (*15,16*) is shown (Figure 11) with the epoxidation typically being conducted in a solvent of low ionization potential (i.e., Benzene). In our case, the reaction was conducted in the less toxic hydrocarbon, toluene. The reaction rate is rapid and the product easy to separate, yet care must be taken to insure that no crosslinking occurs in the system. The epoxidation of LMWGR provides a functional natural rubber derivative with potential utility in the coatings, adhesives and elastomers industries. A commercial epoxidized Hevea rubber is available in the form of ENR-50 (available from Guthrie Latex, Inc., Arizona). Thus, it was of interest to determine the feasibility of producing and evaluating epoxidized LMWGR.

Experimental. Initially two products have been synthesized: a 50 mole percent epoxidized rubber and a 25 mole percent epoxidized rubber. This is accomplished by preparing a 5% solution of LMWGR in toluene. The solution, in a three neck round bottomed flask, is cooled in an ice bath to reduce unwanted side reactions (*11*). After addition of molar amounts of sodium carbonate, to consume the side product m-chlorobenzoic acid, the reaction is allowed to equilibrate with stirring after which m-chloroperbenzoic acid is slowly added to the reaction vessel. The product is precipitated into ethanol and washed with water to remove the benzoic acid salt. Purified by recrystallization from ethanol gives a grey, non-tacky, rubber like product.

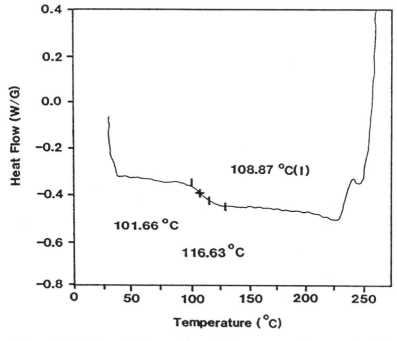

Figure 10. Differential scanning calorimetry curve for guayule CR.

Table III
Guayule CR Formulated Marine Coating

MATERIALS	AMOUNT (grams)
Grind:	
Alloprene R20	28.6
Chlorafin 40	14.3
Chlorowax 70	7.1
Epi-Rez 510	0.425
Ti-Pure R-960	5.0
Zinc Oxide	6.4
International 3X	12.6
Mica, 325-mesh	13.4
Xylene	50.0
Letdown:	
Epi-Rez 510	0.425
Xylene	50.0

Table IV
Comparative Impact Resistance Tests

PAINT	DIRECT IMPACT (in-lbs)	REVERSE IMPACT (in-lbs)	ADHESION
Guayule Marine	160+	160+	5A
Alloprene Marine	160+	160+	5A
Guayule Clear	40	10	5A
Alloprene Clear	10	<10	5A
Guayule Floor	160+	160+	5A
Alloprene Floor	160+	140	5A

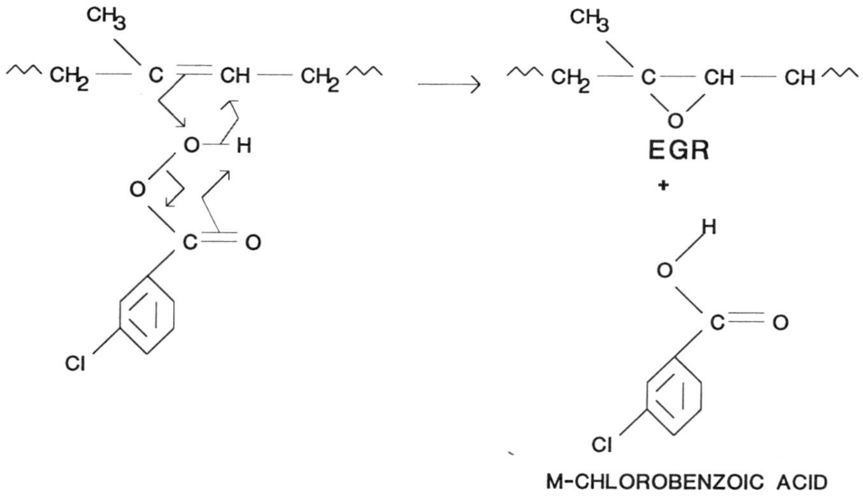

Figure 11. Epoxidation mechanism.

Characterization of EGR. The products of the epoxidation reaction were characterized via ^{1}H and ^{13}C NMR, IR, and DSC. The IR spectrum (Figure 12) confirms the disappearance of alkene bond via diminution of the 3100 cm^{-1} absorption and the presence of the 1261.61 cm^{-1} absorption characteristic of the epoxide ring. Since a strong tendency for ring opening exists, care must be exercised in the reaction workup.

The ^{13}C NMR spectra were consistent with literature values for epoxidized Hevea rubber (Figure 13) with absorptions at 60.7 ppm and 64.7 ppm corresponding to the published values of Burfield and co-workers (17) of 60.8 ppm and 64.5 ppm. Alkene character was retained and confirms the continued presence of unsaturation. Elemental analysis by MHW Laboratories allowed the calculation of epoxidation to be set at 56% of the original alkene content.

The ^{1}H spectrum is likewise in agreement with published values for epoxidized Hevea rubber (Figure 14) (18).

A linear relationship between Tg and mole % epoxidation has been published for epoxidized Hevea rubber; a one degree increase in Tg accompanies the addition of each 1 mole percent epoxidation. In our case, a 56 mole percent epoxidation resulted in a Tg of -20 ± 1 °C, representing a 54 degree increase. The DSC chromatogram is shown in Figure 15.

The Reaction of LMWGR with Maleic Anhydride (MA). Maleic anhydride can be free-radically grafted onto the LMWGR backbone (11) and thus offers an attractive means of functionalizing guayule rubber. This has particular significance as the carboxyl functionality associated with MA allows for water dispersibility and/or solubility. Furthermore, as the anhydride moiety is capable of facile ring opening and subsequent reactions with numerous active hydrogen compounds, chain extension, crosslinking or simply, unique monomer formation can result. The reaction (Figure 16) is initiated by benzoyl peroxide and presumably provides inter- and intramolecular crosslinks to the rubber (11) (Figure 17).

Organic Soluble Resins (OSR)

The OSR fraction, obtained during the processing of guayule shrub, has previously been reported as being effective in the formulation of amine-epoxy strippable coatings and as a termiticide for wood treatment. In addition to small quantities of LMWGR and HMWGR, the organic soluble resin contains a variety of components including parthenoils and their cinnamic esters; fatty acids such as linoleic, linolenic, palmitic, and stearic; high molecular weight alcohols; and guayulin A and guayulin B (21-23). Guayule OSR contains triglycerides of fatty acids (25,26) which when incorporated into epoxy coatings formulations, impart flexibility and plasticization to the resultant coatings (24). Thus, guayule resin was found to be effective as an adhesion modifier in a typical epoxide coatings formulation. Accordingly, strippable coatings with controlled limits of adhesion were prepared. Indeed, bis-phenol A epoxides are reported to have a broad

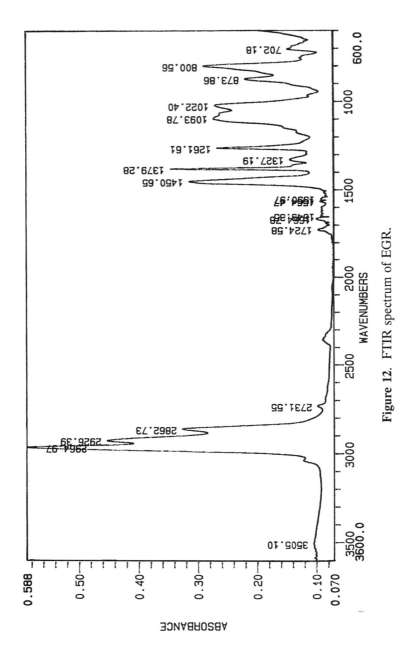

Figure 12. FTIR spectrum of EGR.

16. THAMES & POOLE Polymeric Materials from Agricultural Commodities 289

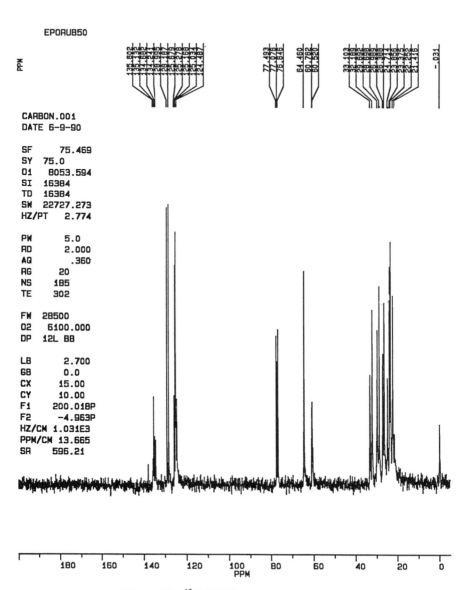

Figure 13. ^{13}C NMR spectrum of EGR.

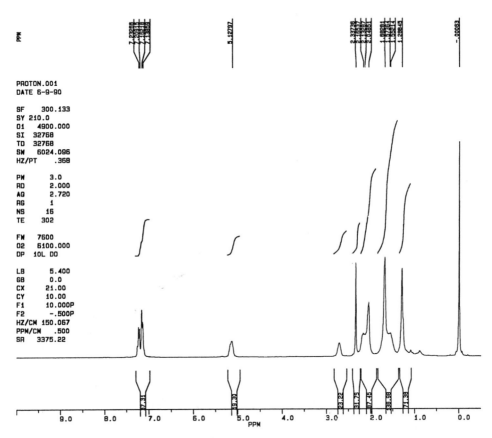

Figure 14. ^{1}H NMR spectrum of EGR.

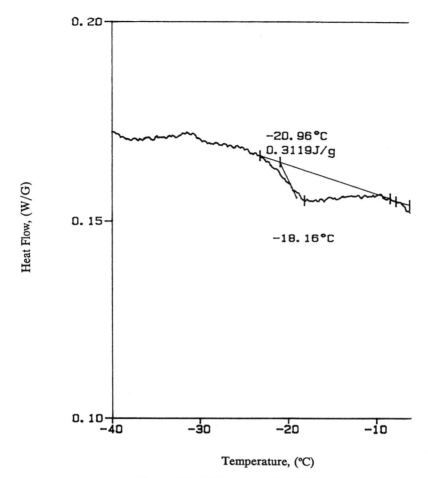

Figure 15. DSC curve of EGR.

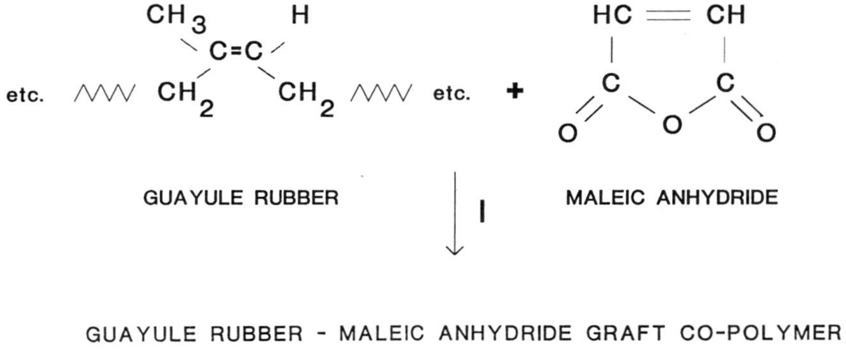

Figure 16. Maleination conditions of GR.

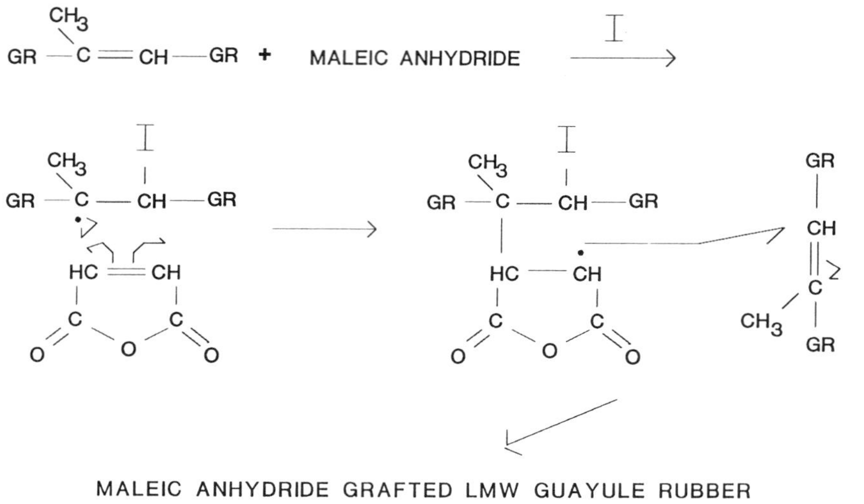

Figure 17. Maleination mechanism of GR.

range of applications including their use as high performance coatings with outstanding adhesion, low shrinkage during cure and excellent chemical resistance. Thus, high performance, yet strippable coatings were prepared. Great latitude in coating formulations is provided by flexibility of cure temperatures including ambient to elevated conditions (27). Thus, epoxides are an excellent polymer type for physical property modification.

Formulating Conditions. Formulations of epoxy strippable coatings include EPON 828, the diglycidyl ether of Bisphenol A from the Shell Chemical Company, and guayule resin dissolved in a solvent mixture of n-butanol and xylene to which Jeffamine D-400, a diamine product of the Texaco Chemical Company, has been added. A flow control agent, Byk-341 supplied by Byk-Chemie USA, is added with subsequent filtration through glass wool to remove traces of insoluble materials. The films were cast on several substrates such as pretreated steel, phosphatized steel, cold rolled steel, and aluminum in order to test its adhesion to a wide variety of substrates. Coatings were cured at 150°C for 3h in an air forced oven. Table V shows typical coating formulations.

Characterization of Films. DSC measurements were used to determine Tg of the films, with the plasticization providing a marked decrease in the glass transitions (Figure 18). The decrease in Tg for guayule resin containing formulations is consistent with the plasticization by the addition of guayule OSR.

In the case of guayule OSR modified epoxy coatings, surface treatment of panels improves adhesion to the substrate. In contrast, epoxy resin control coatings do not show noticeable changes in film performance, regardless of whether the coated panels are treated or untreated. For untreated panels, such

Table V
Formulation of Guayule Modified Epoxy Coatings

		Epon-Amine/Guayule resin (wt/wt)			
		100/0	95/5	90/10	80/20
		(g)	(g)	(g)	(g)
Epon 828 (Epoxy ether resin)	:	10.00	10.00	10.00	10.00
Polyoxy propylene amine (Jeffamine D-400)	:	6.185	6.185	6.185	6.185
Guayule resin (Firestone facility)	:	--	0.855	1.798	4.050
Solvent Mixture (n-Butanol-xylene, 50/50)	:	6.890	7.300	7.710	8.670
Byk - 341 (Byk Chemie)	:	0.180	0.180	0.180	0.180

as cold rolled steel, guayule OSR acts as an effective surface modifier rendering epoxy systems strippable (see Table VI). This characteristic of Guayule OSR and epoxy resins is a unique application for the OSR extracts and offers significant potential for commercialization. The modification with guayule OSR provides a viable and economically feasible approach to formulate strippable epoxy coatings for use in harsh environments (27).

Impact resistance was determined (Table VI) and found to remain quite acceptable with the incorporation of guayule resin modification. Tensile and elongation properties were likewise ascertained to determine the extent of change on physical properties by guayule resin modification (Table VII).

The Efficacy of Guayule OSR as a Pesticide. The United States armed forces employ the use of large quantities of treated wood in marine and terrestrial environments with the annual cost of repair and replacement of these products amounting to approximately 25 million dollars (28,29). Consequently, guayule OSR is currently under evaluation as a wood protectant against termites, fungi and marine borers such as barnacles (29).

Guayule OSR treated pine, exposed to termites for 33 to 44 months in the Panamanian rain forest at Chiva, shows no damage. However, the resin did not act as a repellant since the termites tube over it in order to reach the control baitwood. It is felt that since guayule resin is active against wood fungi, a

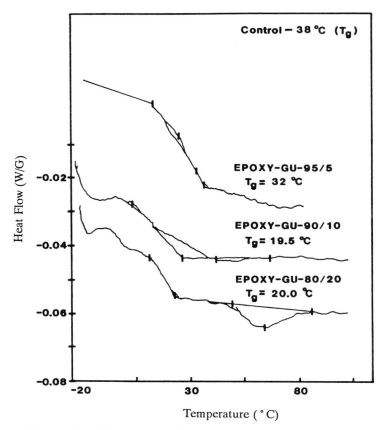

Figure 18. Effect of guayule resin on T_g of epoxy coatings.

Table VI
Film Properties of Guayule Modified Epoxy Coatings on Non-Treated Panels

Substrate	:	* Cold Rolled Steel Panels ** Untreated Aluminum Panels			
Solvent	:	Xylene/Butanol 50/10 (wt/wt)			
Cure Cycle	:	125°C/3 h			
Flow Control Agent	:	Byk 341 (Byk Chemie)			

Epoxy-Amine Guayule (wt/wt)			100	95/5	90/10	80/20
Thickness (mils)	:		1.5	1.5	1.5	1.5
Pencil Hardness		* **	H H	2B 2B	2B 2B	2B 2B
Resistance to boiling water	:			Film came off after 30 mins.		
Impact Resistance (in. –lbs)						
Direct	:	* **	150 60	-- film released at 5 --		
Reverse	:	* **	150 60	-- film released at 5 --		

* Cold Rolled Steel Panels
** Untreated Aluminum Panels

Table VII
Tensile Properties of Guayule Modified Epoxy Films*

Epoxy-Amine/Guayule (wt/wt):	100	95/5	90/10
Tensile Strength (MPa) :	41.37	40.00	39.29
% Elongation At Break :	3.5	3.6	4,000

*Substrate: Cold Rolled Steel Panels

condition required for termite activity, the wood resists termitic attack yet does not act as a termite repellant. Furthermore, the exposed samples do not show weathering patterns characteristic of heavy rainfalls (29).

The majority of Guayule OSR treated wood, exposed to the marine environments of Limon Bay in Panama, remains unattacked after 45 months of exposure. This phenomenon appears to be guayule resin concentration dependent, as the wood most highly impregnated with OSR shows increased resistance to attack. The wood specimens have become lighter in color and some surface leaching of the resin has occurred. All untreated wood panels, however, were heavily damaged. Work is underway to identify the exact components of the resin responsible for this wood protective activity.

Water Soluble Resins

The processing fraction remaining the least explored is the water soluble portion. Fractionation of the water soluble materials is accomplished by membrane separation technology. Dr. Wagner and co-workers at Texas A&M are currently performing separation and characterization of fraction components. Characterization techniques include NMR, IR, DSC, and GPC.

Bagasse

Guayule bagasse (GB) has been mentioned as a potential fuel feedstock, a cellulosic provider, and a source of fermentable sugars or fibers. Direct combustion of GB gives a fuel value of 18,200 kJ/kg or 7,838 Btu/lb (30). It has been reported that a gas containing olefins, hydrogen and carbon monoxide can be formed when GB is passed into a fluidized bed gasification system (31).

Various GB cellulosic derivatives have been prepared; these include but are not limited to cellulose acetates, cellulose nitrates and regenerated cellulosics (rayon). Pulping and bleaching of GB are necessary prior to derivation. The authors utilized contemporary wood pulping processes and therefore concluded that the process was efficient for GB(32).

Pulping. Lignin and other non-cellulosics must be removed from the GB in order to purify the cellulosics for subsequent use. A convenient laboratory method involves the Kraft pulping process carried out in a Parr reaction vessel allowing mild reaction conditions so as to protect the cellulose from degradation. After pulping, the GB is separated by filtration and squeezed with a hand press to remove all pulping liquor. Water and acetone washes are then performed until a Ph of 7 1 1 is reached with subsequent air drying to insure complete removal of residual acetone. The pulping liquor contains tall oil and sulfate turpentine that offer other potential uses (32).

Bleaching. The bleaching process removes lignins and colored matter without adversely affecting the cellulose. Sodium hypochlorite is employed as the bleaching reactant while the reaction mass is monitored for Ph. Upon completion of bleaching, the product is washed with acetone and air dried to insure high purity and high yield (32).

Cellulose Acetate. Acetylation is performed on pulped, bleached GB. Glacial acetic acid, acetic anhydride and concentrated sulfuric acid (catalyst) make up the acetylating solution. After 20h in solution, the remaining solids are removed via filtration and centrifugation. The filtrate is poured into distilled water whereupon a white gelatinous material, cellulose acetate, immediately forms. The infrared spectrum of this product is essentially identical to that of cotton cellulose acetate. Cellulose acetates are used in lacquers, plastics, safety film, and fabrics, and therefore, provide attractive product areas for the guayule industry (*32*).

Summary

Ongoing research and development efforts on guayule are focused to provide a better understanding of the numerous guayule components, their chemistry and ultimately, their commercialization. The success of developing value added products from guayule components will dictate the fate of guayule as a domestic source of natural rubber. By isolating, identifying and utilizing the majority of guayule's processing fractions a domestic, natural rubber industry is likely to become a reality.

Acknowledgements

We acknowledge the continuous support and helpful comments of Dr. Daniel Kugler of the USDA Office of Agricultural Materials and Mr. George J. Donovan of the Department of Defense.

Literature Cited

1. Bragg, D. M.; Lamb, C. W., Jr. *The Market for Guayule Rubber*; Center for Strategic Technology, Texas Engineering Experiment Station, Texas A & M University: College Station, TX, 1980.
2. Hammond, B. L.; Polhamus, L. G. *Research on Guayule (Parthenium Argentatum Grey): 1942-1959*; U.S.D.A. Agr. Res. Ser. Tech. Bill. No. 1327, U. S. Dept. of Agriculture: Washington, DC, 1965; p. 143.
3. Backhaus, R. A.; Nakayama, F. S. *Rubber Chem. and Tech.* **1986**, *61*, 78-85.
4. *Guayule: An Alternative Source of Natural Rubber*; National Academy of Science, 1977.
5. Hager, T. A.; et al. *Rubber Chem. and Tech.* **1979**, *52*, 693.
6. Soltes; et al. Report on NSF Grant No. PFR 78-12713; Texas A & M Univ.; College Station, TX, 1979; pp 77-99.
7. Wagner J. P.; et al. Presentation to Fourth Int. Conf. on Guayule Research and Development, Tucson, AZ.
8. Thames S. F.; Kaleem, K. In *Agricultural and Synthetic Polymers*; Glass, Edward; Swift, Graham, Eds.; ACS Symposium Series; ACS: Washington, DC, 1990; pp 230-241.
9. Bloomfield, G. F. *J. Chem Soc.* **1943**, 289.

10. *Rubber Technology*; Morton, Maurice, Ed.; Robert Krieger Publishing Co.: Florida, 1981; p 165.
11. *Rubber Chemistry*; Brydson, J. A., Ed.; Applied Science Pub. LTD.: London, 1978; pp 172-3, 187.
12. March, Jerry *Advanced Organic Chem.*, Third Ed.; John Wiley & Sons: New York, 1985; p 735.
13. Emmons; Pogano *J. Am. Chem. Soc.* **1955**, *77*, 89.
14. Rastetter, R.; Lewis, J. *J. Org. Chem.* **1978**, *42*, 3163.
15. Bartlett *Rec. Chem. Prog.*, *18*, 111.
16. Dryuk *Tetrahedron* **1976**, *32*, 2855-2866.
17. Burfield, D.; Lim, K.; Law, K. *J. Appl. Poly. Sci.*, *29*, 1661-1673.
18. Bradbury, J. H.; Perera, M. C. *J. Appl. Poly. Sci.*, *30*, 3347-3364.
19. Davies, C; Wolfe, S.; Gelling, J. R.; Thomas A. G. *Polymer* **1983**, *24*, 107-113.
20. *Chemical Reaction of Polymers*; Fettes, E. M., Ed.; High Polymer Series, Vol. XIX.; Interscience Pub.: New York; pp 125-132, 188.
21. *Guayule: An Alternative Source of Natural Rubber*; Contract No. KS1C14200978, Bureau of Indian Affairs: Washington, DC, 1977; Chapter 9.
22. Dorado, E. B. *Chim. Ind.* **1962**, *87(5)*, 617.
23. Banigan, T. F.; Meeks, J. W. *J. Am. Chem. Soc.* **1953**, *75*, 3829.
24. Bauer, R. S. In *Applied Polymer Science*; Tess, R. W.; Poehlin, G. W., Eds.; ACS Symposium Series, 285; ACS: Washington, DC, 1985; pp 931-961.
25. Belmares, H.; Jiemenez, L. L.; and Ortega, M. *Ind. Eng. Chem. Prod. Res. Dev.* **1980**, *19*, 107.
26. Schloman, W. W., Jr.; et al. *J. Agric. Food Chem.* **1986**, *34*, 177-179.
27. Thames S. F.; Kaleem, K. In *Guayule Natural Rubber*; 1990, pp 338-346.
28. Pendleton, D.; O'Neill, T. Naval Civil Engineering Laboratory Technical Report N-1773, 1987.
29. Bultman, J. D.; et al. "The Efficacy of Guayule Resin as a Pesticide," *Biomass*, in press.
30. Kuester, J. L.; et al. In *International Conference on Fundamentals of Thermochemical Biomass Conversion*; Overend, R. P., et al., Eds.; Elsevier Applied Science: 1982; pp 875-895.
31. Schloman; Wagner *Guayule Rubber*; 1990; Chapter 12.
32. Bultman, J. D. *Possible Uses for Guayule Bagasse*; Report to Naval Research Laboratory, El Guayulero; 1989, Vol. 11, No. 3/4, pp 35-45.

RECEIVED May 22, 1991

Chapter 17

Emerging Polymeric Materials Based on Soy Protein

Thomas L. Krinski

Protein Technologies International, Checkerboard Square, St. Louis, MO 63164

This chapter will describe the chemical modification of native soy protein, a globular protein primarily used as a food source. A review of the past technology will describe the early industrial uses of soy protein and how the protein was modified to alter its function as an adhesive. Current and future directions will show how chemical modification and alteration of protein chain association can further enhance soy protein polymers function. The past unfavorable attribute of biodegradation of the soy polymer chain may now guide the future of soy polymer derivatives as the need increases for biodegradable polymers from renewable resources.

Industrial Soy Protein of the Past

Industrial soy protein used are primarily in the coating of paper and paperboard, with minor areas in water based inks and water based adhesives. The water based adhesives include bottle label adhesives, foil laminating adhesives, and cone/tube winding adhesives.

Protein Composition and Isolation. The protein present in the soy bean is not a simple singular globulin but a mixed assortment of different types of protein fractions. These fractions consist of:

Bowman-Birk Trypsin Inhibitor. A globular protein of 24M molecular weight that can form dimers and trimers due to high cystine content.

Kunitz Trypsin Inhibitor. This is another low molecular weight protein (molecular weight of 21M) which possesses disulfide linkages.

Hemaglutinin. A glycoprotein with a molecular weight of 100-110M without any disulfide crosslinks.

Lipoxygenase. An enzyme with a molecular weight of 102M which can be disassociated into two near equal fragments using guanidine hydrochloride.

7S Globulin. A large associated polymer of 180-210M molecular weight consisting of nine polypeptide chains. These peptide chains can be disassociated into 2S and 5S subunits at pH 2.0. At neutral and slightly alkaline pH (pH of 7.6), the 7S can dimerize into the 9S fraction. Fully disassociated, the polypeptide chains that make up the 7S have molecular weights in the 20-25M range.

11S Globulin. A large 330-350M molecular weight associated polymer which easily disassociates into 7S size fragments. These fragments can likewise further disassociate into various smaller subunits capable of associating with one another to form dimers and trimers.

Urease. An enzyme having a sedimentation fraction equivalent to an 18S form (1).

Soy proteins are isolated by aqueous alkaline extraction of defatted soybeans. The native protein can be isolated prior to any additional treatment by precipitation at their average isoelectric point of pH 4.5 and separated from the soy sugars and salts. The molecular weight distribution by HPLC in 6 Molar guanidine hydrochloride is represented in Figure 1. The native protein has an average molecular weight (Mw) of 192,000 daltons.

Native soy protein is difficult to work with as an adhesive because the many different globulins disassociate and reassociate as the protein is solubilized on the alkaline side. Even though the molecular weight of the native protein is quite high and it has good adhesive qualities, the problem of solution rheology control usually discourages its use (2).

Physico-Chemical Treatment of Soy Protein (Caustic Treatment).
Native soy protein has been modified using a controlled alkaline heat treatment of the protein while in solution. This processing utilizes three factors to control the protein reorganization: pH, temperature and time (3).

Table 1 shows the primary events which occur during physico-chemical treatment. Protein globulins slowly unfold with a minimum of backbone chain cleavage. As the chains unfold, they reorganize or reassociate, this time by hydrophobic/hydrophilic regions. this exposes more potential hydrophilic groups to aqueous contact.

A portion of the asparagine and glutamine residues are hydrolyzed to the free acids. This increases the anionic character of the protein as seen in the titration curve of Figure 2.

Finally, we also achieve some internal chemical crosslinking through the formation of lysinoalanine (LAL). This occurs when the cysteine or serine residues undergo a β elimination reaction (Figure 3) in the presence of elevated ph and temperature, forming a dehydroalanyl residue. This residue can then react with lysine amine through a base catalysed reaction achieving potential crosslinking of two independent protein chains.

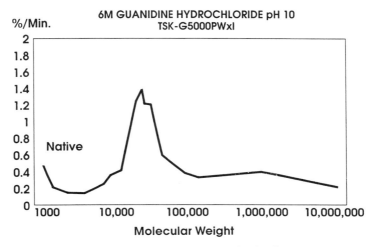

Figure 1. Native protein Mw distribution.

Table I. Caustic Treatment (pH>10.5, 40°C<Temp<70°C, Time)

1. Encourages unfolding of native globular structures.
2. Encourages Reorganization of unraveled globules by hydrophilic/hydrophobic regions.
3. Partial hydrolysis of asparagine and glutamine primary amides.
4. Protein chain crosslinking by formation of lysinoalanine.

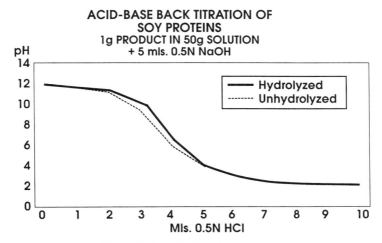

Figure 2. Protein titration curve.

Figure 4 shows the molecular weight distribution of a caustic treated protein compared with the native. With an average molecular weight of 150,000 daltons, there is a marked shift to lower molecular weights as compared with the native protein.

The physico-chemical treatment step achieves reorganization of the protein globules/chains. Hydrophilic residues are exposed to more aqueous contact, while hydrophobics are buried. Combined with increased anionic character, the treatment greatly reduces the protein globules tendancy to self associate, which changes their rheological function and how they interact when used in coating paper (4).

Paper Coating Applications The major use of soy protein polymers is in the coating of paper and paperboard. Paper is coated (Table 2) to improve its surface appearance and to make it whiter and more uniform in color. The coatings impart a smoother surface which accepts ink uniformly without feathering of the ink.

A paper coating consists of three classes of components (Table 3). pigment makes up most of the coating weight. The combined use of different levels of protein, latex, and starch provides the adhesive which holds the pigment on the paper surface.

Coating rheology influences how well the coating can be applied to the paper surface. The hercules rheogram (52% solids, 4400 RPM) of the native protein (Figure 5) has a much higher viscosity than the caustic treated protein rheogram (Figure 6). Physico-chemical treatment reduces not only the internal reactivity of the protein chains but also the reactivity of the protein with pigments such as clay. This leads to more uniform application of the coating to the paper surface with improved overall surface properties.

Chemical Modification of Soy Proteins

Recent technology indicates that chemical modification of some of the amino acid residues can provide the ability to design new protein polymers with more specific functionality changes than simple caustic treatment can achieve. Both of these protein modification tools can be combined to further enhance the protein modification and extend the application.

In chemical modification of the proteins, we must keep in mind that the hydrophilic amino acids (Table 4) due to their reactivity provide the best targets considering the aqueous conditions under which these proteins are processed (5).

Specific amino acid residues can be targeted (figure 7), by choosing the proper reaction pH (6).

Anhydride Reactions. Acid anhydrides reacting with pendant free amine groups can be used to modify the protein. A primary anhydride such as acetic (Figure 8), can be used to react with the amine groups negating their cationic charge. This tends to decrease solution and paper coating viscosities of the soy protein (7).

The use of di and tricarboxylic anhydrides enables the modification not only to negate the cationic amines, but also to

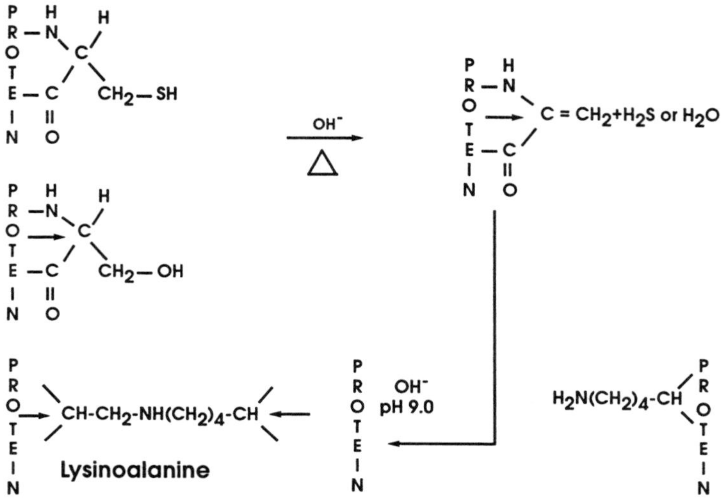

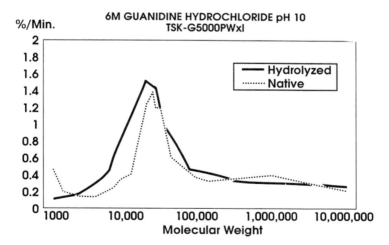

Figure 3. Protein crosslinking by lysinoalanine formation.

Figure 4. Protein Mw distribution.

Table II. Primary Use Paper Coatings
General Coating Applied to Paper

- Improved appearance
- Improved color
- Improved printing surface

Table III. Primary Use Paper Coatings Composition (Aqueous)

- Pigment
  Clay
  Calcium Carbonate
  Titanium Dioxide
- Adhesive
  Soy Protein
  Latex
  Starch
- Minor Additives

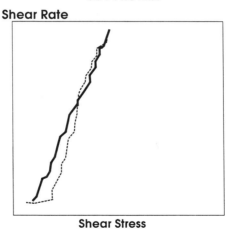

Figure 5. Native protein rheogram.

17. KRINSKI *Emerging Polymeric Materials Based on Soy Protein*

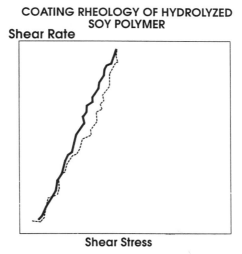

Figure 6. Soy polymer rheogram.

TABLE IV. AMINO ACID COMPOSITION OF SOY PROTEINS

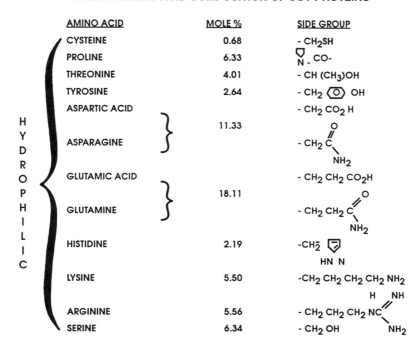

replace them with additional acid groups (Figure 9 & 10). The increased anionic character is evident in the titration curves of Figure 11 where di and tricarboxylic anhydrides were used to modify the soy protein. This type of modification in coatings tends to decrease the viscosity of soy protein even further (Hercules Rheogram in Figure 12). Increased anionic charge on the protein further reduces the associative nature of the protein and increases its dispersant nature. A shift to lower average molecular weight (133,000 daltons - Figure 13) occurs due to increased processing time and increased dissociation of the soy globulins.

Crosslinking Reactions. Soy proteins can be mildly crosslinked through the use of various simple or complex reagents. Using epichlorohydrin (Figure 14), the soy protein molecular weight and solution viscosity can be increased by varying the amount of reagent used.

A more complex monomer such as methyl acrylamidoglycolate methyl ester (MAGME) which has the ability to crosslink both during initial reaction and later during application (Figure 15) can be used (8). Under basic conditions, the primary reaction with the protein can occur either at the double bond or at the activated ester. At elevated temperatures, the proton on the amide group is acidic enough to encourage further crosslinking at the amide nitrogen. This second reaction could occur during the soy polymers use in the coating-drying-calendering of paper.

Base Catalysed Reactions. Due to the alkaline processing conditions, nucleophilic addition reactions are excellent fits for modification of soy protein and can be tailored even to selective amino acid residues. water soluble monomers such as acrylamide (Figure 16) and acrylonitrile provide excellent reactivity with lysine amine groups and sulfhydryl of cysteine.

Hydroxyacrylates can also be used to modify any free amine groups and add additional hydroxyl residues to the protein (9). In paper coating applications, this allows the protein to hydrate more thus adding water holding to the paper coating as it is applied. The runnability of the coating as it's applied to the paper is improved.

Exotic monomers, such as glycidyl oxypropyl trimethoxy silane, can add a silane ester to the protein amines (10). This modification greatly increases the viscosity of the protein with clay pigments or causes total destabilization of the clay suspension.

Interpolymer Technology. When soy polymers are used in the coating of paper, a styrene butadiene latex is also used. The increased dispersant action and lower viscosity of the anionic modified soy protein has been used in the preparation of various latices (Figure 17). Monomers can be emulsified by using the soy polymer as the only surfactant. These miscelles can then be polymerized in the presence of the soy polymer resulting in a soy latex interpolymer whose core is stabilized by a "surfactant shell" of soy protein. Figures 18 and 19 show the particle size analysis of such a styrene butadiene latex with an average size of 0.2 micron. This is

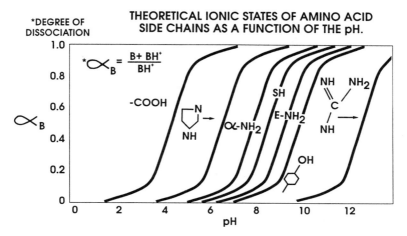

Figure 7. Ionization of side residues.

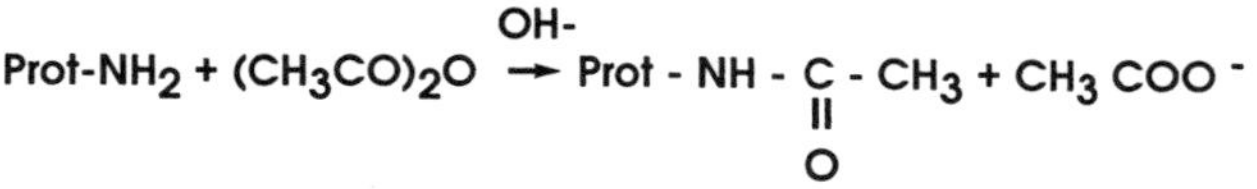

$$\text{Prot-NH}_2 + (CH_3CO)_2O \xrightarrow{OH^-} \text{Prot - NH - }\underset{\underset{O}{\|}}{C}\text{ - CH}_3 + CH_3COO^-$$

Figure 8. Anhydride reaction with amine groups: monocarboxylic anhydride.

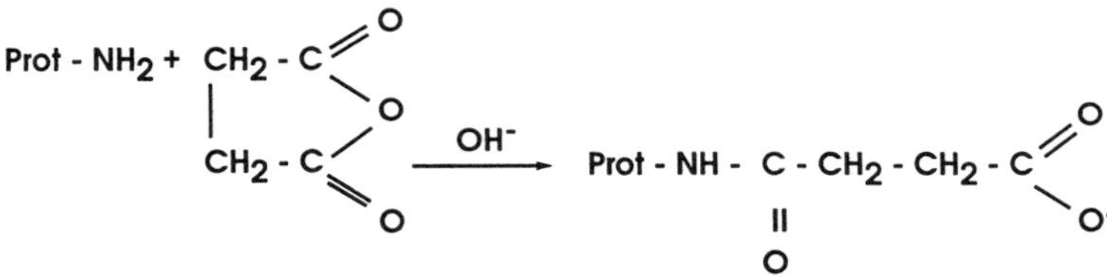

Figure 9. Anhydride reaction with amine groups: dicarboxylic anhydride.

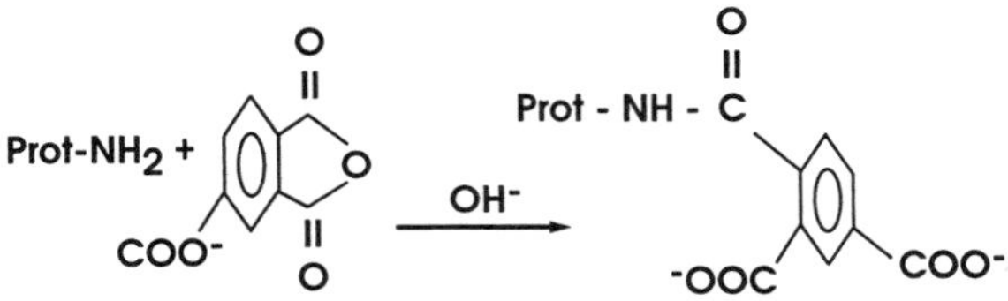

Figure 10. Anhydride reaction with amine groups: tricarboxylic anhydride.

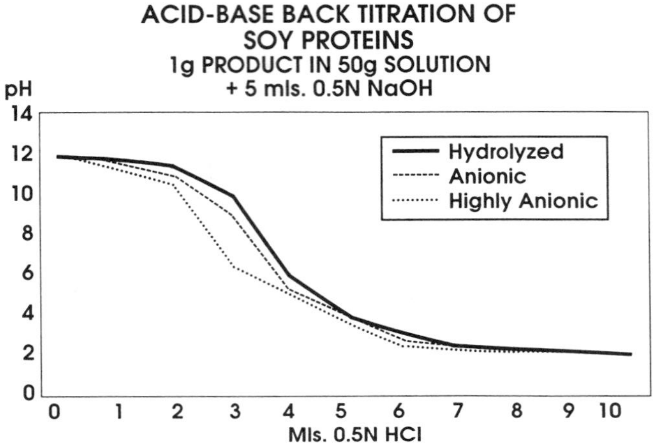

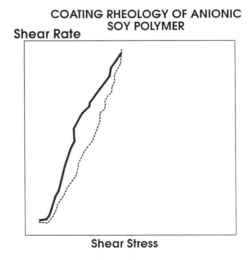

Figure 11. Protein titration curve.

Figure 12. Soy polymer rheogram.

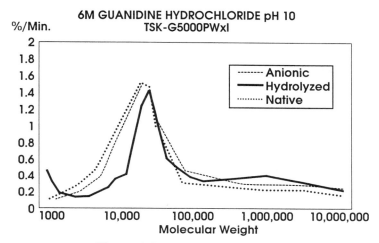

Figure 13. Protein Mw distribution.

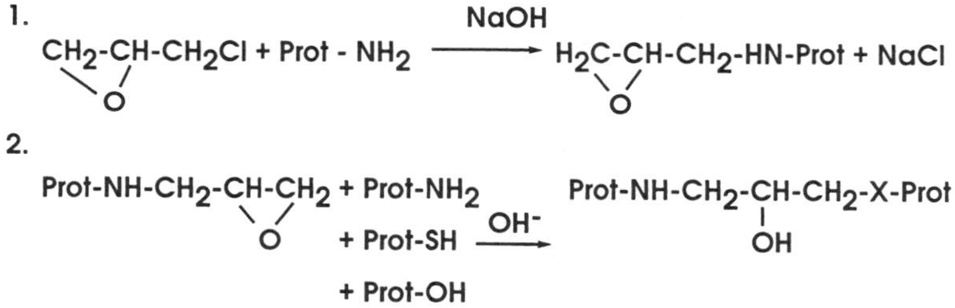

1.

$$CH_2\text{-CH-CH}_2Cl + Prot\text{-}NH_2 \xrightarrow{NaOH} H_2C\text{-CH-CH}_2\text{-HN-Prot} + NaCl$$
$$\quad\quad\underset{O}{\diagdown\diagup}\quad\quad\quad\quad\quad\quad\quad\quad\quad\quad\quad\underset{O}{\diagdown\diagup}$$

2.

$$Prot\text{-}NH\text{-}CH_2\text{-CH-CH}_2 + Prot\text{-}NH_2 \quad\quad Prot\text{-}NH\text{-}CH_2\text{-CH-CH}_2\text{-X-Prot}$$
$$\quad\quad\quad\underset{O}{\diagdown\diagup}\quad + Prot\text{-}SH \xrightarrow{OH^-}\quad\quad\quad\quad\quad|\quad\quad\quad\quad\quad\quad$$
$$\quad\quad\quad\quad\quad\quad\quad + Prot\text{-}OH\quad\quad\quad\quad\quad\quad\quad OH$$

Figure 14. Crosslinking reactions using epichlorohydrin.

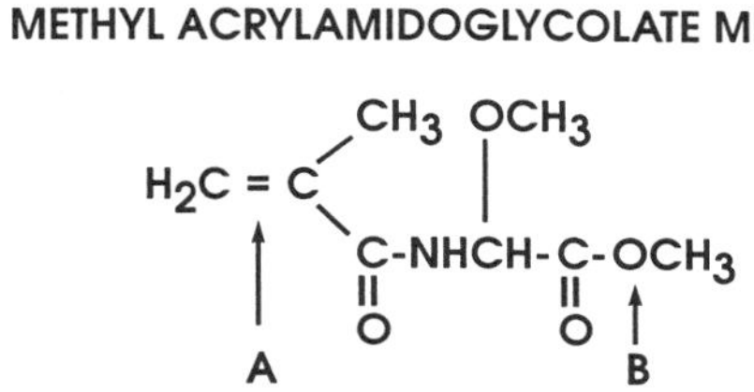

METHYL ACRYLAMIDOGLYCOLATE METHYL ESTER

$$H_2C = C\underset{\underset{A}{\uparrow}}{\overset{CH_3}{\diagup}}\underset{\underset{O}{\|}}{\overset{OCH_3}{|}}\text{C-NHCH-C-OCH}_3$$

Figure 15. Crosslinking reactions using MAGME.

UNSATURATED CARBONYLS CAN REACT WITH UNPROTONATED AMINES

$$CH_2 = CH - \underset{}{\overset{NH_2}{\overset{|}{C}}} = O + Prot - NH_2 \xrightarrow{pH\ 9.0} Prot - NH - CH_2 - CH_2 - \overset{NH_2}{\overset{|}{C}} = O$$

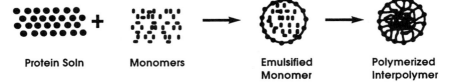

ARCYLAMIDE

Figure 16. Base catalyzed additions (nucleophilic).

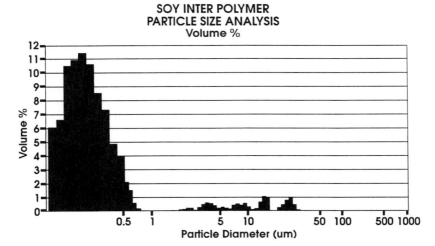

The Interpolymer is prepared by the free radical polymerization of styrene and butadiene in the presence of modified soy protein. The monomers are first emulsified by the protein. These emulsified particles are spherical in shape with the hydrophillic soy protein covering the surface and the hydrophobic monomers buried in the interior. The soy protein is polymerization.

Figure 17. Interpolymer preparation.

Figure 18. Interpolymer particle size.

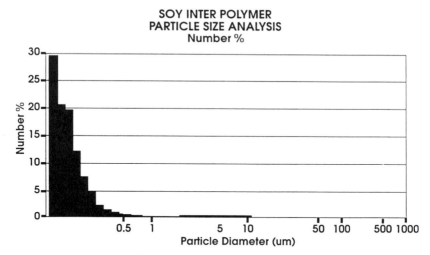

Figure 19. Interpolymer particle size.

comparable to commercial latices yet, when used in the coating of paper and paperboard, possesses the advantages of both soy polymer and S/B latex (11). Through the use of chemical modification, we have brought the old caustic treatment technology of soy protein out of the dark ages. Combining the caustic treatment step with chemical modification has enabled us to change substantially the functionality of the protein polymer by modifying a small number of amino acid residues. Using these "tools", we can alter the reactivity/associativity of the globular protein of the native soy with itself, alter its molecular weight and generally provide a more uniform protein.

Future Soy Polymers

Previously, most industries had considered soy proteins strictly as adhesives (12). However, this attitude has dramatically changed as the chemistry of the soy protein has been modified. Modern soy polymers are now being used as paper pigment structuring agents and flow modifiers. Their amphoteric nature has shown they can be used as protective colloids and even surfactants in the stabilization of latices.

The use of chemical modification to change soy protein function is still in its infancy. As we learn how to use more reagents to modify the protein in conjunction with other treatments which can control protein reorganization, we can further specialize the functionality of the finished soy polymer. In doing so, we can expand the use of soy polymers into other industries.

Use of soy polymers in aqueous based inks has increased due to the environmental concerns of hazardous fumes from solvent based inks. As environmental concerns of biodegradability, waste disposal, and sewer effluent treatment become issues, new soy polymers could be offered as solutions.

Creation of new soy interpolymers, solution viscosity modifiers, flocculating agents and ingredients for biodegradable plastics, are potential avenues for future soy polymer technology.

References

1. Smith, A. K., and Circle, S. J., Soybeans - Chemistry and Technology; 1978, Vol. 1; Proteins, pp. 114-128.
2. Coco, C. E., Krinski, T. L., Tappi Coating Binders Seminar, 1986.
3. Ishino, K, Okamoto, S., "Molecular Interaction in Alkali Denatured Soybean Proteins"; Cereal Chemistry 52, 9, 1975, p 19.
4. Coco, C. E., Preprints Tappi Coating Conference, 1987.
5. IBID.
6. Means, G. E., Feeney, R.E., Chemical Modifications of Proteins, 1971, p 14.
7. Coco, C. E., Krinski, T. L., Graham, P. M., U.S. Patent #4474694.
8. Steinmetz, A. L., Krinski, T. L., U.S. Patent #4554337, 11/19/85.
9. Krinski, T. L., Steinmetz, A. L., U.S. Patent #4687826, 8/18/887.
10. Krinski, T. L., Steinmetz, A. L., U.S. Patent #4713116, 12/15/87.
11. Coco, C. E., Preprints Tappi Coating Conference, 1987, pp. 133-140.
12. Strauss, R. W., Protein Binders in Paper and Paperboard Coating; Tappi; Monograph No. 36.

RECEIVED July 9, 1991

Chapter 18

Enzymatic Treatments of Pulps

Thomas W. Jeffries

Institute for Microbial and Biochemical Technology, U.S. Department of Agriculture Forest Service, Forest Products Laboratory, One Gifford Pinchot Drive, Madison, WI 53705−2398

The pulp and paper industry processes huge quantities of lignocellulosic biomass every year; the technology for pulp manufacture is highly diverse, and numerous opportunities exist for applying microbial enzymes. Enzymes have been used to increase pulp fibrillation and water retention and to reduce beating times in virgin pulps. With recycled fibers, enzymes have been used to restore bonding and increase freeness. Specialized applications include the reduction of vessel picking in tropical hardwood pulps and the selective removal of xylan from dissolving pulps. The single most important application is in the removal of lignin from kraft pulps. Both lignin-degrading and hemicellulose-degrading enzymes have been shown to be effective for this purpose, but many other microbial enzymes remain to be examined. Eventually, microbial enzymes in conjunction with the appropriate chemical of thermochemical extractions might enable the fractionation of cellulose, hemicellulose, and lignin from pulps and other lignocellulosics.

The conversion of wood into pulp for use in paper comprises one of the largest industries in the United States today. In 1988, the United States produced almost 71 million metric tons of paper and paperboard (*1*) with a wholesale value of over $47 billion (*2*). Paper consumption increases with both population and gross domestic product, and it amounts to almost 315 kg per capita per year. The industry is highly diverse. It produces many different types of pulp, the most important being kraft and thermomechanical. Because pulp production consumes large quantities of wood and has a number of environmental effects, the industry has come under increasing pressure to reduce discharges and increase use of secondary (post-consumer) fibers; recycled fibers are rapidly increasing in importance.

Several biological treatments have been developed to reduce environmental effluents or to alter pulp properties. They can be classified as either **microbial**, in which growing cells are used, or **enzymatic**, in which pre-formed enzymes are employed. These two areas are related, but the applications and process considerations are quite different. Microbial and enzymatic processes can be combined. In such instances, enzymes are formed *in situ* by growing cells, and the culture conditions are then altered

This chapter not subject to U.S. copyright
Published 1992 American Chemical Society

to favor specific enzymatic activities over cell growth. Enzymatic digestion takes place in the latter stage. For example, a solid substrate culture might be flooded, and the temperature could be elevated and enzyme cofactors added. Enzymes secreted by microbes within the pulp could act effectively. Cell growth and *in situ* production of selected enzymes could combine the best characteristics of both approaches. At present, no such two-stage processes are known for pulps, but endogenously produced enzymes have been studied extensively.

Enzymes are widely employed in commercial processes, particularly in food manufacture, so the technology for large-scale enzyme production and application is fairly well-established. Many industrial enzymes are simply concentrated, stabilized, and clarified extracellular microbial broths or plant extracts that are used once and disposed of. Others are immobilized preparations that are used repeatedly. The activities are often defined in terms of empirical units, such as the amount of color removed or the degree of viscosity attained under given conditions, but they can sometimes be related to more fundamental changes, such as particular bond cleavages.

The objective of this review is to provide examples of enzymatic treatments of pulps and to discuss their efficacy. Most of the information has been obtained from patents and from the proceedings of specialized meetings. Relatively few studies have been published in refereed journals. Claims are presented here essentially as reported, so the nature of the source material should be kept in mind.

Microbial treatments of pulps such as biopulping, biobleaching, and bio-oxidation have been dealt with elsewhere (*3*). Related applications that will not be elaborated on include the use of pectinolytic or xylanolytic enzymes for pulping or retting bast (*4,5,6,7*), flax (*8,9*), or other non-wood materials (*10,11*), the selective cleavage of lignin–carbohydrate bonds (*12,13,14*), and the use of isolated ligninases to decolorize bleach plant effluent (*15,16*) or to modify lignin (*17,18*).

Several applications have been investigated for enzymatic treatments of pulp. These include the modification of pulp properties, specific removal of pulp components, and enhanced bleaching. Enzymatic treatments have been shown to improve fiber flexibility, fibrillation and drainage, and to decrease vessel picking. Xylanases can improve the quality of dissolving pulp and facilitate bleaching of kraft pulps.

Factors Determining Enzymatic Activity

The extent of product removal is largely a function of the substrate characteristics, but enzyme specificity also plays an important role. Some purported xylanases, for example, appear also to have activity against substituted (carboxymethyl) cellulose. When present, they can reduce cellulose viscosity by attacking the amorphous regions.

Microbial enzymes actually have relatively little effect on wood and thermomechanical pulps. Particle size and substrate surface area play large roles in determining enzyme accessibility, but substrate reactivity is also critical. The resistance of wood to attack is apparently attributable in part to the presence of alkali-labile bonds such as acetyl esters. Acetylated xylan is essentially resistant to attack by most xylanases. Acetyl groups are removed and other ester linkages are broken by dilute alkali. Such treatment greatly facilitates enzymatic attack. *Aureobasidium pullulans* xylanase, for example, will remove only small amounts of reducing sugars from untreated aspen thermomechanical pulp. If the pulp is treated with 4% to 8% (w/w, oven-dry basis) sodium hydroxide, as much as 40% of the xylan can subsequently be removed by xylanases. The kinetics of xylanase adsorption also shift. With alkali-treated thermomechanical pulp (TMP), xylanase is taken up rapidly, and the amount remaining in the supernatant solution is essentially unchanged with time. With untreated TMP, less xylanase is taken up initially, but the remaining enzyme gradually disappears from free solution. The enzyme has no discernable effect on the release of reducing sugars during that time (*19*).

Numerous substances can interfere with enzyme activity. In addition to contaminants and heavy metals, the nonspecific adsorption or inactivation of xylanases by cellulose can reduce the effect of the enzyme on xylan. Senior et al. (20) studied the adsorption kinetics of xylanases by low-xylan content cellulosic materials. They found that with highly crystalline, purified cellulose, nearly all of the xylanase could be recovered. When the crystallinity was reduced by swelling in acid, however, significant quantities of xylanase were adsorbed by the substrate. They also found that unbleached pulps contain inhibitory compounds that leach into solution to inactivate xylanases. Similar effects of lignin could inactivate or adsorb lignin-degrading enzymes. For these reasons, great care must be taken in evaluating enzyme activities in the presence of heterogeneous substrates.

Selective Modification of Fiber Properties

Fibrillation. Enhanced fibrillation was the first application of cell-free enzymes to pulps. Fibrillation occurs during refining wherein threadlike microfibrils are exposed and the overall surface area of the fiber increases. Fibrillation enhances inter-fiber bonding. It is normally induced by refining or beating. Beating increases the strength of paper to a maximum value, after which it decreases as a result of fiber breakage. Cellulases enhance fibrillation.

As early as 1942, a patent claimed that microbial hemicellulases from *Bacillus* and various *Aspergillus* species could aid in the refining and hydration of pulp fibers (21). In 1959, Bolaski et al. (22) patented the use of cellulases from *Aspergillus niger* to separate and fibrillate pulp The technology was principally applied to cotton linters and other nonwood pulps. Crude or purified cellulases from *Aspergillus niger* were placed in solution at pH 4.0, 37°C, and fibers were added with agitation to a consistency of about 6%. The enzyme concentration used was 0.01% to 2% (weight of crude enzyme to weight of oven-dry pulp). The treatment took about one-half hour before microfibrils were observed.

A process patented in 1968 used cellulases from a white-rot fungus, *Trametes suaveolens*, to reduce refining or beating time (23). The enzyme was applied at a concentration of 0.1% to 1% by weight. In addition to enhancing beating, the enzyme selectively removed fines, thereby facilitating drainage. In other applications of this sort, cellulases have been used to remove fines from pits and felts in the papermaking machinery.

One of the disadvantages of using cellulases on pulps is that they will attack the fiber and can diminish pulp properties. Particularly, cellulases decrease the degree of polymerization (DP) of cellulose and thereby reduce viscosity. To circumvent this difficulty, a number of "cellulase-free" xylanase preparations have been investigated. A group of French researchers have described the actions of xylanases on chemical pulp fibers and the details of their analytical techniques in two scientific papers (24,25), in the proceedings of a conference (26) and in a patent (27).

These researchers employed enzymes from mutants of *Sporotricum pulverulentum* and *Sporotricum dimorphosphorum* to fibrillate pulps. They used relatively low xylanase loadings (0.001%–0.1%, weight of enzyme/weight of dry pulp) and carried out treatments at 40°C for up to 20 h in the presence of 1 mM $HgCl_2$. This procedure removed less than 2% of the total dry weight of the pulp. The $HgCl_2$ inhibited cellulase activity and thereby avoided viscosity losses.

Enzyme treatment in the presence of $HgCl_2$ increased the Schopper–Riegler index (Table I). This index is a measure of freeness; larger values indicate slower drainage. When enzyme-treated pulps were compared to untreated controls, the time required to obtain the same degree of freeness decreased by about 60%. In keeping with the slower drainage, water retention increased about 40% following enzyme treatment and

Table I. Effects of *Sporotricum Pulverulentum* Xylanase
Treatments on Bleached Bisulfite Softwood Pulp (27)

Enzyme treatment	Pulp property					
	SR[a]	WRV[b] (g/g)	DP[c]	LR[d] (km)	LRo[e] (km)	MV[f] (g/cm³)
None (control)	13	1	1500	0.9	6.6	0.52
Same, refined 51 min	57	unch.[g]	unch.	unch.	unch.	unch.
0.02% enz, 20 h, 40°C[h]	10	1.36	1400	2.8	10.7	0.61
Same, refined 23 min	57	2.2	1400	6.8	11.7	0.93
0.1% enz, 6 h, 40°C	21	1.40	1400	3.00	11.0	0.62
Same, refined 18 min	55	unch.	unch.	unch.	unch.	unch.

[a]SR = Schopper–Riegler degree   [b]WRV = water retention value
[c]DP = degree of polymerization   [d]LR = length of rupture (tensile)
[e]LRo = zero-span breaking   [f]MV = volumetric weight
[g]Authors state that properties were not significantly modified.
[h]All reaction mixtures contained 1 mM $HgCl_2$; activity of enzyme was not given.

more than two-fold following refining. The tensile strength of the enzyme-treated, refined pulp increased, as did the zero-span breaking length.

In the presence of 1 mM $HgCl_2$, hydrolysis of cellulosic substrates is essentially blocked at enzyme concentrations of up to 500 mg/L; by comparison, the rate of xylan hydrolysis decreases only slightly with addition of $HgCl_2$. Likewise, molecular weight distribution studies show that the cellulose DP is little changed even after 88 h of incubation when Hg(II) is present. Hg(II) inactivates sulfhydryls; however, the basis for selectivity and efficacy of this additive is not understood, because at least some microbial xylanases are inactivated by Hg(II) (28).

The amount of $HgCl_2$ required can be reduced by adding it to the enzyme prior to pulp treatment. Even though only small amounts of mercury could be employed, disposal of the element in the waste streams would probably be environmentally unacceptable. Comtat *et al.* (27) claimed similar results were obtained using xylanases that had been produced by cloning the DNA for the enzyme into a bacterium.

In addition to the reported increase in water retention, Mora *et al.* (24) showed that the mean pore radius of aspen wood is reduced by a factor of ten following treatment with xylanases. Presumably this results from the opening of small cracks in the walls of pores. Electron microscopy showed increased fibrillation in enzyme-treated as compared to control pulps. In a companion paper, Noé *et al.* (25) reported the characteristics of enzyme-treated pulps of birch and spruce. The Schopper–Riegler index, water retention, breaking length and apparent density all increased with treatment, but viscosity decreased by over 30% and the wet zero-span breaking length decreased significantly. The authors concluded from these studies that enzyme-treated pulps show enhanced beatability and better bonding as a result of increased fiber flexibility, but the intrinsic fiber strength decreases as a result of the loss of the xylan.

Enzymatic Treatments of Recycled Fibers. Treatments with xylanases and cellulases have been used to improve the drainage of recycled fibers. Recycled fibers contain a larger fraction of fines than do virgin fibers, and fines account in large part for the slow drainage of pulps. Rapid drainage is desirable because to increase the efficiency of the paper machine, the drying phase must be as short as possible. Fuentes and Robert (29) found that treating recycled fibers with cellulases and xylanases increased drainage by

18% to 20%. Commercial enzymes such as Maxazyme 2000 (Rapidase), Cellulase 250 P (Genencor), and SP 249 (Novo) were used at levels of 0.07% to 2% of the dry weight of the pulp. The preparation employed had a C_1 cellulase activity of 0.17 SI units, a C_x activity of 3.9 SI units, and a xylanase activity of 31 SI units per mg of powder. The reactions were carried out at 50°C for 30 to 90 min. A mixture of 40% recycled corrugated and 60% recycled paper was treated enzymatically. Following enzyme treatment, drainage improved while strength was reduced. The latter could, however, be improved by refining the pulp before enzyme treatment. Starch sizing likewise improved mechanical properties.

Many different cellulases and hemicellulases can be used to improve freeness, but good results have been obtained with commercial cellulases from *Trichoderma reesei* (*30*). The lower the initial freeness, the greater the gain following treatment. Freeness shows rapid initial increases, with over half the observed effect occurring in the first 30 min. Extending treatment times beyond 2 h is not necessary. A relatively small amount of enzyme is required; 0.1% (w/w) is about 70% as effective as 0.5%. A 3% pulp consistency is nearly optimal. While the initial effects are largely beneficial, extending the reaction time with high enzyme concentrations is detrimental. These studies have been extended to pilot studies in industrial paper mills (*31*).

Cellulases have been used to accelerate digestion of pulps. Cellulase (0.05%–0.5%) plus cellobiase (0.005%–0.015%) when added to pulps in water, at pH 4–7, 60°C has been reported to accelerate digestion of pulp without damaging fibers (*32*). The effect of such treatment on viscosity or paper strength properties was not reported.

Recently, Karsila *et al.* (*33*) showed that the drainability of mechanical pulp is enhanced by the addition of hemicellulases rather than cellulases. Xylanase (5 U per g of pulp) improves the Canadian standard freeness of deinked, recycled pulp from 148 to 164 ml while it has no detrimental effect on strength properties. By comparison, the tear indices of recycled pulps treated with cellulases decrease.

In addition to the use of pre-formed enzymes, mixed microbial cultures have been used to facilitate the recovery of recycled fibers. Papers containing polyethylene and aluminum laminates and coated papers used in packaging are particularly difficult to recycle. However, these papers contain 70% to 90% high quality fibers, so they are of considerable interest. Enzymatic treatments are probably not well-suited to such substrates because of their heterogeneous nature. In a patented bioprocess, fibrous materials and laminates are introduced into an aqueous suspension of microorganisms capable of growing on starch, cellulose, carboxylic acids, flotation media, and inks that are found in these fibers. Mechanical agitation and shear combined with microbial growth loosens fibers from laminated paper, including juice boxes and cigarette packets. The freed fibers are then separated by filtration or flotation (*34*).

Cellulase Treatment to Reduce Vessel Picking. The use of tropical hardwood, such as eucalyptus, for pulp production has increased in recent years. The trees grow rapidly, so the chip supply is plentiful, and the pulps are useful for many applications. The vessel elements of tropical hardwoods are, however, large and hard, and they do not fibrillate during normal beating. As a consequence, they stick up out of the surface of the paper. During printing, the vessels are torn out, leaving voids. This characteristic reduces the value of tropical hardwood pulps. Although increased beating can eventually increase vessel fibrillation and flexibility, it also can result in poor drainage.

A patent of disclosure from Honshu Paper Co. (*35*) describes the use of commercial cellulases to enhance the flexibility of hardwood vessels. A 5% slurry of bleached kraft hardwood pulp was treated with 0.5% (w/w) of cellulase at 50°C, pH 5, for 4 h. After dilution to 3% consistency and refining to a Canadian standard freeness of 450 ml, enzyme treatment reduced vessel picking by 85% from 160/cm^2 to 21/cm^2. At the same time, smoothness increased from 50 to 129 s, and tensile strength increased from 4.45

to 5.50. Drainage time increased slightly with enzyme treatment, but the difference was negligible.

Selective Removal of Xylan. Certain applications of pulps require purified cellulose having a high degree of polymerization. Rayon, for example, is a regenerated cellulose formed by extruding a cellulosic-dissolving pulp through a fine nozzle. When xylan is present, bulking occurs, and the quality of the rayon fiber is diminished. Xylan can be removed by partial acid hydrolysis, but this also degrades the cellulose, leading to diminished strength properties. By using a specific xylanase to remove xylan, the polymeric properties can be preserved.

Some xylanases can remove xylan from pulps without affecting other components. The purified cellulose can then be used in making dissolving pulps. Xylanase from *Schizophylum commune* reduced hemicellulose 22% in a delignified mechanical aspen pulp (*36*). Purified xylanase from *Trichoderma harzianum* reduced xylan content of unbleached kraft 25% (*37*), and a cloned xylanase from *Bacillus subtilis* reduced xylan content of bleached hardwood pulp 20% (*38*). More recently, an extracellular xylanase from a thermophilic actinomycete, *Saccharomonospora viridis*, was shown to selectively remove about 20% of the xylan present in a birch wood pulp (*39*), and a crude xylanase from *Aureobasidium pullulans* was shown to reduce the xylan content of alkali-treated thermomechanical pulp by 33% to 50% (*19*).

Roberts *et al.*(*39*) sought to elucidate the roles of xylans in pulp properties by selectively removing them through enzymatic hydrolysis. Following enzymatic treatments in which at least 20% of the residual xylan was removed from a bleached kraft birch pulp, two strength properties, burst and zero-span breaking length, both decreased about 25% to 30%. Jeffries and Lins (*19*) likewise enzymatically removed about 33% of the xylan from aspen kraft and TMPs using a specific microbial xylanase. In the case of the TMP, the burst decreased about 30% while tear decreased about 11%. Tensile strength was essentially unchanged; brightness and opacity increased. In the case of the kraft pulp, burst was unchanged while tear decreased 17% and tensile strength decreased 10%.

Enzymatically Enhanced Bleaching

The most studied application of enzymes to pulps is to enhance bleaching. At least three enzyme-based approaches have been investigated: one is to use low molecular weight biomimetic agents, another is to use extracellular oxidative enzymes, and a third is to use hemicellulases.

Production of bleached kraft pulp is one of the most important segments of the pulp and paper industry. The kraft process results in excellent pulp from a wide variety of wood species. It removes large quantities of hemicellulose and lignin and disrupts ester linkages between lignin and carbohydrate and between hemicellulose chains. Even so, significant amounts of lignin and color bodies remain after kraft pulping. During the phase of bulk delignification, kraft pulping removes lignin without greatly degrading the cellulose. As the reaction progresses, however, residual lignin becomes harder and harder to remove. Lignin remaining in the kraft pulp cannot be removed without unacceptably large yield losses. Extending digestion removes cellulose and results in unacceptably low pulp yields, so more specific methods of removing residual lignin and color must be employed.

Kraft pulp is commonly bleached with chlorine and chlorine dioxide. Alkali extraction then removes chlorinated aromatic derivatives of lignin. Some of these derivatives include dioxins and other toxic compounds. They are recalcitrant and pass through conventional biological waste treatment. Moreover, residual chlorine compounds left in the paper are released during incineration. The chlorine bleaching of paper is a major concern to the industry.

Residual Lignin in Kraft Pulp. Chemical delignification during pulping may be retarded by the association of lignin with carbohydrates. Even though lignin can be solubilized under alkaline conditions, covalent alkali-stable bonds between lignin and cellulose (LC bonds) can bind lignin in the matrix. Yamasaki *et al.* (*40*) found evidence of LC bonds in characterizing the residual lignin left after digesting a bleachable-grade kraft pulp from loblolly pine.

The pulp was exhaustively digested with cellulases and hemicellulases, leaving an insoluble residue that consisted chiefly of lignin. About 10% to 20% consisted of hemicellulosic sugars that could not be removed by further digestion. The residual, enzymatically isolated lignin had a higher molecular weight than that of either native (milled wood) lignin from pine or lignin solubilized during the kraft process. However, it was fully soluble in dilute sodium hydroxide or sodium carbonate. The residue was somewhat more condensed than lignin extracted from wood meal, but it was less condensed than kraft lignin that had been solubilized in the pulping process. The molecular weight, although higher than that of most material extracted during pulping, was lower than that of the lignin extracted during the final phase. After the isolated lignin was mixed with holocellulose, the two components could easily be separated by a simple extraction. From these results, the authors concluded that covalent linkages between lignin and carbohydrate were the most probable reasons for lignin to remain in the kraft pulp. More recently, Hortling *et al.* (*41*) arrived at a similar conclusion.

Based on the nature of the residual sugars bound to the enzymatically liberated lignin, Yamasaki *et al.* (*40*) proposed that the most likely bond is an α phenyl ether linkage to the *O*-6 of galactoglucomannan. Although this structure is in accord with what is known about native LC bonds, the bonds in kraft pulp can be very different. The nature of the covalent linkages in kraft pulp have not been fully characterized because many rearrangements occur during pulping. At least some changes appear to result from the formation of alkali- and acid-stable carbon–carbon bonds between lignin and carbohydrate (*42*). A number of rearrangements are possible. Particularly, primary hydroxyls of glucose and mannose can react with the α, β, or γ carbons of phenylpropane units to form ether linkages. The reducing-end groups can also react (*43*). Glucose is the most prevalent sugar bound to residual lignins from kraft pulps, and it seems probable that this results from the reaction of cellulose with lignin during the pulping process. This conclusion is supported, but not proven, by the observation that the glucose content of residual lignins from pulps is much higher than the glucose content of lignins from wood (*41*).

Minor (*44*) used methylation analysis to determine the characteristics of polysaccharides attached to residual lignin in cellulase- and hemicellulase-digested loblolly pine kraft pulps. The total carbohydrate content of the residual MWEL was only about 8%, as compared to the 12% obtained with MWEL from native wood, but methylation data indicated that the carbohydrate bonding was similar in kraft and native wood. The primary *O*-6 position was most frequently found for hexans and the primary *O*-5 for arabinan. Xylan was bonded to lignin at *O*-3, with a small amount at *O*-2. The predominant methylated derivatives obtained from galactose and arabinose indicated 1 → 4 and 1 → 5 linkages, respectively. The apparent DP ranged from 4 for xylan to almost 13 for galactan. Because of the small differences in methylation patterns between carbohydrates from MWELs of pine and pine kraft pulp, Minor (*44*) was not able to confirm the possible formation of LC bonds during pulping. Reactions with the β and γ hydroxyls of the phenyl propane units are particularly acid- and alkali-stable and may escape detection in methylation analysis (*45*).

We may conclude from these studies that lignin–carbohydrate bonds probably exist in kraft pulps, but they are not necessarily similar to those found in native wood. They are stable to alkali, so they likely consist of ether (C–O–C) or carbon–carbon linkages; they are heterogeneous, involving both cellulosic and hemicellulosic polymers.

Biomimetic Oxidative Agents. Paice and Jurasek (46) examined horseradish peroxidase, catalase, laccase, singlet oxygen, superoxide radical, and hydroxyl radical for their abilities to catalyze oxidative bleaching of TMPs. Glucose oxidase in the presence of glucose increased brightness two points, probably as a result of *in situ* generation of hydrogen peroxide, but none of the other treatments was effective when compared to appropriate controls (Table II).

Table II. Effects of Biochemical Bleaching Agents on Spruce Thermomechanical Pulps (46)

Additive[a]	Proposed reactive species	Brightness ISO
Blank	None	59.8
H_2O_2 (1%)	HO_2-	65.6
Peroxidase + H_2O_2 (1%)	$HO_2\cdot$	65.2
Catalase + H_2O_2 (1%)	HO_2-	61.8
Laccase	$O_2\cdot$	59.6
Glucose oxidase + glucose	H_2O_2 (HO_2-)	61.9

[a]All experiments performed at 10% consistency, 40°C, pH 7.0.

Petterson et al. (47) examined hemoglobin, cytochrome C, and horseradish peroxidase in the presence of solvents such as dioxane and palmitoyl chloride for their ability to enhance chlorine bleaching. In this study, hemoglobin in 90% dioxane reduced the kappa number of kraft pulp by 27 to 40% within 24 h. The authors also observed an increase in brightness. At the same time, ISO brightness increased 25% to 50%. Viscosities were somewhat reduced. Horseradish peroxidase in 90% dioxane reduced kappa by almost 70%, but cut viscosity in half. Dioxane alone did not have a significant effect on kappa. Other porphyrin structures such as vitamin B_{12}, hematoporphyrin, and hemin chloride also bleached kraft pulp when dissolved in either 90% dioxane or water. Yang et al. (48) found that 0.75% hemoglobin (w/w, oven-dry pulp basis) at a temperature of 40°C to 60°C and a reaction time of 60 to 90 min are optimal for bleaching with hemoglobin.

The difficulty with using native heme for bleaching is that unusual solvents or extreme conditions are required, and the porphyrin structures are unstable. Chemically modified, porphyrins have therefore been examined for their ability to bleach kraft pulp (49). Although significant delignification occurs, the yields obtained in making these pulps are low, and specialized oxidative agents need to be employed for the reaction.

Lignin-Degrading Enzymes. Chlorination is essentially an oxidative process, so a number of biological oxidative agents have been examined for their ability to break down lignin in pulps. The principal enzymes of interest are lignin and manganese peroxidases (LiP and MnP, respectively) from white-rot fungi. These enzymes are believed to be responsible for breaking down and oxidizing lignin in nature.

The published scientific literature concerning the effect of ligninases on pulps is sparse. Farrell (50) pointed out the potential applications of lignin-degrading enzymes. These include the bleaching of chemical pulps, the partial delignification of coarse thermomechanical pulps, decolorization of bleach plant effluents, and enhancement of the utility of kraft lignin. Initial results demonstrated only marginal effects (51). In part, however, these early findings simply reflected how little was actually known about the production of such enzymes in culture and about the conditions necessary to achieve their action on native substrates.

Viikari et al. (52) applied crude LiPs and MnPs from *Myrothecium graminaceum* and *Phlebia radiata* to birch and pine peroxy acid pulps. *M. graminaceum* produced high MnP but relatively little LiP. Initial studies indicated that enzyme treatments would reduce the kappa numbers of peroxyacid pulps, but that the lignin contents were not affected.

There are at least two U.S. patents (53,54) and one European patent application (55) pertaining to the use of ligninases from white-rot fungi for bleaching pulps. In addition, there is a European patent application on the production of recombinant ligninase (56). All of these patents focus primarily on lignin peroxidase from *Phanerochaete chrysosporium*.

Olsen et al. (55) presented data showing the effects of LiP and MnP on pulps. This work focuses on the development of an enzymatic bleaching procedure using partially purified MnP and Lip isoenzymes from *P. chrysosporium*. The enzymes were applied in three stages, with gradually decreasing amounts of enzymes at each stage. Several replicates of buffer controls and several levels of enzymes were employed with a northern hardwood and a southern pine kraft pulp. Low levels of hydrogen peroxide were supplied by adding glucose oxidase and glucose to the reaction mixtures, but in a commercial process, hydrogen peroxide would be added by an automated metering device. The amounts of LiP and MnP added were relatively high. Cumulative enzyme additions amounted to as much as 32 IU of LiP and 88 IU of MnP per oven-dry grams of hardwood pulp. Similar but slightly lower levels were used for the softwood pulp. Reductions in kappa were significant, and they appeared to be dose-responsive. In the case of the hardwood, the kappa of enzyme-treated pulp was less than half that of the controls, and similar reductions in kappa were achieved with the softwood. Viscosities were also reduced by the enzyme treatments, albeit to a slightly lesser extent. These findings indicated that under the conditions employed, fungal lignin-degrading enzymes can facilitate the removal of lignin from pulp.

Table III. Conditions for Comparing the Effects of Xylanases, Ligninase, and Oxidase on Birch and Pine Kraft Pulps (51)

Enzyme	Source	Concentration (IU/g pulp)
Hemicellulases	*Aspergillus awamori*	30–240
(endo xylanase)	*S. olivochromogenes*	1.7
Ligninase	*Phlebia radiata*	0.4–0.75
(veratryl alcohol oxidase)	*Phanerochaete chrysosporium*	0.3–0.55
Laccase (ATBS)	*Phlebia radiata*	6

Viikari et al. (51, 52) compared the effects of two hemicellulase and two ligninase preparations for their efficacy against hardwood and softwood pulps. The enzymes and pulp loadings were as shown in Table III (51). Different buffers were used for separate enzyme preparations because of the various characteristics of these reagents (Table IV). The nature of the buffer is important since it can often affect an increase in brightness or a reduction in kappa even in the absence of enzyme. Citric acid, gluconic and tartaric acids are efficient chelators, and they can increase brightness of the pulp. However, they are reported to have little effect on kappa (57).

Table IV. Conditions Used in Enzymatic Treatments of Pulps (51)[a]

Enzyme	Buffer	pH	Temperature	Incubation
Hemicellulase	Citrate	5.0	45°C	24 h
Ligninase	HCl-glycine + 0.07 mM H_2O_2	3.0	RT	24 h
Laccase	2,2'-dimethylsuccinate	4.5	RT	24 h

[a]Enzyme treatments were followed by peroxide bleaching at 12% consistency, 80°C, and a reaction time of 60 min.

The effects that Viikari et al. (51) observed on a pine kraft pulp are summarized in Table V. All the treatments showed decreased kappa numbers in comparison to that of the initial unbleached kraft pulp, but the effects of enzymes were not as significant when compared to the buffer controls. The hemicellulase treatments showed slight brightening and some decrease in kappa in comparison to the buffer controls, but the effects were less apparent with ligninase treatments, and the *P. radiata* oxidase actually decreased brightness. A significant reduction in the consumption of chlorine in a subsequent (DC)EDED bleaching sequence was observed with the hemicellulase-treated pulps, and the final properties of enzyme-treated pulps and controls were similar.

Table V. Effects of Xylanases, Ligninases, and Laccases on Brightness of Pine Kraft Pulp (51)

Enzyme	Brightness ISO	Kappa number	Viscosity (dm^3/kg)
Initial unbleached kraft pulp	—	33.7	1,190
Hemicellulase treatment			
Hemicellulase buffer alone	52.9	19.8	1,010
B. subtilis	54.0	19.3	1,030
S. olivochromogenes	55.8	16.7	990
A. awamori	55.8	17.0	990
Ligninase treatment			
Ligninase buffer alone	43.5	22.1	1,040
P. radiata	42.2	22.1	990
P. chrysosporium	44.8	22.1	1,060
Laccase treatment			
Oxidase buffer alone	46.9	21.1	950
P. radiata	38.8	22.1	1,060

Bleaching With Hemicellulases. Several reports from separate research groups have shown that bleaching is enhanced by various hemicellulase preparations. Microbial hemicellulases include various xylanases, acetylmethylglucuranosidases, arabinofuranosidases, and glucomannanases. How hemicellulases enhance bleaching is not well understood. It is possible that they break LC bonds and open the pulp matrix to bleaching chemicals. It is also possible that they remove carbohydrate-derived (or carbohydrate-bound) color bodies. Lignin–carbohydrate bonds occur between the lignin and hemicellulose fractions of wood (13,58,59). Xylanases generally enhance bleaching of hardwood pulps more than that of softwood pulps, probably because xylan comprises a larger fraction of hardwoods. Purity of the xylanase appears to be important. Some preparations contain appreciable amounts of cellulase and therefore reduce the viscosity of pulps, so cellulases should not be present. However, mixtures of hemicellulases with different actions can be more effective than single enzyme activities. Glucomannanases appear to be more useful with softwood pulps. The only reliable method to evaluate an enzyme is to determine its effects on pulp properties and bleachability. Pulps have been treated effectively with cloned xylanases, and it seems likely that products will eventually be developed that will contain a mixture of enzymes to enhance and affect bleaching.

Xylanase treatments can in some instances release lignin from kraft pulp. More commonly, they reduce chemical consumption in subsequent bleaching and extraction steps while attaining greater brightness than pulps that are not enzymatically treated. The mechanical strength of fibers is not necessarily affected; however, inter-fiber bonding often decreases. Xylan does play an important role in fiber strength properties, so excess removal of hemicellulose can be detrimental. If xylanases alone are used to reduce residual lignin content, the negative effects can be significant. The key to success lies in obtaining enzymes that break the bonds between lignin and hemicellulose while opening the pulp structure. It is critical to avoid destroying the cellulose. The latter is sometimes difficult because many microbial enzyme preparations contain cellulases or the xylanases themselves possess activity against amorphous cellulose. Moreover, since the kraft pulping process can introduce chemical bonds between cellulose and lignin, some cellulase activity might be necessary to enhance bleaching.

Table VI. Reduction of Chemical Demand Following Xylanase Treatments on the Bleachability of Softwood and Poplar Kraft Pulps (60)

Bleaching stage	Factor	Softwood[a]		Poplar[b]		Poplar[c]	
		Std.	Xyl.	Std.	Xyl.	Std.	Xyl.
Chlorine	Cl_2 conc. (%)	98.4	95.1	95.9	91.9	93.9	89.3
	Brightness (%)	25.3	29	49.6	49.9	47	50.5
NaOH extract.	Kappa number	4.43	3.3	—	—	2.22	1.53
	Brightness (%)	34.7	42.5	51.9	56	51.5	54.8
Chlorine dioxide	ClO_2 conc. (%)	99.7	94.0	89.5	84.4	70.6	68.8
	Brightness	72.7	81.7	85.8	86.9	85.2	85.7
NaOH extract.	Kappa number	1.29	0.74	—	—	0.69	0.69
	Brightness	70	80	81.8	83.9	79.9	82.1
Chlorine dioxide	ClO_2 conc. (%)	90	87	89.5	84.4	70.8	70.8
	Kappa number	0.5	0.41	—	—	—	—
	Brightness	88	90.6	90.8	90.9	89.1	89

[a] 0.67% enzyme (w/w, oven-dry basis) for 96 h.
[b] 0.33% enzyme for 96 h.
[c] 0.67% enzyme for 48 h.

Chauvet et al. (60) showed that fungal xylanase lowers the lignin content of unbleached kraft pulps, and that the addition of small amounts of Hg(II) can minimize cellulase activity. Xylanase from extracellular filtrate of *Sporotricum dimorphosum* that had been treated with $HgCl_2$ was used to treat never-dried, unbleached softwood and hardwood kraft pulps. Small amounts of lignin were released by the enzyme treatment alone. More importantly, however, a higher brightness was attained with enzyme-treated pulps even though chemical consumption was less (Table VI). Despite the use of Hg(II), pulp strength properties such as breaking length and burst were reduced by the enzyme treatment, as one would expect from the removal of xylan (Table VII).

Typical enzyme loading rates used are in the range of 30 to 150 nkat/g of pulp ($\approx$2–10 IU/g) (58). However, useful effects have been observed with some commercial enzyme treatments using as little as 0.5 to 5 IU of xylanase per gram dry wt of pulp (61).

Roberts et al.(62) used a thermophillic actinomycete xylanase to remove 20% of the xylan from birch wood kraft pulp. The enzyme preparation was essentially free of cellulase activity. They found that the burst and long-span breaking length decreased significantly. However, the zero-span breaking length, i.e., that which is attributable largely to the inherent strength of the cellulosic fiber, was essentially unchanged.

Table VII. Effects of Xylanases on Kraft Pulp Properties (60)

Evaluative index	Bleached softwood[a]		Bleached poplar[b]	
	Std.	Xyl.treated	Std.	Xyl.treated
Shopper–Riegler index (°SR)	21	28	41	37
Apparent density (kg·m^{-3})	720	740	900	690
Breaking length (km)	7.3	5.9	8.5	7.9
Burst index (kPa·m^{2}g^{-1})	4.5	3.8	5.7	4.6
Tear index (mN·m^{2}g^{-1})	15.3	18.4	7.5	7.6
Air porosity	—	—	0.37	0.62
Z- span breaking length (km)				
Dry	6.8	15.3	16.9	17.1
Wet	12.6	11.5	11.9	11.9

[a]Beaten in a JOCKRO mill, 35 min.
[b]PFI mill, 2000 revolutions.

Results from several groups of researchers have indicated that the purity of xylanase is very important. Cellulase should not be present. In the absence of cellulase, xylanase treatment can actually increase viscosity. Probably the best published example of this is the treatment of (aspen) pulps with a xylanase cloned from *Bacillus subtilis* (63,64). Lignin removal was facilitated while pulp retained viscosity and strength properties (65) (Table VIII). Many other microbial (chiefly prokaryotic) xylanases have been cloned, but the properties of relatively few have been tested for their action on kraft pulps. Enzymes that will effectively enhance the bleaching of softwood kraft pulps are still needed.

Kantelinen et al. (66) showed that different hemicellulases applied in a sequential enzyme–peroxide delignification can remove lignin from a pine kraft pulp. Peroxide alone reduced kappa by 35%, and the citrate buffer accounted for an additional 5% reduction, whereas hemicellulase treatment in combination with peroxide and citrate reduced kappa by greater than 50%. Hemicellulases having different specificities for substrate DP and side groups were used. Relatively little cellulase was present in the preparations, but the greater the cellulase activity, the greater the loss of pulp viscosity. Only limited hemicellulose hydrolysis was necessary to enhance lignin removal. Utilization of xylanase and mannanase did not further enhance lignin removal, but enzyme preparations having a number of different hemicellulolytic activities were somewhat more effective.

Viikari et al. (58) evaluated the capacity of five different microbial xylanases and a bacterial mannanase to enhance the alkaline peroxide delignification of a pine kraft pulp. The endo xylanases were the most effective. Only slight differences in brightness or pulp strength properties were observed with the fungal and bacterial xylanases. In keeping with other studies (e.g., 19; 67) only a portion of the xylan was susceptible to enzyme hydrolysis following repeated enzyme treatment and lignin extraction.

Relatively short treatments with low quantities of enzyme are sufficient to remove the susceptible fraction of xylan present. These findings have led to the hypothesis that the efficacy of xylanase treatments is largely limited by substrate accessibility or structure.

Table VIII. Effects of a Cloned *Bacillus Subtilis* Preparation on Properties of a Kraft Pulp (65)[a]

Pulp property	Original	Control	Enzyme-treated
Canadian standard freeness (ml)	527	423	450
Basis weight (g/m^2) (oven dry)	61.9	61.7	61.7
Bulking caliper (μm)	106	100	104
Bulk (cm^3/g)	1.71	1.63	1.68
Burst index (kPa m^2)	2.32	2.49	2.27
Tear index (mN m^2/g)	6.94	6.53	6.84
Breaking length (km)	4.41	4.75	4.21
Stretch (%)	2.2	2.33	2.23
Z-span breaking length (km)	14.6	15.2	13.8
ISO brightness (% absolute)	27.3	30.0	33.9
TAPPI opacity (%)	97.8	98.1	96.9
Scattering coefficient (mm)	22.3	23.5	23.5

[a]Unbleached hardwood pulp

Clark et al. (59) have also examined the effects of mannanases and xylanases on the properties and bleachability of chemical pulps. They employed relatively high enzyme loading levels—on the order of 10 to 100 IU/g—and they treated pulps for 24 h at 2% consistency. When wood chips were pulped to different extents, the lignin content (kappa no.) had surprisingly little influence on the ability of *Bacillus subtilis* mannanase and *Trichoderma harzianum* xylanase to solubilize residual mannan and xylan. The fraction of polysaccharide solubilized was maximal at a kappa of about 50. In the case of xylan, significantly less solubilization (when expressed as percentage of solubilization of the residual material) was observed at kappas less than 50, and with mannan, little solubilization (<5%) was observed with any sample.

The difference observed between xylan and mannan solubilization might be attributable to the locations of these polymers in the cell wall. Xylans are more prevalent in the outer layers, while glucomannan is found predominantly in the S2 layer. The enhancement of bleaching was not proportional to the amount of glucomannan removed. The *B. subtilis* mannanase was much more efficient at enhancing bleaching than was a mannanase from *T. harzianum* mannanase, even though the latter enzyme removed much more sugar. Mannanase treatments had little effect on handsheet properties. Xylanase treatments increased the beatability of unbleached kraft pulps and substantially increased the tear strength of bleached pulps when both were beaten to the same degree of freeness.

At least two patent applications have been filed concerning the hemicellulase-facilitated bleaching of kraft pulps. One application indicates that a mixture of cellulases, hemicellulases and esterases might be used (68). The other patent (69) makes specific reference to bleaching sequence employing the xylanase from *Aureobasidium pullulans*. The usefulness of the latter enzyme derives from the fact that the wild-type strain produces the enzyme easily in very high yield without the co-producing significant amounts of cellulase. Neither of these preparations nor other presently available commercial preparations have very much activity at neutral or alkaline pH.

Other Microbial Enzymes and Applications

Broda and Martin (70) described the use of an enzyme from *Streptomyces cyanus* to solubilize the lignocellulose from wheat. Cells were grown for two days, and the enzyme that was harvested was applied to ^{14}C lignocellulose that had been prepared from wheat. After a reaction period, 30% of the ^{14}C was sound in the supernatant liquid. In other publications (71,72) this enzyme has been reported to have a molecular weight of about 20 kDa. It does not appear to be a polysaccharidase (73). A similar enzyme was identified by Ramachandra *et al.* (74). In this latter instance, a protein with an apparent molecular weight of 17.8 kDa exhibits peroxidase activity and has an associated heme adsorption (75). The assay that Ramachandra *et al.* used to determine peroxidase activity was not the assay routinely used to determine lignin peroxidase activity in white-rot fungi. Moreover, lignin solubilization and veratryl alcohol oxidation do not appear to correlate well when assayed in various actinomycetes (76). A patent exists on the use of extracellular protein from *Streptomyces cyanus* to solubilize lignin (77).

Conclusions

Cell-free enzymes have been used with varying success to treat pulps. The earliest applications were to increase fibrillation and fiber flexibility and to promote drainage. The specific enzymes responsible for these effects have not been identified, but cellulases and xylanases have been implicated in different studies. Purified enzymes are useful for some specific applications, but they are often not essential to achieve the desired effects. Cellulase is usually considered detrimental, but some organisms form xylanases and other hemicellulases in the absence of cellulase activity. Highly specific preparations are essential for some applications. The production of dissolving pulp, for example, requires the use of enzyme preparations that have little activity against amorphous cellulose. The most common application of enzymes to pulps is to enhance bleaching. Again, a number of different enzymes have been shown to be effective, including lignin peroxidase and manganese peroxidase. Most of the research has, however, been performed with microbial hemicellulases—particularly xylanases. Glucomannanases and other debranching hemicellulases have been examined, but we still know little about the relative efficacy of these various treatments. Many novel enzymes remain to be examined for their effects on pulps.

Disclaimer

The Forest Products Laboratory is maintained in cooperation with the University of Wisconsin. This article was written and prepared by U.S. Government employees on official time, and it is therefore in the public domain and not subject to copyright.

The use of trade or firm names in this publication is for reader information and does not imply endorsement by the U.S. Department of Agriculture of any product or service.

Literature Cited

1 Anon. *Statistics of Paper, Paperboard and Wood Pulp*; American Paper Institute: New York, NY, 1989.
2 U.S. Department of Commerce, Bureau of the Census. *Current Industrial Reports: Pulp, Paper, and Board.* November 1989.
3 Kirk, T.K.; Eriksson, K.-E. In *The Principles of Biotechnology: Engineering Considerations;* Cooney, C.L.; Humphrey, A.E., Eds.; Pergamon Press: New York, NY, 1985; pp. 271–294.

4. Kobyashi, Y.; Konae, K.; Tanabe, H.; Matsuo, R. *Biotech. Adv.* **1988**, *6*, 29–37.
5. Kobayashi, Y.; Tanabe, H. *Tappi* **1988**, *42(4)*, 330–338. (Japanese review).
6. Agency for Industry and Science. *Japanese Patent*. JP 63,042,988. 1988.
7. Tanabe, H., Matsuo, R.; Kobyashi, Y. *Japanese Patent Application* 86/187098. 1988.
8. Sharma, H.S.S. *Int. Biodet.* **1987**, *23*, 329–342.
9. Sharma, H.S.S. *Int. Biodet.* **1987**, *23*, 181–186.
10. Agency of Industrial Sciences and Technology. *Japanese Patent* J5 4968-402. 1979.
11. Nazareth, S.; Mavinkurve, S. *Int. Biodet.* **1987**, *23*, 343–355.
12. Watanabe, T.; Koshijima, T.; Azuma, J. *4th International Symposium Wood and Pulping Chemistry;* Poster Presentations: Paris, France, 1987; Vol. 2, pp. 45–48.
13. Jeffries, T.W. *Biodegrad.* **1990**, *2*, 163-176.
14. Koshijima, T.; Watanabe, T.; Yaku, T. In *Lignin Properties and Materials*; Glasser, W.G.; Sarkanen, S. Eds.; ACS Symp. Ser. 397, American Chemical Society: Washington, DC, 1989; pp. 11–28.
15. Paice, M.G.; Jurasek. L. *Biotechnol. Bioeng.* **1984**, *26*, 477-480.
16. Farrell, R.L. *U.S. Patent* 4,692,413. 1987.
17. Reid, I.D. *Polymer* (preprints) **1988**, *29(1)*, 618–619.
18. Sadownick, B.A.; Farrell, R. *Biomass* **1988**, *15*, 77–92.
19. Jeffries, T.W.; Lins, C.W. In *Biotechnology in Pulp and Paper Manufacture*; Kirk, T.K.; Chang, H.-M., Eds.; Butterworth-Heinemann: Boston, MA, 1990; pp. 191–202.
20. Senior, D.J.; Mayers, P.R.; Breuil, C.; Saddler, J.N. In *Biotechnology in the Pulp and Paper Industry;* Kirk, T.K.; Chang, H.-M., Eds.; Butterworth-Heinemann: Boston, MA, 1990; pp. 169–182.
21. Diehm, R.A. *U.S .Patent* 2,289,307. 1942.
22. Bolaski, W.; Gallatin, A.; Gallatin, J.C. *U.S. Patent* 3,041,246. 1959.
23. Yerkes, W.D. *U.S. Patent* 3,406,089. 1968.
24. Mora, F.; Comtat, J.; Barnoud, F.; Pla, F.; Noe, P. *J. Wood Chem. Technol.* **1986**, *6(2)*, 147-165.
25. Noé, P.; Chevalier, J.; Mora, F.; Comtat, J. *J. Wood Chem. Technol.* **1986**, *6(2)*, 167-184.
26. Barnoud, F.; Comtat, J.; Joseleau, J.P.; Mora, F.; Ruel, K. Proc. *3rd. International Conference Biotechnology Pulp Paper Industry:* Stockholm, Sweden, 1986; pp. 70–72.
27. Comtat, J.; Mora, F.; Noé, P. *French Patent* 2,557,894. 1984.
28. Keskar, S.; Srinivasan, M.C.; Deshpande, V.V. *Biochem. J.* **1989**, *261*, 49–55.
29. Fuentes, J.L.; Robert, M. *European Patent* 262040 (translation available). 1988.
30. Pommier, J.-C.; Fuentes, J.-L.; Goma, G. *Tappi.* **1989**, *72 (6)*, 187–191.
31. Pommier, J.-C.; Goma, G.; Fuentes, J.L.; Rousset, C. *Tappi.* **1990**, 73, 197-202.
32. Nomura, Y. *Japanese Patent Application* 126,395/85. 1985.
33. Karsila, S.; Kruss, I.; Puuppo, O. *European Patent* 351655 A 900124 9004. 1990.
34. Doddema, H.J.; van der Meulen Bosma, F.O.J. *European Patent Application* 0 291 142. 1988.
35. Uchimoto, I.; Endo, K.; Yamagishi, Y.; *Japanese Patent* 135,597/88 (translation available). 1988.
36. Paice, M.G.; Jurasek, L. *J. Wood Chem. Technol.* **1984**, *4(2)*, 187–198.
37. Senior, D.J.; Mayers, P.R.; Miller, D.; Sutcliffe, R.; Tan, L.; Saddler, J.N. *Biotechnol. Lett.* **1988**, *10*, 907–912.
38. Jurasek, L.; Paice, M.G. *Biomass* **1988**, *15*, 103-108.
39. Roberts, J.C.; McCarthy, A.J.; Flynn, N.J.; Broda, P. *Enzyme Microb. Technol.* **1990**, *12*, 210-213.
40. Yamasaki, T.S.; Hosoya, C.L.; Chen, J.S.; Gratzl, J.S.; Chang, H.-M. Proc. *The Eckman-Days International Symposium on Wood and Pulping Chemistry*; Stockholm, Sweden, June 9–12, 1981; Vol 2, pp. 34–42.
41. Hortling B.; Ranua B.; Sundquist J. *Nordic Pulp Paper Res J.* **1990**, *1*, 33-37.
42. Gierer J.; Wännström S. *Holzforschung* **1984**, *38*, 181–184.

43 Iversen T.; Westermark U. *Chem. Technol.* **1985**, *19*, 531–536.
44 Minor J. *J. Wood. Chem. Technol.* **1986**, *6(2)*, 185-201.
45 Iversen T.; Wännström S. *Holzforschung* **1986**, *40*, 19-22.
46 Paice, M.G.; Jurasek, L. *J. Wood Chem. Technol.* **1984**, *4(2)*, 187-198.
47 Petterson, B.; Yang, J.-l.; Eriksson, K.-L. *Nordic Pulp and Pap. Res. J.* **1988**, *4(3)*, 198-202.
48 Yang, J.-L.; Pettersson, B.; Eriksson, K.-E. In *Biotechnical Approaches to Analysis of Mechanical Pulp Fiber Surfaces and Pulp Bleaching*, Yang, J.-L. Ed.; Swedish Pulp and Paper Research Institute: Stockholm, Sweden, 1989; p. 115.
49 Skerker, P.S.; Farrell, R.L.; Dolphin, D.; Cui, F.; Wijeskera, T. In: *Biotechnology in Pulp and Paper Manufacture*; Kirk, T.K.; Chang, H.-M. Eds.; Butterworth-Heinemann: Boston, MA, 1990; pp. 203–210.
50 Farrell, R.L. Proc. *3rd International Conference Biotechnology Pulp and Paper Industry*: Stockholm, Sweden, 1986; pp. 61–63.
51 Viikari, L.; Ranua, M.; Kantelinen, A.; Linko, M.; Sundquist, J. Proc. *4th International Symposium Wood and Pulping Chemistry*: Paris, France, 1987; Vol. 1, pp. 151–154.
52 Viikari, L.; Ranua, M.; Kantelinen, A.; Sundquist, J.; Linko, M. Proc. *3rd International Conference Biotechnology Pulp and Paper Industry*: Stockholm, Sweden, 1986; pp. 67–69.
53 Farrell, R.L. U.S. Patent 4,690,895. 1987.
54 Farrell, R.L. U.S. Patent 4,687,745. 1987.
55 Olsen, W.L.; Slocomb, J.P.; Gallagher, H.P.; Kathleen, B.A. *European Patent Application* No. 89110176.8; Publication No. 0345 715. 1989.
56 Farrell, R.L.; Gelep, P.; Anillonis, A.; Javaherian, K.; Malone, T.E.; Rusche, J.R.; Sadownick, B.A.; Jackson, J.A. *European Patent Application* No. 87810516.2 Publication No. 0 216 080. 1987.
57 Kantelinen, A.; Rättö, M.; Sundquist, J.; Ranua, M.; Viikari, L.; Linko, M. Proc. *1988 TAPPI International Pulp Bleaching Conference*: Orlando, FL, 1988; pp. 1–9.
58 Viikari, L.; Kantelinen, A.; Poutanen, K.; Ranua, M. In *Biotechnology in Pulp and Paper Manufacture;* Kirk, T.K.; Chang., H.-Min., Eds.; Butterworths-Heinemann: Boston, MA, 1990; pp. 145–151.
59 Clark, T.A.; McDonald, A.G.; Senior, D.J.; Mayers, P.R. In *Biotechnology in Pulp and Paper Manufacture;* Kirk, T.K.; Chang, H.-Min., Eds.; Butterworths-Heinemann: Boston., MA, 1990; pp. 153-167.
60 Chauvet, J.-M.; Comtat, J.; Noe, P. Proc. *4th International Symposium Wood Pulping Chemistry;* Poster Presentations: Paris, France, 1987; Vol. 2, pp. 325–327.
61 Pedersen, L.S. *On the Use of Pulpzyme ™HA for Bleach Boosting*. Tech. Rep., Novo-Nordisk Process Division, Novo-Nordisk a/s, Novo Allé, 2880 Bagsvaerd, Denmark, 1989.
62 Roberts, J.C.; McCarthy, A.J.; Flynn, N.J.; Broda, P. *Enzyme Microb. Technol.* **1990**, *12*, 210–213.
63 Jurasek, L.; Paice, M.G. Proc. *TAPPI International Pulp Bleaching Conference;* Orlando, FL, 1988; pp. 11–13.
64 Paice, M.G.; Bernier, R.; Jurasek, L. *Bleaching Hardwood Kraft With Enzymes From Cloned Systems*. CPPA Annual Meeting; Montreal, Canada, 1988; preprints 74A:133-136.
65 Paice, M.G.; Bernier, R.; Jurasek, L. *Biotechnol. Bioeng.* **1988**, *32*, 235-239.
66 Kantelinen, A.; Rättö, M.; Sundquist, J.; Ranua, M.; Viikari, L.; Linko, M. Proc. *Tappi International Pulp Bleaching Conference*: Orlando, FL, 1988; pp. 1–9.
67 Puls, J.; Poutanen, K.; Lin, J.J. In *Biotechnology in Pulp and Paper Manufacture*; Kirk, T.K., Chang, H.-M., Eds.; Butterworth-Heinemann: Boston, MA, 1990; pp. 183–190.
68 Salkinoja-Salonen, M.; Vaheri, M.; Koljonen, M. *International Patent Application*, Publication No. WO 89/08738. 1989.
69 Farrell, R.L. *European Patent Application*., Publication No. 0 373 107. 1989.
70 Broda, P.; Martin, A. *International Patent* WO 87/06609. 1987.
71 Mason, J.C.; Richards, M.; Zimmerman, W.; Broda, P. *Appl. Microbol. Biotechnol.* **1988**, *28*, 276-280.

72. Mason, J.C.; Birch, O.M.; Broda, P. *J. Gen. Microbiol.* **1990**, *136*, 227-232.
73. Zimmerman, W.; Broda, P. *Appl. Microbiol. Biotechnol.* **1989**, *30*, 103-109.
74. Ramachandra, M.; Crawford, D.L.; Pometto,III, A.L. *Appl. Environ. Microbiol.* **1987**, *53*, 2754–2760.
75. Ramachandra, M.; Crawford, D.L.; Hertel, G. *Appl. Environ. Microbiol.* **1988**, *54*, 3057–3063.
76. Ball, A.S.; Betts, W.B.; McCarthy, A.J. *Appl. Environ. Microbiol.* **1989**, *55*, 1642–1644.
77. Broda, P.; Martin, A. *International Patent* WO 87/06609. 1987.

RECEIVED July 9, 1991

CHEMICALS AND FUELS FROM BIOMASS AND WASTES

Chapter 19

Chemicals and Fuels from Biomass
Review and Preview

Irving S. Goldstein

Department of Wood and Paper Science, North Carolina State University, Raleigh, NC 27695-8005

> The history and technology of producing chemicals and fuels from biomass are briefly reviewed. Technical feasibility does not guarantee economic feasibility, however, and the lower cost and ample availability of chemicals and fuels from fossil reserves have continued to favor that source despite occasional brief dislocations. Looking toward the future, there are indications that other factors than traditional market forces are beginning to enter the equation, and may foreshadow a greater future role for biomass as a resource for producing chemicals and fuels. Among these are concerns about air pollution, global warming and disposition of municipal solid waste.

It is now 15 years since I first addressed this topic at the ACS meeting in Philadelphia in 1975 (*1*), and 10 years since I revisited it at the Las Vegas meeting in 1980 (*2*). My earlier presentations were before the Division of Chemical Marketing and Economics and the Division of Petroleum Chemistry. Both were prompted by the oil crises then in effect. Since then there has been a dramatic change; we experienced an oil "glut", followed by escalating and then receding oil prices asociated with the war in the Persian Gulf.

The times have obviously changed, and so has the audience. I feel at home speaking of biomass before the Cellulose, Paper and Textile Division. But what of the message? In most respects it remains unchanged. There have been no dramatic breakthroughs in technology, which is not to say that there have been no incremental improvements. But it is not my charge to catalog these. Rather, I have been asked to discuss such generalities as chemical feedstock needs and how they can be met from biomass, past history of biomass conversion and its special problems, and present challenges and opportunities. Most of these generalities fall in the review category of the title. The preview aspects represent my personal perspective of future developments.

Composition and Availability of Biomass

For the purpose of this discussion biomass refers to material of terrestrial plant origin. The continental productivity of the world has been estimated to be as high as

170 billion t/year, with about 70% in forests. Wood, therefore, is the most important biomass component. Agricultural residues such as cereal straws, cornstalks and cow manure, and municipal solid waste are also potential biomass sources. Projections of potential fuel material have included over-optimistic extremes based on maximum yields per acre for special crops on productive sites extrapolated to total acreages, as well as overly conservative minimum values based on residues from present harvesting and manufacturing practices. Realistic appraisals fall somewhere in between. A rough estimate of the biomass potentially available in the U.S. for conversion to fuel or chemicals has been placed at about 2 billion tons annually (3). If only 20% of this material were actually made available, it would provide an energy equivalent of 6.5×10^{15} BTU or less than 10% of our energy needs. However, this would be more than adequate for the total energy consumption for chemicals including both feedstocks and process fuel.

A detailed description of the composition of biomass should not be necessary for this audience. Plant cell walls comprise as much as 95% of the plant material. So the polymers of which they are made --- cellulose, hemicelluloses, and lignin --- provide a small number of starting materials for biomass conversion. Other components which are more variable in structure or quantity are the soluble compounds termed extractives, and bark. The reserve carbohydrate starch is concentrated in grains and tubers, but otherwise negligible.

Of these biomass components cellulose is most abundant, comprising about 50% of the material. For conversion into chemicals the important features of cellulose are that it is a glucose polymer, and that unlike starch, which is also a glucose polymer, the highly ordered crystalline structure of cellulose limits the accessibility of reagents and enzymes. Hemicelluloses are also sugar polymers of both pentoses and hexoses. They are amorphous, usually branched, and have much lower degrees of polymerization than cellulose. Lignins are three-dimensional network polymers of phenylpropane units. Unlike the carbohydrate polymers they are not readily broken down into their precursors because of resistant ether and carbon-carbon bonds. The aromatic and phenolic character of lignin is important in chemical conversion.

Historical Chemical Products

Extractives from pine trees provided the raw material for the naval stores industry, the oldest chemical industry in North America. Gum naval stores in the U.S. have been replaced by the recovery of oleoresins from the kraft pulping process. Turpentine is recovered from digester relief gases, while the fatty acids and resin acids are recovered from pulping liquors in the form of tall oil.

Phenolic acids extracted from bark are used as extenders for synthetic resin adhesives. Bark waxes can be used for general wax applications. High purity chemical cellulose from wood pulp is the starting material for rayon, cellophane, cellulose esters for fiber, film and molding applications and cellulose ethers for use as gums.

Natural rubber latex was for many years the only source of rubber, and natural rubber is still preferred for many applications. Tannins extracted from bark and heartwood were traditionally used in leather processing. Pulping liquors also yield lignosulfonates (dispersants,adhesives, etc.) and alkali lignin (extender, stabilizer, reinforcing agent). Vanillin is produced by oxidation of lignosulfonates, and dimethyl sulfoxide by oxidation of dimethyl sulfide from kraft pulping.

Volatile products of wood pyrolysis such as acetic acid, methanol and acetone are no longer obtained from that source. However, wood hydrolysis to simple sugars is commercially practiced in the USSR. Acid pulping liquors also contain sugars that can be fermented to ethanol, yielding also yeast.

Potential Future Chemical Products

Past and present derivation of chemicals and fuels from biomass has relied to a great extent on extractives and by-products. Future chemical products will come from plant cell walls. This source of raw material far exceeds the extractive components or chemical by-products in volume.

There are three mechanisms by which chemicals from biomass may replace those from fossil sources. These include direct substitution of natural polymers such as rubber and cellulose derivatives for synthetic polymers, conversion of biomass into the same intermediates that are now derived from fossil sources, and production of different but equally useful intermediates that are more readily obtained from biomass.

Two approaches to the chemical conversion of biomass are possible. The non-selective gasification and pyrolysis processes are applicable to all carbonaceous materials. These processes can be applied to the plant cell wall polymers in their natural mixed state just as they are applied to coal. Any fuel or chemical derivable from fossil fuels can alternatively be derived from biomass. There are no inherent barriers to the use of biomass for these purposes. All the constraints are relative in comparison to the cost or ease of processing of the various raw materials.

Alternatively, in contrast to the drastic high temperature non-selective processes, selective processing of the individual biomass components can provide substantial yields of a wide variety of chemicals and liquid fuels. In an energy economy based on carbon, chemicals and fuels are interdependent. The same chemical may serve either as a material or as a fuel.

Non-Selective Processes

Gasification of biomass is its conversion by thermal reactions in the presence of controlled amounts of oxidizing agents to provide a gaseous phase containing principally carbon monoxide, hydrogen, water, carbon dioxide and methane or other hydrocarbons. The oxidizing agent may be oxygen, water, carbon dioxide or a mixture of these. Since gasification is generally carried out at temperatures of $1000^{\circ}F$ and above, where the reaction rates are so fast that heat and mass transfer become controlling, gasifier design is critical.

The gasification of biomass compares favorably with the gasification of coal as far as thermal efficiency is concerned (60-80%). Furthermore, biomass gasification offers a number of advantages over coal. These include: (a) substantially lower oxygen requirements, (b) essentially no steam requirements, (c) lower cost for changing the already higher hydrogen/carbon monoxide ratios, and (d) lower desulfurization costs. Offsetting these advantages are the economies of scale available in coal gasification systems not available for biomass because of raw material procurement and solids handling limitations.

The gas produced may be used directly for energy or for the synthesis of organic compounds usable as liquid fuels or chemicals. By enrichment in hydrogen a synthesis gas is formed which is suitable for the production of ammonia, methanol, or the higher hydrocarbons obtainable by the Fischer-Tropsch synthesis. Methanol may be further processed to formaldehyde, so important in adhesive technology.

Pyrolysis of biomass by thermal reactions in the absence of added oxidizing agents provides a volatile phase and a solid char. The volatile phase can be further separated into condensible liquids and noncondensible gases. Pyrolysis is carried out at lower temperatures than gasification, which passes through a pyrolytic stage before completion.

The condensed liquids from wood pyrolysis consist of the so-called pyroligneous acid, which is an aqueous acidic layer, and the heavier wood tar. Both phases contain many compounds, with the tar much more complex since its components have not been selectively extracted on the basis of water solubility.

Major water-soluble components include methanol, acetic acid and acetone, with smaller quantities of other carboxylic acids, aldehydes, ketones, alcohols and esters. The yields of even the major components are low, with methanol (once produced commercially by pyrolysis) in only 2% yield and acetic acid 3-7%.

The wood tar contains at least 50 identified phenolic compounds, which account for up to 60% of the tar. There are also higher acids, aldehydes, ketones, esters, furans and hydrocarbons. The low yields of volatiles make them suitable only for use as internal fuel in the carbonization process. Phenols from the pyrolysis oils have potential value, but the complicated nature of the mixtures has made applications difficult. The corrosivity caused by acetic and formic acids can be removed to yield a high energy liquid fuel suitable for use in fuel oil boilers.

Selective Processes

In selective processing of biomass the individual cell wall components are either first separated from each other before further conversion to chemicals or fuels, or the components are sequentially removed from the polymeric composite. Hydrolysis in organic solvents, autohydrolysis or steam explosion are examples of the first type. The products in each case are an aqueous solution of hemicellulose sugars, an alkaline or organic solvent solution of lignin and a cellulose residue. In the second type the readily hydrolyzed hemicelluloses are first removed from the biomass and converted into simple sugars under mild conditions that leave the cellulose and lignin essentially unaffected. Subsequent hydrolysis of the cellulose provides almost pure glucose and a lignin residue that can be further processed to phenols and other aromatic compounds or used in polymeric form.

Whether the cellulose is first isolated as a residue or removed to leave a lignin residue, its ultimate fate in the selective processing of biomass is hydrolysis to glucose. This may be effected by dilute or concentrated acids or by enzymes. Inasmuch as the biomass is approximately 50% cellulose, its hydrolysis to glucose is the most critical step in the chemical utilization of biomass.

Despite many decades of research, each of the three recognized methods of cellulose hydrolysis still suffers from a severe limitation that has prevented its economic use on a commercial scale. For hydrolysis with dilute acids at elevated temperatures the limitation is the reduced yield of glucose (50-60%). Hydrolysis with concentrated acids at moderate temperatures can provide quantitative yields of glucose, but the recovery of the acid is expensive. Enzymatic hydrolysis also provides high glucose yields, but it is not effective on lignified cellulose without an expensive pretreatment.

Once a glucose solution is available from the hydrolysis of cellulose, its further conversion to useful chemicals or fuels is effected chiefly by fermentation, although under acidic conditions first hydroxymethylfurfural and then levulinic acid are formed. In addition to ethanol, fermentation of glucose can yield such other organic chemicals as acetic, butyric, citric and lactic acids, acetone, butanol, glycerine or isopropanol. Lactic acid may be further processed into acrylic acid. The greatest potential for useful products from glucose, however, is through the familiar fermentation to ethanol. Alternatively to its use in motor fuels, ethanol could be converted into ethylene, the most important organic chemical of commerce, which can then be processed to a multitude of chemicals and synthetic polymers. Oxygenated organic chemicals are of special interest in glucose utilization, because they do not suffer the loss in mass which accompanies removal of oxygen to form hydrocarbons (4). By oxidation of ethanol to acetaldehyde an intermediate is obtained for the further production of acetic acid and acetic anhydride, acrylonitrile, butadiene and vinyl acetate.

Hemicelluloses are readily hydrolyzed to give principally mannose and xylose. Mannose may be processed along with glucose to yield the same products. Fermentation of xylose to ethanol is still not as effective as the hexose fermentation. Chemical conversion of xylose to xylitol by reduction, and to furfural under acidic conditions, can be carried out in high yield. Furfural has many potential applications as a chemical intermediate.

Under various hydrogenation and hydrogenolysis conditions lignin can be converted into mixtures of alkylated and polyhydroxyphenols in yields of up to 50%. Etherification yields aryl ethers which are useful as octane enhancers in motor fuels. Depending on their origin the lignin residues or solutions may be modified, etherified, esterified and condensed to change their physical properties and form useful adhesives and resins.

Biological Processing of Lignocellulose

It is appropriate to consider the potential use of micro-organisms and enzymes in industrial processes for the breakdown and conversion of lignocellulosic materials. Biotechnology is a buzzword that has dramatically encompassed the application of the procedures and methodology of the biological sciences to novel methods of manufacturing new as well as traditional products. After a euphoric period of promotion and hyperbolic promises this segment of biotechnology may be considered to have come down to earth. Most successful applications have been in health care products, many have been in agriculture, while few have been in chemicals and materials. This is not surprising when it is recognized that bioactive compounds have high value and need only be produced on a small scale, in contrast to chemicals and materials which are of much lower value and require large scale production.

Obviously biomass is susceptible to biological transformation as an inherent part of the natural carbon cycle. However, it requires a giant leap from the slow biological conversion of biomass into low concentrations of intermediate metabolites and ultimately carbon dioxide by mixed populations of organisms in nature to the production of high concentrations of useful chemicals and materials in a short time under controlled conditions.

I have already alluded to such biological processes as hydrolysis of cellulose by enzymes and the fermentation of glucose and xylose. But biotechnology alone is no magic formula for biomass conversion. Although the depolymerization of plant cell wall material and the subsequent metabolism of the fragments are entirely effected by biological processes in nature, practical considerations of rates, concentrations, selectivity and susceptibility will require chemical or physical processing as well for the production of chemicals and fuels from biomass. These processes will break the polymers down into small molecules or at the very least serve as pretreatments to confer accessibility to the polymers in lignified cell walls. Further conversion of the small molecules into useful chemicals and fuels may then be carried out by either biological or chemical processing as desired (5).

Economic and Other Considerations

The technical feasibility of converting biomass into a variety of useful chemicals and fuels has been long established, and where economically feasible in a free market or desired in a controlled economy chemicals from biomass have been produced on a commercial scale.

One economic principle that warrants special emphasis for a multicomponent raw material like biomass is the need to maximize the yield of products for greatest economic efficiency. Schemes for biomass conversion that consider only a single product carry a higher raw material cost. The raw material cost of a cellulose-derived chemical such as ethanol, for example, can double when it is a single product, compared to its cost when coproducts are derived from the remaining components, hemicelluloses and lignin. Even established industries such as the petroleum industry could not thrive on a single product. A number of integrated schemes for converting each of the components of biomass have been suggested. In addition to those directed at the production of chemicals only, others include coproduction of fibers, food and fuels.

The most important considerations affecting the ultimate large scale conversion of biomass into chemicals and fuels, and at the same time the most imponderable, are the cost and availability of the fossil fuels, especially petroleum and natural gas for petrochemicals production. At some time depletion of these resources will lead to higher prices and make renewable raw materials for chemicals and fuels like biomass economically attractive. This has been a moving target, however, and exactly when costs of petrochemicals will converge with and eventually exceed the cost of chemicals from biomass cannot be predicted. Meanwhile, the lower cost and ample availability of chemicals and fuels from fossil reserves have continued to favor those sources despite occasional brief dislocations.

Future Portents

However, the classic economic principles of supply, demand and profitability are not the only important factors in determining a change in a resource base. Political and social decisions may also have an important role, as has been shown during the historical shifts from wood to coal and then to oil and gas as primary energy resources. Looking toward the future, there are indications that factors other than traditional market forces are beginning to enter the equation, and may foreshadow a greater future role for biomass as a resource for producing chemicals and fuels.

The first of these involves the oxygenated fuels provisions in the Clean Air Act of 1990. Concerns about air pollution in congested metropolitan areas have focussed attention on alcohol fuels as potential substitutes to improve air quality. Mandated levels of oxygen of 2.7% in gasoline fuels for transportation are required by 1992 in 41 regions of the country. Using biomass-derived ethanol to fuel automobiles in place of fossil hydrocarbons can help to moderate atmospheric carbon dioxide increases to the extent that a replanted and renewable biomass resource uses as much carbon dioxide as is produced.

Growing concerns about the disposition of municipal solid waste may also impact chemicals and fuels from biomass. The U.S. generated a total of 160 million tons of municipal solid waste in 1988(6). Existing landfills are approaching capacity, and proposals for new sites encounter strong local opposition. As a result disposal fees have been rising rapidly, leading to increasing pressures for recycling. Inasmuch as the solid waste may be as much as 50% cellulosic in nature, its processing into chemicals and fuels by biomass conversion technology would help ameliorate both air pollution and solid waste disposal.

In this connection it is of historical interest that the U.S. Army Natick Laboratory's pioneering work in the enzymatic hydrolysis of cellulose by *Trichoderma*, although begun to understand the degradation of cotton materials in the tropics, had as its first proposed application to the civilian economy the treatment of municipal solid waste, long before the oil crisis of 1973 diverted attention to the need for alternative fuels. It would be altogether fitting that these two objectives should now be combined.

My reference to the Natick work should not be construed as enthusiasm for the enzymatic route to cellulose hydrolysis. I still remain a supporter of concentrated acid hydrolysis, especially now that acid separation and recovery by electrodialysis has been shown to be technically feasible (7). But no matter which method of cellulose hydrolysis ultimately proves to be most suitable, biomass will make an important contribution to the problems of organic chemicals supply that will be caused by the depletion of fossil hydrocarbons.

Literature Cited

(1). Goldstein, I. S., Potential for Converting Wood into Plastics, *Science* . **1975,** *189,* 847-852.
(2). Goldstein, I. S., Chemicals from Biomass: Present Status, *For. Prod.J.* **1981,** *31(10),* 63-68.
(3). *Organic Chemicals from Biomass;* Goldstein, I. S., Ed.; CRC Press: Boca Raton, FL, 1981.
(4). Goldstein, I.S., Oxygenated Aliphatic Chemical Feedstocks from Biomass, *Proc. BIOENERGY '80 Conference* . **1980,** Bioenergy Council, Washington, D.C., pp. 296-297.
(5). Goldstein, I. S., Implications of Chemical and Physical Factors on the Biological Processing of Lignocellulose, *Biomass* . **1988,** *15,* 121-126.
(6). Cook, J., Not in Anybody's Backyard, *Forbes.* **1988,** (Nov. 28),*142,* 172-82.
(7). Goldstein, I.S., Bayat-Makooi, F., Sabharwhal, H.S., and Singh, T.M., Acid Recovery by Electrodialysis and its Economic Implications for Concentrated Acid Hydrolysis of Wood, *Applied Biochemistry and Biotechnology* . **1989,** *20/21,* 95-106.

RECEIVED April 9, 1991

Chapter 20

The Promise and Pitfalls of Biomass and Waste Conversion

Helena L. Chum[1] and Arthur J. Power[2]

[1]Chemical Conversion Research Branch, Solar Energy Research Institute, 1617 Cole Boulevard, Golden, CO 80401
[2]Arthur J. Power and Associates, Inc., 2360 Kalmia Avenue, Boulder, CO 80304

>Technologies are continuing to evolve for converting wood, wood waste, agricultural residues, and the organic portion of municipal solid waste into a variety of chemicals. Although many chemicals can be made in this way, this paper considers only the production of phenolics from lignocellulosic biomass. In particular, examples will be given of technologies that are evolving to replace phenol used in phenol-formaldehyde thermosetting resins. Under consideration are (*1*) the high-temperature, short-residence-time pyrolysis of waste wood and bark coupled to solvent fractionation into an inexpensive phenolic and neutrals (P/N) product; (*2*) the direct liquefaction of wood in phenolic solvents at milder temperature, but longer residence time; and (*3*) the demethylation of kraft lignin at moderate temperature and long residence time followed by solvent extraction. A condensed, preliminary technoeconomic assessment of each technology is given, based on process flow sheets and mass and energy balances. The potential penetration of these types of products is discussed relative to the situation in the established petrochemical industry.

Many excellent books on the subject of chemicals from biomass are available (*1-4*), making it unnecessary to introduce this topic. In addition, Professor I. S. Goldstein reviews the subject and previews its future in the previous chapter. This chapter contains a different message, based on the recent advances in one particular area of biomass and waste conversion to chemicals. This specific area is the substantial replacement of phenol, to be used in phenol-formaldehyde (PF) thermosetting resins. Our approach to chemicals from biomass rests largely on an understanding of today's chemical industry, and of the biomass conversion technologies, including

0097–6156/92/0476–0339$06.00/0
© 1992 American Chemical Society

the nature of the overall market and projections for future growth areas. The understanding of the conversion technologies has arisen mostly from the experience of pulp and paper companies, except in countries where other chemicals and materials are produced.

If industry is to adopt any new process, it will require that the new process be reliable, cost effective (and therefore simple), and environmentally benign. The product must also be reproducible, regardless of possible fluctuations in the quality of the incoming feedstocks. Finally, the risk of the new technology must be reduced, through joint funding between government and industry, to a point where industry might invest in a new chemical venture that has the potential for success in returning adequate profit from plant operation. Returns on the government's investment include the implementation of a technology for replacing fossil-derived petrochemicals with those derived from renewables, energy savings, and decreased impact on the carbon dioxide cycle, all leading to a sustainable future for chemicals from renewable resources.

A deep understanding of the chemical industry in the particular country/region is essential, because an overall assessment of the economic and environmental effects of the new process/product needs to be performed. It is pointless to try to replace a chemical/material that is extremely low cost but operates in an acceptably profitable way, in an integrated industry. Most of the plant capital investment has been depreciated in existing plants which can continue to operate profitably for a reasonable time. One needs to search for significant cost-reduction opportunities in the industry, such that the products we want to produce from biomass or waste, can either be substantially less expensive (both in capital and operating costs) or can provide some special market advantages to the investing company. There are also intangible, long-term growth of business reasons for a desire of the industry to switch.

The petroleum/chemicals industry is quite complex and international in nature. National barriers are no longer as important as they were 20 years ago, making our job of finding opportunities to penetrate the market with biomass-derived chemicals (or materials) more difficult than ever. This is especially true if our goal is to improve economic competitiveness by introducing innovative technologies in one particular country, in our case, the United States. Many other countries have labor and infrastructure costs that make plant operating costs more favorable, and therefore lower cost products. Because petroleum resources are largely in the hands of OPEC countries, some of these industries may shift commodity chemical operations to places where the feedstock and the labor are both available and inexpensive in the near future, provided that the product price can absorb the transportation costs.

The Phenol Case

Phenol producers in the United States now operate near 100% capacity and could add capacity overseas. The cumene process, which co-produces phenol and acetone, is the process used most often in the United States. There is a current mismatch in capacities for use of acetone compared to that of phenol. As is generally true for most petrochemicals, in the manufacture of phenol, the benzene and propylene feedstock costs drive the product cost. About 3.5 billion lb/yr of phenol production supplies both the merchant market and internal uses for the production of higher profit margin, lower volume engineered plastics. A substantial fraction of the phenol produced is used in the production of phenolic resins, which are produced at a current rate of roughly 3.1 billion lb/yr. Whereas the engineered plastics or higher value chemicals/materials, such as polycarbonates, polyarylates, nylon 6, and the pharmaceuticals based on nonyl and decyl phenols require a very high-purity phenol, phenolic resins can tolerate a lower purity. Growth of the engineered thermoplastics will continue to increase at a much faster rate than that of the more mature and established phenolic resins, unless additional R&D, low-cost feedstocks, and new properties in the materials allow these commodity resins to penetrate additional markets or substantially expand existing ones. Engineered plastics growth will compete with established materials markets of the wood products industry, which has been driven more by cost than by performance.

Can Biomass-derived Phenolics Replace Phenol in PF Thermosetting Resins?

This area has fascinated researchers in the forest products industry for some time. The goal has been elusive because it has been difficult to obtain a reproducible technical product at the necessary low cost. Phenol replacements that cure slower than phenol require changes in press times or temperatures, which are not easily accepted in the United States because they decrease production rates. Process changes are often not desired because of the relatively high capital investment in the manufacturing process. Therefore, the ideal replacements for phenol should not change panel production rate. They should also be cheaper than phenol; abundantly available, preferably from a renewable resource; and reproducible. The literature has excellent reviews of the state of the art, with comprehensive citations (5-9). Most of these references include lignin-derived replacements, which have been and continue to be investigated because of their potential low cost. However, few have reached commercial applications (10-11) without changing one or more parameters from operation with phenol alone. Most of the successful operations with replacements are outside the United States, where performance standards for the products are less demanding.

Examples of three technologies that have been under investigation over the past five years in various laboratories follow. Table I summarizes the results of the preliminary technoeconomic assessment for these examples.

Fast Pyrolysis Coupled with Solvent Fractionation

Wood pyrolysis processes offer a different route to produce low-molecular-weight phenolic compounds from lignins or from other sources within wood, a renewable resource (12-14). In particular, fast pyrolysis processes offer the advantage of high yields of liquid products from a variety of feedstocks (15-18). In these processes,

Table I. Comparative Production Costs For Phenol Replacements From Biomass

Cases I and II correspond to 1000 dry tons of wood per day plants and Case III the production facility is coupled to a 2000 tons/day kraft mill using 10% of its black liquor stream

	Capital Investment $MM	Annual Phenolics Production MM lb	Contributing to Amortized Production Cost[5]				
			Feedstock ¢/lb	Capital ¢/lb	Operating ¢/lb	Total ¢/lb
I. Fast Pyrolysis/ Solvent Extraction	37.7	140	9.4[1]	3.2	4.2	16.8
			2.4[2]	3.2	4.2	9.8
II. Wood Solvolysis (Phenolation)	33.8	140	16.6[1,3]	4.8	3.6	25.0
			15.3[2,3]	4.8	3.6	23.7
III. Demethylation Kraft Lignin	2.3	19	8.4[4]	1.9	8.7	19.0

Notes:
(1) Wood chips purchased at $40/dry ton
(2) Wood chips purchased at $10/dry ton
(3) Phenolation agent is 20% distilled phenolics fraction (or phenolics/neutrals oils) at transfer cost of $0.20/lb
(4) Black liquor solids cost at equivalent fuel value, $0.03/lb
(5) 20-year plant life, 15% ROI after taxes

Conversion Factors:	To Convert	To	Multiply By
	¢/lb	¢/kilogram	2.205
	$/ton	$/tonne	1.12
	$/lb	$/kilogram	2.205

the pyrolysis temperatures are 450°-600°C, and the residence times are short, on the order of seconds. A schematic of the fast pyrolysis process coupled with solvent fractionation is shown in Figure 1.

On the fast pyrolysis process, a synergistic interaction between basic and applied research has been fostered to allow basic research results to be translated into applied R&D within the Solar Energy Research Institute (SERI) and collaborating institutions. Reference 19 contains details of some of the technical work, from which the solvent extraction process has been patented (20). The activities include process and product screening on a micro scale using the molecular beam/mass spectrometer (MBMS) (21-22), coupled to a collision-induced dissociation (CID) detector. This instrument can detect, in real time, the products of pyrolysis or fractionation, allowing for fast screening of processes for a wide variety of feedstocks (23-25). These studies permit a deep mechanistic understanding of processes and products. In addition, by coupling the data acquisition to multivariate analyses the number of variables can be reduced to a few that carry significant chemical information on the processes investigated (26-27); this knowledge can be expanded by conventional chemical and spectroscopic data analyses.

Closely coupled to this small-scale experimentation is the work at an engineering bench scale using a bench-scale vortex reactor for ablative fast pyrolysis to develop a variety of processes. This reactor has been operated at dry feed rates as high as 70 lb/h. A key feature of this reactor is the ability to decouple the residence times of solids and vapors. It does this through a solids recycle loop and a mechanism that guides the solids on the reactor surface with raised ribs, so that fast heat transfer from the reactor wall into the solids can be exploited (28-30). It can be operated under conditions suggested by the small-scale technique. A continuous liquid extraction system to produce the phenol-rich extractive from the pyrolysis oil was been developed at SERI at the small batch scale (19-20). It was subsequently set up and operated on a continuous basis by a local engineering research firm, Hazen Research, Inc., Golden, Colorado. These experimental efforts are guided by technoeconomic evaluations of the various processes and their modifications to focus on the key research issues that affect cost or technical feasibility.

More than 60% of the dry weight of the feedstock is converted into a mixture of controllable products by fast pyrolysis (15-18). SERI's vortex reactor has been used to generate oils from a variety of feedstocks. Using sawdust ($10-$40/dry ton), the cost of the resulting pyrolysis oil is projected at about $0.02-$0.08/lb (28-29). Then the oils are chemically fractionated (19-20) and the various fractions are subjected to a variety of analytical techniques (19, 31-32) to determine the best extracts for replacement of phenol in PF resins or other high-value applications. The fractionation method developed for fast pyrolysis oils from lignocellulosic materials removes water-soluble carbohydrates and derived polar compounds, and ethyl-acetate-soluble, strong organic acids from the remaining ethyl-acetate-soluble phenolic and neutrals (P/N) fraction through water and aqueous bicarbonate extractions, respectively. The ethyl-acetate-soluble combined fraction of phenolic and neutrals compounds, the P/N product, is useful in the making of PF thermosetting resins. The P/N products can be separated further into phenolics. From these simple fractionation schemes, about 30% of the pine sawdust oil has been converted into useful P/N products. Bark pyrolysis products are much richer in phenolic compounds, and thus appear to be very good candidates for the manufacture of phenolic resins.

Technoeconomic evaluation of this process indicates that waste sawdust can be converted into phenol replacements at $0.10-$0.27/lb (amortized production cost), compared to a purchase price of $0.42/lb of phenol (October 1990 list price). In this evaluation, feed cost varied from $10-$40/dry ton, plant life from 10-20 years, and return on investment from 15%-30%. The capital equipment investment was estimated at less than $10 million for a 250 tons wood per day plant, producing 35 million pounds of phenolics per year. Simple payback calculations show break even in about one year if the difference between phenol price and phenolics amortized production cost is maintained at $0.20/lb (*19, 30*). This plant size corresponds to an average phenolic resin production plant. Data for a four times bigger facility, using 1000 tons of wood per day, are shown in Table I. This plant size corresponds to a relatively large resin production plant. Phenolic resin plant sizes vary from very small coupled with user factilities to larger integrated plants.

Currently, other fast pyrolysis processes are being scaled up in various parts of the world, as reported in various chapters of this book. The fact that three fast pyrolysis concepts are in such an advanced stage of engineering bodes well for the future of the area. Several fractionation schemes are also under development by other groups worldwide.

Technology Transfer. The potential cost and energy savings of the fast pyrolysis method have been sufficient to attract industrial interest and cost sharing in this adhesives development program sponsored by the U.S. Department of Energy, Office of Industrial Technologies, and conducted at SERI. (*19-20*). Technology transfer started through a consortium of industries that was formed by MRI Ventures, Inc., in July 1989, with a five-year lifetime. The title of the patent issued in the fractionation process (*20*) and applied for in the base formulations technologies resides with Midwest Research Institute (MRI), the not-for-profit company that manages SERI for DOE. MRI has assigned the title of these patents to MRI Ventures, Inc., its for-profit subsidiary, to help expedite the transfer of this technology from the government laboratory to the private sector.

The Pyrolysis Materials Research Consortium (PMRC) was created to further the development of this technology and allow its timely transfer to and commercialization by industry. The consortium companies were selected after extensive publicity through press releases announcing that this technology was available for continued R&D and commercialization. Preference was given to U.S. companies because the U.S. government has partially funded the development of the technology (about 33%) since 1986.

Consortium members are Allied-Signal Corp.; Aristech Chemical Corp.; Georgia Pacific Resins, Inc.; Interchem Corp.; and Plastics Engineering Co. through MRI Ventures, Inc. Both Interchem Corp. and Plastics Engineering Co. are small businesses. The companies collaborate with the government-funded adhesives program through MRI Ventures, Inc., with direct monetary contributions and in-kind work to expedite technology development and transfer to industry. An example of in-kind work being performed by a PMRC member is the scale-up of the vortex reactor now in progress by Interchem. Because intellectual property is being generated in the program, current work is not reported until the patents are filed. Intellectual property generated exclusively by the individual company's R&D resides with that company. Only these companies will be able to license the technologies developed by the program with U.S. government funds on a non-exclusive basis. Companies will be added to the PMRC if additional R&D work is needed that the present companies cannot perform, but only if all participating companies unanimously agree.

The Case of Wood Solvolysis (Phenolation)

Solvolysis of wood is an alternative route to substitute phenolics. If carried out in a range of high severities, at moderate temperatures, and with a long residence time, solvolysis yields wood tars or pastes that are reactive, principally if phenol is used as a solvent. At the lower severity end of these types of processes, one would have the phenol pulping process (33), and at the high severity end, wood liquefaction, as explored by Shiraishi and coworkers (34-36). These processes have in common the introduction of phenols into the α- or β-position of the side chains of the phenylpropane units, in addition to acid-catalyzed ether cleavages at both α- and β-aryl ether linkages of the lignin units with or without free phenolic structures. A more complete description of the wood liquefaction studies by the Kyoto University group is given in this Symposium Series, by Professor Shiraishi. The schematic of the wood phenolation process is shown in Figure 2.

We attempted to calculate a rough amortized production cost of the product of this process based on some of the published results, and also on some of D. K. Johnson's experimental findings (37). Definition of the process chemistry in the literature was found to be sketchy. The following conditions were assumed: wood phenolation is operated with a mixture of 20% of the low-molecular-weight phenolics recovered from the pyrolysis/fractionation process, instead of the more expensive purchased phenol. The liquor-to-wood ratio is 5:1; water is found to be necessary for the reaction to occur. The temperature is 250°C for 120 min (600 psi steam). About 70% soluble oils are obtained. The incorporation of phenolics is assumed at about 60%. An alternative would have been to use the low-molecular-weight phenolics fraction distilled from the phenolation process.

The reaction must be carried out in vessels that are constructed of 316 stainless steel. The addition of catalysts may expedite the reaction and lower its operating temperature some, and thus offsetting some of these corrosion-resistant equipment

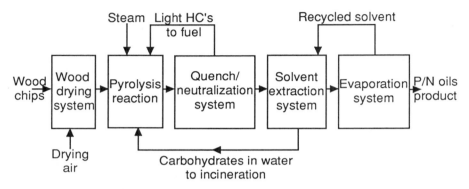

Figure 1. Fast Pyrolysis/Solvent Extraction

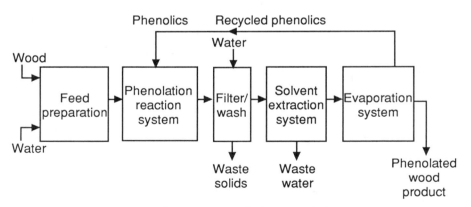

Figure 2. Wood Solvolysis (Phenolation)

requirements. Certain polyhydric alcohols may be substituted for phenol. However, the use of polyhydric alcohols will increase the cost of the product, because of their higher raw materials cost. The phenolation process has potential environmental impediments that could lead to expensive plant equipment, high operating costs, and personnel hazards. Phenol is toxic on skin contact and breathing, with an 8-h TLV of about 5 parts per million. Phenol must be scrupulously removed from waste water before it is discharged to conventional biological treatment; a solvent-extraction system is necessary. The plant would have to be designed to prevent discharge of organic contamination to any part of the biosphere -- land, air, or water. Such requirements translate into high investment and operating costs, especially in this phenolation process that appears to require handling a cycle of about 5 pounds of phenol solution per pound of wood fed. In the design envisioned, a progressing batch operation was chosen that requires the frequent opening and closing of reaction vessels. This would make containment of phenol especially difficult. It is possible that other process designs might somewhat decrease the capital equipment costs. With the amount of published data in the literature, and the need to minimize solids-handling problems, the progressing batch reactor design was a reasonable choice.

The amortized production cost of this wood phenolation product is estimated at $0.25/lb (see Table I). This estimate is approximate because the published data do not contain sufficient detail for a more complete technoeconomic assessment at this point. However, this production cost value, with an uncertainty of about 40%, is close enough to that of the products derived from the fast pyrolysis and fractionation above to indicate that the phenolation process is potentially competitive. The published data on the reactivity and resin formulations suggest that the process has economic promise. There are companies in Japan working with Professor Shiraishi's group to take this process into a commercialization stage.

The Case of Demethylated Kraft Lignin

Although the literature contains many examples of recent investigations of waferboard, oriented strand board (OSB) and other wood composites employing lignins from steam explosion (*38-39*) and organosolv processes (*40-41*), these areas are not included in this assessment. This case focuses on a process for modification of kraft lignins that has not received much attention in the literature, but have a very high reactivity towards formaldehyde, which makes them, in principle, very good candidates for the replacement of phenol in PF resins.

Demethylation of kraft lignin is commercially practiced for the production of dimethyl sulfide. The proposed recovery of demethylated/isolated lignin from kraft black liquor is related to overall kraft mill operations as shown in the simplified block diagram of Figure 3.

About 10% of the mill's concentrated black liquor would be diverted to the proposed recovery system. This fraction is believed to be a practical upper limit to avoid disruption of normal black liquor cycle operations. It is indeed an advantage for mills which are recovery-boiler furnace limited.

By demethylation, it is possible to increase the number of lignin sites that are reactive toward formaldehyde. Schroeder and coworkers have demonstrated that following demethylation with an ethyl acetate solvent extraction, it is possible to isolate about 90% of the lignin as a material containing a substantial amount of low-molecular-weight phenolics (42-43). The demethylation is carried out with the black liquor concentrated to 50% solids, to which 1.5% (weight basis) of sulfur is added. The mixture is heated at 230°C for about 30 min, with a heating rate of about 6°C/min. The rapidly cooled mixture is then diluted with water and ethyl acetate. After neutralization with acid, the organic layer contains the solubilized lignins. Solvent evaporation produces the adhesive starting material. The solid material can be used, probably as an alkaline solution together with the required amount of formaldehyde, directly in an adhesive formulation. A block diagram of this process is shown in Figure 4.

A key problem with this technology is the failure to date to demonstrate adequate reproducibility of the quality of the adhesives produced. Although excellent results have been obtained in many cases, the batch-to-batch variability is still too high, which suggests the need for a deeper chemical understanding of the process and correlations between process variables and product reactivity with formaldehyde.

The technoeconomic feasibility of producing demethylated lignin as described above, as a 10% slip stream of the black liquor of a typical 2000 tons per day kraft mill, was investigated. The amortized production cost was estimated to be $0.19/lb, in the same range as the other two products. The intellectual property of this DOE/OIT/SERI Adhesives Program-sponsored technology resides with Colorado State University Research Foundation, which is discussing funding possibilities with various companies, for the continuation of the development of this process.

Conclusions

The three new technologies described have technoeconomic potential for the replacement of a substantial portion of the phenol in PF resins. Which of these technologies will actually reach the marketplace will depend on a variety of factors as their research, development, and demonstration continues. These factors include performance of the product, the reproducibility, overall environmental impact of the process at a larger pilot scale, followed by the demonstration of the actual cost of these products in commercial plants. The efforts of today and those of 10-20 years ago differ in the degree of chemical understanding of these processes, coupled with substantially improved analytical techniques for the assessment of product reproducibility, which should make it easier to penetrate the market with mixtures of phenolics in replacement of pure phenol.

Future penetration of these products in the United States markets is strongly linked with their materials performance of which cost is only one component.

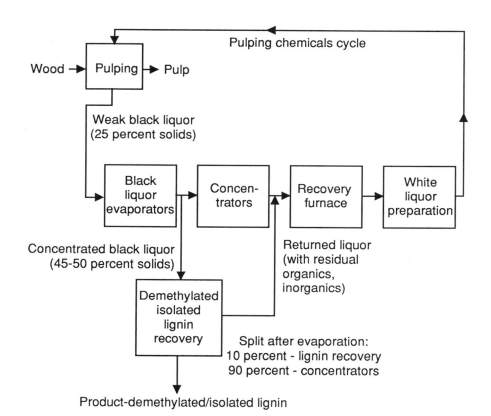

Figure 3. Kraft Pulping Block Diagram

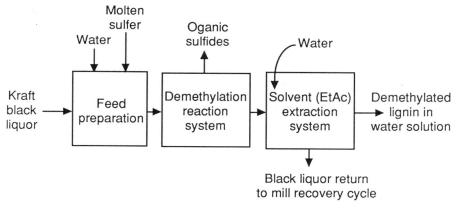

Figure 4. Demethylation of Kraft Lignin

Products with equal or better properties than the conventional ones have a good chance to advance in the marketplace. It is essential that the products are reproducible or have predictable reactivity and performance.

Although there is significant promise in these three routes, the journey to commercialization is a long one. Many factors will continue to play a role, including the sustainability of the efforts (both of government and industry) to decrease the risk of implementing a new technology. The risk in development of any new technology is high; these three processes have somewhat decreased risks by the early demonstration of some quality products and a potential of a reasonably quick return on the investments. They also have in common the need for demonstration at a larger scale to verify the validity of the production cost projections from the current laboratory scales, and that these processes operate as described. Moreover, all of them need continued chemical understanding at the process and resin chemistry level, so that the true benefits of the reactivity of the wood-derived products can be enjoyed.

Acknowledgments

The work described (pyrolysis/fractionation and demethylation) was sponsored by the U. S. Department of Energy's Office of Conservation and Renewable Energy through T. Gross and A. Streb. We gratefully acknowledge the support by the Office of Industrial Technologies Division of Waste Materials Management of the Adhesives program by J. Collins, A. Schroeder, and D. Walter. Thanks are also due to Mr. Paul Kearns of the DOE/SERI area office for his help in the formation of the Consortium. The PMRC representatives are: C. Gilpin and W. Fischer of Allied-Signal; M. Fields, W. Krayer, J. Aiken, T. Smeal, and K. Henry of Aristech; A. Ahmad, J. Outman, A. Gibson, R. Currie, R. McDonald, and K. Wirtz of Georgia-Pacific; W. Ayres, D. Johnson, G. Tomberlin, and L. Derr of Interchem; P. Waitkus, J. Mohr, B. Lepeska, W. Kleine, and R. Brotz of Plenco; and J. Dinwiddie and D. Bodde of MRIV. Special thanks are due to R. Muir and K. Howe for their efforts in the formation of the PMRC. These PMRC representatives are gratefully acknowledged for their support of the R&D and commitment to the program. The staff members of the Chemical Conversion Research Branch at SERI are gratefully acknowledged by the authors for their contribution to the work described, in particular, the efforts of J. Diebold, S. K. Black, J. Scahill, D. K. Johnson, R. E. Evans, which were supported by B. Hames, J. Bozell, K. Tatsumoto, F. Posey Eddy, G. Noll, D. Gratson, M. Echeverria, J. Fennell, C. Elam, J. Fodor, and P. Adam of SERI. Support of the SERI management is greatly appreciated: S. Bull, H. Hubbard, R. Stokes, and D. Sunderman. Thanks are also due to our consultant R. Kreibich and to Hazen Research, Inc. staff members C. Kenney and D. Gertenbach. Thanks are also due to H. A. Schroeder of Colorado State University for his contribution with the demethylated kraft lignins.

Literature Cited

1. *Organic Chemicals from Biomass*; Goldstein, I., Ed.; CRC Press: Boca Raton, FL, 1981; and references therein.
2. Goldstein, I.; *Forest Products J.*, **1981**. *31*(10), 63-68.
3. Goldstein, I.; *Science*, **1975**, 189, 847-852.
4. *Wood and Agricultural Residues - Research on Use for Feed, Fuels and Chemicals*; Soltes, E. J., Ed.; Academic: New York, 1983; and references therein.
5. Goldstein, I., *Appl. Polym. Symp.*, **1975**, *28*, 259-267.
6. Gratzl, J. "Potential of Technical Lignins as Extenders in Phenolic Resins: an Assessment of Reactive Sites," In *Proceedings of Weyerhaeuser Symposium on Phenolic Resins-Chemistry and Applications*, Tacoma, WA, 1979.
7. *Adhesives from Renewable Resources*; Hemingway, R.; Conner, A., Eds.; ACS Symposium Series 385; American Chemical Society: Washington, DC, 1989, and references therein.
8. *Adhesives for Wood -- Research, Applications, and Needs.* Gillespie, R. H., Ed.; Noyes Data Corp:, Park Ridge, NJ, 1983; and references therein.
9. *Wood Adhesives, Chemistry and Technology*; Pizzi, A., Ed.; Marcel Dekker, Inc.: New York, 1983; and references therein.
10. Forss, K.; Fuhrmann, A. *Forest Products J.*, **1972**, *29*, 39.
11. Forss, K.; Fuhrmann, A. U.S. Patent 4,105,606, 1979; Finn. Patent 51105, 1972.
12. Elder, T. J. Ph.D. Dissertation, Texas A&M University, 1979.
13. Soltes, E. *Tappi J.*, **1980**, *63*(7), 75.
14. Soltes, E.; Wiley, A.; Lin, S. *Biotech. Bioeng. Symp.* **1981**, *11* 125.
15. *Research in Thermochemical Biomass Conversion*. Bridgwater, A. V.; Kuester, J., Eds.; Elsevier Applied Science: London, 1988; and references therein.
16. Soltes, E.; Milne, T. A., Eds.; *Pyrolysis Oils from Biomass - Producing, Analyzing and Upgrading*; ACS Symposium Series 376; American Chemical Society: Washington, DC, 1988.
17. *Proc. 7th Canadian Bioenergy Seminar;* Hogan, E. N., Ed.; National Research Council of Canada: Ottawa, Ontario, Canada, 1990; pp 669-720.
18. *Fundamentals of Thermochemical Biomass Conversion*, Overend, R. P.; Milne, T. A.; Mudge, L. K., Eds.; Elsevier Applied Science: London, 1985.
19. Chum, H. L.; Diebold, J. P.; Scahill, J. W.; Johnson, D. K.; Black, S. K.; Schroeder, H. A.; Kreibich, R.; In *Adhesives from Renewable Resources*; Hemingway, R.; Conner, A., Eds.; ACS Symposium Series 385; American Chemical Society: Washington, D.C., 1989; p 135.
20. Chum, H. L.; Black, S. K. U.S. Patent 4,942,269, 1990.
21. Evans, R. J.; Milne, T. A.; Soltys, M. N. *J. Anal. Appl. Pyrol.* **1986** *9*, 207.
22. Milne, T. A.; Soltys, M. N. *J. Anal. Appl. Pyrol.* **1983**, *5*, 93.
23. Evans, R. J.; Milne, T. A. *Energy & Fuels* **1987**, *1*, 127.
24. Evans, R. J.; Milne, T. A. *Energy & Fuels* **1987**, *1*, 311.

25. Evans, R. J.; Milne, T. A. In *Pyrolysis Oils from Biomass - Producing, Analyzing, and Upgrading*; Soltes, E.; Milne, T., Eds.; ACS Symposium Series 376; American Chemical Society: Washington, DC, 1988; p 311.
26. Evans, R. J.; Milne, T. A. In *Energy from Biomass and Wastes XI*; Klass, D., Ed.; Institute of Gas Technology: Chicago, IL, 1988; p 807.
27. Evans, R. J.; Milne, T. A. In *Research in Thermochemical Biomass Conversion*; Bridgwater, A. V.; Kuester, J., Eds.; London: Elsevier Applied Science, 1988; p 264.
28. Diebold, J. P.; Scahill, J. W. In *Pyrolysis Oils from Biomass - Producing, Analyzing, and Upgrading*; Soltes, E.; Milne, T., Eds.; ACS Symposium Series 376; American Chemical Society: Washington, DC, 1988, p 31.
29. Diebold, J. P.; Scahill, J. W. In *Pyrolysis Oils from Biomass - Producing, Analyzing, and Upgrading*; Soltes, E.; Milne, T.,; Eds.; ACS Symposium Series 376; American Chemical Society: Washington, DC, 1988; p 264.
30. Diebold, J. P.; Power, A. J. In *Research in Thermochemical Biomass Conversion*; Bridgwater, A. V.; Kuester, J., Eds.; London: Elsevier Applied Science, 1988; p 609.
31. Johnson, D. K.; Chum, H. L. In *Pyrolysis Oils from Biomass - Producing, Analyzing and Upgrading*; Soltes, E.; Milne, T., Eds.; ACS Symposium Series 376, 1976; p 156.
32. McKinley, J.; Barras, G.; Chum, H. L. In *Research in Thermochemical Biomass Conversion*; Bridgwater, A. V.; Kuester, J., Eds.; London: Elsevier Applied Science, 1988; p 236.
33. Lipinsky, F. S. *Tappi J.* **1983**, *66*(10), 47.
34. Shiraishi, N. In *Lignin: Properties and Materials*; Glasser, W. G.; Sarkanen, S., Eds.; ACS Symposium Series 397; American Chemical Society: Washington, DC, 1989; p 488.
35. Shiraishi, N. *J. Appl. Polym. Sci.* **1986**, *32*, 3189.
36. Shiraishi, N.; Ito, H.; Lonikar, S. V.; Tsujimoto, N. *J. Wood Chem. Technol.* **1987**, *7*(3), 405.
37. Johnson, D. K., unpublished results at SERI, 1990.
38. Ono, H.-K.; Sudo, K. In *Lignin: Properties and Materials*; Glasser, W. G.; Sarkanen, S., Eds.; ACS Symposium Series 397; American Chemical Society: Washington, DC, 1989; p 334.
39. Cyr, N.; Ritchie, N. G. S. In *Lignin: Properties and Materials*; Glasser, W. G.; Sarkanen, S., Eds.; ACS Symposium Series 397; American Chemical Society: Washington, DC, 1989; p 372.
40. Cook, P. M.; Sellers, T., Jr. In *Lignin: Properties and Materials*; Glasser, W. G.; Sarkanen, S., Eds.; ACS Symposium Series 397; American Chemical Society: Washington, DC, 1989; p 324.
41. Lora, J. H.; Wu, C. F.; Pye, E. K.; Balatinecz, J. J. In *Lignin: Properties and Materials*; Glasser, W. G.; Sarkanen, S., Eds.; ACS Symposium Series 397; American Chemical Society: Washington, DC, 1989; p 312.

42. Schroeder, H. A.; Thompson, G. E. In *Proc. 24th International Particleboard/Composite Materials Symposium*, Maloney, T. M., Ed.; Washington State University: Pullman, WA, 1990.
43. Thompson, G. E.; Schroeder, H. A. Presented at the Wood Adhesives 1990 Symposium, 1990; U.S.D.A. Forest Products Laboratory and Forest Products Research Society meeting. Proceedings in press.
44. St. Pierre, L. E.; Brown, G. R., Eds.; *Future Sources of Organic Raw Materials, CHEMRAWN-I*; Pergamon: Oxford, 1979; and references therein.

RECEIVED March 20, 1991

Chapter 21

Potential for Fuels from Biomass and Wastes

K. Grohmann[1], C. E. Wyman, and M. E. Himmel

Biotechnology Research Branch, Fuels and Chemicals Research and Engineering Division, Solar Energy Research Institute, 1617 Cole Boulevard, Golden, CO 80401

Current uncertainty about petroleum supplies and the third sharp rise in petroleum prices in the last 20 years have returned us to the fact that all modern, and many developing, countries have become utterly dependent on petroleum imports for their energy needs. No country has used fully the lessons of the recent past to develop an industrial base for the production of alternative fuels from domestic resources. Many liquid and several gaseous fuels can be produced from the fermentation of treated biomass. The commercial success of this approach appears increasingly assured due primarily to the aggressive application of research and engineering resources to key steps in conversion processes.

The annual consumption of petroleum in 1988 amounted to almost 50% of the total consumption of fossil fuels in the United States (*1*), with approximately 50% of this quantity imported from various countries around the world. The historical trend in annual production and consumption of petroleum in the United States does not provide hopeful trends for the future, because steadily decreasing domestic production and increasing consumption are balanced by ever-increasing imports. An even more ominous domestic trend is the sharp decrease of operating drilling rigs in the last decade, which reached their peak in 1980 and steeply plunged to a postwar low in 1986 (*1*). At the same time, Arab countries stepped up their production and depressed the world petroleum prices. While estimated current reserves can supply petroleum at current rates of production for another 20-30 years (*1*), most fields are past their peak primary production stage, and their output rate cannot be easily increased. Short of discoveries of new large petroleum fields in frontier areas, such as offshore or in Arctic regions, the lower 48 states do not offer great hopes for the abrupt reversal of the slowly declining domestic production trend. Because of complete dominance of petroleum-derived liquid fuels in the transportation and some heating markets, alternative sources of liquid fuels with similar properties need to be developed at an accelerated pace.

In any event, it must be noted that any alternative fuels industry will be a very large enterprise, and will take decades to develop. Its full production potential will be achieved after the research stage is largely completed and pilot and demonstration stages are initiated. For example, the spectacular growth of the fuel alcohol industry based on the fermentation of corn starch to alcohol, from almost zero production in 1980 to approx-

[1]Current address: U.S. Citrus and Subtropical Products Research Laboratory, 600 Avenue S, N.W., P.O. Box 1909, Winter Haven, FL 33883

imately 800 million gal/yr in 1988 (2), still accounts for less than 1% of the annual consumption of gasoline in the United States. Additionally, it did not require an extensive R&D effort to start. Commercial plants for the manufacture of ethanol from starch or molasses have been marketed for decades, and only minor modifications or improvements can convert them to production of fuel alcohol. While our supplies of petroleum do not appear adequate for our current and future needs, the United States has ample supplies of fossil or renewable solid fuels in the form of coal, oil shale, tar sands, and biomass (1). All of these resources can be converted to liquid fuels, but conversion of each resource has its unique aspects. Only biomass conversion falls within the scope of this chapter.

This chapter provides brief analyses of biomass resources and liquid fuels, which can be readily produced by combined biological and chemical processing steps. The key conversion steps, such as depolymerization of cellulose and related carbohydrates, are followed by brief overviews of fermentation processes, which are applicable to the production of liquid and gaseous fuels and their precursors. This chapter is closed with a summary of key engineering issues.

The Nature and Magnitude of Biomass Resources

It has been estimated (3) that 2×10^{11} tons of carbon are fixed annually around the world by photosynthesis of higher plants. This renewable resource could, theoretically, supply approximately 10 times our energy needs and 100 times our food needs (3). While the plant kingdom encompasses tens of thousands of plants, only a few have been developed for large-scale cultivation by man (4). The world's agriculture is dominated by cultivation of grasses (4) (e.g., corn, wheat, rice, and sorghum), which produce starchy seeds necessary for our sustenance. These dense and storable sources of starch are supplemented by starchy tuberous crops (e.g., cassava and taro), which thrive in tropical lowlands unsuitable for grain production, except for rice. Potatoes are an analogous crop for temperate climate. Our sweet tooth is generally satisfied by sucrose crystallized from purified and evaporated juices of yet another grass (sugar cane), or sugar beets, cultivated in temperate zones. Starchy staples are supplemented by a few legumes (e.g., soybeans and peanuts), which produce seeds rich in both oil and protein, or in only protein (e.g., other beans). A few other plants (e.g., sunflowers and rapeseeds) are produced for their oily seeds, and a very small portion of land is devoted to specialty crops, such as vegetables, fruits, herbs, and spices. The major fiber crop is cotton, which is holding its own against synthetic fibers, while other fiber crops (e.g., flax, hemp and sisal) have declined. The remaining large tracts of agricultural land are divided into pasture, dominated again by grasses and herbaceous legumes, and forests where a variety of hardwoods and softwoods are grown.

Our inherent inability to digest polysaccharides other than starch, which as a plant storage polymer accumulates in seeds and tubers, leads to a tremendous wastage of total plant biomass during the harvesting of crops. Only seeds or tubers are removed, and the rest of the plant is usually left in the field as an "agricultural residue" of very little value. In the United States, the two major agricultural residues are corn stover and wheat straw with lesser amounts from minor grain crops, soybeans, and cotton (5,6). Because grain crops are harvested after the grasses have reached their maturity, the residues have very low digestibility and nutritional value, even for ruminants (i.e., cattle, sheep, and horses) that normally feed on grasses in younger stages of growth. Rapid changes in the composition of grass cell walls from germination to maturity were identified many years ago. These changes—mainly increases in xylans and lignins, and decreases in pectin content—were accompanied by a decrease in enzymatic digestibility by bacterial consortia in the rumen (7,8). Therefore, the only uses of grain crop residues, marginal at best, are as bedding and roughage for farm animals. Even if baled and collected, which is commonly done with wheat straw, a lot of residues are left to rot, or simply are burned in the fields. Because wasteful burning practices, whether they apply to forests, pastures, or cropland, are contributing to the greenhouse effect, with marginal recycling of nutrients and clearing of the land as the only discernible benefits, there is mounting pressure to discourage plant burning practices in agriculture. Such a shift in our attitudes provides a strong incentive for the development of new end uses for agricultural residues.

Arguments have been made in the past that the removal of straw and corn stover will rapidly deplete the soil of needed nutrients and accelerate soil erosion. Such arguments represent an extreme view that is not strongly supported by facts. Current harvesting machinery for grain crops does not pull the plants by the roots (9). The stalks are cut at a predetermined height above the ground. Straw or stover, therefore, refers only to the upper, aerial parts of the plant that are cut and removed. Roots and the lower portions of the stems, stubble, are always left in the ground. Because the plants are harvested after maturity, many nutrients (e.g., nitrogen) are translocated to the grain and roots, and only small amounts are left in the stems and leaves (10).

It has been estimated (10) that wheat straw contains approximately 15–30 lb of nitrogen per acre and 35–60 lb of potassium per acre. Due to higher biomass yields, the removal rates are higher for corn stover, approximately 70 lb of nitrogen per acre and 95 lb of potassium per acre. It should be noted that approximately two-thirds of the total phosphorus and nitrogen are irretrievably removed from the field with grain, while only one-third of potassium is removed (10). There are additional trace elements needed by the plants, but nitrogen, potassium and phosphate are the major fertilizer requirements. The vastly increased yields of grain crops in recent decades are directly correlated with heavier fertilizer usage (10), which vastly exceeds the reservoir in plant residues. Therefore, the complete recycle of mineral nutrients in crop residues can only supply a fraction of the fertilizer needs of modern crop production. Only the mineral components, such as potassium and phosphorus, can be eventually recycled back to the soil. Decomposition of crop residues can actually deplete the soil of nitrogen (10), which is consumed by microorganisms decomposing these residues. While we should conserve present nonrenewable deposits of mineral fertilizers and recycle as much of these nutrients as possible back to the soil, there is no inherent conflict between processing of agricultural residues and mineral nutrient recycling. Thermochemical processing, such as combustion and gasification, produces mineral ash streams that can be readily applied back to the fields. In biochemical processing, recycle would be only slightly more complex. Microorganisms require the same major and micronutrients as plants. Therefore, the mineral nutrients released from plant residues during biochemical processing can support growth and maintenance of microorganisms, and after product separation either the aqueous stream or concentrated mineral nutrients can be recycled back to the soil in the vicinity of the fermentation plant. Fermentation plant processing would save on the mineral nutrients purchased by the plant, and the surplus contained in the incoming feedstock can be supplied back to the farmers. Our own experimental work in anaerobic digestion of wheat straw hydrolyzates indicated that only supplementation with nitrogen is needed for maintenance of healthy populations of anaerobic bacteria. The remainder of the nutrients are released from the wheat straw substrate (11).

Another large source of residues in the United States is forestry. Approximately 50% of the standing tree biomass is actually harvested during a typical logging operation (12,13). Branches, crowns of trees, and crooked, diseased, or juvenile trees are not merchantable and are not generally removed from the forest. Because wood has extremely low (approximately 0.5%) ash content (14) and takes a long time to decompose, the benefits of mineral nutrient recycle are minimal. Additional residues are produced during lumber production. The current emphasis on minimizing landfill disposal fees creates incentives for any utilization of sawmill residues, which will decrease disposal problems.

Thanks to the spectacular growth of the pulp and paper industry during the last hundred years and the equally fast increase in population, we are producing ever-increasing amounts of municipal solid waste, which has wastepaper as its major component (15). Because producers pay rather steep prices for disposal of this waste, this feedstock is available at negative or very low cost. In aggregate, the coproduction of agricultural, forestry, and municipal wastes is the large annual resource that is cellulosic in nature and of very low or no value at the present time. When augmented by harvest from underutilized hardwood forests and energy crops on marginal lands and agricultural lands taken out of production, it has been estimated that a renewable fuel resource base twice as large as the total annual consumption of gasoline in the United States could become available, based on cellulosic biomass (2).

Thus, besides the starchy or sugar crops, part of which can be converted to biofuels during times of surpluses in agricultural production, by far the largest resource for fuel production is lignocellulosic biomass from softwoods, hardwoods, and grasses. Other crops, such as legumes, cotton, or oilseed plants, can be of interest in some localities, but they do not approach the magnitude of the first three resources.

By composition, the cell walls of grasses and hardwoods are similar, but they are significantly different from softwoods, see Table I (16-22; Torget, R., Solar Energy Research Institute, unpublished results). The major component in all substrates is cellulose, except in starchy organs such as corn kernels and cassava roots. Significant chemical differences occur in hemicelluloses and lignins (16,21,23-25). The major hemicelluloses in softwoods are glucomannans, which are replaced in angiosperms by acetylated xylans containing arabinose and glucuronic acid side groups. A significant shift also occurred during the evolution of hardwoods and grasses in lignin composition and content. The guaiacyl-type lignins present in softwoods are modified by syringyl groups in hardwoods and grasses. p-Hydroxyphenyl groups may also be present in the lignins of grasses. There is also a significant increase in the lignin content of softwoods when compared to grasses and hardwoods. Compositions of the major starchy crops are included for comparative purposes in Table I. The composition of newsprint and office paper are included to illustrate a range of paper products that are not uniform in composition, but contain mixtures of numerous pulp fibers from fiberized wood (i.e., mechanical pulp) to delignified and bleached chemical pulps that have a high cellulose content. Inspection of Table I indicates that cellulosic biomass and starchy grains contain very similar amounts of carbohydrates, but due to differences in structure and composition, they pose different challenges to conversion processes (discussed below).

Summary of Fuel Properties of Biomass

Biomass, primarily wood, has been used as fuel from the times of prehistory. Even today, large amounts are used for this purpose around the world. Biomass has a reasonable heat of combustion when dry and it is usually low in sulfur, nitrogen, and ash. (Sulfur, nitrogen, and ash cause problems in the utilization of fossil fuels.) However, biomass is a solid fuel with energy content lower than fossil fuels, both on a weight-basis and especially on a volume-basis (Table II) (26-29). Plant cell walls occupy only a small fraction of total plant volume; therefore, the bulk packing density of biomass is very low (Table II). The low volumetric energy content is very severe for agricultural residues, which together with the fast rate of combustion, makes them undesirable even for a farm fuel source, except in times of emergency. The densification of the energy content of biomass by compression or conversion to liquid fuels should be an important economic driver for the development of biomass fuels industries in rural areas. Increased density and reduced perishability will make transportation to distant markets much more attractive. Another economic driver for the conversion of biomass and other solid fuels into liquid fuels is the significant premium customers pay for liquid fuels, both in transportation and in stationary heating markets.

Alternative Fuels from Biomass

Gasoline and other liquid fuels derived from petroleum are extremely complex mixtures of aliphatic and aromatic hydrocarbons, which are characterized by a set of physical properties, such as boiling point range, volatility, ignition properties, flame propagation (antiknock), and heat of combustion. Because these liquid fuels are already complex blends, they can be blended with other organic liquids or replaced by them as long as these

TABLE I
Approximate Chemical Composition of Selected Biomass Resources[a]

Compounds	Wood of hardwoods (wt%)	Wood of softwoods (wt%)	Wheat straw (wt%)	Newspaper (wt%)	Office paper (wt%)	Corn kernel (wt%)	Cassava roots (wt%)
Anhydrosugars							
glucose	45-55	45-55	40	45-52	71-75	0.07-0.17	82-93
xylose	15-25	5.0-7.0	21	4.9-5.3	6.5-8.9	6.2	0.1-1.1
mannose	0.5-3.0	10-12	1.0-2.0	4.9-6.2	2.7	ND	ND
galactose	0.3-1.0	1.0-1.4	1.7	0.5-1.0	0	1.0	ND
arabinose	0.3-0.5	0.5-1.5	1.0-2.0	0.6-1.1	0	4.2	ND
Glucuronic acids	2.0-5.0	2.0-4.0	1.5	ND	ND	0.8	ND
Acetyl groups	2.0-4.0	1.0-1.5	2.2	ND	ND	ND	ND
Lignins	19-28	27-34	18	25.5	0.5	0.2	ND
Crude protein	ND	ND	ND	ND	ND	8-14	2.1-6.2
Ash	0.5-0.7	0.5-0.7	8.0	0.5-3.5	7.7-15	1.1-3.9	0.9-2.4
Water soluble extractives	2.0-5.0	2.0-5.0	8.0-12	ND	ND	ND	ND

ND is not determined, or not available.
[a] refs. 16-24; Torget, R., Solar Energy Research Institute, unpublished data

TABLE II

Selected Physical Properties of Liquid and Solid Fuels[a]						
	Density		Higher heats of combustion			
Fuels	g/mL	lb/ft³	GJ/ton	Btu/lb x10³	MJ/L	Btu/ft³ x10³
Gasoline	0.74	46.2	48.2	20.75	35.7	959
Diesel	0.85	53.0	46.0	19.81	39.1	1050
Vegetable oil	0.90	56.2	39.6	17.05	35.6	958
Methanol	0.79	49.3	22.3	9.6	17.6	473
Ethanol	0.79	49.3	29.8	12.8	23.5	632
Coal	0.6-0.9	40-58	17-35	7.4-15.2	11-33	300-880
Wood	0.16-0.4	10-25	19-22	8-9.7	3-9	80-240
Agricultural residues	0.05-0.2	3-12	16-19	7-8.2	0.8-3.6	21-99

[a] refs. 26-29; GJ = gigajoule, MJ = megajoule

liquids or their mixtures approximate the physical properties of gasoline or other fuels (e.g., jet and diesel). Gasoline extension or replacement is of highest interest in the United States because it dominates the U.S. refinery output (*30*). However, the higher boiling fractions, i.e., diesel fuel and heating oil, also account for a large fraction of refinery output and their replacement should be addressed as well. Because gasoline components must vaporize in the internal combustion engine, the boiling point of gasoline liquids has to be in the range of 60°–200°C. The boiling points of simple organic compounds (*31*) are compiled in Figure 1. Aliphatic, oxygenated compounds were selected for this figure, because many of them are produced directly by microbial fermentations, and the rest can be produced from the primary products by simple chemical or biological transformations. The single exception is methanol, which is not known to be produced by biological systems. However, methanol is produced as an intermediate during methane oxidation by methylotrophic bacteria, so possibilities for biological production of methanol may exist. Methanol is also a major contender for alternative liquid fuels produced by thermochemical methods and was included in Tables I and II for comparative purposes. This survey indicates that many simple alcohols, ethers, ketones, and esters are compatible with the boiling range of gasoline. Organic acids are included because of their value in the preparation of esters, ketones, and alcohols, but they are corrosive, have low heats of combustion, and are usually produced as nonvolatile salts. They are not, therefore, to be considered serious contenders for direct blending with gasoline. It must also be noted that with the exception of acids and two lower alcohols (i.e., methanol and ethanol), all other liquids are fully miscible with gasoline and the freezing points (*31*) of all of them are well below the lowest winter temperatures (-45°C).

The next important consideration is the energy contents (i.e., heats of combustion) of these liquids. They are summarized in Figure 2. This figure shows that numerous organic liquids that can be derived from biomass have significantly higher energy contents than methanol, with many reaching 80% of the energy content of the hydrocarbons in gasoline. Hydrocarbons can also be produced from fermentation products, but with significant weight loss penalty, which will be discussed below.

Another important fuel property needed for compatibility with current internal combustion engines is resistance to abnormal combustion (knocks), reflected by the octane number, which is usually displayed on the pump as an average of the research and motor octane numbers (R+M/2). Both research and motor octane numbers are determined with specialized CFR (Cooperative Fuel Research) knock-test engines operated under ASTM specified conditions. Because the research octane number (RON) is determined under less stringent operating conditions than the motor octane number (MON), it is higher than MON for the same fuel. The oxygenated liquids are clearly superior to hydrocarbons in gasoline in their antiknock properties (*32-34*). The research octane numbers of modern gasolines ranges from 90–98, but only due to extensive processing of oil fractions (*34*). The oxygenated compounds in Figures 1 and 2 show research octane numbers in the range of 108–125, with motor octane numbers correspondingly higher than those for gasoline. The octane numbers of gasoline were improved for many years by the addition of small amounts of tetraethyl lead, but due to its toxicity and poisoning of catalytic converters, its use was discontinued several years ago. Other newer organometallic compounds (*35*) may never be introduced to gasoline markets for the same reason. The industry is currently trying to balance the drop in octane values of gasoline by increased processing to aromatic hydrocarbons and branched aliphatic ones via isomerization, but even the standard for octane number measurements (i.e., isooctane) has a research octane number (i.e., 100) lower than many oxygenated compounds. Since lower aromatic compounds (i.e., benzene, toluene, and xylene) have been identified as carcinogens, it is only a matter of time before political pressure builds to decrease or eliminate their inclusion in gasoline. Preemptive measures may be behind their decrease in "reformulated gasoline." While many oxygenated compounds are also toxic (*32*), they are biodegradable; and none has been identified as a carcinogen. The universal solution

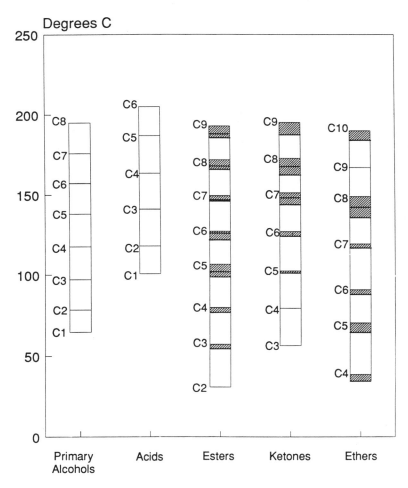

Figure 1. The boiling points of selected organic liquids. Compounds are labeled by the total number of carbon atoms and only straight-chain compounds were considered. Shaded areas depict the boiling point ranges of different compounds possessing the same number of carbon atoms. From reference (*31*). The boiling point range for gasoline is 60°–200°C.

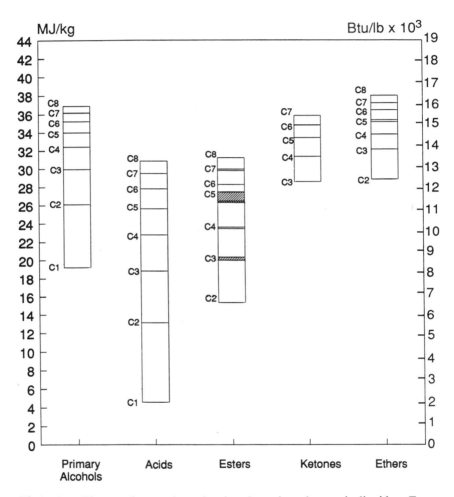

Figure 2. The net heats of combustion for selected organic liquids. From reference (31). The net heat of combustion for gasoline is 46 MJ/kg or 20×10^3 Btu/lb.

to the toxicity problem could be an increase in the lower end of the boiling range of gasoline, or its substitute, to make it less volatile, or the development of better devices for prevention of vapor escape during fuel transfer and from fuel tanks. The higher octane numbers and improved combustion of gasoline are two important benefits of blending oxygenated liquids into gasoline. At least two of them [methyl-t-butyl ether (MTBE) and ethanol] already have been incorporated into commercial premium gasoline production, even though they have not achieved a full market penetration to date. The citywide tests carried out in Denver, Colorado, over the last three years, which involved a mandatory switch of all cars to gasoline blended with MTBE or ethanol during a smog-prone winter season, showed very positive results in terms of air pollution control. The carbon monoxide levels dropped by 30% and smog formation also decreased (*36*). In view of these positive results, it is to be expected that other smog-prone cities and states will try to improve their air quality in a similar fashion, and the demand for oxygenated liquid fuels will rise. California, the largest market in the country, is already showing signs of moving in that direction. Oxygenated liquids shown in Figures 1 and 2 also contain no sulfur or nitrogen. Therefore, their utilization in gasoline blends will decrease sulfate emissions, which are a major component of particulate smog and, as sole fuels, will eliminate sulfate emissions altogether.

There are other important properties of automotive fuels, such as heat of vaporization and others, but their full discussion is beyond the scope of the present chapter and interested readers can find them thoroughly discussed in a recent book (*32*).

Production of Fuels from Cellulosic Biomass

While both thermochemical (e.g., pyrolysis and gasification) and biochemical (i.e., hydrolysis and fermentation) approaches can be used for the production of liquid fuels from biomass, the emphasis in this chapter will be on biochemical conversion, because thermochemical routes are covered elsewhere in the book. However, the current biochemical systems cannot directly convert all the substrate to liquid fuels, because lignin is very resistant to biological depolymerization and conversion (*37*). The range of organic products produced by fermentations is also very limited (*38-40*). These organic products may need to be upgraded to more valuable liquid fuels by using thermochemical routes. Therefore, a combination of thermochemical and biological routes may be necessary for the complete conversion of biomass to liquid and gaseous fuels.

The primary fermentation products produced during anaerobic metabolism by numerous microorganisms include a few simple alcohols, acids, a single ketone (acetone), and two potential gaseous fuels, methane and hydrogen. These products do not normally accumulate during aerobic respiration by microorganisms, yet two groups of highly reduced organic compounds, lipids and poly-β-hydroxybutyrate, can accumulate in storage vesicles during aerobic growth (*38-41*). An interesting feature of some fermentations (see Table III) is that they proceed most efficiently not under strictly anaerobic conditions, but under limited aeration, i.e., in so-called "microaerophilic" mode of operation. Microaerophilic operation does not increase the complexity of fermentation reactors, but does require the accurate monitoring and control of dissolved oxygen levels and of oxygen transfer rates, if cultures are to be prevented from switching to aerobic respiration and production of cell mass instead of the desired products.

The rapid progress in elucidation of metabolic pathways, which occurred after World War II, allowed connection of all these seemingly unrelated products to a single intermediate, pyruvate (*41*). These microorganisms perform anaerobic glycolysis to derive a high energy source, adenosine triphosphate (ATP), for their growth and maintenance. The production of ATP is accompanied by the coproduction of reduced electron carriers (i.e., nicotinamide adenine dinucleotide [NADH] and nicotinamide adenine dinucleotide phosphate [NADPH]), which in the absence of respiration, must be reoxidized for recycle via the reduction of precursors to primary fermentation products. The final metabolic products are often secreted outside of the cell. There are other primary products secreted

TABLE III

Summary of Fermentations for Potential Fuel Production

Reduced Products	Favored Microorganisms	Usual Substrates	Oxygen Requirements	Optimal pH Range	Optimal Temperature Range (°C)	Products Concentration (wt%)	Product Yield (wt%)	Product Formation Rate (g/L/h)
Ethanol	Numerous species of yeasts	C_6 sugars and di- or trisaccharides	Anaerobic to microaerophilic	3-5	30-41	7-12	45-48	2-8 (5-100)
	Pichia stipidis Candida shehatae Pachysolen tannophilus	Xylose and C_6 sugars	Microaerophilic	3-5	30-33	3-4	36-41	0.3-0.5
	Zymomonas mobilis	Glucose, fructose, sucrose	Anaerobic	3-5	35-39	10-12	46-49	6-10 (10-100)
	Transformed Escherichia coli	C_5 and C_6 sugars	Anaerobic	6-8	35-37	3-4	41-48	1-3
Butanol Acetone Ethanol (Isopropanol)	Clostridium acetobutylicum and related Clostridia	Starch, xylan, C_5 and C_6 sugars, disaccharides	Anaerobic	4-5	35-37	1-2	25-44	0.5-1 (1-3)
D(-)2,3-Butanediol	Bacillus polymyxa	C_5 and C_6 sugars, disaccharides, starch, xylan, inulin	Microaerophilic	6-7	30-37	2-2.5	40-46	0.2-1.0
Meso-2,3-Butanediol	Klebsiella pneumoniae	C_5 and C_6 sugars, disaccharides	Microaerophilic	5-6	30-37	9-11	40-46	0.4-2.0 (2-4.5)

TABLE III (CONT'D)

Reduced Products	Favored Microorganisms	Usual Substrates	Oxygen Requirements	Optimal pH range	Optimal Temperature Range (°C)	Product Concentration (wt%)	Product Yield (wt%)	Product Formation Rate (g/L/h)
Salts of acetic acid	*Clostridium aceticum* and related acetogenic bacteria	C_5 and C_6 sugars, syngas (CO_2+H_2 or $CO+CO_2+H_2$)	Anaerobic	6-8	20-70	1.5-4.5	85	0.6-8
Salts of mixed organic acids	Numerous bacteria	C_5 and C_6 sugars, xylan, starch, cellulose, syngas	Anaerobic	6-8	20-70	2-3	66	0.5-2
Lipids	Yeasts, filamentous fungi	C_5 and C_6 sugars, disaccharides, n-alkanes	Aerobic	3-6	25-35	20%-70% of cell dry wt. (0.4%-3% of culture volume)	15-22	0.2-0.4
Methane	Numerous cocultures of anaerobic bacteria (usually undefined)	C_5 and C_6 sugars to polysaccharides and other organic compounds	Anaerobic	6.5-8	25-40 and 50-70	50%-70% CH_4 (v/v) in the gas phase	27	0.02-0.2 (0.2-2) or 0.1-10 v/v/d (10-100 v/v/d)

Note: Production rates in parentheses apply to advanced laboratory systems.

by microorganisms, notably di- and tri-carboxylic acids from the citric acid cycle and polyols, such as glycerol, sorbitol, or xylitol. These high-value chemicals are not of interest for liquid fuel production because they are nonvolatile and highly oxidized. Lactic acid is of marginal interest for the same reasons.

It must be noted that only three primary fermentation products of interest for fuel production are produced in the homo-fermentative mode (i.e., as the sole reduced product). They are acetic acid, ethanol, and butanediol, but butanediol is the sole product produced under only microaerophilic conditions. Ethanol is coproduced with butanediol during strictly anaerobic fermentations. The remaining products are secreted as mixtures (e.g., butanol/acetone/ethanol) of coproducts as in the hetero-fermentative mode. The important features and requirements of microbial fermentations are summarized in Table III. A survey of Table III clearly shows that the primary substrates for practically all these fermentations are sugars. The transport of substrates and other nutrients through cell membranes of microorganisms is tightly controlled and is limited to low molecular weight compounds. In the case of sugars, the largest molecules that can be transported through the cell membranes are di- and trisaccharides. The higher oligomers and polymers have to be depolymerized by chemical or enzymatic means before the microorganisms can ferment the resulting sugars of low molecular weight. Because biomass components are polymeric, the very first step must be their depolymerization to fermentable units.

Depolymerization of Polysaccharides

Polysaccharides are carbohydrate polymers in which individual sugar units are joined by acetal groups (i.e., glycosidic linkages between hemiacetal or ketal) on one sugar unit and numerous hydroxy groups on the other. Since all acetal or ketal linkages are sensitive to acid-catalyzed hydrolysis, it is natural that acid hydrolysis of polysaccharides has been under development for a long time. Acid-catalyzed protonation and enzymatic hydrolysis of glycosidic linkages are common mechanisms of degradation of polysaccharides. Polysaccharides can also be degraded by the action of alkaline solutions, but this so-called "peeling" reaction leads to severe structural changes in cellulose and transformations of sugars to poorly fermentable hydroxy-acids. Therefore, this reaction is to be avoided if one is to produce sugars for subsequent fermentation. There are four families of polysaccharides of interest for large-scale production of fermentable sugars. Two are plant storage polymers, starch and inulin, and two are structural polymers in plant cell walls. There are numerous other polysaccharides produced by plants, but none of these are produced in the amounts needed for full production. Because behavior of storage and structural polysaccharides toward acid- or enzyme-catalyzed hydrolysis is significantly different, they are treated as separate categories.

Hydrolysis of Starch and Inulin. Starch is a storage polymer of α-1,4-linked glucose units, which accumulates in granules in the cells of many organisms. The basic structural unit is the disaccharide, maltose (42). The starch chains are either linear (amylose) or branched (amylopectin), with natural starch granules containing various proportions of amylopectin and amylose. The chains of starch molecules appear to be only loosely bonded by interchain hydrogen bonds, which allow easy penetration of water into starch granules and make partially depolymerized starch soluble in water (i.e., maltodextrins and water soluble starch). The starch granules can also be liberated from plant tissues by simple mechanical or chemomechanical treatments (i.e., wet or dry milling). The properties of high surface area, swelling of starch granules in water, and solubility of large intermediate molecules (maltodextrins) in water make depolymerization and solubilization of starch a relatively easy matter (42). Acid-catalyzed hydrolysis in "starch cookers" has been practiced for many decades, and at high temperatures ($150°-200°C$) it can proceed to completion in seconds to minutes. Dilute acid-catalyzed hydrolysis of starch has been recently supplanted by enzymatic catalysis, both at high ($60°-100°C$) and low ($20°-40°C$) temperatures. The thermophilic α-amylases active at the boiling point of water became commercially available in recent years and allows concurrent depolymerization by enzymes and gelatinization by steaming to occur. The strength of hydrogen bonds between starch molecules decreases rapidly with increasing temperatures; therefore, penetration of water and swelling of starch granules is aided by heating. However, enzyme systems that can depolymerize raw starch at lower temperatures are being developed in the laboratory, because their use could eliminate the energy consumption for steaming (43).

The ease of starch liberation and hydrolysis has led to its widespread industrial use for the production of high-fructose corn syrup and fuel alcohol. It must be pointed out that α-amylase alone does not depolymerize starch to glucose, but a second enzyme, glucoamylase, is needed for liberation of glucose from maltodextrins by endwise (exo-) cleavage. The enzyme or acid consumption needed for starch hydrolysis is very low (< 1:100 by wt) and the cost of hydrolysis is a small portion of the total cost as well. Starch conversion, like many large chemical processes, is dominated by raw material cost.

Inulin, a polyfructosan, is a storage carbohydrate in tubers of numerous species of plants, notably dahlias and jerusalem artichokes (44). Like starch, it is very easy to hydrolyze and solubilize to oligofructosans by both acid- and enzyme-catalyzed hydrolysis. Its hydrolyzates could be easily fermented by microorganisms, but lack of widespread cultivation of inulin-producing crops and repolymerization (reversion) of fructose to unwanted oligofructosans, which interfere with the production of pure fructose syrups, are still important problems (44). However, these issues are not insurmountable and inulin-producing crops may see a brighter future.

Polysaccharides in Plant Cell Walls. The hydrolysis of polysaccharides in plant cell walls is a much more challenging problem than hydrolysis of storage polysaccharides. The plant cell walls can be simplistically depicted as "runs" of cellulose fibers embedded in the crosslinked matrix of lignin-hemicellulose complexes (16,45). These polymers provide not only structural strength to the plants but also protect them against destruction by physical, chemical, and biological agents. Plant cell walls do not turn over during the lifetime of a plant as do storage polysaccharides. Also, in many cases these structures have evolved to last for hundreds or thousands of years in a relatively harsh environment. Due to intimate and strong associations, the major cell wall polysaccharides cannot be liberated by simple mechanical treatments used successfully on starch granules. Very harsh and destructive chemical treatments used in chemical pulping are necessary to liberate even the cellulose fibers from plant cell walls (23-25). These treatments are usually accompanied by the destruction or severe modifications of hemicelluloses and lignin. The hydrolysis of cell walls of plants usually requires "pretreatments" before high yields of sugars can be achieved during the primary hydrolysis. The very first step or series of steps is disintegration of plant tissues into smaller particles to aid penetration of acidic and enzymatic catalysts. These steps are usually accomplished by efficient chipping and shredding machinery, which is commercially available. It must be noted that shredding and milling of herbaceous plants requires much less energy than the same operation for wood (46). The primary size reduction (i.e., through 1/8 in. screen) is quite efficient; only 50 KW·H/dry ton for aspen wood chips and 6 KW·H/dry ton for wheat straw. Milling to finer particles can be much more energy intensive, because energy consumption increases exponentially with a decrease in particle size. The chipping, shredding, and milling steps are an unavoidable part of the process for hydrolysis of polysaccharides in plant tissues and must be carefully optimized for minimal energy consumption while retaining rates and yields. The acid-catalyzed hydrolyses are more tolerant to larger particles because small acid molecules diffuse through cell walls and plant tissues much more rapidly than the relatively large enzyme molecules (47). However, even in acid hydrolysis, the use of large pulping chips will decrease the yields of sugars. An interesting new method for fiberization of large chips via explosive decompression of steam or gases from plant tissues (steam explosion) has been identified. However, the large pressures required for efficient steam or gas explosion increase the cost of the equipment.

While mechanical disintegration is usually the only pretreatment necessary for acid-catalyzed hydrolysis of polysaccharides in biomass, efficient enzymatic hydrolysis usually requires more dramatic changes in porosity and structures of cell walls. These issues will be discussed further below.

There are three basic methods for hydrolysis of polysaccharides in plant cell walls. The simplest one is hydrolysis with dilute mineral acids (usually sulfuric) at elevated temperatures. The second set of methods involves action of very concentrated strong mineral acids at low temperatures (20°–60°C) and the last, most modern methodology involves pretreatment and hydrolysis with enzymes.

Dilute Acid Hydrolysis. Treatment of plant cell walls with dilute aqueous solutions of strong mineral acids releases oligomeric and monomeric sugars from polysaccharides. Due to a relatively high activation energy, hydrolysis is accelerated at elevated temperatures, so treatments are usually performed between 100° and 160°C for hemicelluloses, and between 180° and 220°C for cellulose (*48-51*). The primary drawbacks of this cheap and simple process include the simultaneous decomposition of sugars released from polysaccharides, the high cost of corrosion-resistant equipment, and the low concentrations of sugars produced by the percolation approach. The presence of toxins and microbial inhibitors is another well-documented drawback of utilizing dilute acid hydrolyzates (*51*). The chemistry and kinetics of dilute acid hydrolysis has been extensively studied. Indeed, the process using dilute sulfuric acid was commercialized in Germany before World War II and is in commercial practice in the Soviet Union today.

The dilute acid hydrolysis of polysaccharides is governed by two rate equations. The first equation describes the hydrolysis of polysaccharides to monomeric sugars (formation of intermediate oligomeric sugars is usually omitted) and the second equation describes the decomposition of monomeric sugars to furylaldehydes. The yield of monosaccharides is controlled by the differences between the rates of hydrolysis and decomposition. This difference is quite high for hemicelluloses, and hemicellulosic sugars can be prepared by dilute acid hydrolysis in very high yields (>70%) (*48-51*). The problem appears with cellulose, which is semicrystalline and more resistant to hydrolysis than other polysaccharides. The rates of hydrolysis of cellulose and decomposition of glucose are similar and, therefore, yields of glucose are expected to be quite low (30%–50%). One simple engineering solution to decrease the rate of glucose decomposition has been to decrease the residence time of liquids in high-temperature reactors (*51*). The so-called "percolation reactors," where a stationary bed of biomass chips is leached with percolating dilute acid, are the accepted design of hydrolysis plants. While this stratagem leads to increased yields, the increased volumes of liquids pumped through the reactor dilute the sugars released from biomass. Therefore, dilute acid hydrolysis in percolation reactors produces very dilute (<4%) sugar streams that are unsuitable for fermentation to liquid fuels. The major stream from these reactors is a solid residue of condensed lignins and other polymers, which has little value except as a solid fuel or substrate for thermochemical processing. Two approaches have been followed recently that could potentially overcome the low glucose yields inherent in dilute acid hydrolysis of cellulose. The approach pursued by Chen and co-workers (*52*) attempts to exploit differences in activation energies for cellulose hydrolysis and glucose decomposition by increasing the temperature above 200°C. However, the reaction times become so short (a few seconds) that scale-up of this approach to industrial size may present difficulties. Another potential improvement is replacement of water with organic solvents, first studied by researchers in the Soviet Union (*53*). These results indicate that impressive increases in the rates of cellulose hydrolysis and in glucose yields can be achieved by replacing water with some organic solvents such as acetone (*53-55*). Unfortunately, this research is in a preliminary stage, and important issues such as decomposition and recovery of expensive organic solvents have not been thoroughly addressed.

with a high surface area accessible to cellulase enzymes. Small particles can be produced by mechanical milling or by explosive decompression, but the formation of cellulose surfaces accessible to cellulase enzymes requires dissolution and extensive modification of other components in plant cell walls. Lignin-hemicellulose complexes (59-63) and acetyl groups (20,65) in xylans were all identified as major barriers to enzymatic hydrolysis. These complexes can be broken by alkaline or acidic treatments, however. Solutions of alkali metal hydroxides are known to break lignin-hemicellulose bonds and dissolve both lignin and hemicelluloses. At very high concentrations (5%-20%) they also swell cellulose. The degradative and dissolving action of hot sodium hydroxide solutions is used in well-known soda and Kraft pulping processes. During these treatments acetyl groups are also hydrolyzed. The action of hydroxide solutions in pretreatments must be adjusted to lower the severity from that found in pulping in order to preserve hemicellulosic carbohydrates. This goal seems to be achievable with grass residues (e.g., wheat straw and corn stover) and some hardwoods, but has not been achieved for softwoods. While alkaline pretreatments can be performed with many substrates at ambient temperatures, they are not truly catalytic, and significant amounts of chemicals are consumed for neutralization of acidic carboxylic and phenolic groups in biomass. The recycle and regeneration of hydroxides are thus critical research issues. These pretreatments do not hydrolyze polysaccharides, at least not without simultaneous destruction of sugars. The enzyme mixtures for subsequent hydrolysis must contain both cellulase and hemicellulase enzymes and thus are more complex than enzyme mixtures required for hydrolysis of pure cellulose.

An alternative base used for alkaline pretreatment is ammonia. A whole variety of pretreatment techniques ranging from treatment with ammonia gas to treatment with liquid ammonia has been developed over the past 90 years. Agricultural residues seem to be the most susceptible to this type of pretreatment, with hardwoods giving a variable response and softwoods being very resistant. Treatment with liquid ammonia is rather unique, because cellulose fibers are swollen and recrystallized by this liquid.

Because hemicelluloses can be easily hydrolyzed to monomeric sugars by hot, dilute solutions of mineral and organic acids, the next set of pretreatment methods have evolved around acid-catalyzed prehydrolysis of hemicelluloses in biomass (19,66-73). Depending on hydrogen ion concentration, these pretreatments can be performed at temperatures between 100° and 220°C, and with reaction times ranging from minutes to hours. At very high temperatures (>160°C) no acidic catalysts are required, because acidic compounds released from biomass provide the catalytic effect (68-69). This family of very simple pretreatments, because they involve only high-temperature steaming of biomass, is called "autohydrolysis" or, when explosive decompression is involved, "steam explosion." Conditions and equipment for steam explosion are similar to the commercial masonite process developed in the 1930s for the production of fiberboard. Hemicelluloses are partially hydrolyzed and solubilized during autohydrolysis, and lignin condenses into spherical particles. The pretreatment is very effective with some substrates, namely grasses and hardwoods, but softwoods are again resistant. The reaction times are short at temperatures higher than 190°C, and corrosion-resistant equipment is unnecessary. The main drawback of autohydrolytic pretreatments is the relatively low yield of hemicellulosic sugars (~50%) due to partial hydrolysis and pyrolytic decomposition at high temperatures (69). Baugh and coworkers (69) obtained evidence that alkaline degradation of sugars can extend to acidic pH values at high temperatures. The very high temperatures can be avoided by the addition of small amounts of inorganic acids, such as sulfuric acid or sulfur dioxide. These so-called "dilute acid pretreatments" (19,70) proceed rapidly anywhere between 140° and 200°C. Hemicelluloses are hydrolyzed to monomeric sugars at yields exceeding 70%. The removal of hemicelluloses and concurrent condensation of lignin creates numerous large pores in hardwoods and grasses, which allow penetration of cellulase enzymes to cellulose fibers (19,71). As in autohydrolysis, the creation of large pores is insufficient in softwoods to allow high enzymatic digestibility of cellulose. Reaction with sulfur dioxide seems to improve susceptibility of softwoods, however (72,73). The mechanism of its action is not quite clear, because extensive sulfonation of lignin does not occur. Lignins are retained in pretreated solids, which shows that its removal is unnecessary for efficient enzymatic hydrolysis of some biomass substrates. The retention of lignins, however, dilutes the cellulose stream intended for enzymatic hydrolysis and fermentation. Lignins can also adsorb significant amounts of cellulase enzymes (74,75). Its removal can thus have beneficial effects for downstream processing. More or less selective removal of lignins can be accomplished by the addition of organic solvents (usually lower alcohols) to acidic or alkaline aqueous solutions in numerous variants of organosolv pulping and pretreatments (58-63), or by sulfonation of lignins with sodium bisulfite, as in bisulfite pulping. Lignins and lignosulfonates are not fermentable, so they should be removed from sugar solutions by precipitation or similar means. The high cost of organic solvents, in comparison to water, limits the choice of solvents in organosolv pretreatments and makes solvent recovery an important consideration. The acidic bisulfite pretreatment, in contrast, would be easy to scale up because it is very similar to commercial acid sulfite pulping processes.

All the efficient pretreatments listed above provide cellulosic substrates that are hydrolyzable by potent cellulase enzymes to glucose and cellobiose in very high yields, i.e., 80%–100%. However, these pretreatments seem to be limited to certain substrates, such as grasses and the woody tissue of hardwoods. Softwoods are responsive to delignifying pretreatments or require the partial acid hydrolysis of cellulose before high yields in enzymatic hydrolysis of cellulose can be achieved (72,73). The few reported attempts (76; Torget, R.; Himmel, M.E.; Grohmann, K. *Bioresource Technol.*, in press) at pretreatment of softwood and hardwood barks indicate that some barks are resistant to pretreatment methods that are effective with wood. Because bark is a very important component (10%–40% wt) of branches and juvenile stems, pretreatment methods need to be developed for enzymatic hydrolysis of the carbohydrates in bark, or selection of trees for fast rotation cultivation should take this factor into account. Herbaceous legumes also appear to be more resistant to some pretreatments (*18*; Torget, R.; Werdene, P.; Himmel, M.E.; Grohmann, K., *Appl. Biochem. Biotechnol.*, in press). Ultimately, the substrates, pretreatment methods, and cellulolytic enzymes need to be closely matched in the overall processes for biochemical conversion, because they show a strong interaction in terms of rates, yields, and enzyme consumption. Results from our own series of dilute acid pretreatments of various plants indicate that significant variations in rates of enzymatic hydrolysis can be observed for different substrates milled to the same particle size and pretreated under identical conditions (*18,19,46,78*). Cellulose in pretreated corn cobs hydrolyzes the most rapidly, followed by cellulose in various pretreated grasses. Cellulose in pretreated hardwoods was hydrolyzed at the lowest rates. Whether these differences in the rates of enzymatic hydrolysis reflect inherent differences in cellulose fibers and cell walls, or are a reflection of different rates of enzyme penetration into pretreated plant tissues, is unknown at the present time.

The need for careful matching of some pretreatment methods and cellulase enzyme systems used for subsequent hydrolysis results from the significant differences among cellulase enzyme complexes produced by various microorganisms. Microcrystalline cellulose is never hydrolyzed by a single enzyme. Cooperative hydrolysis by one or more endoglucanases (EC 3.2.1.4) and exoglucanases (EC 3.2.1.91) is needed for efficient hydrolysis of cellulose to soluble sugars, usually a mixture of glucose and cellobiose (*77*). Conversion of cellobiose to glucose requires the presence of β-D-glucosidases (EC 3.2.1.21) or exoglucosidases (EC 3.2.1.74). The complexity of cellulase systems is increased in many anaerobic bacteria that produce very large ($>2 \times 10^6$ daltons) complexes called cellulosomes that are attached to outer cell walls. The penetration of individual enzymes to cellulose fibers and the rates and yields in enzymatic hydrolysis can thus be affected by selection of individual enzyme components. The commercial cellulase enzymes available today are mainly of fungal origin, and primarily from mutants of various *Trichoderma* and *Aspergillus* strains. The crude commercial preparations are complex mixtures of two or more endoglucanases and exoglucanases with additional hemicellulases, β-D-glucosidases and exoglucosidases, acetylxylan esterases, and other hydrolytic enzymes. There are numerous other cellulase systems that have been investigated in recent years by biochemists and molecular biologists, but these cellulases have not been introduced to the industry yet. Their potential for cellulose saccharification remains undetermined.

Therefore, the following discussion of enzymatic hydrolysis will apply to *Trichoderma* cellulases, which have received the most R&D attention over the last 20 years and are available in commercial quantities. The various *Trichoderma* mutants are prolific producers of cellulases and other hydrolytic enzymes. The secreted protein concentrations of about 50 g/L have been observed (78), which is an unusually high concentration of secreted microbial enzymes. The *Trichoderma* enzyme complex is active in the pH range of 3-6 and at temperatures up to 55°C. The high retention of activity over a period of several days has been observed at a temperature of 45°C, and the stability increases as the temperature is decreased below this limit. The problems with *Trichoderma* cellulase systems, and perhaps with cellulases in general, are relatively low specific activity, sensitivity to end product (i.e., glucose and cellobiose) inhibition, and low levels of β-D-glucosidase enzymes (78). The low specific activity leads to high enzyme loading requirements. Even with newer commercial preparations of higher specific activity, approximately 1 kg of enzyme is needed for hydrolysis of 50 kg of cellulose fibers. By comparison, the consumption of more active amylases in starch hydrolysis is much lower. The severe end-product inhibition of *Trichoderma* exoglucanases by cellobiose makes this enzyme system unsuitable for the production of concentrated (10%-25%) sugar solutions. This inhibition can be overcome by the application of simultaneous saccharification/fermentation systems, where yeast or other fermenting microorganisms remove the sugar as soon as it is formed. The development of SSFs requires careful matching of components, or a compromise between the physical and chemical requirements of enzymes and microorganisms.

The low levels of β-D-glucosidase secreted by many mutants of *Trichoderma reesei* lead to a requirement for supplementation by β-D-glucosidases from other microorganisms (78). This expensive supplementation can be decreased or eliminated by utilization of microorganisms that ferment cellobiose.

The microcrystalline nature of cellulosic fibers, the restriction of hydrolysis to the surfaces of cellulose fibers, and changing porosity of pretreated materials all combine to decrease the rates of both enzymatic and dilute acid hydrolysis of cellulose. The tenfold decline between initial and final rates of cellulose hydrolysis has been frequently observed (62) and enzymatic hydrolysis of pretreated cellulosic substrates usually takes one to five days. Austrian and other researchers (62,79) obtained results indicating that perhaps two fractions of cellulose with different rates of enzymatic hydrolysis exist in pretreated cellulose fibers, but other explanations, such as enzyme inactivation and decreased surface area, are also possible. Diversity of cellulose fibers in cell walls and plant tissues may also manifest itself in different rates of enzymatic hydrolysis. The slow rates of enzymatic hydrolysis of cellulose cannot be simply increased by increasing the enzyme loading, because available surfaces become saturated with enzyme molecules. However, all pretreatments that increase surface area or simultaneously change both crystallinity and surface area (e.g., in ball-milled and reprecipitated celluloses) are effective in increasing the rates of enzymatic hydrolysis of cellulose fibers. While many of these methods are too expensive for industrial use, the preparation of clean cellulose fibers with high surface area remains one of the objectives for pretreatment research.

Due to a short research history, enzymatic hydrolysis of cellulosic materials requires significant R&D investment before it can become as efficient as enzymatic starch hydrolysis and can be adopted by the industry for commercial sugar and alternative fuel production. Key improvements are needed in decreasing enzyme consumption and enzyme cost. Enzyme recycle, increased specific activity, resistance to end-product inhibition, and increased productivity by genetically engineered microorganisms are obvious avenues to achieve this goal. The development of improved pretreatment methods and integration with naturally susceptible substrates needs to be addressed as well. Enzymatic hydrolysis of cellulosic substrates has already achieved significant progress in the most important objective, high yields of sugars for fermentation. Modern methods in enzymology, genetics, and biochemical engineering make achievement of other goals possible as well. The sugars produced by enzymatic or acid hydrolysis can then be fermented by appropriate microorganisms to liquid or gaseous fuels. The fermentations of major interest for liquid fuel production can be grouped according to end products such as ethanol, acetone/butanol/ethanol, or organic acids. The major fermentation of organic materials to a gaseous fuel is an anaerobic digestion producing mixtures of methane and carbon dioxide, often called "biogas." The possibilities in the biological production of hydrogen are being explored as well (*34*).

Ethanolic Fermentations

Numerous yeast species (*80*) and two species of bacteria in the genus *Zymomonas* (*81*) efficiently ferment six-carbon sugars to ethanol and carbon dioxide. A few filamentous fungi, notably within the genera *Fusarium*, *Rhizopus*, and *Paecilomyces* can ferment both five- and six-carbon sugars to ethanol and CO_2, but at lower rates and final concentrations of ethanol. Many bacteria ferment sugars to ethanol as a coproduct with lower organic acids, butanediol, and acetone/butanol (*41*). The yeasts and *Zymomonas* are of primary interest for industrial production of ethanol because the conversion yields and rates are high and ethanol can be accumulated in relatively high concentrations of 5%–12% (w/v). *Zymomonas* have even higher yields and rates of ethanol production than yeast strains, but they suffer from a very limited range of fermentable sugars (i.e., glucose and fructose). *Zymomonas* are also very ethanol-tolerant and produce ethanol in very high concentrations (Table III). The high ethanol concentrations (>4% w/v) are necessary for efficient separation of azeotropic (95%) or anhydrous ethanol by distillation of fermented media (*82*).

Yeast strains can ferment a variety of six-carbon sugars and their oligosaccharides. A few of them (notably *Saccharomyces diastaticus*, *Endomycopsis fibuligera*, and *Schwanniomyces castellii*) secrete amylolytic enzymes and can ferment starch (*80*). However, none of the yeasts (*83*) can ferment five-carbon sugars (i.e., xylose and arabinose) or uronic acids to ethanol under anaerobic conditions, and none of the known fermenting yeast strains produce cellulase enzymes in nature (*80*). A few cellulase producing yeast species were recently identified, but they are strictly aerobic and do not ferment sugars to ethanol. The inability of yeasts to directly ferment five-carbon sugars, cellulose, or hemicelluloses to ethanol poses unique problems for the development of fermentation processes using lignocellulosic biomass as a substrate. Xylose, and to a much lesser extent arabinose, are major building blocks of hemicelluloses in hardwoods and other angiosperms. Xylose and arabinose are thus very important components of total sugars produced by enzymatic or acid-catalyzed hydrolysis of numerous plants in the angiosperm family. They are of lesser importance only in hydrolyzates from softwoods and wastepaper because major hemicelluloses in softwoods are glucomannans and galactoglucomannans; and wastepaper is enriched in cellulose at the expense of hemicelluloses (Table I). Three major approaches are being pursued that provide systems for fermentation of xylose to ethanol (*83*). Isolation and identification of microorganisms fermenting xylose to ethanol led to identification of several yeast strains such as *Candida shehatae*, *Pichia stipidis*, *Pachysolen tannophilus*, and *Candida utilis*, which ferment xylose to ethanol under microaerophilic conditions (i.e., under limited and controlled supply of oxygen). The yields of ethanol are currently lower (approx. 80%), fermentation rates are significantly slower, and ethanol concentrations are also lower than those obtainable by fermentation of glucose by industrial yeast strains. The performance of these yeasts can be improved by genetic techniques, and research work in this direction is already in progress (*84*).

The second approach relies on isomerization of xylose to the keto-sugar, xylulose, which is fermentable by some yeast strains (*83,85*). The isomerization is performed by bacterial xylose isomerases, some of which are commercially available as glucose isomerases and are used in large quantities for the production of high fructose corn syrups (*85*). The cost of xylose isomerase production can be decreased by overproduction in genetically engineered bacteria, and immobilization can increase pH compatibility between fermenting yeasts and immobilized xylose isomerase enzymes. The immobilized xylose isomerase enzyme and yeasts are usually combined with xylose substrate in one simultaneous isomerization and fermentation system (SFIX) because fermenting yeasts can remove xylulose as it is formed and drive the isomerization reaction to completion. The equilibrium between xylose and xylulose is 85:15, respectively, and is otherwise unfavorable to the production of xylulose. The yields of current SFIX fermentations (approx. 70%–80%) are similar to those obtainable with xylose fermenting yeasts, but the rates of ethanol production appear to be higher (*85,86*) and the ethanol tolerant yeast strains (e.g., *Saccharomyces cerevisiae and Schizosaccharomyces pombe*) can be used.

The third and newest approach involves construction of ethanologenic bacteria by transfer and expression of pyruvate decarboxylase and alcohol dehydrogenase genes from *Zymomonas* to other bacteria, such as *E. coli* (*87-89*). The activity of these two enzymes seems to be sufficient to change the metabolic pathway of *E. coli* from mixed acid production to ethanol as a major product. The reported yields and rates of these fermentations are very high, and arabinose can be used for ethanol production as well. The production of organic acids can be decreased by mutagenesis, and the whole enzyme system can potentially be transferred to other (e.g., cellulolytic) bacteria.

It can then be concluded that major improvements have been achieved in the fermentation of xylose to ethanol during the last 10 years, and the remaining obstacles can be overcome by genetic modifications. The fermentations of pretreated cellulosic substrates cannot be performed directly by yeasts because they are not cellulolytic, but require separate or simultaneous hydrolysis by cellulolytic enzymes produced by other microorganisms. Some fungal cellulases, such as those produced by *T. reesei*, are very compatible with yeast fermentation in terms of pH and fairly compatible with respect to temperature. The combination of *T. reesei* cellulase enzymes and thermotolerant yeast strains helped overcome the sensitivity of enzymes to end-product inhibition and decreased enzyme consumption in simultaneous saccharification and fermentation systems. High yields and reasonable final concentrations of ethanol (4%–7% v/v) have been achieved in this system using both wood pulp and pretreated cellulosic substrates. Measurements of transient sugar and ethanol concentrations clearly shows that the rates of SSF are limited by rates of cellulose hydrolysis, with yeast fermentation being a limiting factor only at the initial stage of SSF. Further improvements in the rates of SSF thus require continued development of more active enzymes and better pretreated substrates.

A significant difference between SSF of biomass and starch hydrolyzates is the presence of solid substrate throughout the SSF of pretreated cellulosic biomass because cellulose does not become solubilized until the very end of SSF. Because the higher rates of enzymatic hydrolysis are achievable by increasing the concentration of solid substrate, the higher rates of SSF should be achievable by increasing the substrate concentrations. This simple solution for increased rates and ethanol concentrations in SSF introduces mixing problems. Pretreated biomass particles, or fibers, can easily be stirred only at concentrations of less than 10% w/w. As concentrations increase above 10% w/w, the particle slurries rapidly change to wet solids and cannot be mixed by conventional impeller mixers. The improvements in SSF of biomass, which can be achieved by increasing the substrate concentration above 8%–10%, will require development of new large reactors equipped with high solids mixers. Our preliminary results in high solids SSF of pretreated wheat straw (*90*) indicate that significant increases in the rate of ethanol production and final ethanol concentration can be achieved by increasing the initial cellulose concentration from 8% w/w to an optimum value between 12.5% and 15% (w/w). Due to the presence of lignin in pretreated wheat straw, these concentrations of cellulose correspond to total solids concentrations between 12% and 23% (w/w). Above 12.5%–15% w/w cellulose concentrations, ethanol yields rapidly drop due to inhibition of yeast by ethanol.

Acetone/Butanol/Ethanol Fermentation (38,39)

The acetone/butanol/ethanol (ABE) fermentation was conceived by Chaim Weirman circa 1904. Demand for industrial solvents during World War I led to its scale-up to industrial production both in the United States and in Europe. After decades of successful operation, the plants were commonly shut down in the 1940s and 1950s because the process could not compete with cheap petrochemical routes. However, under special circumstances, such as in South Africa, the process has survived even to this day.

The ABE fermentation is carried out by numerous bacterial species in the genus *Clostridium*, which are commonly known as "butyl" organisms. Two species were developed for industrial solvent production, namely *C. acetobutylicum* and *C. beijerinckii*. Some *Clostridia* can also reduce acetone and produce minor amounts of isopropanol. The ABE fermentation proceeds in two stages, controlled mainly by hydrogen ion concentration. During the first stage, which starts at neutral pH, organic (mainly acetic and butyric) acids are produced. As the pH drops below pH 5.5 and inhibitory concentrations of undissociated acids start to accumulate, the bacteria respond by switching to the solventogenic phase where neutral solvents are produced, partially by uptake and reduction of preformed organic acids. Strains of *C. acetobutylicum* are amylolytic and, thus, can ferment starch directly. They also ferment all five- and six-carbon sugars that are present in hemicellulose hydrolyzates. Therefore, ABE fermentation does not suffer from the five-carbon sugar problem, which are major obstacles for ethanol production. Numerous hemicellulolytic and weakly cellulolytic strains of *C. acetobutylicum* were recently identified and, after proper genetic improvement, could be used for direct conversion of pretreated cellulosic biomass to ABE solvents. The total solvent yield is approximately 37% (w/w) of sugar consumed and the approximate ratio of butanol:acetone:ethanol is 6:3:1, respectively, on a weight basis. These ratios can be changed by changes in fermentation conditions, strain differences, and mutagenesis; but usually butanol remains a dominant product. The rates and yields in ABE fermentations are comparable or slightly lower than yields in ethanolic fermentations, but high toxicity of butanol limits the final solvent concentrations to approximately 2% (w/w) and initial substrate concentrations to approximately 60 g/L. Ethanol and acetone are much less toxic than butanol. The production of liquid fuels by ABE fermentation requires low energy consumption in all steps of the process, and it is in product recovery (82) where ABE fermentation suffers most in comparison with ethanol production. The boiling point of butanol is higher and, due to its low concentration in fermented media, the energy consumption for its recovery by distillation is much higher than for ethanol. A whole array of methods from membrane separations to solvent extraction, has been investigated in recent years. Decreased energy consumption for butanol recovery and increased solvent production rates were the major aims. While some of these modern approaches appear promising, none have been scaledup and replaced traditional distillation. Attempts to increase the butanol tolerance of *C. acetobutylicum* by mutagenesis have not been highly successful; therefore, the development of energy-efficient separation methods appears to be a most important avenue for the improvement of ABE fermentation.

2,3-Butanediol/Ethanol Fermentation (39)

2,3-Butanediol fermentation is conducted by many strains of bacteria. The bacteria that received the most R&D attention are various strains of *Bacillus polymyxa* and *Klebsiella* (*Aerobacter*) *pneumoniae*. *B. polymyxa* produces optically active D-(-)-butanediol, while strains of *K. pneumoniae* produce racemic meso 2,3-butanediol. This fermentation is strongly controlled by aeration. Under anaerobic conditions, approximately equimolar amounts of ethanol and 2,3-butanediol are produced by all strains. Limited aeration decreases ethanol production and 2,3-butanediol becomes a major, or sole, product. Both five- and six-carbon sugars are fermented, and some strains of *B. polymyxa* are cellulolytic, hemicellulolytic, and/or amylolytic (39). The possibilities thus exist for direct microbial conversion of pretreated biomass to 2,3-butanediol, and all sugars in biomass can be utilized. The final solvent concentration is relatively low (2%–3% w/v) with *B. polymyxa*, but *K. pneumoniae* can accumulate much higher levels (6%–8% w/v). The rates of solvent production are comparable to ethanolic and ABE fermentations, but product recovery presents a major problem. 2,3-Butanediol has a very high boiling point and heat of vaporization. It forms complexes with water molecules and is very hydrophilic. This combination of properties makes efficient recovery very difficult. The only process tested on the pilot-plant level involved evaporation of water and vacuum distillation of 2,3-butanediol from evaporator bottoms. 2,3-Butanediol is also not a valuable liquid fuel. It has a high boiling point and relatively low heat of combustion. However, it can be easily rearranged and dehydrated to 2-butanone, which is an excellent liquid fuel. Experimental work indicates that this acid-catalyzed diol-rearrangement can occur in aqueous solutions acidified with sulfuric acid, while 2-butanone is being distilled off. The recovery of anhydrous 2,3-butanediol may not be necessary for the production of 2-butanone, and difficulties in recovery of 2,3-butanediol may be decreased.

Fermentations to Volatile Organic Acids (*33,41,82,91*)

Several organic acids are produced during fermentation of sugars by numerous strains of bacteria. The acids of interest for fuel production are two- to five-carbon aliphatic acids from acetic to valeric. Lactic acid has a high boiling point and has a low heat of combustion; therefore, its production is not considered here. Some lower acids, namely acetic and propionic, can be produced as major or sole products, but most often mixtures of acids and alcohols are produced in "mixed acid" fermentations. The formation of mixed organic acids from pyruvate is accompanied by the formation of hydrogen and CO_2 if formate is decomposed by formate lyase or part of the carbon is secreted as formate. It must be emphasized that free organic acids are not produced during these fermentations, because they are toxic to microorganisms. The fermentations are conducted near neutral pH, and salts of organic acids with added base are the actual products. Sodium, potassium, magnesium, calcium, and ammonium hydroxides, or carbonates, are usually added with the substrate or during fermentation to maintain the high pH values and neutralize organic acids as they are formed. This inherently high base consumption is an important feature of organic acid fermentations, unlike ethanol and ABE fermentations, where neutral solvents are produced and carbon dioxide is simply gassed off. All recovery schemes for products from organic acid fermentations should include recovery of basic cations which, except for the calcium from limestone, are rather expensive. Production of acetic acid (i.e., vinegar) by strains of *Acetobacter* is a partial oxidation of ethanol to acetic acid and, thus, is of limited value for fuel production. Homo-fermentative conversion of sugars to salts of acetic acid (*82*) is one of the recent additions to fermentation technology. This conversion is carried out by several species of "acetogenic" bacteria. *Clostridium thermoaceticum* and *Clostridium thermoautotropicum* appear to have the highest potential because they can ferment both glucose and xylose to acetate. The acetogens can ferment fructose and a few other six-carbon sugars to three moles of acetate per mole of sugar utilized. Two moles of acetate are produced by decarboxylation of pyruvate, while the third mole is produced by reduction and incorporation of CO_2. These bacteria are quite unique because they can also produce acetate from mixtures of hydrogen and CO or CO_2 (i.e., syngas), formate, methanol, and other one-carbon compounds. The syngas conversion will be discussed later in the chapter. While fermentation of sugars is quite rapid and reasonable concentrations of acetate (i.e., 0.25 M) can be accumulated, the isolates studied so far suffer from a limited range of sugars utilized for acetate production and require rather complex media. Isolation of new strains and species, or transfer of genes for hydrolytic enzymes, may alleviate some of those problems. Propionate can also be produced by "propionic bacteria" and other bacterial species. The propionate is coproduced with acetic acid, and sometimes with smaller amounts of succinic acid, according to the relationship:

$$3 \text{ glucose} \rightarrow 4 \text{ propionate} + 2 \text{ acetate} + CO_2 + 2 H_2O$$

Succinate is a usual precursor of propionate and can be secreted from some strains with minor amounts of other products. Propionic bacteria can use a variety of five- and six-carbon sugars and some disaccharides. Butyric acid is produced by many *Clostridia* and other bacteria according to the relationship:

$$4 \text{ glucose} + 2 H_2O \rightarrow 3 \text{ butyrate} + 2 \text{ acetate} + 8 CO_2 + 10 H_2$$

Large amounts of hydrogen are thus produced in butyric acid fermentations. Some *Clostridia* can carry this fermentation farther and switch to ABE production in the second stage. *Clostridia* and other butyric acid-producing bacteria can utilize a wide variety of sugars, and some are cellulolytic and hemicellulolytic. Furthermore, the direct microbial conversion of cellulose and hemicelluloses to volatile organic acids occurs in rumens of animals and as a first stage in anaerobic digestion.

Many other bacteria, namely *Enterobacteriaceae*, ferment sugars to mixtures of organic acids, such as lactic, acetic, succinic, butyric, etc., with neutral coproducts, such as ethanol and butanediol. The coproduction of several acids is one of the drawbacks of acid fermentations, with the exception of acetic acid production. Another drawback for further conversion of organic acids to liquid fuels is their production in the form of salts, usually combined with inorganic cations. Three chemical methods have been investigated for the conversion of volatile organic acid salts to liquid fuels. A very simple thermochemical process for the production of valuable ketone fuels by pyrolysis of the calcium salts or esters of organic acids was investigated years ago (*33*). Low temperature (~ 300°C) pyrolysis of calcium salts of aliphatic organic acids proceeds according to the following equation:

$$(R\text{-}CO_2) Ca \rightarrow R\text{-}CO\text{-}R + CaCO_3$$

Due to the formation of mixed calcium salts, a mixture of ketones from acetone to heptanone is produced by pyrolysis of calcium salts of mixed organic acids produced by anaerobic fermentations. These ketones have a high energy content (Figure 2), high octane values, and boiling range compatible with gasoline (Figure 1).

Another approach investigated for the production of fuels from organic acid salts was the Kolbe electrolysis, but this method consumed large amounts of expensive electrical energy and produced mainly gaseous lower hydrocarbons (*92*). Kolbe electrolysis has an advantage over other methods for the conversion of salts of volatile organic acids to liquid fuels because it does not require prior concentration of the salts or separation of the acids from the cations. Hydrocarbons produced during electrolysis of the salts of volatile organic acids can be removed by pervaporation. Alternatively, they will separate from the aqueous reaction mixture due to their immiscibility with water. Pyrolysis of the calcium salts will require concentration of the salts, which are quite soluble in water. The formation of esters by the classical reaction of carboxylic acids with alcohols requires not only concentration of the salts of organic acids, but also separation of the acids from the cations, production of highly concentrated organic acids, and preferably the recycle of the cations back to the fermentations. The formation of esters from carboxylic acids and alcohols is a simple dehydration reaction, which is unfortunately readily reversible in the presence of water. Mixed acid fermentations tend to become inhibited by end-products at concentrations of 20–30 g/L; therefore, the efficient separation and concentration of organic acids become the major problems. Salts of organic acids are not volatile and organic acids have high heats of vaporization, which negates their recovery by conventional distillation methods. Innovative separation concepts using solvent extraction, ion exchange, and membrane technology need to be developed (*33*) before these very simple and attractive systems for direct microbial conversion of pretreated cellulosic materials can be utilized for liquid fuels production. Inspection of Figure 2 also indicates that the formation of esters from higher organic acids, such as butyric acid, and lower alcohols (e.g., methanol and ethanol) would be more attractive because the resulting esters will have higher energy contents than either of the starting reactants. The thermochemical formation of esters or ketones from organic acids or their salts will require additional input of thermal energy, but all or part of this energy can be recovered from the higher energy content of the upgraded products (see Figure 2).

Lipid Production (38)

Many microorganisms, namely some aerobic yeasts and fungi, can accumulate large amounts of triglycerides of fatty acids (lipids) during the later stages of aerobic growth on sugars. The accumulation of lipids is usually triggered by exhaustion of nutrients other than the carbon source. The nutrient that is usually allowed to become exhausted is nitrogen, but with some microorganisms, depletion of other nutrients, such as phosphate sulfur, or iron also stimulates lipid accumulation. Very high fat content (50%–66% w/w) was achieved in some strains (38), but lipid production by microbial conversion of sugars is plagued by very low weight conversion yields. Fatty acids are highly reduced compounds, and overall conversions of 15%–24% of the weight of sugar to the weight of lipid were observed. Numerous lipid-producing strains can utilize a variety of sugars from biomass, but very low conversion yields on a weight basis will limit potential substrates to wastes of negative or very low (1¢–2¢/lb) cost. Direct production of lipids by photosynthetic plants and algae may be a better production route because lipid production is aided by photosynthesis (93). It must be noted that lipid production is rather unique among fermentation processes for liquid fuel production because it can accommodate relatively low substrate concentrations on the order of 10–40 g/L. Lipid droplets are sequestered in microbial cells and can be easily recovered by filtration or sedimentation of dilute cell suspensions. Production and recovery of other biofuels, such as ethanol, require much higher (10%–25% w/w) substrate concentrations.

The conversion systems discussed above all require sugars in one form or another as a substrate, with the exception of photosynthetic lipid production. These systems cannot convert phenolic compounds derived from lignins, tannins, and other phenolic components from biomass to liquid fuels. These processes thus impose a pressure on feedstock selection, requiring feedstocks with the highest possible carbohydrate contents and coupling with the chemical or thermochemical processing of lignin for complete substrate conversion. Biomass feedstocks with high carbohydrate content include some agricultural residues (see Table I) and the wood of hardwood trees. Tree bark, the wood of softwoods, and some forms of wastepaper are examples of substrates with high phenolic content (i.e., lignin, tannins, etc.) and correspondingly lower carbohydrate content. Two biological conversion processes, syngas fermentations and anaerobic digestion, can ultimately accommodate a wide range of biomass feedstocks and their components. Syngas conversion refers to a two-stage process where organic matter is first gasified in the presence of steam to syngas, a mixture of hydrogen, carbon monoxide, and carbon dioxide. Biomass, coal, and other organic materials can thus be converted to a fairly uniform gaseous feedstock, which can be biologically converted to liquid fuels or their precursors, disproportionated to hydrogen and carbon dioxide by photosynthetic bacteria, or converted to mixtures of methane and carbon dioxide (biogas) by anaerobic digestion consortia. The thermochemical conversion of biomass and steam to syngas has been investigated for many years and the technology should be adaptable to the gasification of lignin and mixtures of organic solids. Estimated capital costs are low and the gasification processes appear to be quite efficient. Depending on the feedstock cost, syngas can be produced at approximately 2¢–5¢/lb. Anaerobic digestion, an old set of technologies used primarily in waste treatment, can also convert a wide range of carbohydrates and other organic compounds to biogas without prior gasification.

Syngas Conversion to Liquid and Gaseous Fuels (78,82,94-97)

The anaerobic utilization of carbon monoxide or carbon dioxide-hydrogen mixtures was observed many years ago. Many methanogenic bacteria can disproportionate mixtures of H_2 and CO or CO_2 to mixtures of methane and carbon dioxide, usually called biogas. Likewise, a selected group of anaerobic bacteria, called acetogens, can convert mixtures of carbon monoxide or carbon dioxide with hydrogen to acetate, which can be converted by methanogens to biogas. Single- or two-stage microbial conversions of syngas to biogas are well documented in anaerobic bacterial systems. Additional conversions to alcohols and higher fatty acids were identified recently and will be discussed below.

Syngas, a mixture of carbon monoxide, carbon dioxide, and hydrogen, is being produced by the reaction of numerous carbonaceous substrates with steam at very high temperatures (97). Large-scale commercial processes for the production of ammonia and methanol utilize syngas production as a first step. Syngas is not a uniform mixture of gases at a fixed stoichiometric ratio. Depending on the carbon, hydrogen, and oxygen content of individual feedstocks, syngas mixtures of different hydrogen to carbon monoxide plus carbon dioxide ratios are produced. The syngas produced from natural gas is richest in hydrogen. Its formation proceeds according to two reactions:

$$CH_4 + H_2O \rightleftharpoons CO + 3H_2$$

$$CO + H_2O \rightleftharpoons CO_2 + H_2$$

yielding the following result:

$$CH_4 + 2H_2O \rightleftharpoons CO_2 + 4H_2$$

The second reaction is usually called the water shift reaction. The equilibria are adjusted by choice of reaction conditions and the water shift reaction is promoted by decreased temperatures and use of catalysts. The ratios of hydrogen to combined carbon monoxide and carbon dioxide drop significantly if coal or biomass are used as a feedstock, with intermediate values produced by conversion of various petroleum fractions (94-97). Production of syngas from coal or biomass leads to molar ratios of carbon monoxide to hydrogen of approximately 1:1, or after complete water shift, carbon dioxide to hydrogen ratios of approximately 1:2. Because the composition of lignin is very similar to low ranking coals, the syngas composition from lignin is similar to syngas produced from coal. The low hydrogen content of syngas produced from coal or biomass makes it deficient in hydrogen for numerous chemical and biological syntheses (e.g., methanol and methane) and decreases the yields of many ultimate products. Simple solutions to hydrogen deficiency would be cogasification of methane and coal or biomass, or blending of syngas streams produced by separate gasification of hydrogen-poor substrates and methane. Because methane and water are the best natural substrates for hydrogen production and liquid or gaseous fuel production will involve hydrogenation in one fashion or another, we should be conserving natural gas for future fuel production processes.

The biological conversion of syngas to liquid or gaseous fuels received research attention only recently. The biomethanation of syngas using pure cultures of methanogens, mixed cultures of acetogens and methanogens, or enriched cultures of bacteria from natural populations has been studied (82). Very impressive rates and yields of methane production from coal syngas were observed, but methane concentrations are lower than usual due to the low hydrogen content of syngas. The conversions are also limited by low gas-liquid transfer rates of the relatively insoluble hydrogen and carbon monoxide, but rates have been improved by operating under pressure and decreasing aqueous film thickness (98). The biological conversion of syngas to methane proceeds under mild conditions (i.e., low temperatures and pressures) and is resistant to hydrogen sulfide, which poisons practically all chemical catalysts. Preliminary estimates (99) indicate that biological methanation of syngas could be significantly cheaper than the chemical routes.

Some photosynthetic bacteria (i.e., *Rhodospirilum rubrum*) carry out a water shift reaction and convert carbon monoxide and water to CO_2 and hydrogen (98). The two-stage biomethanation reaction can also be converted to acetate production by selectively inhibiting the growth and metabolism of methanogens. The dominant acetogens will then produce acetate by the homo-fermentative conversion of syngas. Because the formation of higher organic acids has been observed in anaerobic bacterial cultures grown on syngas (100), the isolation of new strains that can convert syngas to higher and more valuable organic acids, such as propionic and butyric, should be relatively straightforward. A new species of *Clostridium* (*C. ljungdahlii*) has been recently isolated (101) that produces mixtures of ethanol and acetate from syngas. Researchers from the Michigan Biotechnology Institute (99) also demonstrated the two-stage conversion of syngas to acetone/butanol by converting syngas to organic acids in the first stage and performing ABE fermentation in the second stage.

The results obtained so far indicate that many bacterial fermentations may proceed using syngas as a substrate, and rapid development in this area can be expected. This route could allow biochemical conversion of lignin to additional liquid or gaseous fuels and complete utilization of all biomass components. The recovery problems for liquid fuel products or their precursors will be similar to those described previously under fermentations using sugars as substrate. Biological conversions of syngas must also compete with numerous chemical conversion processes which have been developed over many decades. Other alternatives for biochemical lignin conversion may occur in mixed anaerobic populations that convert depolymerized and solubilized coal and peat to biogas. Anaerobic fermentation of many organic compounds, for example phenols and furfural, is known to occur in anaerobic digesters, but bacteria responsible for these fermentations have not been isolated and identified. Anaerobic digestion is normally allowed to run its course and only the conversion of total organic carbon (TOC) to biogas is usually monitored. Although the direct anaerobic degradation of lignin to biogas does not seem to occur (37), the conversion of many phenolic and aromatic compounds is documented (102), so possibilities may exist for biological conversions of thermochemically depolymerized and solubilized lignins. Partial biological conversion of degraded lignin compounds from peat to biogas has been experimentally documented (82).

Anaerobic Digestion (103)

Anaerobic digestion of biomass or it's components can provide a major source of high quality gaseous fuel (i.e., methane or biogas). Anaerobic digestion is a sequential conversion process in which numerous organic compounds and biodegradable polymers are ultimately converted to biogas, i.e., gaseous mixtures of methane and carbon dioxide saturated with water vapor and containing small amounts of gaseous impurities, such as hydrogen and hydrogen sulfide. Anaerobic digestion processes are carried out by a consortia of anaerobic bacteria adapted to given substrate or mixture of substrates. Many diverse bacteria known to exist in anaerobic digestion systems have not been isolated or studied yet. Most of the information that has been gathered about bacteria involved in anaerobic digestion of cellulosic materials was derived from the rumena of grazing animals, such as sheep and cattle (104). Other important applications of anaerobic digestion (i.e., municipal sewage and wastewater treatments) did not support very extensive bacteriological studies. The broad and incomplete picture of steps involved in anaerobic digestion appears to be a primary fermentation of organic substrates to salts of organic acids and hydrogen plus carbon dioxide. Hydrogen, carbon dioxide, and the lowest organic acids (i.e., acetate and formate) can be directly converted to biogas by methanogens. Other organic acids are disproportionated to acetate, hydrogen, and carbon dioxide by "acetogenic" bacteria, which were mentioned previously in regard to organic acid production and syngas conversion. Some recent results indicate that the majority of biogas is produced by disproportionation of acetate, with minor contribution from syngas (i.e., hydrogen and carbon dioxide) reduction.

Due to the diversity of bacterial consortia, anaerobic digestion is unique among the microbial fermentations in its adaptability to numerous organic compounds and their mixtures, which can be used as substrates. Because these consortia can contain numerous anaerobic bacteria producing hydrolytic enzymes, such as cellulases or hemicellulases, direct microbial conversion of biopolymers is usually practiced. The drawbacks of this simplistic approach are lower yields and rates of the conversions. Both pretreatments and supplementation with hydrolytic enzymes were shown to be effective methods for increasing rates and yields in the anaerobic digestion of some biomass substrates. Other productive approaches include increasing concentration of bacterial cells by immobilization on solid supports or flocculation for liquid feedstocks, and increasing the concentration of solid substrates for insoluble cellulosic feedstocks, such as municipal solid waste. Cellulosic feedstocks, such as municipal solid waste (MSW) and wheat straw, can be digested at concentrations of solids approaching 35%, and a severalfold improvement in the rate of gas production (over stirred tank reactors operated at low [<8% w/w] concentrations of substrate) can be achieved. Anaerobic digestion of crop residues and food and feed waste is practiced on rather large scale in China, India, and other countries (105,106). In these countries, biogas is used for domestic cooking and also for local transportation. Residues from the digesters are recycled back to the fields as a fertilizer and as a solid soil amendment. Anaerobic digestion is a very suitable technology for fuel production from wet biomass and wastes in rural areas because the process is very simple, requires minimal attention and training of operators, and cycling of substrates and products can be easily done in a rural setting. Scaleup of anaerobic digestion in town- or city-size systems requires increased efficiency and attention to waste disposal problems. Similar to lipid production, the main problem of anaerobic digestion is the low weight yield of methane from substrate utilized. The theoretical weight yield of methane from glucose is 27% (w/w). The actual yield is less than that, due to the incomplete utilization of all components in biomass. Low yields on a weight basis and the low current cost of natural gas limit biogas production to waste treatment systems and hamper efforts for large-scale development. Otherwise, anaerobic digestion is very attractive. Product separation occurs automatically by outgassing, and biogas can be directly used as a fuel in the proximity of the plant. The heat of combustion of biogas (~600 Btu/scf) is lower than for natural gas (1,000 Btu/scf), but only minor adjustments are needed for replacement of natural gas by biogas in furnaces, boilers, and internal combustion engines. The acceptance of biogas by natural gas pipelines requires removal of carbon dioxide, water vapor, and hydrogen sulfide. The commercial technology for upgrading biogas to pipeline quality exists because it was developed for the upgrading of natural gas from selected gas fields, but such an upgrade will naturally increase the biogas costs. If excess biogas can be produced at prices competitive with natural gas, numerous liquid fuels can be produced from it by chemical processes; or it can serve as a source of hydrogen via reaction with steam.

The occurrence of anaerobic digestion during burial of organic wastes and its utilization in waste treatment can also lead to serious underestimates in potential productivities (rates) of anaerobic digestion. The usual volumetric rates of anaerobic gas production are summarized in Table III. The comparison of these rates indicates that gas production rates encountered in landfills and waste treatment systems are not even close to the maximal rates achieved in more advanced laboratory or pilot-scale systems. Anaerobic digestion is incidentally developed under some burial conditions, but anaerobic digestion in landfills is not an engineered production of biogas by any means. Likewise, the sewage and other waste treatment processes must be robust and degrade waste under variable conditions. Biogas production in these systems is not optimized because waste degradation and treatment is the main objective. The anaerobic digestion processes aimed at commercial biogas production require changes in emphasis from waste treatments. Dedicated feedstocks that are highly biodegradable with minimal pretreatments must be integrated with anaerobic digestion consortia and bioreactor design. High yields, increased rates, and cheap reactors are key requirements for further development of anaerobic digestion as a fuel production system. Power consumption for mixing can also be a major energy drain, and efficient mixing systems with low power consumption must be developed.

Biological Hydrogen Production (39,107)

Hydrogen and carbon dioxide are coproduced with organic acids during certain fermentations. If hydrogen is simply vented from the system, energy in the substrate is wasted. The hydrogen production and loss need to be suppressed, or hydrogen should be captured, separated from carbon dioxide, and used for reductions elsewhere in the liquid fuel production system. The only biological systems that produce hydrogen as a major primary product are photosynthetic microorganisms, namely some algae, blue-green algae (cyanobacteria), and other photosynthetic bacteria. Even with these systems, hydrogen is produced at the expense of other organic compounds formed by photosynthesis. The biological production of hydrogen thus does not appear to be very efficient at the present time, but can be useful in coproduction systems where hydrogen can find captive use in the reduction of organic compounds produced as a part of biomass conversion processes.

The biological shift reaction of syngas, performed by some photosynthetic bacteria, appears to have the highest potential for the production of pure hydrogen.

Engineering Issues and Future Improvements

The biological conversions of sugars to liquid and gaseous fuels are very efficient (4) because between approximately 90% and 97% of heat of combustion in substrates can be converted to fuel products. Microorganisms derive only small amounts of energy by anaerobic fermentations and thus convert very small portions of the substrate into cell mass. However, significant losses in efficiency and yield can occur in other parts of the conversion system. The inability of current biological systems to convert lignin to liquid or gaseous fuels results in an obvious loss of energy in original substrate. While lignin is only a minor (10%–30% w/w) component of lignocellulosic biomass, it is a more significant component with respect to total energy content because the heat of combustion of lignin is significantly higher than that for carbohydrates.

The development of advanced biological, or preferably chemical, processes for the conversion of lignin to liquid or gaseous fuels is needed for the total conversion of biomass to other fuels. Simple alternative uses for lignin include its use as a boiler fuel to provide process energy or potentially the cogeneration of electricity. However, in these applications lignin must compete with coal, a relatively cheap solid fuel. Lignin streams also require dewatering to increase combustion efficiency.

Another source of potential losses resides in the pretreatment and depolymerization of polysaccharides in biomass to develop fermentable sugars. Mechanical steps for the disintegration of biomass to smaller particles must always be optimized with respect to energy consumption because tremendous amounts of energy can be dissipated in these steps. Mechanical pretreatments interact strongly with changes in plant cell walls and tissues produced by chemical pretreatments, but the effects of chemical pretreatments on the energy consumption in mechanical disintegration of plant tissues have been scarcely investigated to this day. The decrease in total energy consumption must always remain a goal of processes used for the conversion of biomass to other fuels.

The energy losses in pretreatments and hydrolysis are fairly specific to each of the three types one chooses to pursue, but operation at high concentration (>10% w/w) of biomass solids can provide significant improvements in energy efficiency for all of them. Biomass is highly porous and cell walls occupy only about 30% of the total cell volume. Biomass thus has a very low bulk packing density (Table II) and absorbs large amounts of liquids. Biomass particles in liquids cannot be slurried at concentrations > 10%–12%, and free liquid disappears at about 18%–20% w/w. High solids processing thus provides unique challenges to the design and operation of processing equipment, which must handle wet solids rather than liquid slurries. The savings in energy consumption for heating biomass can be very large, because the concentration of biomass increases from 10% to 30% (*108*).

Two systems for hydrolysis of biomass (i.e., dilute acid hydrolysis and enzymatic hydrolysis) have low consumptions of catalyst, on the order of 2%–5% w/w of substrate. However, dilute acid hydrolysis gives low yields of fermentable sugars and enzymatic hydrolysis requires chemomechanical pretreatments, which add to the total consumption of chemicals. Enzymatic catalysts are also an order of magnitude more expensive than cheap inorganic acids, such as sulfuric acid. With the environmental policies shifting to waste minimization at the source, recycle of both chemical and biochemical catalysts, or reagents, is becoming an important part of process considerations. The recycle of enzymes can also be rewarding in an economic sense, because the enzyme consumption and cost is a significant portion of the total conversion cost. There are other obvious avenues for decreasing the enzyme cost: minimal or no purification, overproduction by genetically engineered microorganisms, and selection of enzymes with higher specific activities. However, even with improvements, enzyme production will remain a relatively complex process, consuming part of the substrate and requiring other nutrient input and energy input for mixing and, usually, aeration. Great strides have already been made in the development of cellulolytic enzymes, but a major R&D effort is still needed for the development of improved cellulolytic enzyme systems.

The key aspect of the overall conversion system for biological fuel production is the integration of plant substrates and pretreatments to provide easily hydrolyzable cellulose for subsequent enzymatic hydrolysis. High yields in enzymatic hydrolysis of cellulose have been achieved routinely for pretreated grasses and hardwoods, but little attention has been paid to the abundant softwoods that appear to be more resistant to enzymatic hydrolysis.

The development of efficient pretreatment methods needs to continue if the promise of efficient enzymatic hydrolysis of the carbohydrates in biomass is to be fulfilled. At a minimum, these methods need to yield porous cell walls or cellulose fibers with high surface area accessible to enzymes. The rates of enzymatic cellulose hydrolysis are still rather slow, requiring reaction times on the order of one to three days. Developments in plant substrate pretreatments and enzyme systems may speed up the rates, but the crystallinity of cellulose and the insolubility of cellodextrin intermediates may provide upper limits for improvements that can be achieved. The long reaction times and insolubility of the substrate can lead to significant power consumption for mixing in fermentation vessels. Anaerobic fermentations do not require the intensive mixing for oxygen transfer as aerobic systems do, so the levels of mixing for efficient hydrolysis and fermentations of biomass need to be investigated and optimized.

Many fermentations of biomass to liquid fuels also require high concentrations of product for efficient product recovery. Due to the fixed stoichiometry of substrate to product conversion, correspondingly high substrate concentrations have to be achieved in saccharification-fermentation reactors, which points to the need for the development of new reactors for solid-state fermentations. High concentrations of substrate can also be achieved by fed-batch operation.

All biological processes targeting the conversion of carbohydrates in biomass must address the integration of carbohydrate processing with the conversion and utilization of lignin and other phenolic or nonphenolic components of biomass which are not readily amenable to biological conversion. These components, primarily lignin, contain a significant portion of the total energy content of biomass feedstocks. One obvious route for lignin is its recovery, dewatering, and utilization as a boiler fuel to provide energy for the rest of the conversion process. Such utilization will provide environmental benefits by minimizing emissions of carbon dioxide from fossil fuels input, but this fate for lignin provides only marginal economic benefits (i.e., 1¢–2¢/lb of lignin used as boiler fuel). Valorization of lignin and other organic components of biomass to liquid fuels, polymeric materials, and higher value organic chemicals can have significant economic benefits for the overall production system. Pyrolytic cracking, hydrogenolytic deoxygenation, and dealkylation can be coupled with catalytic methylation to form anisole and other methylarylethers (112), which are excellent antiknocking agents for gasoline. Other large volume products which can be produced from lignin are asphalt and fuel oil replacements. Extensive R&D is needed for development of marketable products from the lignins of hardwoods and grasses, which are favorable feedstocks for biological fuel production. Lignin conversion research needs to be integrated with feedstocks selection and pretreatment methods, because lignins are not uniform polymers in biomass feedstocks and they are severely modified during pretreatment.

Summary

Integrated biochemical and chemical methods are being developed for the conversion of biomass components to valuable liquid and gaseous fuels. These processes can initially tap vast reservoirs of agricultural, forestry, and municipal solid wastes, with additional supplies potentially available from energy crops.

The conversion of agricultural and related wastes is especially attractive because these organic materials are often burned before replanting or the rejuvenation of fields, forests, orchards, and pastures. Mass burning of biomass is an ancient agricultural and pastoral practice which releases large amounts of carbon dioxide and contributes to the greenhouse effect. Permitting the biomass residues to rot by the action of aerobic microorganisms achieves the same carbon dioxide release, only at a slower rate. We are thus needlessly oxidizing carbon fixed by current photosynthesis, while we are extracting energy from the carbon fixed by photosynthesis of ancient plants and other organisms. Liquid and gaseous fuel production from plant residues can capture part of this modern fixed carbon in a valuable fuel form and allows it's utilization before terminal oxidation to carbon dioxide. The combined production and conversion system can also achieve a closed carbon dioxide cycle because carbon dioxide released during the conversion and combustion of biomass and fuels is reduced back to biomass substrates by photosynthesis.

Furthermore, the conversion of biomass to liquid fuels will provide a significant densification of the energy content in the original substrates and, thus, permits long distance export of fuel products from rural areas (see Table II). The development of processes for the conversion of lignocellulosic biomass also alleviates the food versus fuel controversy, because the conversion is concentrated on parts of the total plant biomass which are essentially undigestible by man and many farm animals.

Biomass production systems are also significant net producers of energy. The estimated energy output/input ratios range from approximately 5:1 for the production of corn and similar herbaceous crops, through the range of 10:1–40:1 for silviculture, to ratios of 50:1–150:1 for the harvesting of agricultural and forestry residues (109-111). Conversion processes for the production of liquid and gaseous fuels from biomass, whether thermochemical or biochemical, have to strive for the highest thermal efficiencies and product yields to decrease the dissipation of energy surplus from the production section. Biological conversion processes are eminently suitable for the conversion of moist feedstocks because they can utilize moisture in biomass as a reaction medium. They can also accommodate dry biomass equally well.

Several biological routes for the production of liquid fuels from carbohydrates in biomass have been reviewed in this chapter. Each projected process has it's own unique features and limitations, but they share the common steps of depolymerization of complex carbohydrates, fermentation, and product recovery. Lower alcohols, such as ethanol, can be used directly as a fuel, while some other products, such as butanediol or salts of organic acids, will require additional chemical conversion steps to turn them into liquid fuels.

Great strides have been made over the last two decades in the pretreatment of biomass for enzymatic hydrolysis, especially the development of enzymes for the hydrolysis of cellulose and the development of microorganisms for fermentation of xylose to ethanol. High yields of sugars and fuel products have been achieved from a variety of pretreated substrates. Further improvements are needed in the integration of biomass feedstocks, pretreatment methods, and enzymes for the hydrolysis of polysaccharides. A decrease in the consumption of enzymes and the increased rates of cellulose hydrolysis will also have important benefits for economic viability of biological fuel production.

A serious obstacle to complete biological conversion of biomass to liquid or gaseous fuels is the inability of current biological systems to depolymerize lignin into fermentable units or to convert lignin and other phenolic compounds in biomass to liquid fuel products. Therefore, biological conversions of carbohydrates must be coupled to the recovery and chemical conversion of lignin. At a minimum, lignin must be recovered in a form suitable for boiler fuel, but this application does not provide much coproduct credit. Possibilities for combined chemical and biological conversion of lignin may exist in syngas conversion and anaerobic digestion systems, but both approaches have not been thoroughly investigated to date and will require significant research investment. There are many possibilities in the chemical conversion of lignins to valuable products, such as liquid fuels, polymeric materials, and organic chemicals, some of which are discussed in other chapters of this book. Increased attention should, however, be paid to the conversion of lignins from hardwoods and grasses, which are also modified by various pretreatment methods. Furthermore, the extensive research results from R&D on softwood lignins modified by pulping processes may not always be directly applicable to lignins from other plant families, some of which are different in structure and composition. Finally, the ultimate biomass refinery should operate with the same attention to marketable products as do current petrochemical refineries. R&D investment should also be made in recovery and utilization of such minor components as uronic acids, waxes, and acetic acid, all of which will be coproduced in significant quantities.

Acknowledgments

This work was funded by the Ethanol from Biomass Program of the DOE Biofuels Systems Division. The authors wish to thank C.J. Rivard and R. Torget for technical assistance.

References

1. *Fuels to Drive Our Future*; National Academy Press: Washington, DC, 1990; 223 pp.

2. Wyman, C.E.; Hinman, N.D. *Appl. Biochem. Biotechnol.* 1990, *24/25*, 735.

3. Hall, D.O. *Solar Energy* 1979, *22*, 307.

4. Scherry, R.W. *Plants for Man*; Prentice-Hall: Englewood Cliffs, NJ, 1972; 657 pp.

5. Lynd, L.R. *Adv. Biochem. Eng./Biotechnol.* 1989, *38*, 1.

6. Miller, D.C.; Eisenhauer, R.A. In *CRC Handbook of Processing and Utilization in Agriculture*; Wolff, I.A., Ed.; Chemical Rubber Company Press: Boca Raton, FL, 1982, Vol. 2; p 691.

7. Mowat, D.N.; Kwain, M.L.; Winch, J.E. *Can. J. Plant Sci.* 1969, *49*, 499.

8. Van Soest, P.J. *Agric. & Environ.* 1981, *6*, 135.

9. Smith, H.P.; Wilkes, L.H. *Farm Machinery and Equipment*; McGraw-Hill Books: New York, NY, 1976; pp 286-343.

10. *Fertilizer Technology and Usage*; McVickar, M.H.; Bridges, G.C.; Nelson, L.B., Eds.; Soil Science Society of America: Madison, WI, 1963; 464 pp.

11. Rivard, C.J.; Himmel, M.E.; Grohmann, K. *Biotechnol. Bioeng. Symp.* **1986**, *15*, 375.

12. Sajdak, R.L.; Lai, Y.Z.; Mroz, G.D.; Jurgensen, M.F. In *Biomass as a Nonfossil Fuel Resource*; Klass, D.L., Ed.; ACS Symposium Series No. 144; American Chemical Society: Washington, DC, 1981; pp 21-48.

13. Wahlgren, H.G.; Ellis, T.H. *Tappi* **1978**, *61*, 37.

14. Young, H.E.; Guinn, V.P. *Tappi* **1966**, *49*, 190.

15. Goodman, B.J. *Overview of the Anaerobic Digestion Research Program*; SERI/SP-231-3520; Solar Energy Research Institute: Golden, CO, 1988; pp 1-7.

16. Fengel, D.; Weneger, D. *Wood: Chemistry, Ultrastructure, Reactions*; Walter de Gruyer: New York, NY, 1984; 613 pp.

17. Timell, T.E. *Adv. Carb. Chem.* **1964**, *19*, 247.

18. Torget, R.; Werdene, P.; Himmel, M.E.; Grohmann, K. *Appl. Biochem. Biotechnol.* **1990**, *24/25*, 115.

19. Grohmann, K.; Torget, R.; Himmel, M.E. *Biotechnol. Bioeng. Symp.* **1986**, *15*, 59.

20. Grohmann, K.; Mitchell, D.J.; Himmel, M.E.; Dale, B.E.; Schroeder, H.A. *Appl. Biochem. Biotechnol.* **1989**, *20/21*, 45.

21. Menezes, T.J.B. In *Liquid Fuels System*; Wise, D.L., Ed.; CRC Press: Boca Raton, FL, 1983; p 31.

22. Watson, S.A. In *CRC Handbook of Processing and Utilization in Agriculture*; Wolff, I.A., Ed.; CRC Press: Boca Raton, FL, 1982, Vol. 2; pp 16-17.

23. Timell, T.E. *Adv. Carb. Chem.* **1965**, *20*, 409.

24. Wilkie, K.C.B. *Adv. Carb. Chem. Biochem.* **1979**, *36*, 215.

25. *Lignins: Occurrence, Formation, Structure and Reactions*; Sarkanen, K.V.; Ludwig, C.M., Eds.; Wiley-Interscience: New York, NY, 1971; 916 pp.

26. Domalski, E.S.; Jobe, T.L.; Milne, T.A. *Thermodynamic Data for Biomass Conversion and Waste Incineration*; SERI/SP-271-2839, Solar Energy Research Institute: Golden, CO, 1986; 352 pp.

27. *Chemical Engineers' Handbook*; Perry, R.H.; Chilton, C.H., Eds.; McGraw-Hill Books: New York, NY, 1973; pp 9.3-9.14.

28. Rovenskii, V.T.; Lomova, G.P.; Malinovskaya, N.M.; Palii, I.S.; Lisitskaya, S.M. *Gidroliz. Lesokhim. Prom.* **1984**, 11.

29. *Handbook of Chemistry and Physics*; Hodgman, C.D.; Weast, R.C.; Shankland, R.S.; Selby, S.M., Eds.; 44th edition, The Chemical Rubber Publishing Company: Cleveland, OH, 1961; pp 766-1305 and 1926-1939.

30. Jahnig, C.E. In *Kirk-Othmer Encyclopedia of Chemical Technology*; Grayson, M.; Eckroth, D., Eds.; John Wiley & Sons: New York, NY, 1982, Vol. 17; pp 183-256.

31. *Physical and Thermodynamic Properties of Pure Chemicals*; Daubert, T.E.; Danner, R.P., Eds.; Hemisphere Publishing: New York, NY, 1989.

32. *Present and Future Automotive Fuels: Performance and Exhaust Clarification*; Hirao, O.; Pefley, R.K., Eds.; John Wiley & Sons: New York, NY, 1988; pp 115-139.

33. Datta, R. *Biotechnol. Bioeng. Symp.* **1981**, *11*, 521.

34. Lane, J.C. In *Kirk-Othmer Encyclopedia of Chemical Technology*; Grayson, M.; Eckroth, D., Eds.; John Wiley & Sons: New York, NY, 1981, Vol. 11; pp 652-695.

35. *Chemical Additives for Fuels: Developments Since 1978*; Gillies, M.T., Ed.; Noyes Data Corporation: Park Ridge, NJ, 1982; 306 pp.

36. Livo, K.B.; Gallagher, J. *Environmental Influence of Oxygenates*; Presented at the 1989 AIChE National Meeting, San Francisco, CA.

37. *Lignin Biodegradation: Microbiology, Chemistry, and Potential Applications*; Kirk, K.T.; Higuchi, T.; Chang, H., Eds.; Chemical Rubber Company Press: Boca Raton, FL, 1980; 488 pp.

38. *Primary Products of Metabolism*; Rose, A.H., Ed.; Academic Press: New York, NY, 1978; 470 pp.

39. *Biotechnology*; Rehm, H.J.; Reed, G., Eds.; VCH Publishing: Weinheim, Germany, 1988, Vol. 6b; pp 1-176.

40. Grohmann, K.; Villet, R. In *Bioconversion Systems*; Wise, D.L., Ed.; Chemical Rubber Company Press: Boca Raton, FL, 1984; pp 1-16.

41. Stanier, R.Y.; Adelberg, E.A.; Ingraham, J.L. *The Microbial World*; Prentice-Hall: Englewood Cliffs, NJ, 1976; 871 pp.

42. *Starch: Chemistry and Technology*; Whistler, R.L.; Beniller, J.N.; Paschall, E.F., Eds.; Academic Press: New York, NY, 1984; 718 pp.

43. *Handbook of Amylases and Related Enzymes*; The Amylase Research Society of Japan, Eds.; Pergamon Press: New York, NY, 1988; 274 pp.

44. Kosaric, N.; Wilcrorek, A.; Consentina, G.P.; Duonjak, Z. *Adv. Biochem. Eng./Biotechnol.* **1985**, *32*, 1.

45. Koshijima, T.; Watanabe, T.; Yaku, F. In *Lignin, Properties and Materials*; Glasser, W.G.; Sarkanen, S., Eds.; American Chemical Society, Washington, DC, 1989, Vol. 397; p 11.

46. Himmel, M.E.; Tucker, M.P.; Baker, J.O.; Rivard, C.J.; Oh, K.K.; Grohmann, K. *Biotechnol. Bioeng. Symp.* **1985**, *15*, 39.

47. Torget, R.; Himmel, M.E.; Wright, J.D.; Grohmann, K. *Appl. Biochem. Biotechnol.* **1988**, *17*, 89.

48. Wenzl, H.F. *The Chemical Technology of Wood*; Academic Press: New York, NY, 1970, pp 157-251.

49. Harris, J.F.; Baker, A.J.; Conner, A.H.; Jeffries, T.W.; Minor, J.L.; Pettersen, R.C.; Scott, R.W.; Springer, E.L.; Wegner, T.H.; Zerbe, J.I. *Two-Stage Dilute Sulfuric Acid Hydrolysis of Wood: An Investigation of Fundamentals*; General Technical Report FPL-45, United States Forest Products Laboratory - United States Department of Agriculture: Madison, WI, 1985; 73 pp.

50. Ranganathan, S.; Bakkshi, N.N.; MacDonald, D.G. In *Biomass Conversion Technology: Principles and Practice*; Moo-Young, M., Ed.; Pergamon Press: New York, NY, 1987; pp 11-17.

51. *Technology of Hydrolytic and Sulfite-Alcohol Production*; Sharkov, V.I., Ed.; Goslezbumizdat Publishing: Moscow, U.S.S.R., 1959; 438 pp.

52. Chen, H.C.; Grethlein, H.E. *Biomass* **1990**, *23*, 319-326.

53. Fedarov, A.L.; Korolkov, I.I. *Khimiya i Tekhnol. Tsellyulozy* **1978**, *5*, 40.

54. Pazzner, L.; Chang, P.C. United States Patent No. 4,409,032; October 11, 1983.

55. Ward, J.P.; Grethlein, H. *Biomass* **1988**, *17*, 153.

56. Wright, J.D.; Power, A.J.; Bergeron, P.W. *Evaluation of Concentrated Halogen Acid Processes for Alcohol Fuel Production*; SERI/TR-232-2386, Solar Energy Research Institute: Golden, CO, 1985; 74 pp.

57. Gauss, W.F.; Suzuki, S.; Tagaki, M. United States Patent No. 3,099,944; November, 1976.

58. Millett, M.A.; Baker, A.J.; Satter, L.D. *Biotechnol. Bioeng. Symp.* **1976**, *6*, 125.

59. Chang, M.M.; Chou, T.Y.C.; Tsao, G.T. *Adv. Biochem. Eng.* **1981**, *20*, 15.

60. Fan, L.T.; Lee, Y.H.; Gharpuray, M.M. *Adv. Biochem. Eng.* **1982**, *23*, 157.

61. Datta, R. *Process Biochem.* **1981**, 16.

62. Schurz, J. *Holzforschung* **1986**, *40*, 225.

63. Horton, G.L.; Rivers, D.B.; Emert, G.H. *Ind. Eng. Chem. Prod. Res. Dev.* **1980**, *19*, 422.

64. *Pulp and Paper Manufacture*; Stephenson, J.N., Ed.; McGraw-Hill Books: New York, NY, 1950, Vols. 1 and 2; 1043 pp.

65. Mitchell, D.J.; Grohmann, K.; Himmel, M.E. *J. Wood Chem. & Technol.* **1990**, *10*, 111.

66. Algeo, J.W. United States Patent No. 3,817,786; 1974.

67. DeLong, E.A. Canadian Patent No. 1,141,376; 1983.

68. Foody, P. Canadian Patent No. 1,163,058; 1984.

69. Baugh, K.; Bachmann, A.; Beard, V.L.; Levy, J.; McCarty, P.L. *Thermochemical Pretreatment of Lignocellulosic Biomass for Increasing Anaerobic Biodegradability to Methane*; SERI/STR-231-2458; Solar Energy Research Institute: Golden, CO, 1985; 164 pp.

70. Grethlein, H.E. United States Patent No. 4,237,226, 1980.

71. Grethlein, H.E. *Bio/Technology* **1985**, *3*, 155.

72. Mackie, K.L.; Brownell, H.H.; West, K.L.; Saddler, J.N. *J. Wood Chem. Technol.* **1985**, *5*, 405.

73. Clark, T.A.; Mackie, K.L.; Dare, P.H.; McDonald, A.G. *J. Wood Chem. Technol.* **1989**, *9*, 135.

74. Tatsumoto, K.; Baker, J.O.; Tucker, M.P.; Oh, K.K.; Mohagheghi, A.; Grohmann, K.; Himmel, M.E. *Appl. Biochem. Biotechnol.* **1988**, *24/25*, 159.

75. Sutcliff, R.; Saddler, J.N. *Biotechnol. Bioeng. Symp.* **1986**, *17*, 749.

76. Vazquez, G.; Antorrena, G.; Parajo, J.C.; Francisco, X.L. *Wood Sci. Technol.* **1988**, *22*, 219.

77. Wood, T.M.; McCrae, S.I. In *Hydrolysis of Cellulose: Mechanisms of Enzymatic and Acid Catalysis*; Brown, R.D.; Jurasek, L., Eds.; American Chemical Society: Washington, DC, 1979, Vol. 181; pp 181-210.

78. Eveleigh, D.E. *Phil. Trans. R. Soc. Lond.* **1987**, *A321*, 435.

79. Chum, H.L.; Johnson, D.K.; Black, S.; Baker, J.O.; Grohmann, K.; Sarkanen, K.V.; Wallace, K.; Schroeder, H.A. *Biotechnol. Bioeng.* **1988**, *31*, 643.

80. Stewart, G.G.; Russell, I. In *Yeast Biotechnology*; Berry, D.R.; Russell, I.; Stewart, G.G., Eds.; Allen & Unwin: London, UK, 1987; pp 277-310.

81. Montenecourt, B.S. In *Biology of Industrial Microorganisms*; Demain, A.L.; Solomon, N.A., Eds.; The Benjamin/Cummings Company: Menlo Park, CA, 1985; pp 261-290.

82. *Organic Chemicals from Biomass*; Wise, D.L., Ed.; The Benjamin/Cummings Company: Menlo Park, CA, 1983; 465 pp.

83. Magee, R.J.; Kosaric, N. *Adv. Biochem. Eng./Biotechnol.* **1985**, *32*, 61.

84. Chang, S.F.; Ho, N.W.Y. *Appl. Biochem. Biotechnol.* **1988**, *16*, 316.

85. Lastick, S.M.; Tucker, M.P.; Beyette, J.R.; Noll, G.R.; Grohmann, K. *Appl. Microbiol. Biotechnol.* **1989**, *30*, 574.

86. Lastick, S.M.; Mohagheghi, A.; Tucker, M.P.; Grohmann, K. *Appl. Biochem. Biotechnol.* **1990**, *24/25*, 431.

87. Tolan, J.S.; Finn, R.K. *Appl. Environ. Microbiol.* **1987**, *53*, 2033.

88. Tolan, J.S.; Finn, R.K. *Appl. Environ. Microbiol.* **1987**, *53*, 2039.

89. Ingram, L.O.; Conway, T.; Clark, D.P.; Sewell, G.W.; Preston, J.F. *Appl. Environ. Microbiol.* **1987**, *53*, 2420.

90. Mohagheghi, A.; Tucker, M.P.; Grohmann, K.; Wyman, C.E. Presented at the Annual Meeting of the American Chemical Society, Miami Beach, FL, September 1989.

91. *Bacterial Metabolism*; Doelle, H.W., Ed.; Academic Press: New York, NY, 1975.

92. Levy, F.P.; Sanderson, J.E.; Ashare, E.; de Riel, S.R. In *Liquid Fuels Development*; Wise, D.L., Ed.; Chemical Rubber Company Press: Boca Raton, FL, 1983; p 159.

93. Neenan, B.; Feinberg, D.; Hill, A.; McIntosh, R.; Terry, K. *Fuels from Microalgae: Technology Status, Potential, and Research Requirements*; SERI/SP-231-2550, Solar Energy Research Institute: Golden, CO, 1986; 149 pp.

94. Wan, E.I.; Fraser, M.D. In *Proceedings of the 1989 IGT Conference*; Klass, D.L., Ed.; Institute of Gas Technology: Chicago, IL, 1989.

95. Barik, D.; Corder, R.E.; Clausen, E.C.; Gaddy, J.L. *Energy Progress* **1987**, *7*, 157.

96. Clements, L.D.; Beck, S.R.; Heintz, C. *Chem. Eng. Progress* **1983**, *79*, 59.

97. Hahn, A.V.G.; Williams, R.; Zabel, H.W. *The Petrochemical Industry, Market, and Economics*; McGraw-Hill Books: New York, NY, 1970; pp 19-185.

98. Vega, J.L.; Clausen, E.C.; Gaddy, J.L. *Resources, Conservation and Recycling*; **1990**, *3*, 149.

99. Srivastava, R.D.; Cambpell, I.M.; Blaustein, B.D. *Chem. Eng. Progress* **1989**, *85*, 45.

100. Gold, D.S.; Goldberg, I.; Cooney, C.L. *Prepr. Amer. Chem. Soc., Div. Petr. Chem.* **1980**, *25*, 575.

101. Klasson, K.T.; Elmore, B.B.; Vega, J.L.; Ackerson, M.D.; Clausen, E.C.; Gaddy, J.L. *Appl. Biochem. Biotechnol.* **1990**, *24/25*, 857.

102. *Microbial Degradation of Organic Compounds*; Gibson, D.T., Ed.; Marcel Dekker: New York, NY, 1984.

103. Hungate, R.E. *The Rumen and It's Microbes*; Academic Press: New York, NY, 1966.

104. *The Rumen Microbial Ecosystem*; Hobson, P.N., Ed.; Elsevier Applied Science: New York, NY, 1988.

105. *Fuel Gas Production from Biomass*; Wise, D.L., Ed.; Chemical Rubber Company Press: Boca Raton, FL, 1981, Vol. I.

106. *Anaerobic Digestion and Carbohydrate Hydrolysis of Waste*; Ferrero, G.L.; Ferranti, M.P.; Naveau, H., Eds.; Elsevier Applied Science: New York, NY, 1984.

107. Weaver, P.F.; Lien, S.; Seibert, M. *Solar Energy* **1980**, *24*, 3-45.

108. Grohmann, K.; Torget, R.; Himmel, M.E. *Biotechnol. Bioeng. Symp.* **1986**, *17*, 135.

109. Smith, D.L.O.; Rutherford, I; Radley, R.W. *Agr. Engineer.* **1975**, *30*, 70.

110. Ledig, T.F. In *Biomass as a Nonfossil Fuel Source*, Klass, D.L., Ed.; ACS Symposium Series No. 144, American Chemical Society: Washington, D.C., 1981; pp 447-461.

111. Bagnaresi, U.; Baldini, S.; Berti, S.; Minotta, G. In *Biomass Energy: From Harvesting to Storage*, Ferrero, G.L.; Grassi, G.; Williams, H.E., Eds.; Elsevier Applied Science: New York, NY, 1987; pp 75-79.

112. Chum, H.L.; Johnson, D.K.; Black, S.; Ratcliff, M.; Goheen, B.W. *Adv. Solar Energy* **1987**, *4*, 91.

RECEIVED April 8, 1991

Chapter 22

Chemicals from Pulping
Product Generation from Pulping Residuals

David E. Knox and Philip L. Robinson

Westvaco Corporation, 5600 Virginia Avenue, North Charleston, SC 29411–2905

> Recent advances in producing improved materials from pulping byproducts are reviewed. Terpenes and their derivatives are examined in flavors and fragrance and adhesive applications. Rosin derivatives of similar quality to products from non-pulping sources are now available. Derivatives of fatty acids are useful in high-solids coatings, corrosion inhibitors, and soaps and detergents. The synthesis of these derivatives from pulping byproducts is discussed. Also, benefits provided to finished goods by the unique characteristics of these derivatives are examined.

For many centuries the softwood tree has been recognized as a renewable resource that provides a readily available source of chemicals. The constituents of softwoods which are the largest source of industrial chemicals are terpenes, rosins, fatty acids, and some tannins. Uses for these materials range from food applications to oilfield corrosion inhibitors. This wide application spectrum has enabled the industry to maintain a consistently broad market for these chemicals.

This chapter considers some recent developments in the uses of turpentine, rosin, fatty acid and tannins. The emphasis is on areas where these materials and their derivatives have shown recent benefits. The chemistry that differentiates these materials is described. A full review of Naval Stores has been published recently (1).

TERPENES

Flavors and Fragrances

Terpenes have classically been employed in the production of pine oil for the soap and detergent industry. This remains the chief use of pinenes. Pinenes are C-10 structures consisting of α-pinene, β-pinene, and dipentene; each of these structures is ultimately derived from the C-5 isoprene unit. The reaction of α-pinene with aqueous acid is known to give α-terpineol, which under appropriate conditions, can add a second water molecule to give terpin hydrate (Figure 1). Pine oil may be considered a textbook example of how terpenoid compounds are employed in fragrance chemicals (1).

Crude sulfate turpentine, a sulfur-rich byproduct of the pulping operation, contains the terpene mixture. Use of this material in flavor and fragrance applications requires desulfurization. Desulfurization of turpentine feed sources may be performed by the use of peroxide or hypochlorite, although recent work has described the use of catalytic cobalt and molybdenum oxides for sulfur removal (2,3). This step is necessary to insure appropriate aroma characteristics for flavor and fragrance production. Subsequent distillation provides the major isomeric terpenes which include α-pinene, β-pinene, and dipentene. Classically, β-pinene was pyrolyzed to myrcene which is a precursor to a number of important flavor and fragrance chemicals including linalool, geraniol, and nerol (4). An alternative route to these products has been developed from a mixture of hydrogenated α- and β-pinenes. This method overcomes industry reliance on β-pinene, which constitutes only about one-quarter of the cyclic terpenes in crude turpentine. Hydrogenation, oxidation, reduction of the hydroperoxy group, and thermolysis yields linalool, which under acidic rearrangement, gives nerol and geraniol. These latter alcohols may subsequently be converted to desired flavor and fragrance compounds (Figure 2).

Recent literature has described the use of terpenoids as hygiene chemicals. An application described by Colgate-Palmolive employed α-ionone in non-cariogenic mouthwashes (5). Ionones are well-known derivatives of terpenoid compounds. The use of ionones for decreasing mouth odor had previously been described with cariogenic sugar solutions (6,7). It was found that the use of α-ionone in conjunction with zinc salts and sodium gluconate gave systems that effectively reduced tartar. The systems were also able to eliminate orally generated volatile sulfur compounds from food decay for extended periods of time. The ionone, by itself, does not cause a large decrease in volatile sulfur compounds over time. Presumably, a suitable complex forms with the zinc salt to provide oral time release

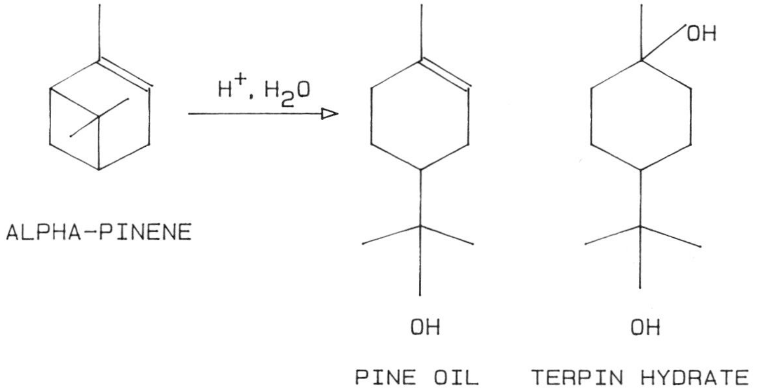

Figure 1. Hydrolysis of α-Pinene to Give Pine Oil and Terpin Hydrate

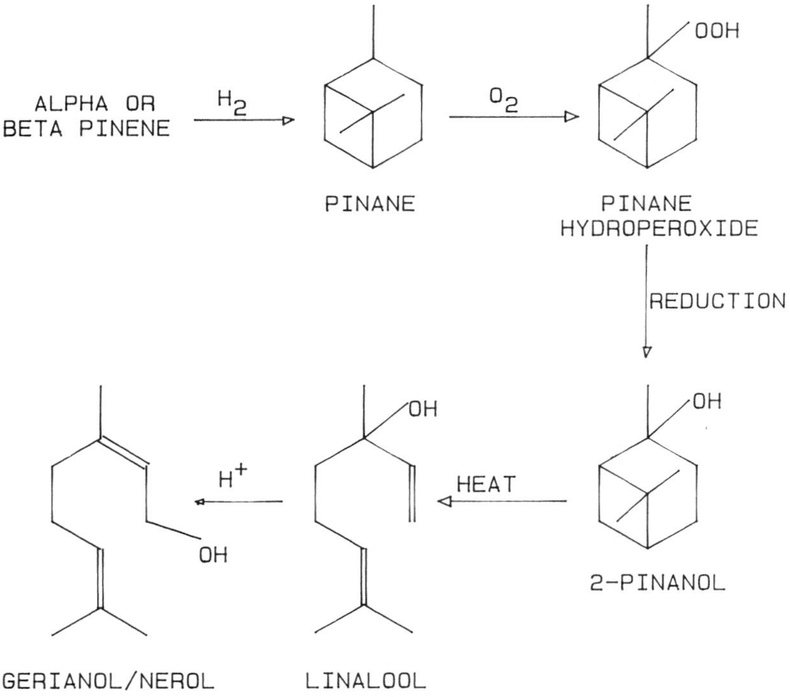

Figure 2. Pinane-Based Flavor and Fragrance Chemicals

properties for the ionone. This type of complex formation of zinc salts for oral hygiene has been described for diketones (8); non-chelating species such as α-ionone should have similar characteristics with different zinc release properties due to the single ketone group.

An interesting use of terpenic compounds has been described based on the reaction of dipentene (d,l-limonene from tall oil sources) with sulfur and thionyl chloride (9) (Figure 3). Thio-terpenes are known to occur in trace quantities in such compounds as rose oil. In this reaction, monosulfurized and polysulfurized compounds are generated that have unique and desirable aroma enhancing characteristics. The chief monosulfurized components were trans-(-)-2,8-thiomenthylene, 6,8-thiomenthene-1-ylene, and the achiral epithiomenthane. When used in trace quantities, these compounds enhance the aroma of a variety of personal use items including hair sprays, toiletry articles, and perfumes.

Terpenoids in Polymer Applications

Although the use of terpenoid compounds in flavor and fragrance applications offers exciting potential for different types of chemistry, a much larger use for terpenoid compounds is in hot-melt and pressure-sensitive adhesives. This area has been fairly well developed (1). In these applications, α- or β-pinene and a variety of acyclic terpenoid compounds are cationically polymerized to give tackifiers for a variety of random and block copolymers. Cationic polymerization of α-pinene is shown in Figure 4.

The main adhesive copolymers employed in hot-melt adhesive formulations include ethylene-vinyl acetate copolymers (EVA's) and styrene-butadiene-styrene (SBS) or styrene-isoprene-styrene (SIS) based materials. In the case of block copolymers, the polyterpenes are presumably compatible with the aliphatic (butadiene or isoprene) section of the polymer. In the case of EVA's, the structures cited in the literature appear to be mostly random copolymers indicating that the tackifier is compatible with the resin as a whole. Generally, when used as tackifiers, polyterpenes are simply blended with the polymer and formed into a rod which may be extruded from a suitable apparatus such as an adhesive gun. These resins generally have very good "open time", the amount of time available for an adhesive prior to bonding to another substrate. In addition, good "set times" have been observed. This is the time it takes for the system to adhere after two substrates have been placed in contact. More recently, the use of hydrogenated resins has been described (10). These hydrogenated resins have desired color and oxidative stability for use in more demanding applications.

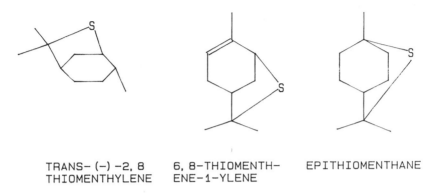

S
S
S
TRANS- (-) -2, 8
THIOMENTHYLENE
6, 8-THIOMENTH-
ENE-1-YLENE
EPITHIOMENTHANE

Figure 3. Sulfur-Containing Aroma Enhancers From Limonene

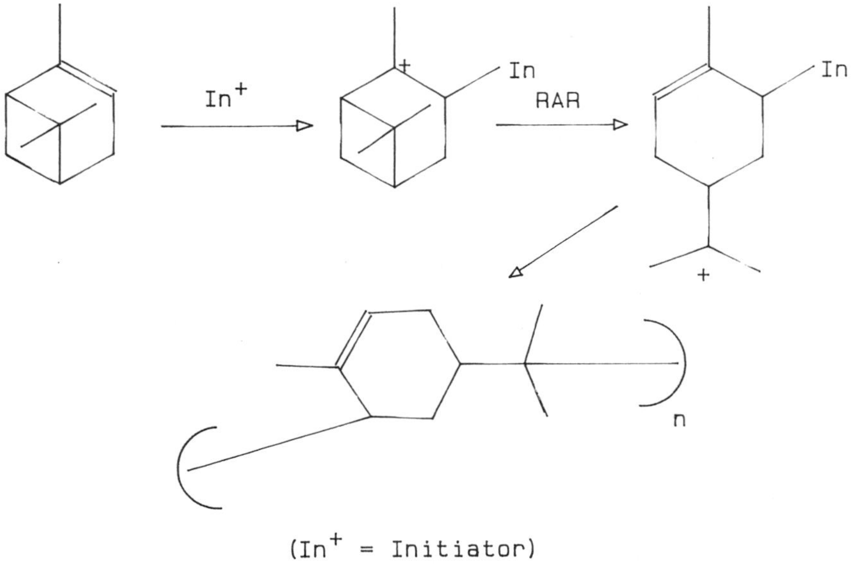

In+
+
In
RAR
In
+
n
(In+ = Initiator)

Figure 4. Cationic Polymerization of α-Pinene

The use of pinenes in water-borne adhesives has been described recently (11). The copolymerization of α-pinene with dicyclopentadiene and styrene, followed by maleinization and subsequent neutralization, yields resins that have excellent adhesive qualities. These polymers are also described with tall oil rosin as a water-borne composition. In either of the above cases, the adhesive has a fairly low softening point of about 65°C to 75°C; this would probably need to be raised for many commercial applications. However, the fact that pinenes (and tall oil rosins) may be readily employed in water-borne formulations indicates that these materials can play an important role as technology shifts due to environmental legislation from solvent-based systems to water-borne.

In conclusion, terpenes are viable commercial products in flavor and fragrance and tackifier resin applications. The fact that current use of these materials exceeds domestic production and turpentine importation is rising indicates the industrial importance of these products.

ROSIN AND ITS DERIVATIVES

Rosin Esters in Adhesive and Ink Applications

For many years, gum rosin was obtained from the wounding of trees and wood rosin from the extraction of stumps. Procurement of rosin from both techniques is labor intensive and the wounding process for gum rosin results in a premature loss of trees. Wood rosin has become scarcer with the harvesting of younger trees that have less resinous material in their stumps. Although the use of the paraquat (Hercules PINEX® process) held promise for alleviating this situation, the use of this chemical has been largely abandoned due to unfavorable economics. Although gum is still the primary rosin source, both gum and wood rosins have become increasingly expensive. As a result, there has been an push to use tall oil rosin in many applications that have previously been reserved for the wood and gum grades. Research has focused on two chief areas: the decolorization and the deodorization of tall oil rosin so that it is, in effect, equivalent to gum or wood rosin. The use of different decolorization aids during the synthesis of rosin esters and the use of various antioxidants for maintaining color stability during processing have been developed (12-15). Deodorization may be accomplished by a number of means including masking of odors and chemical treatment. These processes are designed to give tall oil rosin the characteristics of the other rosin grades.

The main rosin esters that are employed in the industry consist of either glycerol or pentaerythritol esters. Generally, the degree of esterification with pentaerythritol is between two and four. These esters are compatible with the "soft" blocks of SBS and SIS copolymers and have suitable softening points for use in hot melt and pressure sensitive applications.

It is very difficult to esterify rosin due to the quaternary carboxylic acid group; therefore, very high temperatures must be employed. These temperatures, which are frequently in the range of 240°C to 260°C, lead to darker products. Decolorization during the synthesis of rosin esters is accomplished by using either phosphinic (12) or phosphorous (13) acid; these materials also act as catalysts for the esterification reaction. Color improvements for the different systems are noted in Table I. The use of the phosphorous-containing acids resulted in color reductions on the order of four Gardner color units.

TABLE I

Color Improvement of Pentaerythritol Containing Rosin Esters with Phosphorous Containing Acids (12, 13)

Sample	Initial Rosin Color	Final Ester Color
Control	4	8
H_3PO_2	4	4
H_3PO_3	4	4

An alternative method for color control involves the use of disulfide resorcinols in conjunction with phosphinic acid during esterification (14); no catalytic activity or color reduction was noted with the use of alternative acids such as para-toluene sulfonic acid. Hydrogenated rosin and its derivatives are still employed in applications where essentially water white materials are required (16).

Many of the rosin esters that have been mentioned for use in adhesives also find use in ink applications. These esters and other derivatized products, have found use in all four ink resin types: letterpress, lithography, flexography and gravure. An extensive review of these areas is given in reference 1.

FATTY ACID DERIVATIVES AND THEIR USES

Fatty acid derivatives are noted extensively in the literature and have found many applications including soaps and detergents (17), coatings (18), and corrosion inhibition (19). The extensive uses of these materials in commerce have shown their versatility in meeting specifications for many product applications.

The structures of the two fatty acid derivatives to be discussed are shown in Figure 5. Synthesis of these adducts involves either a Diels-Alder adduction of acrylic acid with linoleic acid or an ene adduction of maleic anhydride with oleic acid (20).

Applications for these products have been in the above mentioned areas of detergents, corrosion inhibitors, and coatings applications. However, the properties of the materials obtained are frequently different than those obtained from other fatty acid derivatives such as C-36 dimer acid. Each of these areas are addressed in turn.

For the uses that are addressed, different purity materials are frequently required. The purity needs of the soap and detergent and the corrosion inhibitor industries are considerably different from the higher purity requirements (95%) demanded of coatings.

Acrylated Fatty Acid in Soap and Detergent Applications

The C-21 dicarboxylic acid generated from the cycloaddition of linoleic and acrylic acids has been found to have several unique properties in soap and detergent applications (21). Some of the unique properties of the systems include improved hydrotroping in detergents, formulations at higher solids, and lower pH formulations for cleansers.

Hydrotroping in detergents is defined as "the ability of a material to make a slightly soluble product more water soluble" (22). Formulations using an acrylated linoleic system have excellent compatibility with both anionic and nonionic fatty surfactants. This gives increased detergency with a lower loading of surfactant and hydrotrope. For instance, the hydrotroping capabilities of acrylated linoleic acid are demonstrated in its ability to produce clear pine oil formulations (Figure 6). The use of an acrylated linoleic acid requires one-quarter to one-half the amount of product than competitive hydrotropes. Improved performance on a weight (and cost) basis coupled with biodegradability (by BOD and TOC), indicates that this fatty acid derivative would be a preferred product for hydrotropes in detergent formulations.

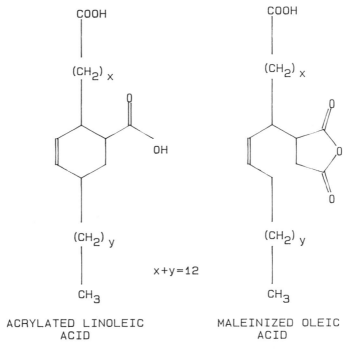

Figure 5. Acrylated Linoleic and Maleinized Oleic Acid

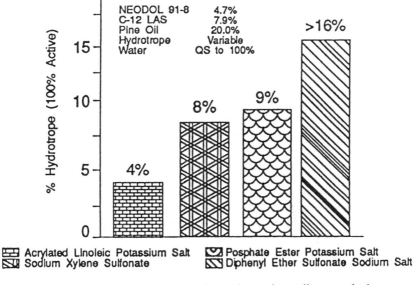

Figure 6. Hydrotrope Needed to Clear Pine Oil Formulation

A second area where C-21 acrylated linoleic acid has shown promise is in preparing high-solids soap solutions. Comparative data in Figure 7 show viscosity versus solids curves for potassium oleate and C-21 acrylated linoleic acid soaps, and a blend of the two (23).

In this figure, the ability of C-21 acrylated linoleic acid to maintain low viscosity at increased solids of a soap solution is demonstrated. Potassium oleate viscosity rises rapidly at about 20% solids; for a combination of 90% oleic acid and 10% C-21 acrylated linoleic acid, an analogous viscosity rise occurs at about 33% solids level. In these cases, the vertical rise in viscosity indicates a maximum effective concentration at which these soaps could be used. With pure C-21 acrylated linoleic acid, the major increase in viscosities does not occur until about 65% solids. Formulations would still be possible at much higher solids levels. A combination of effects, including interference of liquid crystal formation (24-29) with the straight chain C-18 soaps, and the large number of diastereomeric pairs available in a C-21 acrylated linoleic acid, are postulated as the reason for the system's fluidity to a much higher solids level. This ability to formulate to a much higher solids is significant since it is possible to make high-solids detergents which require less packaging and lower transportation and manufacturing costs.

Another practical advantage of C-21 acrylated linoleic acid is the ability to make clear hard-surface cleaning formulations in soaps and detergents. Generally, hard-surface cleaners are formulated at a pH of approximately 9. With normal straight-chain fatty acid soap cleaners, this pH is necessary to obtain solubility and to overcome liquid crystal formation that occurs closer to neutrality. With C-21 acrylated linoleic acid, the cleaning formulation may be mixed at a pH of about 7, which is less aggressive to sensitive surfaces such as wood. Again, C-21 acrylated linoleic acid can be used at lower pH's since liquid crystal formation is not occurring.

Acrylated linoleic acid has been demonstrated to give superior performance in the area of hydrotroping, high-solids formulations, and low pH cleansers. The dual functionality and amorphous nature of this system make it superior to standard fatty acid soaps as well as a number of synthetic hydrotropes that are either petroleum based or contain phosphorous.

Maleinized Fatty Acid in Corrosion Inhibitor Applications

A second area of fatty acid utilization, where new findings are significant, is the use of maleinized fatty acid (MFA) in corrosion inhibitor applications. Maleinized fatty acid has received attention as a potential aqueous corrosion inhibitor in a number of patent publications (30,31) and, more recently, as a product for use in downwell oilfield applications (32).

In the presence of a fatty imidazoline, maleinized fatty acid forms more tenacious films than the industrial standard 80% C-36 dimer/20% C-54 trimer acid mixes. These dimer/trimer mixes have been the workhorse products of the oilfield corrosion inhibitor industry for the last thirty years. Comparative test data is given in Figures 8 and 9.

Minimum performance requirements are considered to be 90% protection under testing conditions. In both sweet (carbon dioxide) and sour (hydrogen sulfide) environments, maleinized fatty acids perform about seven to ten times better than commercial dimer/trimer.

In sweet environments, only 200 ppm of the inhibitor is required versus a standard of about 2000 ppm for dimer/trimer. In sour environments, about 700 ppm of inhibitor is required versus a standard of nearly 5000 ppm for dimer/trimer. Although the cost of the maleinized fatty acid derivative is about two to three times that of commercial dimer/trimer products, it is still considerably more cost effective because of superior performance.

Of particular interest are the data for sour wells. Domestic production requirements have necessitated the drilling of deeper wells to tap oil and gas reserves. These wells tend to have tremendous temperatures, pressures, and high levels of hydrogen sulfide. The use of maleinized fatty acid based corrosion inhibitor formulations in these wells has been found to be particularly beneficial since greater protection is realized.

The mechanism of corrosion inhibition by maleinized fatty acid is not known. Infrared analyses of films on metal coupons subjected to corrosive environments show that imides are present in the film. Imides are formed when the maleinized fatty acid is combined with a fatty imidazoline derivative to give the final corrosion inhibitor package. The presence of this group may influence the film-forming capability and the bonding of the molecule to a metal surface.

Acrylated and Maleinized Fatty Acids in Coatings

Recent work has demonstrated advantages for using either acrylated or maleinized fatty acids in coating applications (33,34). These materials have been found to give viscosity reductions in solvent- and water-borne alkyds when substituted for either phthalic anhydride or trimellitic anhydride.

Using high-purity C-21 acrylated linoleic acid in alkyds based on phthalic anhydride up to substitution levels of about 50% gives viscosity reductions enabling higher solids paints to be formulated. High-solids coatings are

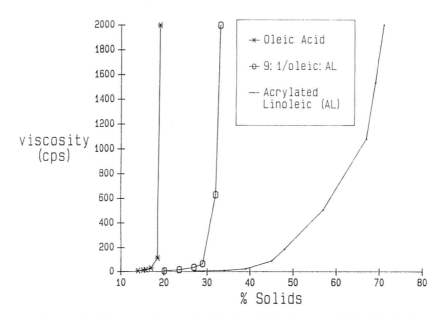

Figure 7. Viscosity Versus Solids for C-21 Acrylated Linoleic Acid and Oleic Acid Potassium Salts

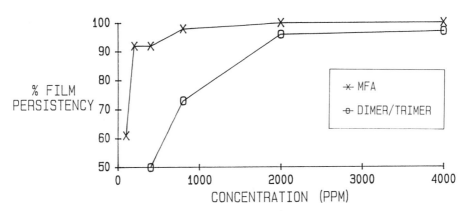

Figure 8. Corrosion Inhibition of Maleinized Fatty Acid Derivatives vs. Dimer/Trimer Mixes - Sweet Wells

becoming increasingly important as clean air regulations require the use of less solvent in paint formulations. In Figure 10, viscosity curves for a standard phthalic anhydride, C-21 acrylated linoleic acid, and C-36 dimer acid alkyds are given. Results clearly indicate that the C-21 acrylated linoleic acid gives the desired low viscosity compared to either phthalic anhydride or dimer acid and would be preferred in low-viscosity formulations. The reduced viscosity can presumably be attributed to the diastereomeric nature of the product. Dimer acid would theoretically have the same number of diastereomers as C-21 acrylated linoleic acid, but dimer acid can presumably undergo chain extension more readily than C-21 acrylated linoleic acid and gives systems of intermediate viscosity. This performance readily demonstrates the advantage of C-21 acrylated linoleic acid in alkyd resins.

Maleinized oleic acid has been evaluated in water-borne coatings as a means of reducing organic cosolvent. Generally, water-borne coating systems are formulated with organic cosolvents to maintain solubility. Maleinized oleic acid is capable of reducing the amount of organic cosolvent required for solubilization. Viscosity reductions for typical alkyd paints are shown in Figure 11. Maleinized oleic acid-containing resins show a three to fourfold decrease in viscosity compared with a trimellitic anhydride-based system. Paint formulations containing maleinized oleic acid can be formulated into water from a solids level of about 80% versus a normal of 75% with trimellitic anhydride. Although this five percent difference may seem small, it is significant for manufacturers attempting to comply with volatile organic emissions from alkyd paints. Presumably, the aliphatic nature of the maleinized oleic acid makes less coupling solvent necessary.

TANNINS AND THEIR DERIVATIVES

There has recently been considerable interest in the area of tannins and their derivatives. A recent book edited by Hemingway and Karchesy gives an extensive overview of work in the tannin area (35). We will briefly mention some of the latest developments in tannin utilization. Tannins are very broadly classified as hydrolyzable or condensed, depending on the source and structure. Hydrolyzable tannins are generally based on gallic acid, and the non-hydrolyzable systems are based on flavonoids. The gallic acid is generally esterified with a sugar moiety, whereas the flavonoids are generally condensed in a number of positions on the skeleton. A more detailed description of these condensates and their biosynthetic pathway has been given (36).

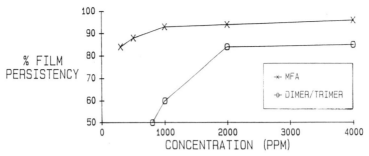

Figure 9. Corrosion Inhibition of Maleinized Fatty Acid Derivatives vs. Dimer/Timer Mixes—Sour Wells

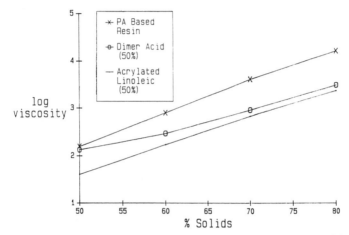

Figure 10. Log Viscosity vs. Solids for Alkyd Resins in Mineral Spirits

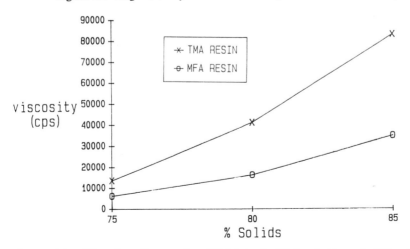

Figure 11. Viscosity Curves for Maleinized Oleic Acid and Trimellitate Formulations

Condensed tannins have received considerable interest in conjunction with the defense function in trees (37,38). In light of this function, there has been considerable interest in the use of condensed tannins in fungicide applications, although no commercial applications appear to have been made (39). An interesting proposed use of proanthocyanadin has been described in its use as an anti-radical agent for preventing tumor formation (40). This compound has shown evidence as a radical scavenger that presumably could serve this function in humans and prevent diseases that are believed to be caused from free radical generation. This use is only speculative. In fact, some studies have suggested that oral ingestion of condensed tannins found in vegetable sources and teas may lead to an increased risk of esophageal cancer (41). In either case, further studies on these compounds are necessary since they are obviously present in vegetable sources and in high fiber compounds.

A final note on tannins involves their use in adhesive applications. The wattle tannin industry has developed in South Africa and the South Pacific. These tannins have found use in particleboard adhesives as a substitute for phenolics (42-44). A similar industry in North America does not appear to be developing with great impetus, although the prices of phenolic compounds may drive the development of such an industry at a future date. Tannin molecules give phenolic resins special advantages such as faster cure cycles (35) (an advantage in the bonding of hardwoods); advances such as this may accelerate procurement and development of these compounds. Over the course of this century, uses for tannins, other than leather tanning, have been slow to develop on a concerted basis compared with other tree extracts such as terpenes, fatty acids, rosins, and lignins. As more emphasis is put on development of technology from biomass, this situation will undoubtedly change with a renewed emphasis on tannin chemistry.

CONCLUSIONS

The use of tall oil terpenes, rosins, fatty acids, and tannins continues to play an important role in the development of many commercial products. Recent advances have enabled the use of tall oil-based products where only wood or gum derived products would have originally been acceptable. Tall oil derivatives have been shown to be especially good in environmentally-friendly systems such as water-borne adhesives and coatings, high solids soaps, and high-solids alkyd formulations. Continued development of new uses for terpenes, rosins, fatty acid, and tannin derivatives is expected as industries become acquainted with the ability of these materials to give excellent properties at a reasonable cost.

ACKNOWLEDGEMENTS

We wish to thank Dr. John Alford for helping compile some of the information in this paper, Dr. Carl Bailey for his reviews and helpful comments, and Drs. Ben Ward and John Glomb for their editing and inputs. Mr. Gene Fischer has offered many helpful suggestions in clarifying certain points for oilfield area and thanks are also extended to him. Thanks are also due to Judy Smith for her efforts in the preparation of this chapter.

REFERENCES

1. Naval Stores; Zinkel, D. F.; Russell, J., Eds.; Pulp Chemicals Association: New York, N. Y., 1989
2. Casbas, F.; Duprez, D.; Ollivier, J. Appl. Catal., 1989, 50, 1, 87-97
3. Cabas, F.; Duprez, D.; Ollivier, J.; Rolley, R. European Patent Application 267,833; 1987
4. Zinkel, D. F. In Organic Chemicals from Biomass; I.S. Goldstein, Ed.; CRC Press: Boca Raton, Fla, 1980
5. Barth, J. U. S. Patent 4,814,163; 1989
6. McNamara, T. F.; Rubin, H. French Patent 2,127,005; 1971
7. McNamara, T. F.; Rubin, H. Canadian Patent 987,597; 1971
8. Eilberg, R. G.; Mazzinobile, S. German Patent 2,229,466; 1972
9. French Patent (to P. Robertet and Co.) 2,338,037; 1976
10. Vanhaeren, G. European Patent Application 271,254; 1987
11. Komitsky, Jr. F.; Thompson, K. L.; Evans, J. M.; Ocampo, E.; Patel V. European Patent Applicaton 300,624; 1988
12. Duncan, D. P.; Cameron, T. B. U. S. Patent 4,548,746; 1985
13. Johnson, Jr. R. W. European Patent Application 257,622; 1987
14. Lampo, C. S.; Turner, W. T. U. S. Patent 4,650,607; 1987
15. Knoblock, G.; Martin, H.; Patel, A.; Graziosi, P. R. TAPPI. J., 1989, 72, 71-78
16. Hicks, J. P.; European Patent Application 342,808; 1989
17. Schwartz, A. M.; Perry, J. W.; Berch, J. Surface Active Agents and Detergents, Interscience: New York, N. Y., 1949, Vol. 1 and 2
18. Fulmer, R. W. In Fatty Acids and Their Industrial Applications; E. S. Pattison, Ed.; Marcel Dekker: New York, N.Y., 1968, 187-208.
19. McSweeney, E. E.; Arlt, Jr. H. G.; Russell, J.; Tall Oil and Its Uses, Pulp Chemicals Assoc. Inc.: New York, N. Y., 1987, 92-93.
20. Danzig, M. J.; O'Donnell, J. L.; Bell, E. W.; Cowan, J. C. J. Am. Oil Chem. Soc., 1957, 34, 136-138
21. Ward, B. F., Jr.; Force, C. G.; Bills, A. M.; Woodward, F. E. J. Am. Oil Chem. Soc., 1975, 52, 219

22. Hawley's Condensed Chemical Dictionary; Sax, N. I. and Lewis, R., Eds.; Van-Nostrand: New York, N. Y., 1987, 620.
23. Robinson, P. L. J. Am. Oil Chem. Soc., In preparation
24. Cox, J. M. and Friberg, S. E. J. Am. Oil Chem. Soc., 1981, 58, 743
25. Flaim, T. and Friber, S. E. J. Colloid Interface Science, 1984, 97, 26
26. Friber, S. E.; Flaim, T. D. In Structure/Performance Relationships in Surfactants; M. J. Rosen, Ed.; Symposium Series, No. 253; ACS: Washington D. C., 1984, 107
27. Friberg, S. E.; Rananavare, S. B.; Osborne, D. W. J. Colloid Interface Science, 1986, 109, 487
28. Friberg, S. J. Am. Oil Chem. Soc., 1971, 48, 578-581.
29. Bell, A.; Birdi, K. S. In Structure/Performance Relationships in Surfactants; Rosen, M. J., Ed.; ACS: Washington DC, 1984, 117-128
30. Kindscher, W.; Oppenlaender, K. German Patent 2,357,951; 1975
31. Desai, N. B. German Patent 2,651,438; 1978
32. Knox, D. E.; Fischer, E. R. U. S. Patent 4,927,669; 1990
33. Cosgrove, J. P. In Proceedings 17th Water-Borne and Higher Solids Coatings Symposium, 1990, 488-498
34. Knox, D. E. In Proceedings 17th Water-Borne and Higher Solids Coatings Symposium, 1990, 262-278
35. Chemistry and Significance of Condensed Tannins; Hemingway, R. W., Karchesy, J. J., Eds.; Plenum Press: New York, N. Y., 1989
36. Lewis, N. G.; Yamamoto, E. In Chemistry and Significance of Condensed Tannins; Hemingway, R. W.; Karchesy, J. J., Eds.; Plenum Press: New York, N. Y., 1989
37. Scalbert, A.; Haslam, E. Phytochemistry, 1987 26, 3191-3195
38. Stafford, H. A. Phytochemistry, 1988 27, 1-6
39. Rao, S. S.; Rao, K. V. Indian J. Plant Physiol., 1986, 29, 278-280
40. Masquelier, J. U. S. Patent 4,698,360; 1987
41. Morton, J. F. In Chemistry and Significance of Condensed Tannins; Hemingway, R. W.; Karchesy, J. J., Eds.; Plenum Press: New York, N. Y., 1989
42. Pizzi, A., Orovan, E., Cameron, F. A. Holz Roh-Werkst., 1988, 46, 67-71
43. Collins, J. C., Yazaki, Y. Australian Patent 569,439; 1987
44. Collins, J. C., Yazaki, Y. Australian Patent Application 81663/87; 1987
45. Collins, P. J. Australian Patent Application 50817/85; 1985

RECEIVED April 12, 1991

Chapter 23

Feedstock Availability of Biomass and Wastes

J. W. Barrier and M. M. Bulls

National Fertilizer and Environmental Research Center, Biotechnical Research Department, Tennessee Valley Authority, Muscle Shoals, AL 35660

Biomass is defined as all organic matter including plant material whether grown on land or in water; animal products and manure; food processing and forestry by-products; and urban waste (*1*). Biomass represents a renewable and abundant source of energy, with an annual production on the land area of the world estimated to be between 10^{11} and 10^{12} tons (*1*). Biomass accounts for one-seventh of the worldwide energy consumption and for as much as 43% of the energy consumption in some developing countries (*2*). This worldwide usage is equivalent to 25 million barrels of oil a day (*3*). In the United States, the annual pro duction of biomass is estimated to be 2.1 billion tons (*4*), though not all is available for energy use. Table I gives a breakdown of total energy usage in the U. S. by resource.

Table I. Estimated Energy Consumption in the U. S., 1988

Energy Source	Quads	10^6 BOE/day
Fossil Fuels		
Petroleum	33.62	15.88
Coal	18.86	8.91
Natural Gas	17.22	8.13
Renewables		
Hydropower	3.02	1.43
Biomass	3.28	1.55
Other		
Nuclear	5.74	2.71
Geothermal	0.26	0.12
Total	82.00	38.73

This chapter not subject to U.S. copyright
Published 1992 American Chemical Society

As shown, the U. S. uses about 82 quadrillion British ther mal units, or quads (1 quad = 10^{15} Btu), annually (5). About 4% of this total is supplied by biomass resources, or about 3.3 quads. The U. S. currently imports over 8 million barrels of oil a day (6). As shown in Table I, biomass resources have the potential to make a significant contribution towards the reduction of oil imports.

For biomass to represent a convenient feedstock for energy production, it must be available in adequate amounts and have sufficient energy content so that the energy conversion process can operate efficiently. Cellulosic feedstocks such as wood and woodwastes, crop residues, and municipal solid wastes represent traditional sources of biomass which are abundant in the U. S. These three sources alone account for over 90% of the total biomass currently used for energy production. The average energy content of these feedstocks is about 7000 Btu/lb, making them excellent feedstocks for energy production. Other nontradition al sources such as animal wastes, food processing wastes, energy crops and aquaculture, though significant, contribute to a much lesser extent (either in quantity, energy content, or both).

Biomass has the unique advantage over most energy sources in that it can be converted to liquid and gaseous fuel, as well as being used directly in its solid form (i.e. combustion). Over the years several processes for biomass energy conversion have evolved including anaerobic digestion, pyrolysis, and hydrolysis. Combustion of biomass, however, remains as the most widely used method of energy conversion.

This chapter will focus on the availability of various traditional and nontraditional cellulosic biomass resources and their potential contribution to energy production in the U. S.

Traditional Cellulosic Biomass Feedstocks

Traditional biomass feedstocks include such resources as woody biomass, agricultural residues and municipal solid wastes (MSW). The use of these feedstocks for energy production dates back hundreds of years. In the U. S., wood provided over half of the energy supply up until about 1880 (7). MSW has been used for the past 80 years in the U. S. for the energy production on a commercial basis (8). The combustion of agricultural residues for energy production also dates back many years. The potential energy contribution of these feedstocks is discussed below.

Woody Biomass. Wood for energy has been defined as the biomass on a forest ed area that is unusable, that is available in supplies exceeding anticipated demands for nonenergy products, and/or residue that is associated with the processing of wood harvested for nonenergy purposes (8).

With this general definition as the basis, several studies have been conducted to determine the quantity of wood available for energy production in the U. S. According to Zerbe (9), available woody biomass for energy production totals about 600 million dry tons annually. Zerbe includes in this amount

160 million dry tons of logging residues and cull trees; 20 million dry tons of standing live and dead trees following logging opera tions; 215 million dry tons of growth in excess of cut; 95 mil lion dry tons of trees killed by insects, disease and fire; 70 million dry tons from urban tree removals and wood wastes such as that used for pallets and demolition wastes; and 40 million dry tons from industrial waste and land clearing.

Table II shows estimates of wood resources available for energy production as determined by other studies. As shown, wood resources range from 100 million tons of waste wood to 1 billion tons of excess wood production. The average quantities of collectible waste and excess wood produced are 180 and 483 million ton, respectively, per year (10). Therefore, a total of 663 million tons are available for energy production, which agrees well with Zerbe's (9) estimation.

Table II. Estimated Available Wood Resources For Energy Production (10^6 Tons/year)

Study	Wastes		ExcessProduction	
	Total	Collectible	Total	Collectible
Young et al. (1986)	–	100	–	–
Jeffries (1983)	383	175	417	200
Ferchak and Pye (1981)	–	278	3000	1000
Ng et al. (1983)	230	170	450	269
OTA (1980)	84	–	678-1803	308-616
Hymphry et al. (1977)	60	–	–	–
Soc. Am. Foresters (1979)	–	–	1237	–
Average	189	180	1268	483

Source: Adapted from ref. 10.

Wood currently supplies about 3.7% (approximately 3.0 quads) of the country's energy needs (9). However, the potential exists to generate as much as 20 quads from available excess wood.

Agricultural Residues. Agricultural residues are defined as the plant parts left in the field after harvests, the remains that are generated from packing operations, and the materials discarded during processing (8). The most abundant residues – cornstalks, wheat straw, and soybean field waste – comprise more than 85% of the residues generated each year (11).

As with wood, the literature offers considerable variability on the quantity of crop residues available for energy production in the U. S. These differences have been attributed to two main factors: differences in the estimation of the amount of residue generated and differences in the methods used to determine what quantity of that generated is available for energy production.

Table III gives the results of various studies conducted to determine the amount of residue available for energy production. As shown, collectible residue estimates range from 52 million tons per year to 400 million tons per year with the average being about 213 million tons (*10*).

Table III. Estimated Availability of Agricultural Residues For Energy Production (10^6 Tons/year)

Study	Total	Collectible
Young et al. (1986)	—	400
Jeffries (1983)	385	—
Ng et al (1983)	809	318
Goldstein (1983)	355	—
OTA (1980)	419	82
Vergara and Pimentel (1979)	473	—
Humphrey et al. (1977)	400	—
Tyner and Bottum (1979)	—	52
Average	473	213

Source: Adapted from ref. 10.

Potential energy production from crop residues is estimated to be 0.14 quads. The potential exists to produce about 2 quads of energy from agricultural residues.

Municipal Solid Wastes (MSW). Each year about 250 million tons of residential, commercial, and industrial wastes are generated in the U. S. Residential and commercial MSW account for about 160 million tons. This figure is expected to reach 193 million tons by the year 2000 (*12*).

The composition of MSW can vary considerably based on such factors as season, location, residential vs. commercial, etc. Table IV gives an average composition for MSW as estimated by the Environmental Protection Agency (*12*). Paper products comprise the greatest fraction of the waste stream.

Because of the numerous problems associated with the land filling of MSW, several alternatives have been developed to convert this waste into a useful product such as boiler fuel for steam and electricity production. Combustion of the entire waste stream (called mass burning) was the original method of energy conversion and remains the predominant method to this day. However, environmental concerns have increased the popularity of burning refuse-derived fuel, which is MSW from which recyclables have been removed.

As of 1988, the U.S. had 105 MSW combustion facilities either operating or in the shakedown phase, 29 under construction, and 61 in the advanced planning stage (*13*). The currently operating waste-to-energy facilities annually account for 0.15 quads of energy (*13*). The potential exists to produce 1.4 quads of energy per year from MSW.

Table IV. Average Composition of MSW

Component	Amount (%)
Paper and paperboard	41.0
Glass	8.2
Metals	8.7
Plastics	6.5
Rubber and leather	2.5
Textiles	1.8
Wood	3.7
Food waste	7.9
Yard waste	17.9
Other	1.7
Total	100.0

Sources: Adapted from ref. 12.

Non-Traditional Biomass Sources

According to the Department of Energy, a significant increase in the use of biomass for energy production will be necessary to expand the biomass resource base. Non-traditional sources of biomass and methods of production must be explored, such as short rotation energy crops, food processing and animal wastes, and aquatic resources.

Energy Crops. Energy crops are fast-growing hardwood trees and grasses that can be converted to fuels (*14*). These crops fall into two general categories : short rotation woody crops and herbaceous energy crops. Short rotation woody crops include such species as eucalyptus, hybrid poplars, and sycamore. These crops are se lected for their fast growth, regeneration from stumps, and hardiness on marginal land. They are harvested after 2 to 8 years, rather than the 30 to 60 year rotation common in conventional forestry. The average yield for these woody crops is about 6 dry tons per acre (*14*). Herbaceous energy crops are forage crops, grasses, and legumes that yield the same energy product as wood. These crops are harvested one to three times a year with conventional agricultural equipment. The potential contribution of energy crops to the U. S. energy supply is esti mated to be about 5 quads in the year 2000 (*15*).

Food Processing and Animal Wastes. Food processing and animal wastes represent a relatively small percentage of the total biomass energy potential. On a national basis, there appears to be enough food wastes to consti tute a resource for energy production. However, these materials are so widely distributed that only a few isolated sources have economic potential. The food processing industry in the U. S. produces over 11 million tons of waste products annually in the form of discrete, suspended, and dissolved solids, plus 114×10^9 gallons of waste waters (*16*). The energy potential of these streams is very small (less than 0.005 quads).

Wastes from animals in confinement (about 50 million dry tons annually) account for more than 50% of total livestock wastes generated each year (*17*). The primary sources of animal wastes are cattle feedlots and dairy, hog, and poultry farms. The Department of Energy's Office of Technology Assessment estimates that by the year 2000 as much as 0.3 quad of energy could be derived annually from wastes annually (*17*).

Aquatic Resources. Four primary types of aquaculture are suitable for consideration as energy sources: microalgae, macroalgae, floating plants and emergent plants (*18*). Microalgae can contain up to 90% of their mass in hydrocarbons and can grow in different aquatic environments including fresh, saline, brackish and wastewater (*18*). Macroalgae, such as giant kelp, grow naturally in the open ocean and in coastal areas. Macroalgae have high growth rates and produce significant amounts of carbohydrates which can be converted to methane and valuable byproducts via anaerobic digestion. Floating plants include feedstocks such as the water hyacinth. Considerable research focuses on the water hyacinth because it is one of the most productive plants in the world. In addition to its potential as a biomass energy feedstock, the water hyacinth can also be used as a water purifying agent. Emergent plants include cattails, reeds, and bulrushes and are also highly productive. Because aquatic feedstock production is a relatively new area of biomass research, the energy potential of these feedstocks has not been estimated.

Future Potential of Biomass

Table V summarizes current and potential energy production from biomass. As shown, biomass resources currently account for about 3.28 quads of energy per year, but the potential exists for as much as 19.2 quads, based on current resources. Several other estimates of the future potential of biomass feedstocks have been made by various organizations. Studies conducted by the Congressional Office of Technology Assessment and the U. S. Department of Energy's Energy Research Advisory Board have estimated the contribution of biomass to U. S. energy production at 6.0 to 16.6 quads by the year 2000 (*17*). In a study conducted by the Department of Energy's Office of Policy, Planning, and Analysis, the potential contribution of all renewable resources by the year 2030 was estimated to be 22 quads, based on the current level of expenditures for renewable energy research (*5*).

It is estimated if the funding level for renewable energy research is increased by $3 billion over the next 20 years, the energy contribution of renewables could be as much as 41 quads by the year 2030 (*5*). Biomass currently accounts for about 50% of these renewable resources; therefore, the potential exists for about 20 quads to be supplied by biomass resources.

Table V. Summary of Current and Potential Energy Production From Biomass

Biomass Energy Resource	Annual Production (10^6 T/yr)	Quads Current	Quads Potential
Traditional			
Wood	600	3.0	10
Ag. Residues	200	0.14	2.5
MSW	250	0.15	1.4
Nontraditional			
Food Processing Wastes	11	–	–
Energy Crops	–	–	5.0
Aquatic Resources	–	–	–
Animal Wastes	50	–	0.3
Total	1111	3.28	19.2

Sources: Adapted from refs. 9, 13, 17

Biomass Utilization Processes

Biomass can be converted to energy by using a variety of processes including thermochemical, photobiological, and biochemical methods. These processes produce energy products that can be used by residential, commercial, and industrial consumers for heat, power, and fuel. The processes also produce chemical products that can be used by the chemical industry as petrochemical substitutes.

Thermochemical processes use heat to convert biomass into various products. The four basic types of thermochemical processes include direct combustion, gasification, direct liquefaction, and pyrolysis. Direct combustion produces heat which can be used to produce steam and electricity. Gasification yields low- and medium-BTU gas which can be used directly as fuel. Direct liquefaction converts biomass into crude oil. Pyrolysis yields gases, oils, and solid char. Photobiological hydrogen production involves the manipulation of certain types of hydrogen-producing photosynthetic microorganisms to generate hydrogen efficiently and in large quantities.

Biochemical conversion processes involve converting biomass to fuels and chemicals by the use of chemical and/or biological processes which utilize microorganisms. Anaerobic digestion and fermentation are two biological processes that are widely used to convert biomass to products. Anaerobic digestion yields biogas, a mixture of methane and carbon dioxide. Fermentation is usually used in combination with hydrolysis to produce ethanol from biomass. Hydrolysis processes use either acid or enzymes to convert the carbohydrates in biomass to sugars which can then be fermented to ethanol.

The Tennessee Valley Authority's Biomass Research Program

Considerable research has been conducted by various organizations to optimize these thermochemical, photobiological, and biochemical technologies for commercial application. However, the oil embargo of the 1970's resulted in much emphasis being placed on the development of technologies to convert biomass to liquid fuel such as ethanol. Since 1980, the Tennessee Valley Authority (TVA) has been involved in the development of hydrolysis and fermentation technology for ethanol production. Two processes have evolved from this work — a dilute process and a concentrated sulfuric acid hydrolysis process. Both process are being developed and evaluated in bench-studies. Pilot-scale evaluations of the dilute and the concentrated acid hydrolysis processes have been conducted in 2- and 4-ton-per-day facilities, respectively. Figures 1 and 2 show flow diagrams of the two processes. The dilute process is a two-stage process that uses high temperatures (320 °F) and pressures (10 atm) to convert hemicellulose and cellulose to sugars. The concentrated acid process is a two-stage process that uses low temperatures (212 °F) and pressures (1 atm) during hydrolysis (*19*).

Both processes have been evaluated using a variety of feed stocks including agricultural residues, hardwoods, and MSW. Concentrated acid hydrolysis has been evaluated using mainly agricultural residues such as corn stover and rice hulls. Table VI shows the hemicellulose and cellulose conversion efficiencies of several agricultural feedstocks using the concentrated acid process. These conversion efficiencies correspond to ethanol yields of 60-80 gallons per ton of feedstock (*20*). Hardwoods such as red oak have been the primary feedstock in dilute acid hydrolysis studies. Ethanol yields from oak have ranged from 45-60 gal/ton (*21*).

Because of the problems associated with the disposal of MSW, TVA's current research efforts have focused on ethanol production from this biomass waste product. About 60% of this waste stream is composed of paper products,

Table VI. Concentrated Acid Conversion Efficiencies of Selected Biomass Feedstocks

Feedstock	Conversion Efficiencies (%)	
	Hemicellulose-to-xylose	Cellulose-to-glucose
Alfalfa stems	96	89
Corn hulls	96	93
Corn stover	92	90
Rice hulls	95	89
Sugarcane bagasse	99	92
Sweet sorghum residue	96	88
Average	96	90

Source: Adapted from ref. 20.

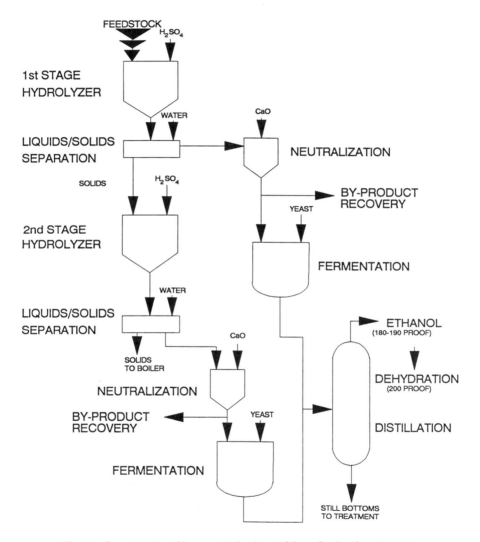

Figure 1. TVA's Dilute Sulfuric Acid Hydrolysis Process

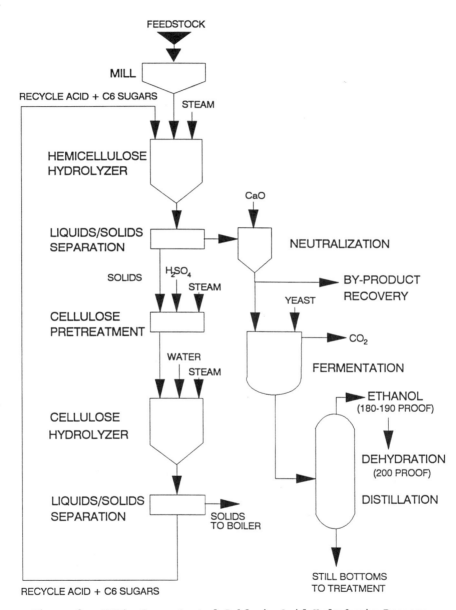

Figure 2. TVA's Concentrated Sulfuric Acid Hydrolysis Process

yard wastes, and food wastes. TVA uses this cellulosic fraction as a waste-derived feedstock (WDF) in the dilute acid hydrolysis process. Preliminary laboratory evaluations using WDF and newsprint have resulted in sugar yields equivalent to 25-40 gallons per ton of feedstock processed. By- products of the process include furfural and carbon dioxide (22).

Preliminary economic evaluations were conducted to determine the commercial feasibility of producing ethanol from MSW using TVA's process. Based on ethanol yields ranging from 25-40 gallons per ton and a required return on investment of 15%, tipping fees are estimated to range from $20-40 per ton for a 500 ton-per-day facility (22).

Pilot-scale evaluations of dilute acid hydrolysis of MSW have begun. Results have been comparable to those achieved in the laboratory. Research and development activities are being conducted to optimize process conditions. Tests will also be conducted to determine optimum methods for by-product recovery and feedstock handling.

Conclusions

The use of biomass-derived energy in the U. S. has increased about 25% since 1980 and is estimated to increase an additional 100-400% over the next 10 years. This projected increase can be attributed to several factors including a heightened awareness of the nation's vulnerability to interruptions in imported oil and to the growing budget deficit, of which 40% can be attributed to oil imports (10). Despite the increased use of biomass resources for energy production, the bulk of the resources remain untapped. This scenario can change, however, based on advances in biomass production and utilization technologies and on implementation of national policies which emphasize the use of biomass resources.

Literature Cited

1. Stout, B. A. In *Biomass Energy Profiles*. FAO Agricultural Services, *Bulletin 54*, **1983**.
2. Jones, M. R. In *Biomass Handbook*; Kitani, O.; Hall, C. W., Eds.; Gordon and Breach Science Publishers: New York, NY, 1989, p. 97.
3. Hall, D. O. In *Biomass Handbook*; Kitani, O.; Hall, C. W., Eds.; Gordon and Breach Science Publishers: New York, NY, 1989, p. 3.
4. Goldstein, I. S. In *Organic Chemicals From Biomass*; Editor, I. S., 1; CRC Press Inc.: Boca Raton, FL, 1981.
5. *The Potential of Renewable Energy: An Interlaboratory White Paper*, U. S. Department of Energy, DE-AC02-83CH100093, 1990.
6. *Geotimes*; 1990.
7. Klass, D. O. In *Biomass As A Nonfossil Fuel Source*, Klass, D. O., Ed., ACS Symposium Series 144, Washington, D. C., 1981.
8. Eoff, K. M. and W. H. Smith. In *Biomass Handbook*; Kitani, O.; Hall, C. W., Eds.; Gordon and Breach Science Publishers: New York, NY, 1989, P. 567.

9. Zerbe, J.; *Forest Residuals-The Universal Energy Source*;. Proceedings of the Southern Biomass Conference, 1988.
10. Wyman, C. E. and N. D. Hinman. *Ethanol: Fundamentals of Production From Renewable Feedstocks and Use As A Transportation Fuel*; 1990.
11. *Biomass Energy Technology Research Program Summary FY 1983*; U. S. Department of Energy, DOE/CE-0032/1. 1983. p. 2.
12. *Solid Waste Disposal Overview*; National Solid Waste Management Association, 1988.
13. Klass, D. L. *Energy From Biomass and Wastes XII*; Institute of Gas Technology: Chicago, IL, 1990, p. 13.
14. *Biofuels Program Summary*; U. S. Department of Energy, 1989.
15. *Biomass Energy Technology Research Program Summary FY 1983*; U. S. Department of Energy, DOE/CE-0032/1, 1983.
16. Weathers, P. J. In *Biomass Handbook*; Kitani, O.; Hall, C. W., Eds.; Gordon and Breach Science Publishers: New York, NY, 1989, p.176.
17. *Biomass Energy Technology Research Program Summary FY 1983*; U. S. Department of Energy, DOE/CE-0032/1, 1983, p.2.
18. *Biomass Energy, An R and D Overview*; U. S. Department of Energy DE-AC01-80ET.
19. Lambert R. O., Moore-Bulls, M. R., and J. W. Barrier; In *Eleventh Symposium on Biotechnology For Fuels and Chemicals*; The Humana Press, Clifton, N. J., 1990, pp773-784.
20. Broder, J. D. and J. W. Barrier. *Producing Ethanol and Coproducts From Multiple Feedstocks*; Proceedings of the American Society of Agricultural Engineers, 1988.
21. Strickland, R. C., R. L. Griffith, M. J. Beck and J. R. Watson. *Conversion of Hardwoods to Ethanol – The Tennessee Valley Approach*; Eleventh Symposium on Energy From Biomass and Wastes, 1985.
22. Barrier, J. W., M. M. Bulls, J. D. Broder, and R. O. Lambert. *Production of Ethanol and Coproducts From MSW-Derived Cellulosics Using Dilute Sulfuric Acid Hydrolysis*; 1990.

RECEIVED April 16, 1991

Chapter 24

Pyrolysis of Agricultural and Forest Wastes

D. S. Scott, J. Piskorz, and D. St. A. G. Radlein

Department of Chemical Engineering, University of Waterloo, Waterloo, Ontario N2L 3G1, Canada

Pyrolysis of biomass materials has been practised for many years to prepare charcoal. However, in recent years, it has been shown that short residence time pyrolysis at high heating rates can give high conversions to liquid products. The liquid obtained in these processes can contain surprisingly large amounts of specific oxychemicals or anhydro sugars which are of industrial interest.

The amounts of liquid products obtainable and the chemical characteristics of typical liquids from various biomass feedstocks are described. Modifications of the fast pyrolysis process can influence selectivity and different products can then be produced. Finally, some results of research on upgrading or conversion of bio-oil fractions to higher value products is described. It is concluded that fast pyrolysis can represent a direct route to the production of a variety of special chemical products.

The pyrolysis of wood to produce gases, liquids and char is an ancient art. Only one hundred years ago, the process was the source of many basic organic chemicals such as acetone and methanol (wood alcohol). Pyrolysis is no longer a source of such chemicals, and has been practiced in recent years on an industrial scale primarily as a method to produce wood charcoal.

By definition, pyrolysis implies decomposition by heating in the absence of oxygen. Traditionally, heating was done slowly at high temperatures, over long periods of time, to give maximum yields of charcoal. Only in recent years has some attention been paid to the effects associated with short time rapid heating of wood particles.

Flash pyrolysis or flash hydropyrolysis (that is, pyrolysis in a hydrogen atmosphere) are terms usually used to describe processes with reaction times of only several seconds, or less. Flash pyrolysis is often carried out at near atmospheric pressures, while hydropyrolysis commonly employs pressures to 20 MPa.

Residence times of less than a few seconds with reaction at high temperatures require a reactor configuration capable of very high heating rates. Among the most appropriate designs are the entrained flow reactor, the fluidized bed reactor and the ablative reactor. Nearly all flash pyrolysis studies have employed one or the other of these reactor types.

It has been found in the past decade that "fast" or "flash" pyrolysis processes can yield a variety of products in concentrations of economic interest. This chapter will describe the nature of these products and discuss some aspects of their production.

Fast Pyrolysis Processes. All fast pyrolysis processes have in common a short residence time at high temperature for the volatile decomposition products which yield gases and condensable liquids. Generally, such processes exclude oxygen or other reactive gases, and it is only processes of this kind which will be described.

In addition to the residence time of the volatiles, the temperature of reaction is a critical variable in fast pyrolysis. At low temperature, below 400°C, decomposition is relatively slow, and gas and char are the major products. At higher temperatures, 450° to 600°C, the amount of condensable liquid product increases to a maximum and then decreases. Gas yield increases steadily as temperature is increased and above 650°C, it will become the major product. Therefore, depending on whether a char, a liquid or a gas is the principal product desired, fast pyrolysis may operate anywhere in the temperature range of 400° to 1200°C.

Any process development must have an economic rationale, even though it is rather tentative, to justify continued development. Novel energy conversion processes for biomass are no exception, and are characterized by a unique dependence on local environments and economics, raw material supply, technological factors and politics. In the present case, the overriding limitation is often the supply of raw material (biomass) which can be economically harvested and transported to a central conversion plant. The source area for harvesting can be estimated roughly as a circle with a radius of about 50 km. For readily gathered forest waste, this might support a plant of 1000 dry tonnes per day throughput. For a single sawmill or pulpwood operation, waste available from logging might be from 100 to 1000 tonnes per day. None of this represents a large plant in energy production terms, although 1000 tonnes of dry biomass per day is a large scale forest or agricultural operation. The reasonable conclusion might be that a biomass conversion process would be most useful if it were a simple process, not capital intensive, that could be operated efficiently on a small scale and which was capable of producing higher value added chemicals rather than alternative fuels. Such a plant could then be sited where raw material could be supplied at reasonable cost. Liquid products might be used or modified on site, if feasible. If not, the liquids could be readily transported to a central upgrading or refining plant.

In the past ten years or so, a number of methods of achieving fast pyrolysis of biomass have been described. Some of these were developed primarily as gasification processes to make synthesis gas or ethylene (1)(2). Biomass as a gasification raw material has the considerable advantage of being nearly sulfur free. However, pyrolytic gasification produces primarily carbon monoxide, carbon dioxide, hydrogen and some methane and ethylene. High temperatures, normally 800° to 1000°C are required. Also, to obtain rapid heating to these high temperatures small particles are also required which usually implies a high consumption of energy for grinding. As a result, pyrolytic gasification without steam or oxygen addition is usually not economically attractive, although a good quality synthesis gas can be made (3) and ethylene yields from 5% to 10% of the dry biomass fed have been achieved. High temperature fast pyrolysis of biomass has not yet, therefore, shown economic potential as a gas-producing industrial process.

A second approach to flash pyrolysis has been described by Scott and Piskorz (4)(5) and Scott et al. (6). In these publications, the development has been outlined of an atmospheric pressure flash pyrolysis process utilizing a fluidized bed of solid as heat carrier. The process studied has as a primary objective the determination of conditions for maximum yield of liquids from biomass, particularly forest materials. Results from a bench-scale unit (15 g/h feed rate) indicated that at apparent vapor residence times of about 0.5 s, organic liquid yields of 60 to 70% on a moisture-free basis could be obtained from hardwoods such as aspen-poplar and maple. Lower but still high yields of organic liquids (40 to 60%) could be obtained from agricultural wastes such as wheat straw, corn stover, and bagasse. Bench-scale reaction conditions used a biomass particle size of -295 +104 μm in a nitrogen atmosphere over a temperature range of 400° to 650°C. In terms of calorific value and hydrogen-to-carbon ratio, the best quality liquid was obtained at conditions which also gave the maximum liquid yield.

In view of the high yields of organic liquids obtained -- among the highest yields reported for a pyrolysis conversion process for biomass -- and the reasonable operating conditions, a larger scale continuous process unit was designed and constructed. The liquid yields attained in this unit were as good as, or better than, those achieved in the bench-scale unit. Detailed results and a description of the process have been given as well as a preliminary economic analysis by Scott and Piskorz (5).

More recently, Knight et al. (7) described the operation of an entrained flow reactor for the production of liquids. A somewhat different upflow entrained pyrolyzer for the production of liquids from wood has been described by Beaumont (8). Kosstrin (9) has also used a fluidized bed for thermal conversion of biomass to liquid. In general, processes for the attainment of high liquid yields operate at much lower temperatures, commonly 450° to 550°C, than do processes to yield gaseous product, but at about the same vapor residence times, about 200-700 ms.

In a publication by Scott et al. (10), it was shown that any type of reactor capable of meeting certain heat transfer rate criteria would be expected to give results comparable to those obtained with the fluidized bed. Indeed, the "ablative" reactor developed by workers at SERI has recently been modified and operated at lower temperatures to produce higher liquid yields (11). It has also been reported that high liquid yields, especially at very short residence times, were obtained in an entrained flow reactor (12). In addition, a vacuum pyrolysis unit which has slow heating rates but short volatiles residence time has been reported to obtain good liquid yields (13).

Detailed analysis of the fast pyrolysis liquid products obtained from wood or agricultural wastes as reported by Piskorz et al. (14)(15) and Radlein et al. (16)(17) have shown the presence of a number of simple organic oxychemicals in significant concentrations, up to 15% concentration in some cases. These same researchers have reported a change in selectivity of the fast pyrolysis process by an appropriate pretreatment which can give high conversions of cellulose to sugars (18). They have also described the upgrading of a lignin fraction (19), and additional products obtainable by use of a catalytic fluid bed (20)(23).

All of these alternative fast pyrolysis methods will yield liquids of different compositions from which high value chemicals could be extracted. With this approach, biomass fast pyrolysis has now become a potentially economically attractive alternative process in the short term. In the next section, the products with possible commercial potential will be discussed.

Products from Pyrolysis Liquids

Pyrolysis of Wood. The atmospheric pressure fluidized bed fast pyrolysis process developed at the University of Waterloo (now known as the Waterloo Fast Pyrolysis Process or WFPP) has well-defined optimal conditions for maximum yields of liquids. As mentioned earlier, any reactor which can achieve these operating conditions together with the required high rates of heat transfer can obtain similar liquid yields. However, discussion will be limited to the fluidized bed reactor as a typical fast pyrolysis process, and specifically to the WFPP for which a good deal of experimental data has been made available. A schematic of the WFPP as operated in a pilot plant is shown in Figure 1. The typical performance of the fluid bed reactor for sawdust is shown in Figure 2. The rather narrow temperature range (about $50°C$) in which maximum liquid yields are obtainable (Figure 2) is typical of all fast pyrolysis processes and for a wide variety of biomass feedstocks. Fast pyrolysis processes are now under active semi-commercial development. For example, two demonstration-scale plants using WFPP technology at a scale of five dry tonnes per day of biomass feed will shortly be in operation. Similar scales of operation are reported to be planned or in operation for other fast pyrolysis processes.

The yields of gas, liquid and solids for two typical woods, a hardwood and a softwood, are shown in Table I together with the gas composition (15). The liquid yields on a moisture-free feed basis, are about 77% for both the spruce and the poplar. The properties of this liquid product are given in Table II. The liquid is typically a dark brown homogeneous fluid of low viscosity, with a high density and low pH. It has a high oxygen content of 37% to 40%, and although it can be used quite satisfactorily as an alternative fuel oil, this would represent, in most circumstances, a low value application.

Analysis of the pyrolytic liquid presents a much more interesting picture. A detailed analysis (by HPLC and GC) of the organic liquid fractions for the spruce and poplar is shown in Table III

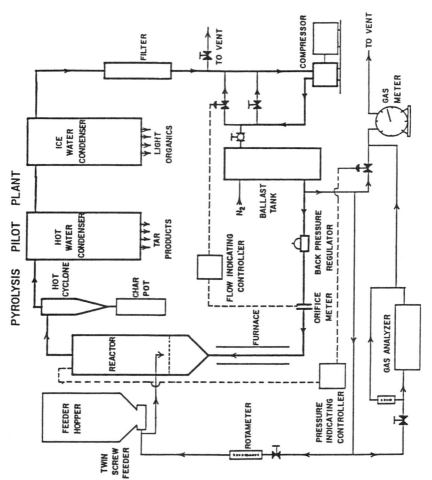

Figure 1. Flow Diagram of a Fluidized Bed Fast Pyrolysis Process [the Waterloo Fast Pyrolysis Process (WFPP)].

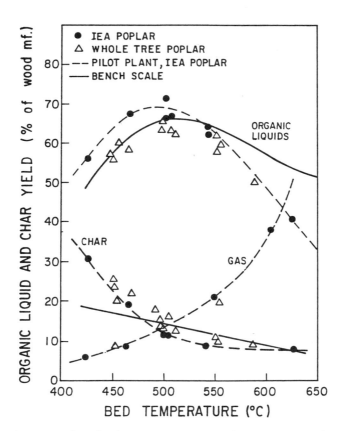

Figure 2. Organic Liquid, Gas and Char Yields from Poplar Wood with the WFPP Fluid Bed Units.

Table I. Pyrolysis Yields from Different Woods

	White Spruce (Softwood)			Poplar (Hardwood)		
Run #	42	43	45	27	59	A-2
Temperature, °C	485	500	520	500	504	497
Moisture content, wt%	7.0	7.0	7.0	6.2	4.6	3.3
Particle Top Size, μm	1000	1000	1000	590	1000	590
Apparent Residence Time, s	0.70	0.65	0.62	0.55	0.48	0.46
Feed Rate, kg/hr	2.07	1.91	1.58	2.24	1.85	0.05
Yields, wt% of m.f. wood						
Organic liquid	63.1	66.5	66.1	65.8	66.2	65.7
Water	10.7	11.6	11.1	9.3	10.7	12.2
Char	16.3	12.2	12.3	12.1	11.8	7.7
Gas:						
H_2	0.4	0.02	0.01	0.02	0.01	--
CO	4.16	3.82	4.01	5.32	4.44	5.34
CO_2	3.38	3.37	2.69	6.30	5.75	4.78
CH_4	0.34	0.38	0.43	0.48	0.37	0.41
C_2H_4	0.16	0.17	0.16	0.20	0.13	0.19
C_2H_6	0.02	0.03	0.05	0.04	0.05	--
C_3	0.03	0.04	0.06	0.09	0.08	0.09
C_4					0.19	3.10
Total Gas	8.1	7.8	7.4	12.4	11.0	10.8
Overall recovery wt%, m.f.	97.8	97.7	96.7	99.7	99.8	96.4

Reproduced from ref. 15. Copyright 1988 American Chemical Society

Table II. Properties of Pyrolytic Liquids

	White Spruce			Poplar
Run #	42	43	45	59
Yields, wt% of wood as fed	75.2	79.2	78.5	77.6
Water content, wt%	21.9	22.4	21.8	18.6
pH	2.1	2.1	2.3	2.4
Density, g/cc	1.22	1.22	1.22	1.23
Elemental analysis, wt%, m.f.				
Carbon	53.5	54.0	56.6	53.6
Hydrogen	6.6	6.8	6.9	7.0

Reproduced from ref. 15. Copyright 1988 American Chemical Society

Table III. Analysis of Liquid Products

	White Spruce	Poplar
Run #	43	A-2
Temperature	500	504
Yields, wt % of feed, m.f.		
Organic Liquid	66.5	62.6
1. Oligosaccharides	--	0.70
2. Cellobiosan	2.49	1.30
3. Glucose	0.99	0.41
4. Fructose & other hexoses	2.27	1.32
5. Glyoxal	2.47	2.18
6. Methylglyoxal	--	0.65
7. Levoglucosan	3.96	3.04
8. 1,6 anhydroglucofuranose	--	2.43
9. Hydroxyacetaldehyde	7.67	10.03
10. Formic Acid	7.15	3.09
11. Formaldehyde	--	1.16
12. Acetic Acid	3.86	5.43
13. Ethylene Glycol	0.89	1.05
14. Acetol	1.24	1.40
15. Acetaldehyde	--	0.02
Water-Solubles - Total Above	33.0	34.2
Pyrolytic Lignin	20.6	16.2
Amount not accounted for (losses, water soluble phenols, furans, etc.)	12.9	11.91
<u>by G.C.</u>		
Methanol	1.11	0.12
Furfural	0.30	--
Methylfurfural	0.05	--

Reproduced from ref. 15. Copyright 1988 American Chemical Society

(15). Note that about 52% to 55% of the liquid has been identified as individual water-soluble components. A further 26% to 31% is characterized as "pyrolytic lignin" (the insoluble fraction remaining after water extraction of the pyrolysis liquid). This fraction represents the compounds which are aromatic and largely phenolic in character, presumably deriving from the lignin fraction of the wood. The outstanding feature of these liquids is the presence of relatively large amounts of a few chemicals such as hydroxyacetaldehyde, formic acid, acetic acid, acetol and glyoxal. Some sugar and anhydrosugars are also present. It is apparent that the pyrolysis liquid has a value as a chemical feedstock that may eclipse that of a liquid fuel if efficient separation and recovery methods can be developed.

Pyrolysis of Pretreated Wood. Early work carried out by several workers including Shafizadeh and co-workers (21)(22) suggested that pretreatment of wood to remove hemicellulose and inorganic ash could result in significant increases in the yields of anhydrosugars, principally levoglucosan, during slow pyrolysis under vacuum. In our laboratory, samples of poplar wood and of steam-exploded (Stake) cellulose were pretreated with a mild acid hydrolysis to remove much of the hemicellulose fraction (18). The pretreated wood was then subjected to fast pyrolysis at optimal WFPP conditions, and the resulting liquids showed a completely different composition from that for untreated wood. Two typical results are shown in Table IV (18). It is clear that the pretreatment changed the thermal decomposition mechanism, from fragmentation of the cellulose polymer to yield carbonyl compounds of low molecular weight in untreated wood or cellulose to a depolymerization mode which yielded mainly monomeric anhydrosugars from the pretreated samples. In this latter case, removal of the pyrolytic lignin followed by mild acid hydrolysis yields glucose which it has been shown recently in our laboratory can be rapidly and quantitatively fermented to ethanol (18). On the other hand, it should be realized that the anhydroglucose, levoglucosan, which is a chiral compound, is in itself a substance of considerable potential commercial interest. Again, methods for recovery of this material on a bulk scale would allow its exploitation as a new and reasonably priced compound.

An important concept arising from this work is the realization that thermal pyrolysis is not necessarily a random and largely uncontrollable reaction. In fact, selectivity to hydroxyacetaldehyde production on the one hand, or for levoglucosan production on the other, can be as high as 80% for the cellulose conversion step. This kind of result suggests that many selected products may be possible in fast pyrolysis given the appropriate reaction conditions or feed treatment. The use of a catalyst in a fluid bed rather than sand, together with a hydrogen atmosphere, is one such modification of the WFPP.

Hydropyrolysis in a Catalytic Bed

If a nickel-alumina catalyst is used in the fluid bed WFPP reactor together with a hydrogen atmosphere, a high conversion of the carbon in a biomass feed to methane can be achieved. Garg et al.

Table IV. Pyrolysis of raw and treated wood

	Poplar Wood		Stake Cellulose	
	Untreated	Pretreated	Untreated	Pretreated
Temperature, °C	497	501	500	490
Vapor residence time, s	0.46	0.45	0.50	0.50
Particle size, μm	-590	-590	-1000	-1000
Moisture, %	3.3	16.5	24.1	3.1
Cellulose, % mf	49.1	62.8	93.9	91.3
DP	---	---	303	149
Ash, % mf	0.46	0.04	2.5	0.32
Yields, % mf wood				
Organic liquid	65.8	79.6	58.2	65.3
Water	12.2	0.9	9.0	7.0
Char	7.7	6.7	15.4	(19.0)*
Gas	10.8	6.5	15.3	3.0
Yields of tar components, % mf feed				
Oligosaccharides	0.7	1.19	ND	(29.4)**
Cellobiosan	1.3	5.68	0	3.1
Glucose	0.4	1.89	0	1.7
Fructose (?)	1.31	3.89	0	2.0
Glyoxal	2.18	0.11	ND	2.5
1,6-Anhydroglucofuranose	2.43	4.50	0	5.5
Levoglucosan	3.04	30.42	0	27.3
Hydroxyacetaldehyde	10.03	0.37	17.1	0.4
Formic acid	3.09	1.42	7.4	0.1
Acetic acid	5.43	0.17	8.5	0.1
Ethylene glycol	1.05			
Acetol	1.40	0.06	5.4	
Methylglyoxal	0.65	0.38	3.3	
Formaldehyde	1.16	0.8		1.0
Pyrolytic lignin***	16.2	19.0		7.0
Totals	51.5	69.9	41.7	71.1
% mf pyrolysis oil	78.3	87.8	71.6	0.4
Sugars, glucose equivalent yield, % cellulose	20.4	83.4	0	67.5

*Over 90% of this water-soluble material held up in the outlet tube of the reactor. See below.
**This includes the 19% reported above as 'char'. Since it was essentially all soluble in water and on hydrolysis with sulfuric acid showed only glucose by the HPLC analysis.
***'Pyrolytic lignin' is the material precipitated by addition of water [5].

Reproduced with permission from reference 18.
Copyright 1989 Elsevier Science Publishers.

(20) showed that 70% to 75% of the feed carbon can be converted to methane at the optimal WFPP pyrolysis conditions. Recent work in our laboratory with an improved catalyst has attained 80% to 85% conversions to CH_4. As the necessary hydrogen can be prepared from biomass, two or three stage overall gasification reactions as shown by equations (1) and (2) (using cellulose as a model compound) can become technically feasible without the use of steam or oxygen as gasifying agents by employing catalytic hydropyrolysis followed by reforming (20).

$$C_6H_{10}O_5 + H_2O \longrightarrow 3\ CH_4 + 3\ CO_2 \quad (1)$$
$$\text{or}\ C_6H_{10}O_5 + H_2O \longrightarrow 6\ CO + 6\ H_2 \quad (2)$$

The use of other catalysts in a hydrogen atmosphere can lead to much larger yields of light hydrocarbons than is possible by high temperature fast thermal pyrolysis. For example, Scott et al. (23) have shown that 21% by weight of biomass fed can be directly converted to C_2-C_7 hydrocarbons with 62% of this amount being C_4+ hydrocarbons. The mechanisms of these light hydrocarbon-forming reactions from wood or cellulose are not yet clear, and much further research is required.

Upgrading of Pyrolysis Products

The pyrolytic liquids obtained from biomass are rich in oxygen. As is apparent from the previous discussion, this oxygen is present in oxychemicals which may have themselves a high value. However, because of the demand for hydrocarbon transportation fuels or octane enhancers, many attempts have been made to "upgrade" the crude pyrolysis liquids to gasoline or diesel oil fractions.

Two general approaches have been used, first to hydrogenate the crude pyrolysis liquid, or secondly, to directly catalytically hydrogenate or reform the pyrolysis vapors (usually by a dehydration catalyst) in a second stage reactor. The first of these, direct hydrogenation, has been developed by Elliott et al. (24) and by Elliott and Baker (25) for pyrolysis oils from peat using hydrotreating catalysts in a two-stage process. About a 30% to 35% yield of hydrocarbon liquids could be obtained. In our laboratory, Piskorz et al. (19) have shown that by use of a non-isothermal plug flow packed bed hydrotreater 64% of the pyrolytic lignin fraction can be converted to liquid hydrocarbons in the gasoline and diesel oil range. Therefore, the technology for production of a hydrocarbon transportation fuel from biomass has been achieved.

The second approach, to process pyrolytic vapors in a second reactor has been described by Marshall (26) and by Diebold and Scahill (27) using nickel catalysts or zeolite catalysts. While some encouraging results have been achieved, the process is not yet judged to be economic, although it has the potential to be so (28).

Use of Agricultural Wastes

Much of the basic research in liquefying biomass by fast pyrolysis or by other techniques has been carried out on wood. It is possible, of course, to use any lignocellulosic material and to obtain similar results to those obtained with wood. However, there will be differences in overall yields of gas, liquid and char as well as

Table V. WFPP Pyrolysis of Agricultural Wastes

Feedstock	Wheat Chaff	Sunflower Hulls	Flax Shives
Moisture, wt%	6.9	11.1	16.3
Ash, wt%	22.5	4.0	2.65
Particle Size, mm	-1.0	-1.0	-2.0
Temp. °C	515	500	500
Yields, wt% mf Feed	*		
Gas	15.9	18.7	17.7
Char	15.7	9.4	13.7
Water	15.7	9.4	13.7
Organic Liquid	51.0	44.4	42.4
Total Recovery	100.2	98.8	96.8

* maf basis

Reproduced with permission from reference 29.
Copyright 1990 D.S. Scott.

in specific compounds because of the wide variations in holocellulose/lignin amounts, as well as due to the presence of other organic or inorganic materials. A very wide variety of agricultural wastes are available in particular local circumstances (e.g. straw, bagasse, nut shells, chaff etc.). Their use in pyrolysis will be dictated by both the supply and by the economics of alternative uses or of disposal.

A considerable number of tests of various agricultural wastes have been reported for both pyrolysis to liquids and pyrolytic gasification. Pyrolytic gasification with partial oxidation to yield a low BTU gas or a char is an old and well-established process with a variety of technologies, and has been developed on a local scale in many third-world countries. Pure pyrolytic gasification of agricultural wastes is much less common, and indeed is not usually practised on an industrial scale.

Fast pyrolysis of agricultural wastes has been carried out in our laboratories on a variety of materials. Some results are given in Table V and Table VI to illustrate the differences and similarities to wood as a feedstock. In general, liquid yields are somewhat lower from agricultural biomass, but the chemical products and their yields relative to the amounts of cellulose, hemicellulose and lignin are similar. Hence, as a generalization, the yields of products obtainable may be roughly estimated from a knowledge of the amounts of each of the three main constituents. However, it must be said that the chemical nature of the lignin in agricultural biomass, for example, grasses or straws, can be very different from that found in woody biomass, and much research needs yet to be done to evaluate this type of pyrolytic lignin.

Table VI. Liquid Product Composition - Agricultural Wastes

Feedstock	Wheat Chaff	Sunflower Hulls	Flax Shives
Yields, wt%, mf Feed	*		
Cellobiosan	0.40	0.06	0.11
Glucose	0.19	--	--
Fructose (?)	0.70	0.12	0.28
Glyoxal	0.70	0.07	0.21
Levoglucosan	1.95	0.33	0.58
Hydroxyacetaldehyde	6.53	0.78	1.44
Formic Acid	ND	1.01	ND
Formaldehyde	1.30	ND	0.41
Acetic Acid	6.11	2.12	2.86
Ethylene Glycol	0.93	0.27	ND
Acetol	3.20	1.16	1.42
Water Insoluble	15.1	38.4	22.5

*maf basis

Reproduced with permission from reference 29.
Copyright 1990 D.S. Scott.

Summary

Fast pyrolysis appears to have a promising future as a method of preparing a liquid feedstock from which a variety of high value products can be recovered. In particular, a number of low molecular weight carbonyl compounds, and sugars, can be produced in sufficiently high yields to make their recovery as individual commercial chemicals economically worthwhile. In addition, some of these compounds for example, levoglucosan, hydroxyacetaldehyde, and pyrolytic lignins, may have unique characteristics which would lead to new market applications.

The pyrolytic liquids produced from biomass can be successfully upgraded to hydrocarbon fuels, but not yet in an economical way. The use of catalysts, or of special pretreatments of biomass, can lead to changes in selectivity of the thermal decomposition process, and fast pyrolysis allows a surprising degree of control over these alternative pyrolysis mechanisms.

Much research needs to be done, partly because of the wide variety of the feed materials available, but also because our knowledge of methods of recovery for the higher valued products of fast pyrolysis is still in its early stages.

Acknowledgments

The authors are pleased to acknowledge the financial assistance of the Natural Sciences and Engineering Research Council and of the

Bioenergy Development Program of Energy, Mines and Resources Canada. The valuable assistance of Peter Majerski with experimental work and analysis was also much appreciated.

Literature Cited

(1) Diebold, J.P.; Scahill, J.W. Proceedings of the 15th Biomass Thermo-Contractors' Meeting; CONF-830323, PNL-SA-11306; Pacific Northwest Laboratory: Richland, WA, 1983.
(2) Mok, L.K.; Graham, R.G.; Overend, R.P.; Freel, B.A.; Bergougnou, M.A. In BioEnergy 84; Egneus, H., Ellegard, A. Eds.; Elsevier: London, 1985; pp 23-30.
(3) Kuester, J.L. In BioEnergy 84; Egneus, H., Ellegard, A., Eds.; Elsevier: London, 1985; pp 48-55.
(4) Scott, D.S.; Piskorz, J. Can.J.Chem.Eng., 1982, 60, pp 666-674.
(5) Scott, D.S.; Piskorz, J. Can.J.Chem.Eng., 1984, 62, pp 405-412.
(6) Scott, D.S.; Piskorz, J. Ind.Eng.Chem.Process Des.Devel., 1985, 24, pp 581-588.
(7) Knight, J.A.; Gorton, C.W.; Stevens, D.J. In BioEnergy 84; Egneus, H., Ellegard, A., Eds.; Elsevier: London, 1985; pp 9-14.
(8) Beaumont, O. Ind.Eng.Chem. Process Des.Devel., 1984, 23, pp 637-641
(9) Kosstrin, H. Proc. Specialists Workshop on Fast Pyrolysis of Biomass; SERI/CP 622-1096; 1980; pp 105-121.
(10) Scott, D.S.; Piskorz, J.; Bergougnou, M.A.; Graham, R.; Overend, R.P. I&EC Research, 1988, 27, pp 8-15.
(11) Diebold, J.; Scahill, J. In Pyrolysis Oils from Biomass; Soltes, E.J., Milne, R.A., Eds.; ACS Symp. Series 376, American Chemical Society: Washington, DC, 1988, pp 31-40. See also Diebold, J.; Power, A. In Research in Thermochemical Biomass Conversion; Bridgewater, A.V., Kuester, J.L., Eds.; Elsevier: NY, 1988; pp 609-628.
(12) Graham, R.G.; Freel, B.A.; Bergougnou, M. In Research in Thermochemical Biomass Conversion; Bridgewater, A.V., Kuester, J.L., Eds.; Elsevier: London, 1988, pp 629-641.
(13) Roy, C.; De Caumia, B.; Pakdel, H. In Research in Thermochemical Biomass Conversion; Bridgewater, A.V., Kuester, J.L., Eds.; Elsevier: London, 1988, pp 585-596.
(14) Piskorz, J.; Radlein, D.; Scott, D.S., J.Anal.Appl.Pyrolysis, 1986, 9, pp 121-137.
(15) Piskorz, J.; Scott, D.S.; Radlein, D. In Pyrolysis Oils from Biomass; Soltes, E.J., Milne, T.A., Eds.; ACS Symposium Series 376, American Chemical Society: Washington, DC, 1988, pp 167-178.
(16) Radlein, D.St.A.G.; Grinshpun, A.; Piskorz, J.; Scott, D.S. J.Anal.Appl.Pyrolysis, 1987, 12, pp 39-49.
(17) Radlein, D.St.A.G.; Piskorz, J.; Scott, D.S. J. Anal. Appl. Pyrolysis, 1987, 12, pp 51-59.
(18) Piskorz, J.; Radlein, D.St.A.G.; Scott, D.S.; Czernik, S. J. Anal.Appl.Pyrolysis, 1989, 16, pp 127-142.
(19) Piskorz, J.; Majerski, P.; Radlein, D.; Scott, D.S. Energy & Fuels, 1989, 3, pp 723-726.
(20) Garg, M.; Piskorz, J.; Scott, D.S.; Radlein, D. I&EC Research, 1988, 27, pp 256-264. See also US Patent 4 822 935, April, 1989.
(21) Shafizadeh, F.; Stevenson, T.T. J.Appl.Polym.Sci., 1982, 27, pp 4577.

(22) Shafizadeh, F.; Furneaux, R.H.; Cochran, T.G.; Scholl, J.P.; Sakai, Y. J.Appl.Polym.Sci., 1979, 23, 3525.
(23) Scott, D.S.; Radlein, D.; Piskorz, J.; Mason, S.L. In Biomass for Energy and Industry; Grassi, G., Gosse, G., dos Santos, G., Eds.; Proc. 5th EC Conference; Elsevier: London, 1990, Vol. 2; pp 2.600-2.605.
(24) Elliott, D.C.; Baker, E.G.; Piskorz, J.; Scott, D.S.; Solantausta, Y. Energy & Fuels, 1988, 2, pp 234-235.
(25) Elliott, D.C.; Baker, E.G. In Energy from Biomass and Wastes X; Klass, D.L., Ed.; Inst. of Gas Technology: Chicago, IL, 1987, pp 765-784.
(26) Marshall, A.J.; MASc Thesis, "Catalytic Conversion of Pyrolysis Oil in the Vapour Phase", Dept. of Chemical Engineering, University of Waterloo, Waterloo, Ontario, Canada, 1984.
(27) Diebold, J.P.; Scahill, J.W. Energy Progress, 1988, 8, 59-65.
(28) Beckman, D.; Elliott, D.C.; Covert, B.; Hornell, C.; Kjellstrom, B.; Ostman, A.; Solantausta, Y.; Tulenheimo, V. Techno-Economic Assessment of Selected Biomass Liquefaction Processes; Final Report of IEA Cooperative Project Direct Biomass Liquefaction, Report 697, Tech. Research Centre of Finland, Espoo, Finland, 1990.
(29) Piskorz, J.; Majerski, P.; Radlein, D.; Scott, D.S. Fast Pyrolysis of Some Agricultural and Industrial Materials; presented at the 8th Canadian Bioenergy R&D Seminar, Ottawa, Canada, October 23 & 24, 1990.

RECEIVED May 20, 1991

Chapter 25

Plants as Sources of Drugs and Agrochemicals

James D. McChesney and Alice M. Clark

Research Institute of Pharmaceutical Sciences and Department of Pharmacognosy, School of Pharmacy, The University of Mississippi, University, MS 38677

> The higher plants are remarkable in their ability to produce a vast number of diverse metabolites ranging in chemical complexity and biological activities. Natural products have historically served as templates for the development of many important classes of drugs. Further, efforts are now being increasingly focused on the potential application of higher plant-derived natural products as agrochemicals. Both areas have enjoyed the development of new bioassays to detect desired biological activity and direct isolation efforts. Such efforts have resulted in the isolation of prototype plant-derived drugs such as the terpene antimalarial qinghaosu (artemisinin), the diterpene anticancer agent, taxol, and the antifertility agent, gossypol. The discovery of the insecticide pyrethrin as a constituent of the *Pyrethrum* daisy led to the development of commercially useful synthetic pyrethroid insecticides. Clearly, natural products are perhaps the best source of **prototypes** to serve as templates for new drug and agrochemical development. All aspects of the discovery and development process will be discussed, as well as a survey of recent developments in each area.

The discovery and development of drugs and agrochemicals from plant sources falls within the domain of the pharmacognosist. Pharmacognosy is the unique discipline of pharmacy which focuses primarily on natural products. Literally, pharmacognosy means a knowledge of drugs and the discipline had its origins in the early healing arts and sciences developed by centuries-old ancient civilizations. From those early beginnings of the *Materia Medica* and the use of concoctions of plant parts as healing agents, the science has developed into a specialized discipline which includes all aspects of the study of natural products and encompasses many different sciences (botany, chemistry, microbiology, biochemistry, pharmacology, molecular biology, agronomy, and biotechnology). Pharmacognosy, as a term to describe "the knowledge of drugs", was introduced by Seydler in 1815. At that time, "drugs" consisted of crude plant parts and extracts. As the discipline has progressed, so, too, has the "definition" of pharmacognosy. Thus, in this day, pharmacognosy generally refers to those scientific endeavors pertaining to natural products utilized to benefit mankind. Such endeavors might include (but are not restricted to) the identification of traditional medicines (ethnobotany), the detection, isolation and structure elucidation of natural products possessing desired biological activities, the synthesis of natural products and their analogs, and the metabolism and/or degradation of natural products by animals and microbes. The focus of this discussion will be on the discovery and development of natural products with potential application as drugs or agrochemicals.

Human survival has always depended on plants. Early man relied entirely on them for food, medicine, much of his clothing and shelter, and his botanical skills should not be underestimated. All of the world's major crops were brought in from the wild and domesticated in prehistoric times. Aside from their value as sources of food, drugs or industrial raw materials, plants are important to man in many other ways. One can hardly imagine modern society without soaps and toiletries, perfumes, condiments and spices, and similar materials, all of plant origin, which enhance our standard of living. The roles of forests and other types of natural vegetation in controlling floods and erosion, and in providing recreational facilities are of immeasurable worth.

An adequate food supply is, and always has been, man's most outstanding need. That food shortages are often due to political, socio-economic, unequal distribution, or even cultural issues is recognized. However, in many cases, shortages are caused by attempts to employ inappropriate Western technology or to grow inappropriate crops. As reported by a recent National Academy of Sciences study, man has used some 3,000 plant species for food throughout history (1). At least 150 of them have been, to some extent, commercially cultivated. However, over the centuries the tendency has been to concentrate on fewer and fewer. Today, nearly all of the people in the world are fed by about 20 crops.

Yet, as the prospect of food shortages becomes more acute, people must depend increasingly on plants rather than animals for the protein in their diet. Reliance on a small number of plants carries great risk, for monocultures are extremely vulnerable to catastrophic failure brought about by disease, variations in climate, or insect infestations. As is well recognized, research to increase the yield of the established high protein food plants must be continued. Increased productivity of agriculture to meet the ever increasing needs of our exploding world population can be attained in several ways. The traditional approach of improving existing crop plants through selection of improved strains and varieties and improvement of soil fertility through increased fertilization practices have been very productive to this point. The exciting prospects of genetic engineering applied to agricultural production hold much promise for the future. However, at the present time, our most effective and immediate methodology for increasing crop productivity is through the utilization of chemical substances which reduce the effects of environmental and biological stresses upon the crop plants. These include selective herbicides, insecticides or repellants, fungicides, plant and animal growth regulators, antibiotics, antiparasitics, and substances which would increase crop tolerance to frost or drought. These agrochemicals must have selective activity and be non-persistent in the environment.

Both the mounting demand for food and the spector of starvation and severe malnutrition have intensified the need to reduce the devastation of food supplies by man's perennial enemy – destructive insects. It is estimated by the World Health Organization that as much as one-third of the agricultural products grown by man worldwide are consumed or destroyed by insects. Insects ravage not only food crops but livestock, timber, fiber crops, etc. Also they are responsible for the transmission of numerous diseases of man and his livestock. All this places a very high premium on effective methods of insect control. The need for new methods of insect control is accentuated by many factors. First, there is no known method to adequately control certain devastating insect pests. Second, government restrictions have been imposed on the use of several effective and previously widely used insecticides which are believed to pose potential threat to human health and/or the environment. Third, there is increasing resistance of many insects to agents which were once effective in control of those insects. Finally, the societal demand that insect control agents, when properly used, must only control target insects and not injure other organisms (humans, livestock, crops, fish, wildlife, beneficial insects, etc.) cannot be ignored. Moreover, residues must not be present in food or feeds or must be demonstrated to cause no harm if they are. Thus, the search for new insect control agents must take into account the following criteria: a) agents must control the target insects at reasonable dosages; b) agents must be selective so that they control the target insects while leaving beneficial insects, natural predators and other organisms unharmed; c) agent(s) must not generate resistance in the population of target insects; d) handling and application methods must be uncomplicated; e) agent(s) should be inexpensive; f) agent(s) must not be overly persistent and should be metabolically degraded reasonably rapidly to harmless compounds by plants, animals or soil microorganisms.

Although a majority of man's cultivated plants and domestic animals as well as man himself are attacked by several insect pests, many plants have been observed to be relatively resistant to insect damage. Today, a wide variety of plants have been shown to be effective not only as insect toxicants but also as repellants, feeding deterrents, attractants, inhibitors of growth and development, and sterilants. The ecology literature is replete with examples of resistance of specific plants to insect damage. Many times this resistance can be demonstrated to result from the presence of specific chemical substances in the plant tissues. Quite often this chemical substance may not be directly toxic (insecticidal) but rather inhibit the feeding of the insect (antifeedant) on the plant. In turn, the insect either moves to a different plant to feed or, if not capable of changing to a different plant, starves. Such an agent has many advantages. For example, such agents often are selective but inhibit feeding by a broad range of phytophagus insects, thereby conferring general protection to the plant. Also, antifeedants are generally not toxic and thereby do not pose an environmental hazard to non-target organisms. Further, there is slow or no development of resistance in the target insect population and applications and handling of such agents should be safer and less complicated than for highly toxic insecticides. Finally, it is possible that, as a natural product, there are already mechanisms in the environment for degrading the agent. An added advantage is that, as a plant derived natural product, the agent could be produced by cultivation and processing perhaps in many cases right at the site of usage so as to be relatively inexpensive.

Paralleling man's need for food is his need for agents to treat his ailments. The practice of medicine by physicians today in the United States and Western Europe is very different from the practice of their predecessors. This is largely because modern doctors have available a large array of medicines with specific curative effect. However, we lack specific curative agents for a number of important diseases. Some 800 million or even 1 billion people, at least one-fifth of the world's population, suffer from tropical diseases: malaria, schistosomiasis, trypanosomiasis, leprosy, leishmaniasis, *etc*. Even in the United States heart disease, cancer, viral diseases (for example AIDS), antibiotic resistant infections, and many others still lack adequate treatment.

Progress in the future toward cures of the serious diseases which still afflict mankind (cancer, AIDS, hypertension, various tropical diseases, etc.) depends upon discovery of new chemotherapeutic agents (drugs) which will be effective in treatment of those diseases. The search for new drugs has traditionally taken the form of evaluations of preparations of organisms (particularly higher plants) for appropriate biological activity and the purification and characterization of the substance(s) which are responsible for the desired activity. This approach to drug discovery became commonplace after the discovery in 1820 of quinine as the active constituent of the antimalarial *Cinchona* bark. Prior to that effort, investigations into the constituents of medicinal plants was done primarily as an "academic" or "scientific" endeavor. With the discovery and utilization of quinine as a pure chemical entity (drug), the course of modern drug discovery was irreversibly altered. Now, of course, new drug discovery is often much more targeted and the concept of rational drug design is one which generates significant interest and effort. However, it should be noted that a vast majority of clinically useful drugs are derived from prototype natural products. DerMarderosian estimates that "25% of all prescriptions dispensed in community pharmacies between 1959 and 1980 contain active ingredients that are extracted from higher plants" (2). Nevertheless, less than 15% of the estimated 200,000 - 250,000 species of higher plants have been investigated for biologically active constituents. Given that the evaluations which have been accomplished have generally focused on specific, restricted biological activities, the opportunities for new drug/agrochemical discoveries from natural sources becomes even more evident. Thus, the discovery of efficacious preparations effective against important diseases still afflicting mankind should be given highest priority.

An additional point should be made regarding traditional medicine. As the costs of health care continue to increase (in the United States at nearly twice the rate of inflation over the last several years), persons of limited incomes are forced more and more to rely on self-medication. These treatments often include imported and locally gathered wild and cultivated medicinal plants. Further, according to a recent study by the World Health Organization, the vast majority of the world's peoples are absolutely dependent on traditional medicine to meet their daily health care needs (3). Clearly, we must have increased scientific knowledge and information about traditional medicines upon which to develop and recommend their rational use as the basis for health care of 65-70% of the population of the world.

The areas of research and development need, outlined above, represent facets of the study of chemical substances found in plants (natural products) and the exploitation of those substances to meet the needs of mankind. If these research and development needs are to be met adequately, it will only be as a result of a broadly based, multidisciplinary effort coordinated in philosophy and endeavor. Such an endeavor must enlist the expertise of many disciplines in a systematic and coordinated effort.

The potential for developing new sources of valuable plant chemicals is largely unexplored and the benefits from doing so unexploited. Plants are known sources of insecticides, herbicides, medicines, and other useful substances; developing new industries and crops based upon plant extracts and extraction residues provides opportunities for agricultural and industrial expansion that will benefit farmers, consumers and industry, both in the developing as well as the developed nations. For example, new-crop and plant-product development will:

- potentially provide less environmentally hazardous pesticides.
- provide consumers with new products, including new drugs to treat diseases.
- diversify and increase efficiency of agricultural production.
- improve land resource use.
- offer increased economic stability to farmers.
- create new and improve existing agriculturally related industries.
- increase employment opportunities.
- provide industries with alternative and sustainable sources of raw materials.

The common thread of all research endeavors is identification and development of specific plant-derived substances having a desired selective biological activity. In this regard, it is important in the discussion of drug and/or agrochemical discovery from natural sources to appreciate fully the importance of the "prototype" discovery, for it is this that paves the way for significant new developments. In general, it should not be viewed as necessary to discover an agent which is safer and more efficacious than known agents (although this is always desirable). The compound discovered as a natural product will not likely be the final drug in the marketplace or agrochemical used in the field (although there are numerous examples of such cases). Rather, it is expected that such **prototype** agents will define a new "class" which would be subject to refinement through standard structure-activity relationship studies, improved formulation and delivery, etc. Thus, one need only demonstrate the desired biological activity and thus the potential for future development. Clearly, the further this process is carried, the more interesting the compound.

As noted earlier, the discovery of plant chemicals with potential benefit to mankind involves a multidisciplinary approach, generally utilizing a tiered screening approach. What follows is a general description of our multidisciplinary approach to the discovery of useful natural products from plants.

Preliminary Economic Assessment

Preliminary economic assessment consists primarily of examinations of the magnitude of various market segments, the market potential of classes of products, and the potential for commercialization of classes of products.

Identify Plants Likely to Possess Desired Biological Activity

Plant genera and species likely to possess a desired biological activity can be identified through one or more of several sources. Databases (*e.g.*, NAPRALERT, USDA EBL DATABASE, DIALOG, etc.) can be searched for specific biological activities, reports of phytochemical constituents, geographical sources, etc. Additional information useful in plant selection can come from research conducted by programs such as the Rural Health Research Program of The University of Mississippi Research Institute of Pharmaceutical Sciences and the World Health Organization's Collaborative Programs for Study in Traditional Practices. Further, collaborative arrangements with colleagues throughout the world are a fruitful source of materials which have been selected for evaluation based upon a local history of traditional use. Finally, after most plants with a history of traditional use have been examined, a systematic evaluation of plants can be undertaken based upon a combination of chemotaxonomic and ecology data. Such information can identify species which are closely related and, thus, likely to contain similar constituents. This is particularly useful if an active agent has already been identified and one wishes to search for other related natural products.

Collect Plant Material

Plant material should be collected by a Field Botanist or by arrangement with other laboratories and groups throughout the world. It is helpful to obtain sufficient quantities in the initial collection to be utilized for replicate biological assays. It is also important to record the location, season, local common name, etc. for use in future recollections of larger quantity.

Verify Plant Identification and Prepare Voucher Specimen

This function is **extremely important** because plants are often misidentified by folk users and in the literature. Since initial collections are generally smaller in quantity, a recollection of active plants is usually required. Failure to have properly identified the plant can lead to recollection of the wrong material and, consequently, the inability to reproduce encouraging data.

Prepare Extracts of the Plant Material

Standard procedures for the preparation of extracts are generally followed. However, some flexibility in this aspect is encouraged, particularly when investigating "folklore" or traditional medicine remedies which are prepared or processed in a specific manner.

Conduct Biological Evaluation of the Extracts

Biological evaluation of the extracts are conducted using standardized bioassays such as molecularly based *in vitro* assays as well as intact organisms. Confirmation of biological activity is extremely important. It saves the effort and expense of fractionation and purification of those compounds whose activity cannot be verified (replicated). Many drug discovery programs have failed because sufficient attention was not given to reconfirmation and verification of biological activity in plant material.

Perform Bulk Collection, Extraction, Fractionation & Purification

Plant species showing reconfirmed activity require bulk recollection. Large scale extraction, fractionation, and purification of the active principals should provide sufficient quantity for complete chemical characterization. This fractionation and purification should be guided by careful biological evaluation of all fractions; fractions continuing to show biological activity are carried on to isolation and purification of the active chemical constituent(s). Since the biological assays to be utilized are dependent upon the target therapeutic drug class or agrochemical application, a greater diversity of expertise is generally required for this activity.

Identify the Active Principals

The chemical structure of the purified, biologically active principals are determined utilizing, as appropriate, various spectroscopic, chemical or x-ray crystallographic means.

Define the Toxicity and Pharmacology of the Active Principals

The pure, chemically characterized active principals should be systematically evaluated for toxicity and biological activity using specific *in vivo* and *in vitro* bioassays. These evaluations, even if only preliminary in nature, can provide significant impetus for further development of active, relatively nontoxic products.

Determine Structure-Activity Relationships

As noted earlier, in most cases the active chemical constituent of the plant material serves as a "prototype" or "lead" for a more desirable molecule having greater efficacy or fewer or less severe side effects. Typically, the lead compound is modified chemically and the efficacy and toxicity of the analogs are determined. The utilization of molecular modeling can also provide significant support in this endeavor by identifying chemical analogs that possess the greatest therapeutic potential and the most probable efficient pathways to their syntheses.

Conduct Clinical Trials or Field Trials

Clinical trials (pharmaceuticals) or field trials (agricultural products) should be conducted in accordance with regulatory guidelines promulgated by the appropriate agency.

Commercialization

Commercialization is actually much more than a final step in the development of a natural product. It actually represents the beginning of the utility of the discovery, i.e., its delivery to the potential marketplace in a manner which makes it accessible to its users.

Conduct Agronomic Studies

Agronomic studies should be conducted to determine optimal growth conditions and to increase the yield of the desired chemical constituent of the plant. The ultimate goal of this activity is to develop alternative cash crops.

Selected Examples of Drugs and Agrochemicals From Plants

Cocaine. The prototype local anesthetic cocaine was discovered as the active principle in the leaves of the South American shrub *Erythroxylon coca*. Cocaine, which in addition to its ability to produce a sense of euphoria (which has led to its abuse) also is a local anesthetic, vasoconstrictor, and CNS stimulant. Introduced by Koller in 1884 as a local anesthetic for ophthalmic surgery, cocaine is considered the prototype local anesthetic which served as a model for the synthesis of clinically useful local anesthetics acting by inhibition of the neural uptake of amines. These include the well known anesthetics novocaine and benzocaine.

Antimalarials. The discovery of quinine as the active chemical constituent in the antimalarial *Cinchona* bark irreversibly altered the course of drug discovery and led to the use of pure natural products as drugs. The report by Pelletier and Caventou which appeared in 1820, was followed by Francois Magendie's treatise, *Formulaire* in 1821, which for the first time described single chemicals rather than plant mixtures. The inability to obtain *Cinchona* bark in World War II sparked one of the largest and most elegant synthetic challenges, *i.e.* the total synthesis of quinine. Quinine subsequently served as the prototype antimalarial for the synthesis of a series of simpler aminoquinolines from which arose two principal antimalarial drugs used today: primaquine and chloroquine.

Resistant strains of malaria persist and given that malaria claims more than 1 million lives each year in Africa alone, significant effort continues to focus on the development of new, safer agents effective against drug-resistant strains. For this researchers turned once again to natural products. For centuries, extracts of the plant known as Qinghao were used in Chinese traditional medicine for the treatment of malaria, including cerebral malaria. Chinese investigators reported the isolation and identification of the active constituent of Qinghao as the unusual sesquiterpene endoperoxide artemisinin (*4*). Numerous studies since the initial discovery of artemisinin have investigated the production, efficacy, chemical and microbial transformation, mammalian and microbial metabolism, and pharmaco-

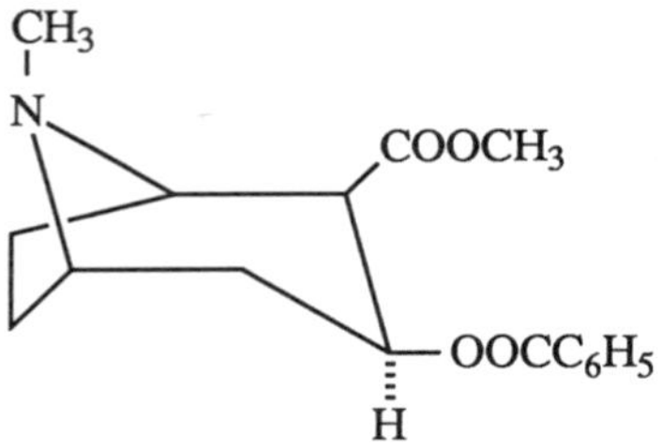

Cocaine

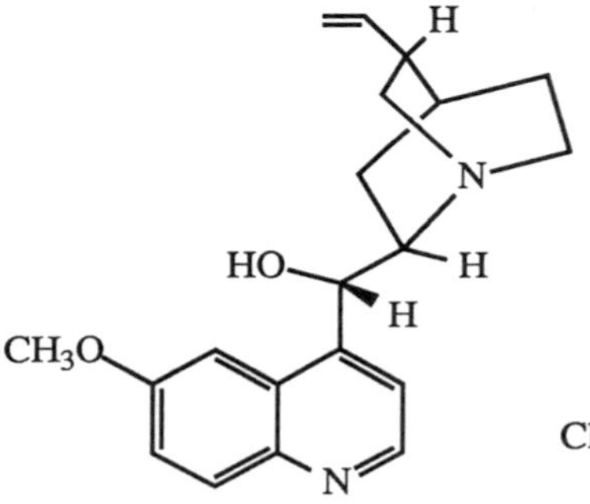

Quinine

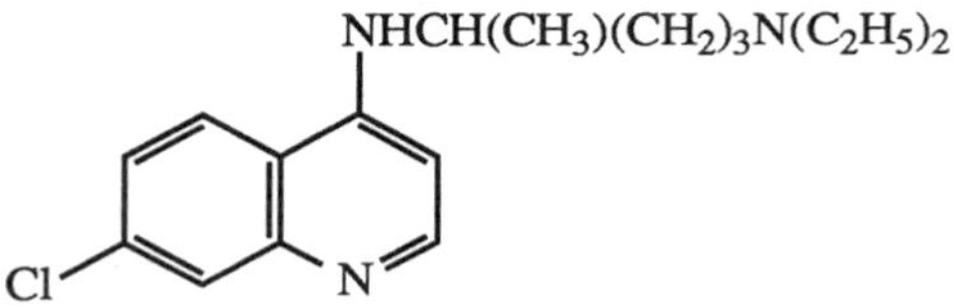

Chloroquine

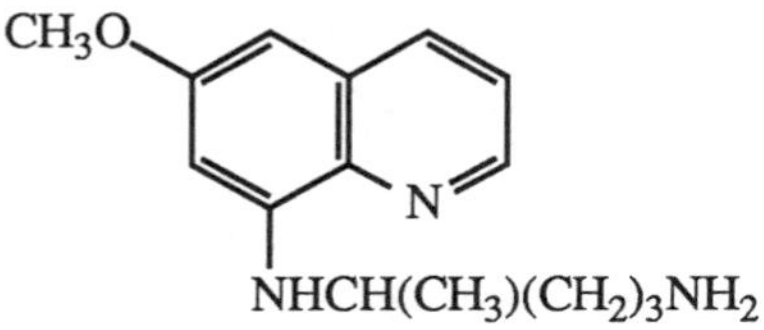

Primaquine

kinetics of this drug (*5 -7*). Further, it has now been shown that certain semisynthetic agents may hold more promise for development and the World Health Organization has concentrated its development on the semisynthetic derivative arteether.

Anticancer. Taxol, a diterpene isolated from the stem bark of the Western yew, *Taxus brevifolia* (*8*), is a promising new anticancer agent. Although taxol was discovered in 1971, its development has been impeded due to the low yield (0.004-0.016%) from the plant. However, the discovery that taxol is effective against advanced ovarian cancer has generated significant interest in the development of the compound and it is now in phase II clinical trials. Taxol is an antimicrotubule drug, which is unique among mitotic poisons in that it **enhances** tubulin polymerization.

Podophyllotoxin, a lignan constituent of the roots of the may apple, *Podophyllum peltatum*, is also a specific inhibitor of mitosis which acts by irreversible binding to tubulin (*9-14*). The discovery of the anticancer activity of podophyllotoxin led to the development of two new synthetic anticancer drugs, etoposide and teniposide, both of which are currently marketed as anticancer agents in the U.S. (*15, 16*).

Antiviral Agents. Given the inability to treat adequately viral diseases in general, there has been significant effort to identify effective antiviral agents from higher plants. Two important examples of such agents which can be considered prototypes are castanospermine and hypericin, both of which exhibit *in vitro* activity against Human Immunodeficiency Virus (HIV), *i.e.*, the AIDS virus. Castanospermine is an alkaloid isolated from the horse chestnut, *Castanospernum australe* (*17*). As a prototype agent, castanospermine led to deoxynojirmycin, a microbial product which is currently in phase I clinical trials for the treatment of AIDS (*18*). Hypericin and pseudohypericin are constituents of a common herb known as St. Johnswort (*Hypericum*) which have demonstrated significant *in vitro* anti HIV activity (*19*). These compounds are in very early stages of development and investigation into their antiviral properties.

Gossypol. Gossypol is a dimeric polyphenol obtained from cottonseed oil (*Gossypium*). It was noted in China that consumption of large quantities of cottonseed oil led to male infertility. Isolation and identification of gossypol as the active constituent led to the use of this agent in China as a male contraceptive (*20*).

Antifungals. In our laboratories, a comprehensive program to discover and develop prototype antifungal antibiotics from higher plants has produced a number of active natural products. Using bioassy-directed fractionation the oxoaporphine alkaloid, liriodenine was isolated as the active antifungal constituent of the heartwood of the tulip poplar tree, *Liriodendron tulipifera* (*21*). Quaternization to the methiodide salt significantly enhanced activity (*21*) and both compounds have been shown to be effective in the treatment of experimental murine candidiasis (*22*).

The azafluoranthene alkaloid, eupolauridine, was found to be the active antifungal component of the West African shrub *Cleistopholis patens* (*23*). It has also been shown to be effective *in vivo* in the treatment of experimental candidiasis (*24*). Recently, an additional antifungal component of *C. patens* was identified as the novel copyrine alkaloid, 3-methoxysampangine (*25*). Related synthetic studies have now yielded the parent compound sampangine (previously reported as a constituent of *Canaga odorata* (*26*)), 4-bromosampangine, 4-meth-

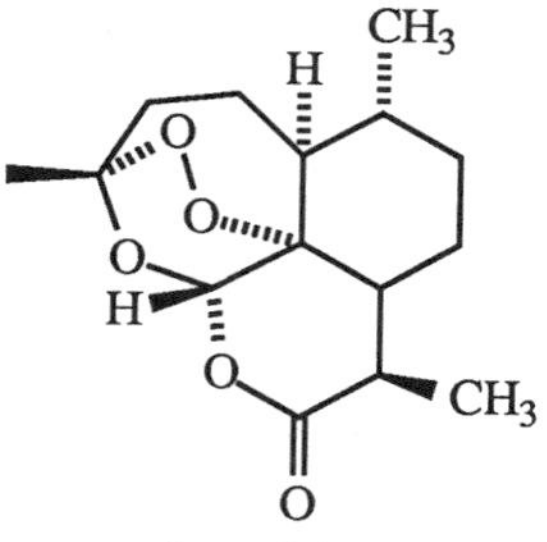

MATERIALS AND CHEMICALS FROM BIOMASS

Artemisinin

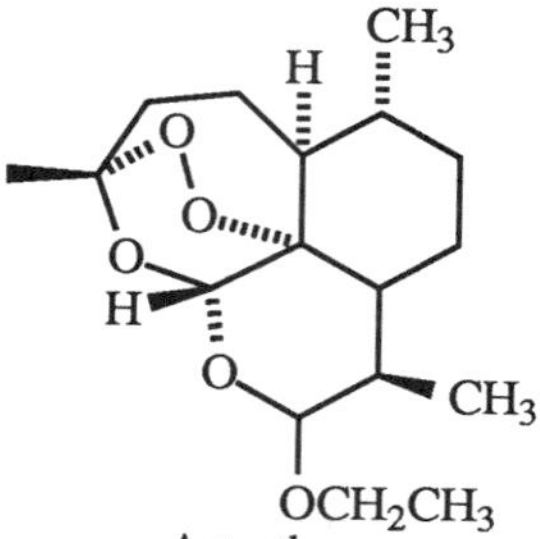

Arteether

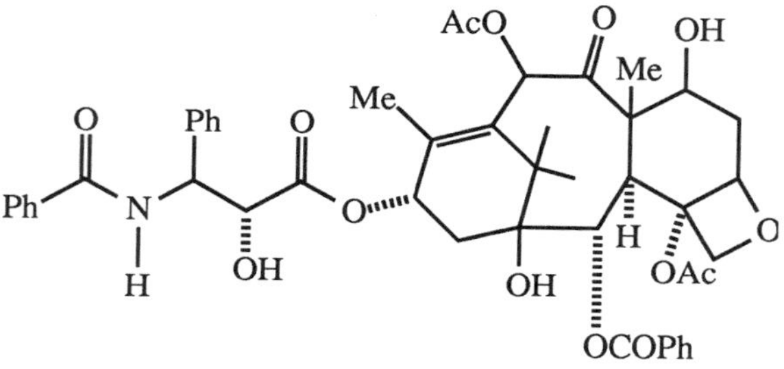

Taxol

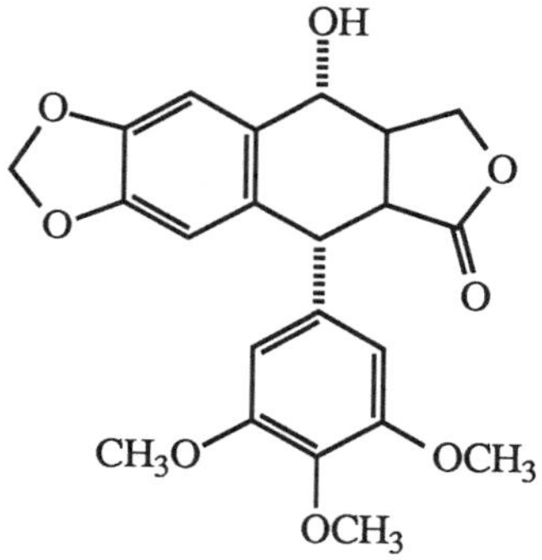

Podophyllotoxin

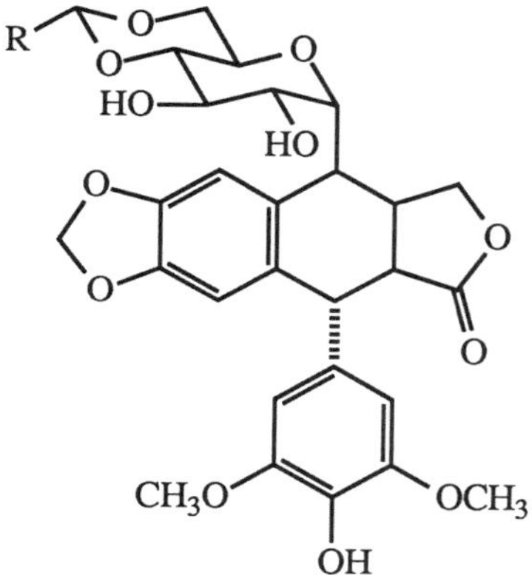

Etoposide (R=CH3)
Teniposide (R=thiazolyl)

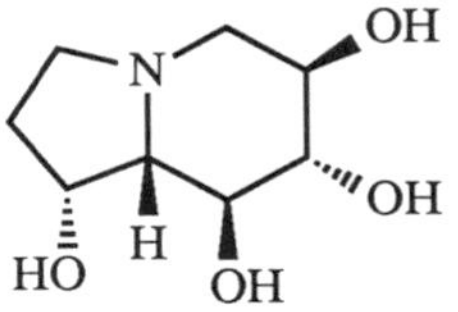

Castanospermine

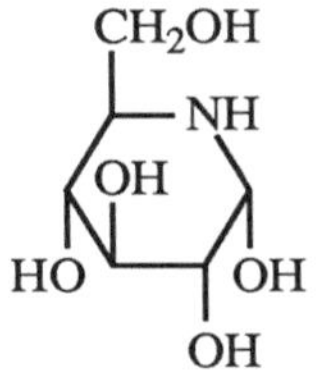

Deoxynojirmycin

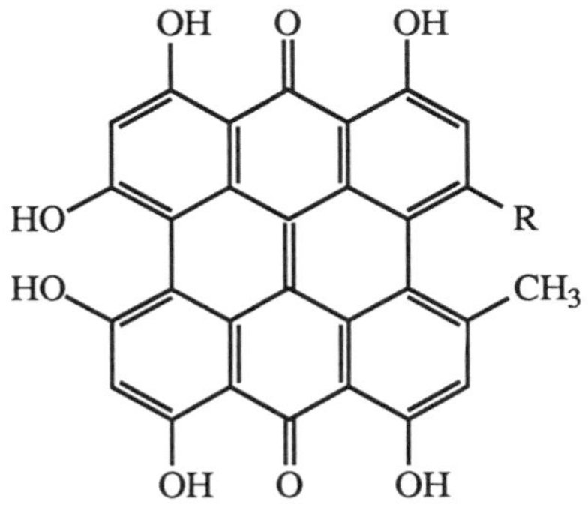

Hypericin (R=CH₃)
Pseudohypericin (R=CH₂OH)

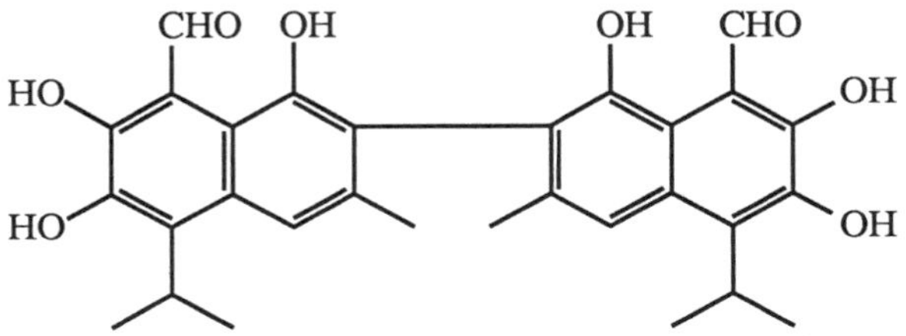

Gossypol

Liriodenine   Liriodenine methiodide

Eupolauridine   Sampangine ($R_1=R_2=H$)   Benzosampangine

oxysampangine, and benzosampangine, all of which also exhibit significant antifungal activity and are being developed as antifungal drugs (27).

Agrochemicals from Plants. While there are fewer examples of commercially available agrochemicals from plants, this is probably a reflection of the lack of research focus in this area to date rather than a lack of occurrence of such agents in higher plants. Certainly, it is well known that higher plants produce allelochemicals, natural herbicides, pesticides, growth regulators, etc. and the potential application of such agents to the improvement of agricultural productivity is obvious.

One of the first and perhaps most well known examples of a natural biodegradable insecticide is the isoflavonoid, rotenone, which can be obtained from several sources (2).

The natural insecticidal pyrethrins were originally obtained from *Chrysanthemum pryrethrum* and served as the prototype for the synthesis and development of the highly effective pyrethroid insecticides (2).

Azadirachtin is an insect antifeedant obtained from the seeds of *Melia azadirachta* and the Indian neem tree, *Azadirachta indica* (28-32). It is a prototype agent for a new generation of insect control agents as a result of its phagorepellent and growth disruption activities.

Finally, strigol, a well known growth regulator (33-35), serves as a prototype for a class of agents which may find utility in their application to improve crop productivity.

Rotenone

Pyrethrins

Azadirachtin

Strigol

Future Directions

It has already been noted that significant opportunity exists for the discovery of useful natural products from higher plants. Several specific areas of needed research focus can be identified. For example, it is likely that great potential can be found in the products of woody plants. However, given the inherent difficulties of collecting the roots and trunks of large woody plants, these are not as routinely investigated. Along these lines, one can speculate as to the possible utility of residual material from related wood industries. By the same token, other nonagricultural, nonmedicinal uses of the "marc" (extracted plant material) of plants should be investigated. Finally, the areas of biotechnology, agronomy, and tissue culture, hold much promise for the concerted, multidisciplinary approach to mass production of useful plant products.

Literature Cited

1. "Underexploited Tropical Plants with Promising Economic Value", Report of an Ad Hoc Panel of the Advisory Committee on Technology Innovation. National Academy of Sciences, USA, Washington, D.C. 1975.
2. DerMarderosian, A., and L. E. Liberti, 1988. *Natural Product Medicine: A Scientific Guide to Foods, Drugs, Cosmetics*, George F. Stickley Company, Philadelphia, PA, p. 3.
3. "The Conservation of Medicinal Plants, An International Consultation." Sponsored by World Health Organization, The International Union for Conservation of Nature and Natural Resources and The World Wide Fund for Nature. Chiang Mai, Thailand, March 21-27, 1988.
4. China Cooperative Research Group on Qinghauso and Its Derivatives as Antimalarials. 1982. *J. Trad. Chin. Med.* **2** (1), 17.
5. Klayman, D. L., 1985. *Science*, **228**:1049.
6. Luo, X.D., and C.C. Shen. 1987. *Medicinal Res. Rev.* **7**:29.
7. Lee, I.-S., and C.D. Hufford. 1990. *Pharmacology and Therapeutics*, in press.
8. Wani, M.C., H.L. Taylor, M.E. Wall, P. Coggon, and A.T. McPhail. 1971. *J. Am. Chem. Soc.* **93**:2325
9. Weiss, S.G., M. Tin-Wa, R.E. Perdue, Jr., and N.R. Farnsworth. 1975. *J. Pharm. Sci.* **64**:95.
10. Carter, S.K., and R.B. Livingston. 1976. *Cancer Treat. Rep.* **60**:1141.
11. Jardin, I. in *Anticancer Agents Based on Natural Product Models*; eds. Cassady, J.M. and J.D. Douros. 1980. Academic Press, New York, p. 319.
12. Wilson, L. 1975. *Life Sci.* 303.
13. Loike, J.D., and S.B. Horwitz. 1976. *Biochemistry* **15**:5435.
14. Brewer, C.F., J.D. Loike, S.B. Horwitz, H. Sternlicht, and W.J.Gensler. 1979. *J. Med. Chem.* **22**:215.
15. Stahelin, H., and A. von Wartburg, in *Progress in Drug Research*, ed. Jucker, J. 1989. Birkhauser Vertag, Berlin, p. 169.
16. Showalter, H.D.H., and R.T. Winters, and A.D. Sercel in *Wenkert Symposium on Natural Product Research*, The University of Mississippi, October 10-13, 1990.
17. Hohenschutz, L. D. 1981. *Phytochemistry* **20**:811.
18. *AIDS Update*, ed. K.B. Goldberg. 1990. **3** (36):3.
19. Meruelo, D., G. Lavie, and D. Lavie. 1988. *Proc. Natl. Acad. Sci. USA* **85**:5230.

20. Davie, G., F. Valentine, B. Levin, Y. Mazur, G. Gallo, D. Lavie, D. Weiner and D. Meruelo. 1989. *Proc. Natl. Acad. Sci. USA* **86**:5963.
21. Liu, Z.-Q., G.-Z. Liu, L.-S. Hei, R.-A. Zhang, and C-Z. Yu, 1981. In *Recent Advances in Fertility Regulation*; Editors, C.F. Chang, D. Griffin and A. Woolman; S.A. Atar, Geneva, Switzerland; pp. 160-163.
21. Hufford, C.D., M.J. Funderburk, J.M. Morgan, and L.W. Robertson. 1975. *J. Pharm. Sci.* **64**:789.
22. Clark, A.M., E.S. Watson, M.K. Ashfaq, and C.D. Hufford. 1987. *Pharm. Res.* **4**:495.
23. Hufford, C.D., S. Liu, A.M. Clark, and B.O. Oguntimein. 1987. *J. Nat. Prod.* **50**:961.
24. Unpublished results
25. Liu, S., B. Oguntimein, C.D. Hufford, and A.M. Clark. 1990. *Antimicrob. Ag. Chemother.* **34**:529.
26. Rao, J.U.M., G.S. Giri, T. Hanumaiah, and K.V.J. Rao. 1986. *J. Nat. Prod.* **49**:346.
27. Unpublished results
28. Nakani, K. 1975. *Recent Adv. Phytochem.* **9**:283.
29. Zanno, P.R. 1975. *J. Am. Chem. Soc.* **97**:1975.
30. Bilton, J.N. 1985. *J. Chem. Soc., Chem. Commun.* 968.
31. Kraus, W. 1985. *Tetrahedron Lett.* **26**:6435.
32. Broughton, H.B. 1986. *J. Chem. Soc., Chem. Commun.* 46.
33. Pepperman, A.B., Jr. and E.J. Blanchard, in *The Chemistry of Allelopathy: Biochemical Interactions Among Plants*, ed. A.C. Thompson. 1985. American Chemical Society, Washington, D.C., pp. 427-435.
34. Brooks, D.W., E. Kennedy, and H.S. Bevinakatti, *ibid*, pp. 437-444.
35. Vail, S.L., O.D. Dailey, Jr., W.J. Connick, Jr., and A.B. Pepperman, Jr., *ibid*, pp. 445-456.

RECEIVED March 28, 1991

Author Index

Barrier, J. W., 410
Bulls, M. M., 410
Chum, Helena L., 28,339
Clark, Alice M., 438
Dirlikov, Stoil K., 231
Doane, William M., 197
Fanta, George F., 197
Goldstein, Irving S., 332
Grohmann, K., 354
Harten, Teresa M., 42
Himmel, M. E., 354
Hon, David N.-S., 176
Jeffries, Thomas W., 313
Knox, David E., 393
Kokta, B. V., 76
Krinski, Thomas L., 299
Majewicz, Thomas G., 265
Matsuda, Hideaki, 98
McChesney, James D., 438
Myers, George E., 42
Narayan, Ramani, 1,57
Northey, Robert A., 146
Piskorz, J., 423
Plackett, David V., 88
Poole, P. W., 273
Power, Arthur J., 28,339
Radlein, D. St. A. G., 423
Raj, R. G., 76
Robinson, Philip L., 393
Rowell, Roger M., 12
Sau, Arjun C., 265
Schultz, Tor P., ix
Scott, D. S., 423
Shiraishi, Nobuo, 136
Swanson, Charles L., 197
Thames, S. F., 273
Wyman, C. E., 354
Young, Raymond A., 115
Youngquist, John, 42

Affiliation Index

Agricultural Research Service, 197
Aqualon Company, 265
Arthur J. Power and Associates, Inc., 28,339
Clemson University, 176
Eastern Michigan University, 231
Forest Research Institute, 88
Georgia-Pacific Corporation, 146
Kyoto University, 136
Michigan Biotechnology Institute, 1,57
Michigan State University, 1,57
North Carolina State University, 332
Okura Industrial Company, Ltd., 98
Protein Technologies International, 299
Solar Energy Research Institute, 28,339,354
Tennessee Valley Authority, 410
U.S. Department of Agriculture, 12,42,313
U.S. Environmental Protection Agency, 42
Université du Québec à Trois-Rivières, 76
University of Mississippi, 438
University of Southern Mississippi, 273
University of Waterloo, 423
University of Wisconsin, 12,115
Westvaco Corporation, 393

Subject Index

A

Acetic acid, product of fermentation for fuel production, 366
Acetogens, syngas conversion, 380
Acetone–butanol–ethanol fermentation, 376
Acetylated wood
   commercialization trends, 37–38

INDEX

Acetylated wood—*Continued*
  technoeconomic assessment of process, 34,35*f*,36–37*t*
Acetylation
  advantages and disadvantages, 93
  description, 92
  examples of composite enhancement, 92–94
  long-term color stability, 94,95*f*
Acetylation of wood
  capital investment summary, 34,36*t*
  estimated processing costs, 34,36*t*,37
  process flow diagram, 34,35*f*
  processing cost vs. wood cost and moisture content, 37*t*
  properties of products, 34
Acid sulfite liquors, purification, 151
Acrylated fatty acids
  in coatings
    advantages, 403
    viscosity curves, 403,405,406*f*
  in soap and detergent applications
    hard-surface cleaning formulations, 402
    hydrotroping capabilities, 400,401*f*
    properties, 400
    viscosity vs. solids, 402,404*f*
Activation, description, 121
Adenosine triphosphate, production, 363
Adhesives
  preparation
    chemically modified wood, 138–140
    untreated wood, 141
  use of lignins, 166
Agricultural commodities, source of polymeric materials, 273–297
Agricultural residues
  availability for energy production, 412,413*t*
  definition, 412
  development of end uses, 356
  relationship to mineral nutrient recycling, 356
  waste product, 355–356
Agricultural wastes
  fluidized bed fast pyrolysis, 432,433*t*
  liquid pyrolysis product composition, 433,434*t*
  use, 432,433–434*t*
Agrochemicals from plants
  applications, 448
  structures, 448–449

Air laying, *See* Nonwoven mat thermoformable composites
Alloys
  concept, 60,61*f*
  definition, 58
Alternately adding esterification reactions, carboxyl group bearing esterified woods, 104
Alternative fuels from biomass
  abnormal combustion, resistance, 360
  boiling points of selected organic liquids, 360,361*f*
  net heats of combustion for selected organic liquids, 360,362*f*
  octane numbers of gasoline, improvement, 360,363
  use in gasoline, 357,360
Alternative fuels industry, developmental needs, 354–355
Anaerobic digestion
  adaptability, 382
  description, 381–382,416
  rates of production, 382–383
  source of high-quality gaseous fuel, 381
Anhydride reactions with soy protein
  protein molecular-weight distribution, 306,309*f*
  protein titration curve, 306,308*f*
  reactions, 302,306,307*f*
  soy polymer rheogram, 306,308*f*
3,6-Anhydrogluconic acid 1,4-lactone, *See* 1,4-Lactone of 3,6-anhydrogluconic acid
3,6-Anhydrogluconic acid 1,4-lactone polyurethanes
  applications, 241
  degradation, 239–240
  hydrolysis, 240–241
  preparation, 239
  properties, 239
Animal feed binder, use of lignins, 161–162
Animal feed molasses additive, use of lignins, 161
Anticancer agents, development and structures, 445–446
Antifungals, applications and structures, 445,448
Antimalarials
  development, 443,445
  discovery, 443
  structures, 443–444,446

Antiviral agents, applications and structures, 445,447
Aquatic resources
  availability for energy consumption, 415
  examples, 415
Artemisinin
  application, 443,445
  structure, 443,446
ASA mats, example of composite, 33–34
Autohydrolysis, definition, 371
Automobile weight reduction, use of composites, 33
Azadirachtin, application and structure, 448–449

B

Bagasse, *See* Guayule bagasse
Base-catalyzed reactions of soy proteins, addition reactions, 306,310*f*
Battery expander, use of lignins, 165
Benzosampangine, application and structure, 448
Binders, use of lignins, 156
Biobased materials
  cost-effective production, examples, 34–38
  cradle-to-grave materials cycle analysis, 29–30
  future applications, examples, 38–39
  requirements, 28–29
Biochemical biomass conversion processes, description, 416
Biodegradable plastics
  importance, 200–201
  markets, 202
  role of starch, 201
Biodegradable polymers, advantages, 7
Biodegradation, lignocellulosics, 20–21
Biofuels, advantages, 8
Biofuels Municipal Waste Technology program, pathways for biofuel production, 8
Biological conversions of sugars to liquid and gaseous fuels, efficiency, 383
Biological hydrogen production, description, 383
Biological resistance, lignocellulosic composites, 23

Biological treatments of pulp, classifications, 313
Biomass
  alternative fuels, 357,360–363
  availability, 333,411
  cellulosic biomass feedstocks, 411,412–414*t*
  chemical products, future potential, 334
  composition, 333
  definition, 12,332,410
  economic feasibility of conversion to useful chemicals and fuels, 336–337
  energy consumption, 410*t*,411
  energy conversion, processes, 411
  energy source, future potential, 415,416*t*
  fossil fuels, availability, 337
  fuel properties, summary, 357,359*t*
  future use, 337
  historical chemical products, 333
  lignocellulose, biological processing, 336
  nontraditional sources, 414–415
  productivity, 332–333
  pyrolysis of agricultural and forest wastes, 422–434
  Tennessee Valley Authority's research program, 417*t*,418–419*f*,420
Biomass and waste conversion, potential and pitfalls, 339–349
Biomass conversion technologies, requirements, 340
Biomass-derived materials
  biopolymers and derivatives, 7–8
  lignocellulosic composites, markets, 5,6*f*,7
  production levels, 5
Biomass feedstocks, 9
Biomass renewable resources
  emerging materials, chemicals, and fuels, 3,4*f*
  environmentalism, 3
  global warming, effect of resource use, 3,5,6*f*
  U.S. national policy, 5
Biomass resources, 355–357
Biomimetic oxidative agents, use for bleaching, 320*t*
Biopolymer(s), markets, 7–8
Biopolymer derivatives, markets, 7–8
Biopolymer plastics via starch fermentation
  poly(hydroxybutyrate-*co*-hydroxyvalerate), 203,204*f*
  poly(lactic acid), 202–203,204*f*

INDEX

Bisepoxide, cross-linking with esterified woods, 104–105
Bleached kraft pulp, importance of production, 318
Bleaching, enzymatically enhanced, *See* Enzymatically enhanced bleaching
Bleaching with hemicellulases
  enhancement, 322,323–325*t*
  reduction in chemical demand, 323*t*
Blends
  concept, 60,61*f*
  definition, 58
Block copolymers
  definition, 187
  description, 59
  preparation from cellulose, 187,188*f*
  problems in large-scale use, 59–60
Body fluid absorption, use of starch graft copolymers, 220
Bowman–Birk trypsin inhibitor, 299
Butanediol, product of fermentation for fuel production, 366
2,3-Butanediol–ethanol fermentation
  bacteria used, 376
  description, 376–377
Butyrate, production via fermentation, 378

C

Carbon black and pigments, use of lignins, 161
Carbon fibers
  preparation from untreated wood, 142
  surface modification by plasma, 130*t*
Carboxyl group(s), introduction into wood by esterification, 100–101,102*f*,103*t*
Carboxyl group bearing esterified woods
  addition reaction with epoxides, 103–104
  alternately adding esterification reactions, 104
Castanospermine, applications and structure, 445,447
Cationic monomers, structures, 221,222*f*
Cellobiose
  polymerization, 247
  preparation of polymers, 246–247
Cellobiose polyesters, preparation and properties, 248–249
Cellobiose polyurethanes, 247–248

Cellulases
  enhancement of fibrillation, 315
  treatment of recycled fibers, 316–317
  treatment to reduce vessel picking, 317–318
Cellulose
  accessibility in fiber, 177–178
  activation treatments, 178
  annual production, 176
  conversion to ethers, 265
  derivatives and copolymers, 176
  description, 15,148
  distribution in plants, 176
  heterogeneous reactions, 178
  history of modification, 178–179
  importance, 176
  properties, 265
  structure, 177,180*f*,265*f*
Cellulose acetate, preparation from cellulose, 182
Cellulose acetate–polystyrene alloys, differential scanning calorimetry, 64,66*f*
Cellulose acetate–polystyrene–maleic anhydride alloys
  differential scanning calorimetry, 67,71,72*f*
  grafting reaction, 67,71*f*
  transmission electron microscopic studies, 71,73*f*
Cellulose carbamates, preparation from cellulose, 182
Cellulose ethers, preparation and properties, 183–184
Cellulose ferrocenyl derivatives, preparation from cellulose, 183
Cellulose fiber(s), use as fillers in thermoplastic polymer matrices, 76–77
Cellulose fiber–thermoplastic composites
  additive effect on tensile strength and modulus, 79,80*t*,81*f*,82
  calculation of modulus, 84
  fiber dispersion, effect on properties, 82–86
  fiber effect on Izod strength, 82,83*f*
  fiber orientations in short fiber composites, 84,85*f*
  formation technology, 77–78
  fracture surface
    treated composite, 82,84,85*f*
    untreated composite, 82,83*f*
  future needs, 84,86*f*
  orientation factor, fiber concentration, and tensile modulus, 84,86*f*

Cellulose fiber–thermoplastic composites—*Continued*
  processing aids–coupling agents, effect on properties, 78–83
  short fiber composites, mechanical properties, 78t
Cellulose hydrolysis, limitations of methods, 335
Cellulose phosphate, preparation from cellulose, 182–183
Cellulose sulfates, preparation from cellulose, 181
Cellulosic(s)
  ESCA, 131,132–133f
  surface modification by plasma, 130,131t,132–133f
Cellulosic alloys, synthesis of tailor-made graft copolymers, 60,62–63f,64,65f
Cellulosic biomass, production of fuels, 363,364–365t,366
Cellulosic biomass feedstocks
  agricultural residues, 412,413t
  examples, 411
  municipal solid wastes, 413,414t
  woody biomass, 411,412t
Cellulosic derivatives and copolymers
  future trends, 191
  history and background, 177–179,180f
  preparation
    block and graft copolymers, 187,188f,189–191
    cellulose acetate, 182
    cellulose carbamates, 182
    cellulose phosphate, 182–183
    cellulose sulfates, 181
    deoxycelluloses, 185,186f,187
    ethers, 183–184,186f
    halodeoxycellulose, 185,186f
    iododeoxycellulose, 185,187
    oxycelluloses, 179,180f
    polyalcohol, 178
Cement manufacture, use of lignins, 159
Chemical(s), use as biomass resources, 8–9
Chemical activation, surface modification of fibers, 121–127
Chemical conversion of biomass, approaches, 334–336
Chemical(s) from biomass, *See* Biomass

Chemical(s) from pulping
  fatty acid derivatives, 400–406
  rosin and derivatives, 398,399t
  tannins and derivatives, 405,407
  terpenes, 394–398
Chemically modified wood
  applications of liquefaction, 138–140
  liquefaction, 137
Chemical modification for dimensional stabilization of lignocellulosic composites
  acetylation processes, 92–93
  color stability, long-term, 94,95f
  cross-linking reactions, 93
  equilibrium moisture content vs. bonded acetyl content, 92–93,95f
  polymerization reactions, 93
Chemical modification of soy proteins
  anhydride reactions, 302,306,307–309f
  base-catalyzed reactions, 306,310f
  cross-linking reactions, 306,309f
  hydrophilic amino acids as targets, 302,305t
  interpolymer technology, 306,310–311f
  ionization of side residues, 302,307f
Chemical products from biomass
  examples, 333
  potential future products, 334
Chlorinated guayule rubber
  characterization, 278,280–285
  ^{13}C-NMR chemical shift, 278,280t,281f
  differential scanning calorimetric curve, 284,285f
  Fourier-transform IR spectra, 278,282f
  gel permeation chromatograms, 278,282–283f
  preparation, 278
  utility, 284,285–286t
Chloroquine, structure, 443–444
Cloned *Bacillus subtilis* preparation, effect on properties of kraft pulp, 324,325t
CO_2 in atmosphere, effect of biomass utilization, 3,5,6f
Cocaine, applications and structure, 443–444
Commercial cleaning, use of lignins, 164–165
Composite(s)
  comparison of fiber properties of synthetic and wood materials, 33,35f
  definition, 32
  examples, 32–34
  markets, 33
Composite materials, growth, 115

INDEX

Composition
  kraft lignin, 154
  lignin materials, 152–154
  lignocellulosics, 15,16–17f,18,19t
  lignosulfonates, 152–154
  wood, 147–148
Concentrated acid hydrolyses of polysaccharides, 369
Concrete admixtures, use of lignins, 158–159
Condensed tannins, applications, 407
Corn stover, source of biomass, 356
Corona, surface modification of cellulosics, 130,131t,132–133f
Cross-linking reactions of soy proteins
  epichlorohydrin, 306,309f
  magme, 306,309f
Cyclodextrin(s)
  applications, 258–259
  experimental preparation procedure, 261
  structure, 257
  types, 257
Cyclodextrin membranes
  advantages, 257–258
  permeability rates, 258
  preparation, 258
  properties, 258–259

D

Demethylation of kraft lignin
  schematic representation, 348,349f
  technoeconomic feasibility, 348
Deoxycelluloses
  description, 185
  preparation from cellulose, 185,186f,187
Deoxynojirmycin, applications and structure, 445,447
Depolymerization of polysaccharides
  dilute acid hydrolysis, 368
  enzymatic hydrolysis, 369–374
  hydrolysis
    concentrated mineral acids, 369
    plant cell walls, 367–368
    starch and inulin, 366–367
1,2:5,6-Dianhydro-3,4-O-isopropylidene-D-mannitol
  NMR and IR spectra, 245
  preparation of diepoxy derivatives, 245–246
  preparation of polymers, 243–245

Dicarboxylic acid anhydrides, reaction with hydroxyl groups of wood, 99
1,6-Dichloro-1,6-dideoxy-3,4-O-isopropylidene-D-mannitol, experimental preparation procedure, 260
1,2:3,4-Di-O-isopropylidenegalactopyranose methacrylate, structure, 252–253
1,2:5,6-Di-O-isopropylideneglucofuranose methacrylate, preparation and structure, 252–253
1,4:3,6-Dilactone of D-glucaric acid, preparation, 242–243
1,4:3,6-Dilactone of mannosaccharic acid, forms and preparation, 241–242
Dilute acid hydrolysis of polysaccharides, rate equations, 368
Dilute acid pretreatments, description, 371–372
Dimensional stability of lignocellulosic composites, 21,22f
  chemical modification methods, 92–94,95f
  future stabilization research, 94,96
  heat treatment method, 89–91
Dimethyl isomannide, structure, 236
Dimethyl isosorbide, preparation and structure, 236
Direct combustion, description, 416
Dispersants, use of lignins, 155–156
Dissolution of wood, See Liquefaction of lignocellulosics in organic solvents
Drugs
  need for new discoveries, 439
  search approaches, 439–440
Drugs from plants
  anticancer agents, 445–446
  antifungals, 445,448
  antimalarials, 443–446
  antiviral agents, 445,447
  cocaine, 443
  gossypol, 445,447
Dye manufacture, use of lignins, 160–161

E

Economic assessment of plant chemicals, preliminary, See Preliminary economic assessment of plant chemicals
Electric discharge activation, surface modification of fibers by plasmas,125,128–133

Electron spectroscopy for chemical analysis (ESCA)
  procedure, 120–121
  range of binding energies for C–O bonds, 121,124f
Emergent plants, use as energy source, 415
Emulsifying agents and stabilizers, use of lignins, 156–157
Emulsion(s)
  definition, 156
  use of lignins, 163
Emulsion systems, classifications of breakdown processes, 157
Energy consumption, estimated usage in United States, by resource, 410t,411
Energy crops, 414
Engineering materials
  definition, 12
  properties, 18t
Entrained flow reactor, use for biomass conversion, 424–425
Environmentalism, importance, 2
Enzymatically enhanced bleaching
  approaches, 318–325
  biomimetic oxidative agents, 320t
  lignin-degrading enzymes, 320,321–322t
  residual lignin in kraft pulp, 319–320
Enzymatic treatments of pulps
  cellulase treatment to reduce vessel picking, 317–318
  conditions, 321,322t
  enzymatic activity, 314–315
  fiber properties, selective modification, 315,316t,317–318
  fibrillation, 315,316t
  recycled fibers, treatments, 316–317
  xylan, selective removal, 318
Enzyme(s)
  techniques for production and application, 314
  use in biological treatments of pulp, 314
Enzyme-catalyzed hydrolysis of polysaccharides, 369–370
  microcrystalline cellulose, 372–373
  pretreatment method(s), 370–372
  pretreatment method and cellulose enzyme systems, importance of matching, 372–373
  technoeconomic evaluation, 373–374
  Trichoderma cellulase systems, 373

Epoxides, addition reaction with carboxyl group bearing esterified woods, 103–104
Epoxidized guayule rubber
  characterization, 287,288–291f
  ^{13}C-NMR spectrum, 287,289f
  differential scanning calorimetric curve, 287,291f
  epoxidation mechanism, 284,286f
  experimental procedure, 284
  Fourier-transform IR spectrum, 287,288f
  ^{1}H-NMR spectrum, 287,290f
Epoxy-activated cellulose
  preparation, 184
  reactions, 184,186f
Epoxy resins
  NMR and IR spectra, 245
  preparation of polymers, 243–245
Equilibrium moisture content, lignocellulosic composites, 23,25f
Esterification
  cellulose, 181–183
  introduction of carboxyl groups into wood, 100–101,102f
Esterified woods
  cross-linking with bisepoxide, 104–105
  thermal plasticity, 101,103t
Ethanol, product of fermentation for fuel production, 366
Ethanolic fermentations
  ethanologenic bacteria, construction, 375
  xylose isomerization, 374–375
  yeast, 374
  Zymomonas, 374
Etherification, cellulose, 183–184,186f
Ethylene–vinyl acetate copolymers, adhesive applications, 396
Etoposide, development and structure, 445–446
Eupolauridine, application and structure, 445,448
Extractives, description, 148

F

1990 Farm Bill, development of biomass resources, 5
Fast pyrolysis processes
  biomass conversion
    economic rationale, 423
    methods, 423–424

INDEX

Fast pyrolysis processes—*Continued*
  effect of reaction temperature, 423
  effect of residence time, 423
  fluidized bed process, 424–425
Fast pyrolysis–solvent fractionation for phenolic production
  advantages, 341–343
  description, 343–344
  schematic representation, 341,343,346f
  technoeconomic evaluation, 344
  technology transfer, 344–345
Fatty acid derivatives
  acrylated and maleinized fatty acids in coatings, 403,405
  acrylated fatty acid in soap and detergent applications, 400,401f,402,404f
  applications, 400
  maleinized fatty acid in corrosion inhibitor applications, 402–403,404f
  purity requirements, 400
  structures, 400,401f
Feedstock of biomass and wastes, availability, 410–420
Fermentation, description, 416
Fiber(s)
  surface characterization, 115–121,124
  surface modification, 121–133
Fiber mats, formation using lignocellulosic composites, 24,25–26f
Fiber reactivity, influencing factors, 177–178
Fibrillation
  cellulases for enhancement, 315
  xylanases for enhancement, 315,316t
Fibrit mat, example of composite, 33
Filaments, preparation from chemically modified wood, 140
Fire performance, lignocellulosic composites, 23
Flash hydropyrolysis, 422–423
Flash pyrolysis, 422–423
Floating plants, use as energy source, 415
Flocculation, use of starch graft copolymers, 221
Flotation aids in mining operations, use of lignins, 167
Fluidized bed, use for biomass conversion, 424
Fluidized bed fast pyrolysis process
  analysis of liquid products, 425,429t
  flow diagram, 425,426f

Fluidized bed fast pyrolysis process—*Continued*
  organic liquid, gas, and char yields, 425,427f
  pyrolysis of agricultural wastes, 433t
Foams, preparation from untreated wood, 141
Food processing and animal wastes, availability for energy production, 414–415
Food processing wastewaters, use of lignins in treatment, 167
Food supply, importance of plants, 438
Foreign oil, U.S. dependence, 1–2,4f
Forest products based industry, 31
Fuel(s) from biomass, *See* Biomass
Fuel(s) from biomass and wastes
  acetone–butanol–ethanol fermentations, 376
  alternative fuels, 357,360–363
  anaerobic digestion, 381–383
  biological hydrogen production, 383
  2,3-butanediol–ethanol fermentation, 376–377
  cellulosic biomass fermentations, 363,364–365t
  depolymerization of polysaccharides, 366–374
  engineering, 383–384
  ethanolic fermentations, 374–375
  fermentations to volatile organic acids, 377–379
  fuels production from cellulosic biomass, 363,364–365t,366
  future improvements, 384–385
  lipid production, 379
  nature and magnitude of biomass resources, 355–357
  syngas conversion to liquid and gaseous fuels, 380–381
Fuel properties of biomass, summary, 357,359t
Fuel source, use of lignins, 157

G

Gasification
  description, 416
  method for biomass conversion, 423–424
Gasification of biomass, 334
Gasoline, improvement of octane number with additives, 360,363

Gelatinized starch as component in plastics
  advantages, 208
  elastomers, 206
  films, 206,207t,208
Global warming, effect of biomass utilization, 3,5,6f
7S Globulin, description, 300
11S Globulin, description, 300
D-Glucaric acid 1,4:3,6-dilactone, preparation, 242–243
D-Glucosamine, properties and structure, 249
Glucosamine poly(amide–ester)s, properties and structure, 250
Glucosamine poly(urea–urethane)s, properties, 249–250
D-Glucose methacrylate
  applications of polymers, 252
  preparation and properties, 254
Gossypol, application and structure, 445,447
Graft copolymer(s)
  definition, 187
  description, 59
  preparation, 60,61f
  preparation from cellulose, 187,189–191
  problems with large-scale use, 59–60
Graft copolymerization, methods, 187
Granular starch as filler in plastics
  containers, 205
  embrittled petroleum-based polymers, 203,205
  poly(vinyl chloride) plastics, 205–206
  thin films, 203
Grasses, composition, 357,358t
Guayule
  occurrence, 274
  rubber extraction process, 274–275
Guayule bagasse
  acetylation, 297
  bleaching, 296
  pulping, 296
Guayule rubber
  cost vs. quality, 273–274
  high-molecular-weight rubber, 275
  low-molecular-weight rubber, 275–291
Gypsum wallboard, use of lignins, 158

H

Halodeoxycellulose, preparation from cellulose, 185,186f

Hardwoods, composition, 357,358t
Heat treatment for dimensional stabilization of lignocellulosic composites
  development, 89–90
  effect on color, 91
  heat posttreatment, 90
  Shen binderless bonding process, 91
  steam treatment, 90–91
Hemaglutinin, description, 299
Hemicellulases, bleaching, 322,323–325t
Hemicelluloses
  description, 15,18,148
  hydrolysis, 336
High-molecular-weight guayule rubber, properties, 275
Human survival, dependence on plants, 438
Hydrolytic processes, description, 416
Hydrolytic pulping, development, 150
Hydrolyzable tannins, description, 405
Hydrophobically modified Hydroxyethylcellulose, preparation and properties, 266
Hydropyrolysis in catalytic bed, use of nickel–alumina catalyst, 430,432
Hydrotroping in detergents, definition, 400
Hygiene chemicals, use of terpenoids, 394,396
Hypericin, applications and structure, 445,447

I

Immiscible blends, examples and properties, 58–59
Industrial products and processes
  developmental factors, 1
  shift from oil to biomass renewable resources, 2–6
Industrial soy protein, background, 299–305
Insects
  food supply and disease, 438
  need for control methods, 438–439
  resistance by plants, 439
International Chamber of Commerce, principles, 2
Interpolymer technology for soy protein polymers, 306,310–311f
Inulin, hydrolysis, 366–367
Inverse gas chromatography, characterization of surface properties of fibers, 119–120

INDEX

Iododeoxycellulose, preparation from cellulose, 185,187
α-Ionone, use in mouthwashes, 394,396
Isomannide, applications and structure, 235–236
Isosorbide,
  configuration and preparation, 232–233
  preparation of polyurethanes, 233–234
Isosorbide polyurethanes
  applications and preparation, 233–234
  properties, 234–235

K

Kolbe electrolysis, advantages and disadvantages, 378–379
Kraft lignin
  applications, 154–167
  composition, 154
  demethylation for phenol replacement, 347–348,349f
  isolation, 150
  modification for property enhancement, 150
  purity, 147
Kraft pulp, residual lignin, 319–320
Kraft pulping
  carbohydrate dissolution and degradation, 150
  schematic representation, 347,349f
Kunitz trypsin inhibitor, description, 299

L

Laccases, effect on brightness of pulp, 322t
Lactic acid
  polymers, 202,204f
    market, 202
    properties, 202–203
  production via fermentation, 377
1,4-Lactone of 3,6-anhydrogluconic acid (LAGA)
  configuration, 238–239
  experimental preparation procedure, 260
  preparation, 238
  preparation of polyurethanes, 239–241
  properties, 238
Leather tanning, use of lignins, 165–166
Ligna-Tock mats, example of composite, 33

Lignin(s)
  adhesives, 166
  animal feed binder, 161–162
  animal feed molasses additive, 161
  applications, 146,154–167
  battery expander, 165
  binders, 156,162–163
  carbon black and pigments, 161
  cement manufacture, 158–159
  cleaning applications, 164–165
  definition, 146
  deicers, 167
  description, 18,147–148
  dispersants, 155–156
  dye manufacture, 160–161
  emulsifiers, extenders, and emulsion stabilizing agents, 156–157,163
  flotation aids in mining operations, 167
  fuel source, 157
  future, 168
  gypsum wallboard, 158
  high-value uses, 167
  history, 154–155
  isolation, 150–152
  leather tanning, 165–166
  micronutrients, 165
  nonfuel uses, 155t
  oil well drilling muds, 159–160
  pesticide dispersant, 160
  physiological function, 147
  road dust control, 162
  rubber additive, 166
  sequesterants, 157
  treatment of food processing wastewaters, 167
  vanillin production, 166–167
  water treatment, 164–165
Ligninase
  conditions for comparing effects on pulps, 320,321t
  effect on brightness of pulp, 322t
Lignin-degrading enzymes, use for bleaching, 320,321–322t
Lignin materials
  chemical modification for property enhancement, 151
  composition, 152–154
Lignocellulose, biological processing, 336
Lignocellulosic(s)
  applications, 15,88
  composition, 15,16–17f,18,19t

Lignocellulosic(s)—*Continued*
  definition, 12–13
  dimensions, 18–19,20*t*
  properties, 18,19*t*
  sources, 12
  strength of components, 18
  types, 13
Lignocellulosic composites
  advantages in combining with other
    material, 27
  applications
    inorganics, 26
    metal films, 26
    plastics, 26–27
    possible, 88
  biodegradation, 20
  biological resistance, 23
  comparison of cost, performance, and
    production rates for types, 13,14*f*
  definition, 13
  dimensional stability, 21,22*f*
  equilibrium moisture content, 23,25*f*
  examples, 13
  fire performance, 23
  formation of nonwoven fiber mat, 24,25*f*
  future, 24,25–26*f*,27
  interaction of chemical components,
    20–21,22*f*
  markets, 5,6*f*,7
  methods for dimensional stabilization,
    88–89
  modification, 21,22*f*,23,25*f*
  photochemical degradation, 20–21
  possible dimensions of fiber mats, 24,26*f*
  processing temperatures, 89
  properties, 13,19–23,25
  public acceptance, 13,15
  sorption of moisture, 21,23
  sources, 24
  strength, 23
  types, 13
  UV radiation resistance, 23
Lignocellulosic for composites, thermal
    plasticization, 99–113
Lignocellulosic materials, chemical activation
    for surface modification, 122–127
Lignocellulosic–plastic blends
  cellulose acetate–polystyrene alloys, 64,66*f*
  cellulose acetate–polystyrene–maleic
    anhydride alloys, 67,71–73*f*

Lignocellulosic–plastic blends—*Continued*
  differential scanning calorimetry of graft
    copolymers, 64,65*f*
  miscible blends, 58
  preparation of graft copolymers,
    60,62–63*f*,64
  transmission electron microscopic studies,
    64,67,68–70*f*
  wood–plastic alloys, 71,74*f*
Lignocellulosic–plastic composites from
    recycled materials
  advantages, 42
  extreme cleaning and refinement, 43
  manufacturing techniques, 43
  municipal solid waste as source, 44*t*,45
  potential products, 42
Lignocellulosic–thermoplastic composites
    from recycled materials
  development of nonwoven mat
    composites, 49
  effect of polymer
    flexural strength, 46,49,50*f*
    notched impact energy, 46,49,50*f*
  flexural strength, 46,47*f*
  melt-blended composites, 46–50
  nonwoven mat composites, 49,51–54
  notched impact energy, 46,48*f*
  research and development needs, 54
Lignosulfates, composition, 152–154
Lignosulfonates
  applications, 154–167
  composition and purity, 147
  sugar removal, 151
Lipid production, methods, 379
Lipoxygenase, description, 300
Liquefaction of lignocellulosics
  description, 136
  organic solvents
    chemically modified wood, 137–140
    untreated wood, 137–138,141*f*,142
Liriodenine, application and structure,
    445,448
Liriodenine methiodide, application and
    structure, 445,448
Low-molecular-weight guayule rubber
  characterization, 275,276–277*f*,278
  chlorinated rubber
    characterization, 278,280*t*,281*f*
    proposed structure, 278,279*f*
    utility, 284,285–286*t*

INDEX

Low-molecular-weight guayule rubber—*Continued*
chlorination
experimental procedure, 278,280t
mechanism, 278,279f
epoxidation, 284,286–291
Fourier-transform IR spectrum, 275,277f
^{1}H- and ^{13}C-NMR spectra, 275,276f
reaction with chlorine gas, 278,279f
reaction with maleic anhydride, 287,291–292f

M

Macroalgae, use as energy source, 415
Maleic anhydride, reaction with low-molecular-weight guayule rubber, 287,291–292f
Maleinized fatty acids
coatings
advantages, 403
viscosity curves, 405,406f
corrosion inhibitor applications
mechanism of corrosion inhibition, 402
test data for corrosion inhibition property, 402–403,404f
Mannanase, effect on properties of kraft pulp, 324,325t
Mannosaccharic acid 1,4:3,6-dilactone, forms and preparation, 241–242
Mannose, selective processing, 336
Material, definition, 12
Materials cycle analysis, 29–30
Melt-blended thermoformable composites
composition, 45
formation process, 45
properties, effect of polymer, 46–50
recycled ingredients, 45–46
Microaerophilic mode of operation, description, 363
Microalgae, use as energy source, 415
Microbial enzymes, applications, 326
Microbial treatments of pulps, examples, 314
Microcrystalline cellulose, enzymatic hydrolysis, 372
Micronutrients, use of lignins, 165
Mineral nutrient recycling, relationship to processing of agricultural residues, 356
Miscible blends, 58

Mixed cellulose ethers, 266
Modified cellulose ethers, preparation, 184
Molding, preparation from untreated wood, 141f
Molding applications, use of lignins as binder, 162–163
Molding materials, preparation from chemically modified wood, 140
Monomers and polymers based on mono-, di-, and oligosaccharides
applications, 231–232
cellobiose, 246–249
cyclodextrin, 257–259
1,4:3,6-dilactone of D-glucaric acid, 242–243
1,4:3,6-dilactone of mannosaccharic acid, 241–242
dimethyl isosorbide, 236–237
epoxy resins, 243–246
experimental materials, 259–261
D-glucosamine, 249–250
D-glucose methacrylate, 252–254
isomannide, 235–236
isosorbide, 232–235
1,4-lactone of 3,6-anhydrogluconic acid, 238–241
measurement procedure, 261
nylons based on saccharides, 243
poly(vinyl saccharide)s, 256
saccharide-carrying styrene, 255–256
sucrose, 251–252
trehalose, 250–251
1,4:2,5:3,6-trianhydromannitol, 237
Municipal solid wastes
advantages of recycling, 44–45
availability for energy production, 413–414
composition, 413,414t
estimated distribution of materials, 44t
methods for conversion into useful products, 413
problems with usage, 44
source of biomass, 356–357
source of lignocellulosic fiber and plastics, 44t,45

N

Natural products, interest in use as annually renewable raw materials, 197

Natural rubber
　guayule as source, 273–274
　market demands, 273
　U.S. rubber industry, reasons for development, 273
Net retention volume, definition, 120
Nitric acid, activation of lignocellulosic materials, 122–126
Nonselective processing of biomass, 334
Nontraditional biomass sources
　aquatic resources, 415
　energy crops, 414
　food processing and animal wastes, 414–415
Nonwoven mat thermoformable composites
　development, 49,51
　formation process, 45
　properties vs. density of composites, 51,52–53t
　prototype samples, 51,54
　recycled ingredients, 45–46
Nylons based on saccharides, preparation, 243

O

Oil, environmental concerns, 2
Oilification of lignocellulosics, description, 136
Oil well drilling muds, use of lignins, 159–160
Oligoesterification of wood
　effect of wood content on acetone-soluble and -insoluble parts, 108,109f,110
　reaction, 106–108,109f
Organic-soluble resins
　applications, 287,292
　characterization of films, 292–293,294f,295t
　components, 287
　efficacy as pesticide, 293,296
　film properties of guayule-modified epoxy coatings, 292–293,295t
　formulation conditions, 292,293t
　tensile properties of guayule-modified epoxy film, 293,295t
　thermal glass transition temperature, 292,294f
Organic solvents, liquefaction of lignocellulosics, 136–142
Organosolv pulping, development, 150
Oxidase, conditions for comparing effects on pulps, 320,321t
Oxycellulose, preparation from cellulose, 179,180f

P

Paper, consumption, 313
Paper-coating applications of soy protein polymers
　classes of components of paper coating, 302,304t
　coating process, 302,303t
　comparison of rheograms of native vs. caustic-treated proteins, 302,304–305f
Paper strength, effect of starch graft copolymers, 223
Parallam, example of composite, 32
Pesticide dispersant, use of lignins, 160
Petroleum, consumption and production, 354
Petroleum–chemicals industry, criteria for use of biomass-derived chemicals, 340
Pharmacognosy, 437–438
Phenol, replacement by biomass-derived phenolics, 341
Phenolation, *See* Solvolysis of wood
Phenol production, 340–341
Phenol replacements from biomass
　comparative production costs, 341,342t
　demethylation of kraft lignin, 347–348,349f
　fast pyrolysis–solvent fractionation, 341,343–346
　wood solvolysis, 345,346f,347
Photochemical degradation, lignocellulosics, 20–21
Physicochemical treatment of soy protein
　molecular-weight distribution, 302,303f
　primary events, 300,301t
　protein cross-linking by lysinoalanine formation, 300,303f
　protein titration curve, 300,301f
Pinene(s)
　description, 394
　hydrolysis, 394,395f
　structures of flavor and fragrance chemicals, 394,395f
α-Pinene
　adhesive application, 398
　cationic polymerization, 396,397f,398
　hydrolysis, 394,395f

INDEX

Plant(s)
  agrochemicals, 448
  dependence by humans for survival, 438
  drugs, 443–448
  importance in food supply, 438
  resistance to insects, 439
  sources of drugs and agrochemicals
    background, 437–441
    examples, 443–449
    future directions, 450
    preliminary economic assessment
      procedure, 441–443
Plant chemicals
  development of prototype agents, 441
  potential for developing new sources, 440–441
  preliminary economic assessment procedure, 441–443
Plasmas
  description, 125
  surface modification of fibers, 125,128–133
Plastic(s), nonbiodegradability, 201
Plastic composites, markets, 33
Plastic industry
  energy profiles, 31
  market, 30–31
Plasticization of wood, cross-linking, 110,111t,112
Plasticized cross-linked wood
  future research, 112
  properties, 105t,106
  samples, 112,113f
Plastic materials
  gelatinized starch as component, 206,207t,208–209
  granular starch as filler, 203,205–206
  preparation methods from starch, 203
Podophyllotoxin, development and structure, 445–446
Polyalcohol, preparation from cellulose, 179
Polyethylene, surface modification by plasma, 128–129t
Poly(hydroxybutyrate-co-hydroxyvalerate), structure, 203,204f
Poly(isosorbide–hexamethylene diisocyanate), preparation procedure, 260–261
Polymer alloys, 57,59
Polymer blends, 57–59

Polymer(s) containing mono- or disaccharides, applications, 231–232
Polymeric materials, importance of blending with other materials, 57
Polymeric materials from agricultural commodities
  bagasse, 296–297
  guayule, 273–292
  organic-soluble resins, 287,292–296
  water-soluble resins, 296
Polymerizable oligoester chains, introduction into wood via cross-linking, 106–110
Polyolefins, chemical activation for surface modification, 121,122t
Polysaccharides
  definition, 366
  depolymerization, 366–374
  hydrolysis in plant cell walls, 367–368
Poly(vinyl saccharide)s, preparation, 256
Preliminary economic assessment of plant chemicals, 441–443
Pretreated wood, pyrolysis, 430,431t
Primaquine, structure, 443–444
Propionate, production via fermentation, 377–378
Pseudohypericin, applications and structure, 445,447
Pulp(s), enzymatic treatments, 313–326
Pulping processes
  alternative processes, 150
  kraft pulping, 150
  sulfite pulping, 149–150
Pyrethrins, application and structure, 448–449
Pyrolysis
  definition, 422
  description, 416
Pyrolysis liquid products of wood, 425–430,431t
Pyrolysis of agricultural and forest wastes
  agricultural wastes, use, 432,433–434t
  fast pyrolysis processes, 423–425
  history, 422
  hydropyrolysis in catalytic bed, 430,432
  liquid products, analysis, 425,429t,430
  pyrolysis of pretreated wood, 430,431t
  pyrolytic liquids, properties, 425,428t
  upgrading of products, 432
  yields from different woods, 425,428t
Pyrolysis of biomass
  description, 334
  yields of components, 335

Pyrolysis products from biomass, upgrading, 432

Q

Quinine, discovery and structure, 443–444

R

Recycled fibers, enzymatic treatments, 316–317
Recycled materials
  formation of lignocellulosic–plastic composites, 42–54
  use as composites, 42
Renewables
  energy profiles, 31–32
  markets, 31
Residual lignin, kraft pulp, 319–320
Retention aids in manufacture of mineral-filled paper, starch graft copolymers, 221,223t
Road dust control, use of lignins, 162
Rosin, sources, 398
Rosin esters
  adhesive and ink applications, 399
  color improvements, 399t
  description, 398–399
  preparation, 399
Rotenone
  application, 446
  structure, 448–449
Rubber additive, use of lignins, 166
Rubber extraction process for guayule
  constituents other than rubber, 275
  separation of low-molecular-weight rubber, 274–275
  sequential solvent extraction, 274
  simultaneous solvent extraction, 274

S

Saccharide-carrying styrene, 255–256
Sampangine, application and structure, 445
Saponified starch-g-polyacrylonitrile absorbent polymer
  applications, 220–221
  gelatinization of starch, effect on structure, 216,218t
  isolation methods, 219
Saponified starch-g-polyacrylonitrile absorbent polymer—*Continued*
  preparation, 216,217f
  thickening properties, 219
Selective processing of biomass, 335–336
Sequesterants, 157
Shen binderless bonding process, 91
Silated hydroxyethylcellulose
  cross-linking properties, 270
  cross-linking structure, 269,270f
  mechanism of self-cross-linking, 269
  preparation, 267
  self-cross-linking properties, 268
  ^{29}Si-NMR spectrum, 269,271f
  solution properties, 268f
Simultaneous isomerization and fermentation system, description, 375
Sodium hydroxide, activation of lignocellulosic materials, 125,127f
Softwoods
  composition, 357,358t
  constituents, 393
  renewable resource of chemicals, 393
Solvolysis of wood
  amortized production cost of product, 345,347
  schematic representation, 345,346f
Sorbitol, preparation and structure, 232–233
Sorption of moisture, lignocellulosic composites, 21,23
Soy protein
  applications, 299
  chemical modification, 302,305–311
  industrial background, 299–305
  isolation, 300
  molecular-weight distribution of native protein, 300,301f
  physicochemical treatment, 300,301t,f,302,303f
  protein composition, 299–300
Soy protein polymers
  future polymers, 311–312
  interpolymer particle size, 306,310–311f
  interpolymer preparation, 306,310f
  paper-coating applications, 302,303t,304t,f,305f
Specific retention volume, definition, 120
Starch
  chemical bonding to resin, 208–209
  composition, 198,199f

INDEX

Starch—*Continued*
  definition, 198
  hydrolysis, 366–367
  injection molding into material for property modification, 208
  interest as renewable resource, 197
  occurrence, 198
  properties, 198,200
  role in biodegradable plastics, 200–202
  sources, 198
  uses, 200*t*
Starch fermentation, production of biopolymer plastics, 202–203,204*f*
Starch-*g*-poly(methyl acrylate)
  continuous films, formation, 213
  extrusion, 212*t*
  extrusion-blown films, shrinking, 215*t*,216
  tensile properties, 213,214*t*,215
  water, effect on extrusion, 212,213*t*
Starch-*g*-polystyrene, extrusion and preparation, 211*t*,212
Steam explosion, definition, 371
Steam explosion pulping, development, 150
Strength, lignocellulosic composites, 23
Strigol, application and structure, 448–449
Styrene–butadiene–styrene, adhesive applications, 396
Styrene–isoprene–styrene, adhesive applications, 396
Succinate, production via fermentation, 378
Sucrose
  preparation of polymers, 252
  structure, 251
Sulfate turpentine, desulfurization, 394
Sulfite pulping
  definition, 149
  process, 149
  wood criteria, 149–150
Surface characterization of fibers
  electron spectroscopy for chemical analysis, 120–121,124*f*
  instrumentation, 115,116*f*,117,118*t*
  inverse GC, 119–120
  wetting, 117,119
Surface modification by plasma
  cellulosics, 130,131*t*,132–133*f*
  synthetic fibers, 128–130*t*

Surface modification of fibers
  chemical activation of lignocellulosic materials, 122–127
  chemical activation of polyolefins, 121,122*t*
  electric discharge activation, 125–133
Surface science, instrumentation, 115,116*f*,117,118*t*
Syngas conversion
  definition, 379
  liquid and gaseous fuels, 380–381
Synthetic(s)
  energy profiles, 31–32
  markets, 31
Synthetic fibers, surface modification by plasma, 128–130*t*

T

Tall oil rosin, decolorization and deodorization, 398
Tannins, applications and classifications, 405,407
Taxol, development and structure, 445–446
Teniposide, development and structure, 445–446
Tennessee Valley Authority's biomass research program
  concentrated acid conversion efficiencies of selected biomass feedstocks, 417*t*
  concentrated sulfuric acid hydrolysis process for ethanol production, 417,419*f*,420
  dilute sulfuric acid hydrolysis process for ethanol production, 417,418*f*,420
Terpenes
  applications, 394,396,397*f*,398
  preparation, 394,395*f*
  use as hygiene chemicals, 394,396
  use in flavors and fragrances, 394,395*f*,396,397*f*
Thermal plasticity of esterified woods, effect of reaction temperature, 101,103*t*
Thermal plasticization of lignocellulosics for composites
  properties, 105*t*,110,111*t*,112
  reactions, 100–109
  thermal plasticity, 101,103*t*
Thermochemical conversion processes of biomass, 416
Thermoformable composites, 45–46

Thermoplastic(s), undegradability, 7
Thermoplastic starch graft polymers
  applications, 216
  extrusion, 211–215
  preparation, 209,211
  shrinkage, 215t,216
  structure, 209,210f
Thickeners, use of starch graft copolymers, 223–224
Thioterpenes, aroma and enhancement in personal use items, 396,397f
Trehalose, 250–251
1,4:2,5:3,6-Trianhydromannitol, 237
Trichoderma cellulases, enzymatic hydrolysis, 373
Turpentine, desulfurization, 394

U

Untreated wood
  applications of liquefaction, 141f,142
  liquefaction, 137–138
Urease, description, 300
Urethane, chemical bonding of starch to resin, 208–209
U.S. Council for International Business, 2–3
U.S. national policy, development of biomass resources, 5
UV radiation resistance, lignocellulosic composites, 23
UV stabilization of lignocellulosic composites, acetylation, 93–94,95f

V

Vanillin, use of lignin for production, 166–167
Vessel picking, cellulase treatment for reduction, 317–318
Volatile organic acids
  production via fermentations, 377–379
  production via Kolbe electrolysis, 378–379
  pyrolysis reaction, 378

W

Waste, volume in United States, 43
Water-dispersible starch graft polymers
  applications, 220–224

Water-dispersible starch graft polymers—
  Continued
  gelatinization of starch, effect on structure, 216,218t
  preparation, 216,217f,220
  thickening properties, 219
Water removal from organic solvents, use of starch graft copolymers, 221
Water shift reaction, description, 380
Water-soluble cellulose ethers, applications and examples, 265–266
Water-soluble resins, separation and characterization, 296
Water treatment, use of lignins, 164–165
Weathering of wood, influencing factors, 93
Wetting
  dispersion force, 119
  equation, 117
  secondary forces, 117
Wheat straw
  development of end uses, 356
  source of biomass, 356
Wilhelmy principle, determination of work of adhesion, 117,119
Wood
  carboxyl groups, introduction via esterification, 100–101,102f
  chemical modification for property enhancement, 99
  composite, 5
  composition, 147–148
  disadvantages, 15
  polymer components, 15,16–17f
  polymerizable oligoester chains, introduction via cross-linking, 106–108,109f,110
  properties due to plasticization, 110,111t,112
  pulping processes, 149–150
  pyrolysis, 425–430
  reaction of hydroxyl groups with dicarboxylic acid anhydride, 99
  technoeconomic assessment of acetylation, 34,35f,36–37t
Wood fiber, examples of composites, 33–34
Wood fiber–plastic composites, *See* Lignocellulosic–plastic composites from recycled materials
Wood fiber–thermoplastic composites, advantages and disadvantages, 34

Wood–plastic alloys, 71
Wood products, annual production, 88
Wood solvolysis, *See* Solvolysis of wood
Woody biomass
   availability for energy production, 411,412t
   definition, 411
Work of adhesion
   definition, 117
   dispersion force, 119

X

X-ray photoelectron spectroscopy, *See* Electron spectroscopy for chemical analysis
Xylan, selective removal by enzymes, 318

Xylanase
   bleaching of pulps, 322,323t
   brightness of pulp, 322t
   comparing effects on pulps, 320,321t
   fibrillation, enhancement, 315,316t
   kraft pulp properties, 324,325t
   pulp properties, 323,324t
   recycled fibers, treatments, 316–317
   xylan, selective removal, 318
Xylose, selective processing, 336
Xylose isomerization, ethanolic fermentation, 374–375

Y

Yeast, ethanolic fermentation, 374

Production: Donna Lucas
Indexing: Deborah H. Steiner
Acquisition: A. Maureen Rouhi
Cover design: Amy Hayes

Printed and bound by Maple Press, York, PA

Other ACS Books

Chemical Structure Software for Personal Computers
Edited by Daniel E. Meyer, Wendy A. Warr, and Richard A. Love
ACS Professional Reference Book; 107 pp;
clothbound, ISBN 0–8412–1538–3; paperback, ISBN 0–8412–1539–1

Personal Computers for Scientists: A Byte at a Time
By Glenn I. Ouchi
276 pp; clothbound, ISBN 0–8412–1000–4; paperback, ISBN 0–8412–1001–2

Biotechnology and Materials Science: Chemistry for the Future
Edited by Mary L. Good
160 pp; clothbound, ISBN 0–8412–1472–7; paperback, ISBN 0–8412–1473–5

Polymeric Materials: Chemistry for the Future
By Joseph Alper and Gordon L. Nelson
110 pp; clothbound, ISBN 0–8412–1622–3; paperback, ISBN 0–8412–1613–4

The Language of Biotechnology: A Dictionary of Terms
By John M. Walker and Michael Cox
ACS Professional Reference Book; 256 pp;
clothbound, ISBN 0–8412–1489–1; paperback, ISBN 0–8412–1490–5

Cancer: The Outlaw Cell, Second Edition
Edited by Richard E. LaFond
274 pp; clothbound, ISBN 0–8412–1419–0; paperback, ISBN 0–8412–1420–4

Practical Statistics for the Physical Sciences
By Larry L. Havlicek
ACS Professional Reference Book; 198 pp; clothbound; ISBN 0–8412–1453–0

The Basics of Technical Communicating
By B. Edward Cain
ACS Professional Reference Book; 198 pp;
clothbound, ISBN 0–8412–1451–4; paperback, ISBN 0–8412–1452–2

The ACS Style Guide: A Manual for Authors and Editors
Edited by Janet S. Dodd
264 pp; clothbound, ISBN 0–8412–0917–0; paperback, ISBN 0–8412–0943–X

Chemistry and Crime: From Sherlock Holmes to Today's Courtroom
Edited by Samuel M. Gerber
135 pp; clothbound, ISBN 0–8412–0784–4; paperback, ISBN 0–8412–0785–2

For further information and a free catalog of ACS books, contact:
American Chemical Society
Distribution Office, Department 225
1155 16th Street, NW, Washington, DC 20036
Telephone 800–227–5558